T0269222

LONDON MATHEMATICAL SOCIETY LECTURE NOTE SERIES

Managing Editor: Professor J.W.S. Cassels, Department of Pure Mathematics and Mathematical Statistics, University of Cambridge, 16 Mill Lane, Cambridge CB2 1SB, England

The books in the series listed below are available from booksellers, or, in case of difficulty, from Cambridge University Press.

London Mathematical Society Lecture Note Series. 191

Finite Geometry and Combinatorics
The Second International Conference at Deinze

Edited by

A. Beutelspacher
University of Giessen

F. Buekenhout & J. Doyen
University of Brussels

F. De Clerck & J.A. Thas
University of Ghent

and

J.W.P. Hirschfeld
University of Sussex

CAMBRIDGE
UNIVERSITY PRESS

CAMBRIDGE UNIVERSITY PRESS
Cambridge, New York, Melbourne, Madrid, Cape Town, Singapore, São Paulo

Cambridge University Press
The Edinburgh Building, Cambridge CB2 2RU, UK

Published in the United States of America by Cambridge University Press, New York

www.cambridge.org
Information on this title: www.cambridge.org/9780521448505

First published 1993

A catalogue record for this publication is available from the British Library

ISBN-13 978-0-521-44850-5 paperback
ISBN-10 0-521-44850-6 paperback

Transferred to digital printing 2005

Contents

v

PREFACE

This book contains articles based on talks at the Second International Conference on Finite Geometry and Combinatorics, which took place from 31 May to 6 June 1992, at the Conference center *De Ceder* in Astene–Deinze, Belgium. There were 76 participants and 52 talks.

There are 35 articles in these proceedings.

The editors, who were also the conference organisers, are grateful for the financial support of the National Fund for Scientific Research (NFWO) of Belgium and of the University of Ghent.

Above all we are grateful to Zita Oost, the conference secretary, and to Ruth Lauwaert, who helped a great deal with the production of the copy.

Albrecht Beutelspacher
Francis Buekenhout
Frank De Clerck
Jean Doyen
James W. P. Hirschfeld
Joseph A. Thas

April 1993

INTRODUCTION

Discrete mathematics has had many practical applications in recent years and this is only one of the reasons for its increasing dynamism. The study of finite structures is a broad area which has a unity not merely of description but also in practice, since many of the structures studied give results which can be applied to other, apparently dissimilar structures. Apart from the applications, which themselves generate problems, internally there are still many difficult and interesting problems in finite geometry and combinatorics, and we are happy to be able to demonstrate progress.

It was a great pleasure to see several Russian colleagues participating both because they were able to do so, some for the first time, and because this is an area of Mathematics not as diffuse in Russia as elsewhere. It was also good to see the participation of a significant number of talented, younger colleagues, but at the same time sad to note the difficulty they are having in finding permanent positions.

The conference papers are here divided into themes. The division is somewhat artificial as some papers could be placed in more than one group. The style of mathematics is very much resolving problems rather than the construction of grand theories. There are still many puzzling features about the sub-structures of finite projective spaces, as well as about finite strongly regular graphs, finite projective planes, and other particular finite diagram geometries. Finite groups are as ever a strong theme for several reasons. There is still much work to be done to give a clear geometric identification of the finite simple groups. There are also many problems in characterizing structures which either have a particular group acting on them or which have some degree of symmetry from a group action.

Generalized polygons

Bader and *Lunardon* together and *Lunardon* alone give new constructions of classical hexagons. *Buekenhout* and *Van Maldeghem* show that there are no surprises in the action of a classical group on a hexagon or octagon. *De Smet* and *Van Maldeghem* give a new geometrical characterization of the finite classical hexagons and characterise some finite Moufang hexagons, and *van Bon*

1

determines some extended generalized hexagons having certain geometric and group-theoretical properties. *Payne* constructs a coherent configuration inside a translation generalized quadrangle. *Brouwer* shows that the subgeometry of a polygon induced on the objects in general position with respect to a given flag is connected.

Graphs and their groups

Brouwer, *Fon-der-Flaass*, and *Shpectorov* find the three graphs which locally are the incidence graph of the unique biplane on 7 points. *Haemers* shows that a strongly regular graph on 76 points with consistent parametric conditions cannot exist. *Munemasa*, *Pasechnik*, and *Shpectorov* characterize the graphs of alternating and quadratic forms over $GF(2)$, whereas *Munemasa* and *Shpectorov* characterize alternating forms over all larger fields. In a similar vein, *Pasechnik* describes the graph of a certain $GF(3)$-geometry, which leads to a characterization of Fischer's sporadic simple groups. *Soicher* shows that the Lyons simple group has no distance-transitive representation and hence determines all faithful multiplicity-free representations of this group.

Finite Desarguesian planes

A spread of a Hermitian curve in a Desarguesian plane is a set of non-tangent lines partitioning the points of the curve. *Baker*, *Ebert*, *Korchmáros*, and *Szőnyi* study such spreads with the property that no line of the spread contains the pole of any other line in the spread and deduce a result on the linear code of the plane. A minimal blocking set in $PG(2, q)$ has size b satisfying $q + \sqrt{q} + 1 \leq b \leq q\sqrt{q} + 1$, with the bounds being achieved for square q by a Baer subplane and a Hermitian curve. Related to this, *Blokhuis* and *Metsch* study strong representative systems and show that, for $q \geq 25$, one cannot have $b = q\sqrt{q}$. A nucleus of a set of $q + 1$ points in $PG(2, q)$ is a point not in the set such that every line through it meets the set; it is known that the number of nuclei is at most $q - 1$. *Blokhuis* and *Mazzocca* use a mapping to $PG(3, q)$ to deduce more about the structure of sets in $PG(2, q)$ with the maximum number of nuclei. *Glynn* relates the code of $PG(2, q)$ with q even to the study of nonics, where a nonic is defined to be either a conic plus its nucleus or a line pair, real or imaginary, less the point of intersection. *Gordon* obtains a formula for the number of projectively distinct k-arcs in $PG(n, q)$ and also finds an efficient algorithm for determining this number; this is applied to finding the number of projectively distinct k-arcs and projectively distinct complete k-arcs in $PG(2, 11)$ and $PG(2, 13)$. *Hirschfeld* and *Voloch* obtain further results on the characterization of sets of points in $PG(2, q)$ with at most three points on a line as cubic curves.

2

Higher-dimensional projective spaces

Storme and *Szőnyi* examine k-arcs having many points in common with a normal rational curve. For odd q, it is elementary that a plane k-arc not contained in a conic has at most $(q + 3)/2$ points on the conic. The problem is much more difficult in higher dimensions; it is shown that, for large q and bounded n, if an arc has more than $(q+1)/2$ points in common with a normal rational curve, it is contained in the curve. A partial flock of a quadric cone in $PG(3, q)$ is a set of disjoint conics on the cone and a flock is a set of q conics forming a partition of the cone less its vertex. *Thas, Herssens* and *De Clerck* survey the known flocks and construct a new flock for $q = 11$ which generalizes to a partial flock of size 11 for any $q \equiv -1 \pmod{12}$. An ovoid of a quadric is a set of points meeting every generator (subspace of maximum dimension lying on the quadric) precisely once. *Moorhouse* describes a 9-dimensional lattice which defines simply an ovoid on a hyperbolic quadric in $PG(7, p)$, p prime. These ovoids had previously been constructed by the author from the E_8 root lattice.

Non-Desarguesian planes

Ho investigates Singer groups S acting on a projective plane of order n and proves, among other things, that, if the multiplier group $M(S)$ has even order, each subgroup of S is invariant under the involution of $M(S)$, except possibly if $n = 16$ and S is non-abelian. *Jha* and *Wene* find the number of central units of a commutative semifield plane. *Johnson* characterises certain translation planes of order q^2 as equivalent to particular subsets of cardinality q of the collineation group $P\Gamma L(2, q)$. *Wettl* extends results on nuclei in a Desarguesian plane to the non-Desarguesian case.

Block designs

A $t - (v, k, \lambda)$ design is an incidence structure of v points and b blocks with k points on each block and the essential property that through t points there are precisely λ blocks. Such structures exist for all t by Teirlinck's theorem, but the known examples mostly have very large λ. *Cameron* and *Praeger* investigate block-transitive designs with $5 \leq t \leq 8$. Among other results they determine the possible automorphism groups of block-transitive 6-designs and flag-transitive 5-designs.

Polar spaces

Buekenhout presents an old result of Parmentier which axiomatizes a pair (P, π), where P is a projective space and π a polarity, so that the axioms defining P alone are weaker than usual. *Shult* develops the axiomatization of Veldkamp spaces for point-line geometries.

3

Diagram geometries

Buekenhout and *King* study flag-transitive diagrams of rank 3 such that the residues of 0-elements are dual Petersen graphs, of 1-elements are generalized digons, and of 2-elements are finite linear spaces. The linear space turns out to be either a projective plane, in which case there are precisely two geometries, or the complete graph on four vertices. In the latter case there are no geometries for which the group acts primitively on the 2-elements, but examples are given of the imprimitive case.

A grid is an incidence structure of points and lines such that any line has at least two points and any two points are incident with at most one line, with the additional property that the lines fall into two classes such that two lines intersect if and only if they belong to different classes; these intersections give all the points. *Meixner* and *Pasini* describe all known extensions of grids. *Ghinelli* classifies the flag-transitive rank 3 geometries in which the planes are linear spaces with constant line size and the point residues are classical generalized quadrangles other than grids. *Huybrechts* gives a new proof of the commutativity of the division ring for a thick, residually-connected D_n-geometry.

A generalized Fischer space is a partial linear space in which any two intersecting lines generate a subspace that is an affine plane or its dual. Subject only to some non-degeneracy conditions, *Cuypers* gives a complete classification of these spaces. *Mühlherr* describes a geometric method of constructing, from the diagrams, Coxeter groups as subgroups of other Coxeter groups.

4

Generalized hexagons and BLT-sets

L. Bader   G. Lunardon[*]

Abstract

An alternative construction for the dual $G_2(q)$-hexagon is given for q odd and different from 3^n.

1. Introduction

In [4], W.M. Kantor has constructed the generalized quadrangle associated with the Fisher-Thas-Walker flock as a group coset geometry starting from the dual $G_2(q)$-hexagon. Analyzing Kantor's construction, the following question arises in a natural way: is it possible to define new points and new lines in a generalized quadrangle Q associated with a flock of the quadratic cone, in such a way that the new point-line geometry H is a generalized hexagon?

For q odd, we prove that the only possibility is that Q is the Kantor generalized quadrangle constructed in [4] and H is the dual $G_2(q)$-hexagon. If $q \neq 3^n$, using a twisted cubic of $PG(3,q)$ we obtain an alternative construction of the dual $G_2(q)$-hexagon similar to the construction of a generalized quadrangle using a BLT-set ([6] or [11]). For q even, we are able to prove a strong connection between the existence of H and the $(q+1)$-arcs of $PG(3,q)$ but the answer is not complete due to difficulties of the same type that arise when studying BLT-sets in even characteristic.

We would like to express our thanks to S. E. Payne, J. A. Thas and H. Van Maldeghem for critical remarks on earlier versions of this paper, and to W. M. Kantor for useful discussions during his visit in Rome. In particular, Theorem 2.1 generalizes a result of W. M. Kantor (private communication).

2. Generalized hexagons as group coset geometries

Let $s, t > 1$ be natural numbers. We denote by $\hat{F}$ a set of $s+1$ elements. Let G be a group of order $s^2 t^3$. For any u in $\hat{F}$, fix the subgroups $A_1(u), A_2(u), A_3(u), A_4(u)$ of G such that $A_1(u) \leq A_2(u) \leq A_3(u) \leq A_4(u)$, where $|A_1(u)| = t$, $|A_2(u)| = st$, $|A_3(u)| = st^2$ and $|A_4(u)| = s^2 t^2$. Define a point-line geometry $H = (P, L, I)$ as follows:

$$P = \{\mathcal{I}, A_4(u)g, A_2(u)g, g : g \in G, u \in \hat{F}\}$$

[*]. The authors are members of G.N.S.A.G.A. of C.N.R. and have partial financial support by Italian M.U.R.S.T.

5

$$L = \{[u], A_3(u)g, A_1(u)g : g \in G, u \in \hat{F}\}$$

where $\mathcal{I}$ and $[u]$ are symbols and the incidences are $\mathcal{I}I[u]$, $A_4(u)I[u]$ and $gIA_1(u)g$ for all $g \in G$ and $u \in \hat{F}$, while $A_i(u)gIA_{i+1}(v)h$ if and only if $u = v$ and $g \in A_{i+1}(v)h$ with $i = 1, 2, 3$.

Theorem 2.1 $H = (P, L, I)$ is a generalized hexagon with parameters (s, t) if and only if, for all distinct i, j, h, m, n in $\{1, 2..., s + 1\}$, the following conditions hold:

1) $A_4(i) \cap A_1(j) = 1$,
2) $A_3(i) \cap A_1(j)A_1(h) = 1$,
3) $A_3(i) \cap A_2(j) = 1$,
4) $A_2(i)A_2(j) \cap A_1(h) = 1$,
5) $A_2(i) \cap A_1(j)A_1(h)A_1(m) = 1$,
6) $A_2(i) \cap A_1(j)A_1(i)A_1(h) = A_1(i)$,
7) $A_1(i) \cap A_1(j)A_1(h)A_1(m)A_1(n) = 1$,
8) $A_1(i) \cap A_1(j)A_1(h)A_1(m)A_1(h) = 1$,
9) $A_1(i)A_1(j) \cap A_1(j)A_1(i) = A_1(i) \cup A_1(j)$,
10) $A_1(i) \cap A_1(j)A_1(h)A_1(j)A_1(h) = 1$.

Proof. With a direct calculation, we can prove that if H is a generalized hexagon, then the ten conditions are satisfied.

Conversely, if we suppose that the conditions 1-10 hold, then no circuit of length less than 12 can exist in H. The point-line geometry H has exactly $(1+t)(1+st+s^2t^2)$ points and exactly $(s+1)(1+st+s^2t^2)$ lines. Each line is incident with exactly $s+1$ points, and each point is incident with exactly $t+1$ lines. Moreover, H contains at least one circuit of length 12, while no circuit of length less than 12 can exist. Consequently H is a generalized hexagon by [12] p. 5.   □

By condition 9 of Theorem 2.1, if H is a generalized hexagon then G cannot be abelian.

Starting from H, we define a new point-line geometry $Q(H)$ in the following way. The points of $Q(H)$ are $\mathcal{I}$, $A_3(u)g$ and g, with $g \in G$ and $u \in \hat{F}$. The lines of $Q(H)$ are $[u]$ and $A_2(u)g$, with $g \in G$ and $u \in \hat{F}$. The line $[u]$ is incident with $\mathcal{I}$ and $A_3(u)g$ for all g in G, while all other incidences are given by inclusion.

Corollary 2.2 Let H be a generalized hexagon, whose parameters s and t are equal. Then $Q(H)$ is a generalized quadrangle with parameters (s^2, s) if and only if $A_2(u)A_2(v) \cap A_2(w) = 1$ for all distinct $u, v, w \in \hat{F}$.

Proof. If $s = t$, the corollary it is an easy consequence of Section 10.1 of [10]. □

The hypothesis $s = t$ is required in Corollary 2.2, because the point $\mathcal{I}$ is incident with $s + 1$ lines, while the point $A_3(u)g$ is incident with $t + 1$ lines. If $Q(H)$ is a generalized quadrangle, then the conditions of Theorem 2.1 can be simplified (see [1]). In [5] W.M. Kantor has given an explicit description of the subgroups $A_1(u)$, $A_2(u)$, $A_3(u)$ and $A_4(u)$ for the known generalized hexagons. In [4] W.M. Kantor has proved that if H is the dual $G_2(q)$-hexagon then $Q(H)$ is a generalized quadrangle when $q \equiv -1 \pmod 3$. Moreover, $Q(H)$ is the generalized quadrangle associated with the Fisher-Thas-Walker flock of the quadratic cone (see [13]).

3. $(q+1)$-arcs

Let $F = GF(q)$ and $\hat{F} = F \cup \{\infty\}$. In the following, we always denote by G the group whose elements are those of $F^2 \times F \times F^2$, and whose product is defined by

$$(\alpha, c, \beta)(\alpha', c', \beta') = (\alpha + \alpha', c + c' + \alpha' \cdot \beta, \beta + \beta')$$

where $\alpha, \beta, \alpha', \beta' \in F^2$, $c, c' \in F$, and $\alpha' \cdot \beta = \alpha' \beta^T$. The center of G is the set $Z = \{(0, c, 0) : c \in F\}$ and the group $\bar{G} = G/Z$ is elementary abelian. Moreover, we can regard $\bar{G}$ as a four dimensional vector space over F. For each element g of G, let g^* be the preimage in G of the 1-space of $\bar{G}$ spanned by $\bar{g} = gZ$. If $\bar{g}, \bar{h}$ are elements of $\bar{G}$, we notice that $(\bar{g}, \bar{h}) = [g, h]$ defines a non singular alternating F-bilinear form on $\bar{G}$; if q is even, $\bar{g} \mapsto g^2$ defines a quadratic form associated with (,). Thus, $\bar{G}$ is equipped with a symplectic or orthogonal geometry.

If $[g, h] = 1$, then $[g^*, h^*] = 1$. Thus maximal elementary abelian subgroups of G are preimages of totally isotropic (or singular) 2-spaces of $\bar{G}$. Therefore a maximal elementary abelian subgroup of G has order q^3.

Let $PG(3, q)$ be the three dimensional projective space associated with the F-vector space $\bar{G}$. For each $u \in \hat{F}$ if $A_3(u) = A_2(u)Z$, then $A_2(u)$ is elementary abelian because it is canonically isomorphic to the subgroup $A_3(u)/Z$ of $\bar{G}$. So $A_3(u)$ is a maximal elementary abelian subgroup of G. Thus, $L_u = A_3(u)/Z$ is a totally isotropic (or singular) line of the projective space $PG(3, q)$.

Theorem 3.1 *Suppose that H is a generalized hexagon, $A_3(u) = A_2(u)Z$. Denote by p_u, L_u, α_u respectively the point $A_1(u)Z/Z$ [1], the line $A_3(u)/Z$ and*

---

1. As Z is contained in $A_4(v)$, $A_1(u) \cap Z = 1$. Therefore $A_1(u)Z/Z$ is a 1-dimensional vector subspace of $\bar{G}$.

the plane $A_4(u)/Z$ of $PG(3,q)$. The set $\Sigma = \{p_u : u \in \hat{F}\}$ is a $(q+1)$-arc of $PG(3,q)$. Moreover, L_u is a tangent line of Σ at p_u and α_u is the osculating plane of Σ at p_u.

Proof. Let u and v be two distinct elements of $\hat{F}$. By property 9 of Theorem 2.1, if $g_u \in A_1(u)$ and $g_v \in A_1(v)$ then $[g_u, g_v] = (\bar{g}_u, \bar{g}_v) = 1$ if and only if $g_u = 1$ or $g_v = 1$. Thus, $p_u^{\perp} \cap \Sigma = \{p_u\}$ where $p_u^{\perp}$ is the polar plane of p_u with respect to the polarity of $PG(3,q)$ defined by the F-bilinear form $(,)$ of $\bar{G}$.

If $p_v \in \alpha_u$, then $A_1(v) \leq A_1(v)Z \leq A_4(u)$. By property 1 of Theorem 2.1, this implies $u = v$. Therefore, if u and v are different elements of $\hat{F}$, the point p_v does not belong to α_u. Thus, $\alpha_u \cap \Sigma = \{p_u\}$.

Let u, v, w be elements of $\hat{F}$. If $v \neq w$, then the plane $< p_v, L_w >$ of $PG(3,q)$ is defined by a subgroup of G of order q^4 containing $A_1(v)$ and $A_3(w)$. By property 1 of Theorem 2.1, this subgroup is $A_1(v)A_3(w)$. If p_u belongs to the plane $< p_v, L_w >$ then $A_1(u) \leq A_1(u)Z \leq A_1(v)A_3(w)$. By property 2 of Theorem 2.1, either $u = v$ or $u = w$. Thus, $< p_v, L_w > \cap \Sigma = \{p_v, p_w\}$. We have proved that each plane through L_w contains at most one point of Σ different from p_w. As Σ contains exactly q points different from p_w, there are q planes containing the line L_w and a point of Σ different from p_w. So there is exactly one plane through L_w, which contains only one point of Σ. This implies $p_w^{\perp} = \alpha_w$. Moreover, three distinct points of Σ are never collinear.

By way of contradiction, we suppose that four points p_u, p_v, p_w, p_x of Σ are coplanar. Let $\bar{K}$ be the 3-dimensional subspace of $\bar{G}$, which defines the plane $\beta = < p_u, p_v, p_w >$. The subgroup K of G, which is the preimage of $\bar{K}$, has order q^4 and contains the subgroups $A_1(u), A_1(v), A_1(w), A_1(x)$ because the points p_u, p_v, p_w, p_x belong to the plane β.

Let $g_u, h_u \in A_1(u)$, $g_v, h_v \in A_1(v)$, $g_w, h_w \in A_1(w)$ and suppose that $g_u g_v g_w = h_u h_v h_w \xi$ where $\xi \in Z$. We have

$$g_u g_v g_w h_w^{-1} h_v^{-1} h_u^{-1}$$
$$= g_u h_u^{-1} g_v h_v^{-1} g_w h_w^{-1} [h_v g_v^{-1}, h_u][h_w g_w^{-1}, h_v][h_w g_w^{-1}, h_u] = \xi.$$

Therefore, $(g_u h_u^{-1})(g_v h_v^{-1})(g_w h_w^{-1}) \in Z$ because G' is contained in Z. Hence $(g_u h_u^{-1})(g_v h_v^{-1})(g_w h_w^{-1})$ defines the 0-vector of $\bar{G}$. Thus, the vectors $g_u h_u^{-1}Z, g_v h_v^{-1}Z, g_w h_w^{-1}Z$ of $\bar{G}$ are linear dependent. As the points p_u, p_v and p_w are never collinear, we have $g_u h_u^{-1} \in Z$ or $g_v h_v^{-1} \in Z$ or $g_w h_w^{-1} \in Z$. This implies $g_u = h_u$, $g_v = h_v$ and $g_w = h_w$ because $Z \leq A_4(x)$ and $A_4(x) \cap A_1(y) = 1$ for all distinct x and y in $\hat{F}$. Then $A_1(u)A_1(v)A_1(w) \cap h_u h_v h_w Z = h_u h_v h_w$. Moreover, for $h_u h_v h_w = 1$, we have also proved $A_1(u)A_1(v)A_1(w) \cap Z = 1$.

With the same argument, we can prove that if $g_u g_v g_w = h_u h_v h_w$, then $g_u = h_u$, $g_v = h_v$ and $g_w = h_w$. Hence, the subset $A_1(u)A_1(v)A_1(w)$ of G has order q^3. As $A_1(u)$, $A_1(v)$, $A_1(w)$ and Z are subgroups of K, we have

$K = A_1(u)A_1(v)A_1(w)Z$.

The subset $A_1(u)A_1(v)A_1(w)A_1(v)$ is contained in K because $A_1(u)$, $A_1(v)$ and $A_1(w)$ are contained in K. Let $g_u \in A_1(u)$, $g_v, h_v \in A_1(v)$ and $g_w \in A_1(w)$ be elements of G different from 1. Then the element $g_u g_v g_w h_v = g_u g_v h_v g_w [g_w^{-1}, h_v^{-1}]$ does not belong to $A_1(u)A_1(v)A_1(w)$ because $[g_w^{-1}, h_v^{-1}] \neq 1$. Thus, the subset $A_1(u)A_1(v)A_1(w)A_1(v)$ of K has order $> q^3$. As $A_1(x) \cap A_1(u)A_1(v)A_1(w)A_1(v) = 1$ by property 8 of Theorem 2.1, the subset $A_1(x)A_1(u)A_1(v)A_1(w)A_1(v)$ has order $> q^4$; also, it is contained in the subgroup K, because $A_1(x)$ and $A_1(u)A_1(v)A_1(w)A_1(v)$ are subsets of K. We have the required contradiction because K has order q^4.

This completes the proof that Σ is a $(q+1)$-arc of $PG(3,q)$.

For each $u \in \hat{F}$, we denote by β_u the osculating plane of Σ at $p_u{}^2$. Let p_u and p_v be two distinct points of Σ. Define

$$C^* = \{ <p_u, p_w> \cap \beta_v : w \in \hat{F}, w \neq u \}$$

As Σ is a $(q+1)$-arc of $PG(3,q)$, C^* is a q-arc of the plane β_v. Let $C = C^* \cup \{L_u \cap \beta_v\}$.

By way of contradiction, we suppose that a line N of β_v incident with the point $L_u \cap \beta_v$ contains three points of C. Then the plane $< N, p_u >$ is incident with three points of Σ. As the line L_u is contained in $< N, p_u >$, this is impossible. Therefore, each line of β_v contains at most two points of C. This implies that C is a $(q+1)$-arc of β_v. By the definition of tangent of a $(q+1)$-arc of $PG(3,q)$, we have proved that the line L_u is tangent of Σ at p_u.

If q is odd, L_u is the tangent line of Σ at p_u. If q is even, L_u is one of the two tangents of Σ in p_u. In both cases, L_u is contained in β_u. As the osculating plane contains exactly one point of Σ, we have $\alpha_u = \beta_u$ because α_u is the unique plane containing L_u such that $\alpha_u \cap \Sigma = \{p_u\}$. □

Corollary 3.2 *If H is a generalized hexagon with $A_3(u) = A_2(u)Z$, and $q = p^n$ with p a prime number, then $p \neq 3$.*

Proof. Suppose $p = 3$. Thus, Σ is a twisted cubic and all the osculating planes of Σ contain a fixed line L ([8] Section 43). If u, v, w are mutually distinct elements of $\hat{F}$, then $< p_u, p_v, p_w >^\perp = \alpha_u \cap \alpha_v \cap \alpha_w = L$. Therefore, the F-bilinear form $(\ ,\ )$ has non-trivial radical. As this is impossible, we have a contradiction. □

In [4] W.M. Kantor has proved that if H is isomorphic to the dual $G_2(q)$-

---

2. For q even, the osculating plane at p_u is the plane defined by the two tangents of Σ at p_u; for q odd, Σ is a normal rational curve and β_u is the osculating plane of the normal rational curve at p_u (i.e. the plane through p_u with intersection multiplicity 3). For more details see [8]

hexagon, then Σ is a twisted cubic; also, there is a canonical way to construct the dual $G_2(q)$-hexagon starting from a twisted cubic (see [4] Remark 2). Thus, we have proved the following corollary

Corollary 3.3 *Let q be odd. If H be a generalized hexagon with $A_3(u) = A_2(u)Z$, then H is isomorphic to the dual $G_2(q)$-hexagon.*

4. The case q odd

In this section we always suppose that q is odd.

Let $W(5, q)$ be the polar space associated with a symplectic polarity of $PG(5, q)$. Embed the symplectic 3-space $W(3, q)$ in $W(5, q)$ and let p be a fixed point of $W(5, q)$ such that $p \notin W(3, q) \subset p^{\perp} \subset W(5, q)$, where "$\perp$" is relative to $W(5, q)$. We can introduce coordinates in $PG(5, q)$ in such a way that $W(5, q)$ is the polar space associated with the alternating bilinear form

$$b((x_0, \alpha, \beta, x_5), (y_0, \gamma, \delta, y_5)) = x_0 y_5 - x_5 y_0 + \alpha \cdot \delta - \beta \cdot \gamma$$

where $\alpha, \beta, \gamma, \delta \in F^2$, $\alpha \cdot \delta = \alpha \delta^T$ and $\beta \cdot \gamma = \beta \gamma^T$. If $p = (0, 0, 0, 0, 0, 1)$, then $p^{\perp}$ has equation $x_0 = 0$. We can embed $W(3, q)$ in $W(5, q)$ in such a way that $W(3, q)$ is the 3-dimensional subspace of $W(5, q)$ with equations $x_0 = x_5 = 0$.

The nonsingular collineation τ defined by the upper triangular matrix

$$M(a, b, c, d, e) = \begin{pmatrix} 1 & a & b & c & d & e \\ 0 & 1 & 0 & 0 & 0 & c \\ 0 & 0 & 1 & 0 & 0 & d \\ 0 & 0 & 0 & 1 & 0 & -a \\ 0 & 0 & 0 & 0 & 1 & -b \\ 0 & 0 & 0 & 0 & 0 & 1 \end{pmatrix}$$

fixes $W(5, q)$ and all the subspaces of $p^{\perp}$ incident with p. We will identify τ with $M(a, b, c, d, e)$. The group $\tilde{G} = \{M(a, b, c, d, e) : a, b, c, d, e \in F\}$ acts sharply transitive on the points of $PG(5, q) \setminus p^{\perp}$. Moreover, the map

$$g = (a, b, e, c, d) \mapsto \tilde{g} = M(a, b, c, d, ac + bd - 2e)$$

is an isomorphism between G and $\tilde{G}$ (see [11]).

Let $o = (1, 0, 0, 0, 0, 0)$. Then $W(3, q)$ lies in the 3-dimensional subspace $S = p^{\perp} \cap o^{\perp}$ of $PG(5, q)$. Let $\Sigma = \{r_u : u \in \hat{F}\}$ be the twisted cubic of S such that $r_u^{\perp} \cap p^{\perp} \cap o^{\perp}$ is the osculating plane of Σ at r_u for each $u \in \hat{F}$. Let L_u be the tangent line of Σ at r_u ($u \in \hat{F}$). We define a point-line geometry $H(\Sigma)$ in the following way:

Points:

(1) the point p;

(2) the points different from p but contained in one of the lines $< p, r_u >$ $(u \in \hat{F})$;

(3) the totally singular planes not contained in $p^\perp$ and meeting one of the planes $< L_u, p > (u \in \hat{F})$ in a line;

(4) the points of $PG(5, q) \setminus p^\perp$.

Lines:

(a) the lines $< p, r_u > (u \in \hat{F})$;

(b) the lines not incident with p, and contained in one of the planes $< L_u, p > (u \in \hat{F})$;

(c) the totally singular lines not contained in $p^\perp$ and meeting $p^\perp$ in a point of the lines $< p, r_u > (u \in \hat{F})$.

Incidences:

Points of type (2) and lines of type (c) are never incident. All other incidences are inherited from $PG(5, q)$.

Theorem 4.1 *If $q = p^n$ and $p \neq 3$, then $H(\Sigma)$ is isomorphic to the dual $G_2(q)$-hexagon.*

Proof. If $r_\infty = (0, 0, 0, 1, 0, 0)$ and $r_u = (0, 1, u, u^3, -3u^2, 0)$, then $\Sigma = \{r_u : u \in \hat{F}\}$ is a twisted cubic of $S = p^\perp \cap o^\perp$. Moreover, we can prove with a direct calculation that the symplectic polarity π of S associated with $W(3, q)^3$ is the unique symplectic polarity of $S = PG(3, q)$ with the property that r^π is the osculating plane of Σ at r for each point r of Σ ([8], Section 43). Then L_∞ has equations $x_0 = x_1 = x_2 = x_5 = 0$ and

$$L_u = \{(0, x_1, x_2, x_3, x_4, 0) \in S : x_4 = 3u^2 x_1 - 6u x_2; x_3 = -2u^3 x_1 + 3u^2 x_2\}.$$

For each $u \in \hat{F}$, let

$$
\begin{aligned}
\tilde{A}_4(u) &= \{\tilde{g} \in \tilde{G} : r_u \tilde{g} = r_u\}, \\
\tilde{A}_3(u) &= \{\tilde{g} \in \tilde{G} : L_u \tilde{g} = L_u\}, \\
\tilde{A}_2(u) &= \{\tilde{g} \in \tilde{G} : < L_u, o > \tilde{g} = < L_u, o >\}, \\
\tilde{A}_1(u) &= \{\tilde{g} \in \tilde{G} : L_u \tilde{g} = L_u, o\tilde{g} \in < o, r_u >\}.
\end{aligned}
$$

With a direct calculation, one can prove that

$$
\begin{aligned}
\tilde{A}_4(u) &= \{M(a, b, au^3 - 3bu^2 - du, d, e') : a, b, d, e' \in F\}, \\
\tilde{A}_3(u) &= \{M(a, b, -2au^3 + 3bu^2, 3au^2 - 6bu, e') : a, b, e' \in F\},
\end{aligned}
$$

---

3. π is defined by the alternating bilinear form $f((0, x_1, x_2, x_3, x_4, 0), (0, y_1, y_2, y_3, y_4, 0)) = x_1 y_3 + x_2 y_4 - x_3 y_1 - x_4 y_2$

$$\tilde{A}_2(u) = \{M(a,b,-2au^3+3bu^2,3au^2-6bu,0) : a,b \in F\},$$
$$\tilde{A}_1(u) = \{M(a,au,au^3,-3au^2,0) : a \in F\}.$$

if $u \neq \infty$, and

$$\tilde{A}_4(\infty) = \{M(0,b,c,d,e') : b,c,d,e' \in F\},$$
$$\tilde{A}_3(\infty) = \{M(0,0,c,d,e') : c,d,e' \in F\},$$
$$\tilde{A}_2(\infty) = \{M(0,0,c,d,0) : c,d \in F\},$$
$$\tilde{A}_1(\infty) = \{M(0,0,0,d,0) : d \in F\}.$$

For $u \in \hat{F}$ and $i = 1,2,3,4$, let $A_i(u)$ be the subgroup of G defined by $A_i(u) = \{g \in G : \tilde{g} \in \tilde{A}_i(u)\}$. For each u in $\hat{F}$ different from ∞, we have

$$A_4(u) = \{(a,b,c,au^3-3bu^2-du,d) : a,b,c,d, \in F\},$$
$$A_3(u) = \{(a,b,c,-2au^3+3bu^2,3au^2-6bu) : a,b,c \in F\},$$
$$A_2(u) = \{(a,b,-a^2u^3+3abu^2-3b^2u,-2au^3+3bu^2,3au^2-6bu) : a,b \in F\},$$
$$A_1(u) = \{(a,au,-a^2u^3,au^3,-3au^2) : a \in F\}.$$

Moreover

$$A_4(\infty) = \{(0,b,e,c,d) : b,c,d,e \in F\},$$
$$A_3(\infty) = \{(0,0,e,c,d) : c,d,e \in F\},$$
$$A_2(\infty) = \{(0,0,0,c,d) : c,d \in F\},$$
$$A_1(\infty) = \{(0,0,0,0,d) : d \in F\}.$$

Let $H = (P,L,I)$ be the point-line geometry defined in Section 2. As the subgroups $A_4(u)$, $A_3(u)$, $A_2(u)$ and $A_1(u)$ of G are those associated with the dual $G_2(q)$-hexagon ([4] or [10] 10.6.2)[4], H is isomorphic to the dual $G_2(q)$-hexagon. Finally the map $\theta : H \to H(\Sigma)$ defined by

$$\theta : I \mapsto p,$$
$$\theta : [u] \mapsto < p, r_u >,$$
$$\theta : A_4(u)g \mapsto r_u\tilde{g},$$
$$\theta : A_3(u)g \mapsto L_u\tilde{g},$$
$$\theta : A_2(u)g \mapsto < L_u, o > \tilde{g},$$
$$\theta : A_1(u)g \mapsto < r_u, o > \tilde{g},$$
$$\theta : g \mapsto o\tilde{g}.$$

is an isomorphism. □

H. Van Maldeghem (private communication) has proved Theorem 4.1 using coordinates.

---

4. The group used by W.M. Kantor is $G(\circ)$ where $(a,b,c,d,e)\circ(a',b',c',d',e') = (a+a',b+b',c+c'+e'a-3b'd,d+d',e+e')$; the map $\phi : G(\circ) \to G, (a,b,c,d,e) \mapsto (a,b,c,e,-3d)$, is a group isomorphism.

5. BLT-sets and generalized hexagons

Let $q = p^n$ with p an odd prime number.

A <u>BLT-set</u> is a set S of $q+1$ totally isotropic lines of $W(3,q)$ such that each totally isotropic line of $W(3,q) \setminus S$ is concurrent with exactly 0 or 2 lines of S.

In [6] N. Knarr has given the following beautiful construction starting from a BLT-set S of $W(3,q)$. Define a point-line geometry $Q(S)$ as follows.

<u>Points:</u>

    (i)   the point p;

    (ii) the lines not containing p but contained in one of the planes $< p, L >$ where L is a line of S;

    (iii) the points of $PG(5,q) \setminus p^{\perp}$.

<u>Lines:</u>

    (A) the planes $< p, L >$ where $L \in S$;

    (B) the totally isotropic planes of $W(5,q)$ not contained in $p^{\perp}$ and meeting some $< p, L > (L \in S)$ in a line not through p.

<u>Incidences:</u>

The incidences are just the natural incidences inherited from $PG(5,q)$.

The point-line geometry $Q(S)$ is a generalized quadrangle with parameters (q^2, q).

For the relation between BLT-sets and flocks of the quadratic cone of $PG(3,q)$ see [2] and [11]. Moreover, S.E. Payne and J.A. Thas have proved that for each BLT-set S of $W(3,q)$, there is a flock $\mathcal{F}$ of the quadratic cone of $PG(3,q)$ such that $Q(S)$ is isomorphic to the generalized quadrangle associated with $\mathcal{F}$ and vice versa (see [11]).

Suppose $q \neq 3^n$. Let $\Sigma = \{r_u : u \in \hat{F}\}$ be a twisted cubic of $PG(3,q)$ and let L_u be the tangent of Σ at p_u. We suppose that $W(3,q)$ is the symplectic 3-space associated with the polarity π of $PG(3,q)$ such that r_u^{π} is the osculating plane of Σ at the point r_u for each u in $\hat{F}$.

Let $H(\Sigma)$ and H be the generalized hexagons constructed in Section 4 starting from Σ. If $S = \{L_u : u \in \hat{F}\}$, we can define a point-line geometry $Q(\Sigma, S)$ in the following way. The points of $Q(\Sigma, S)$ are the elements of $H(\Sigma)$ at distance zero, three or six from the special point p. The lines of $Q(\Sigma, S)$ are the elements of $H(\Sigma)$ at distance either one or four from p. If x is an element at distance 1 from p and y is an element at distance 3 from p, then $x = < p, r_u >$ for some $u \in \hat{F}$ and y is a line of $p^{\perp}$ contained in some plane $< L_v, p >$. We say that x and y are incident in $Q(\Sigma, S)$ if and only if x and y have a common point, i.e. $< x, y > = < p, L_u >$ for some u in $\hat{F}$. If x' is an element at distance 4 from p and y' is an element at distance 6 from p, then x' is a totally singular plane and y' is a point not in $p^{\perp}$. We say that x' and y' are incident in $Q(\Sigma, S)$ if and only if y' is a point of the plane x' of $PG(5,q)$.

All other incidences are inherited from those of $H(\Sigma)$.

The point-line geometries $Q(H)$ (see Section 2) and $Q(\Sigma, \mathcal{S})$ are isomorphic via the map $\tilde{\theta} : Q(H) \to Q(\Sigma, \mathcal{S})$ defined as follows

$$\tilde{\theta} \ : \ \mathcal{I} \mapsto p,$$
$$\tilde{\theta} \ : \ [u] \mapsto <p, L_u>,$$
$$\tilde{\theta} \ : \ A_3(u)g \mapsto L_u\tilde{g},$$
$$\tilde{\theta} \ : \ A_2(u)g \mapsto <L_u, o>\tilde{g},$$
$$\tilde{\theta} \ : \ g \mapsto o\tilde{g}.$$

The geometry $Q(\Sigma, \mathcal{S})$ is a generalized quadrangle if and only if $\mathcal{S}$ is a BLT-set of $W(3, q)$ (see [11] Section III). The partial spread $\mathcal{S}$ is a BLT-set of $W(3, q)$ if and only if $q \equiv -1 \pmod 3$ (see [7]). In this case, $Q(\Sigma, \mathcal{S})$ is the generalized quadrangle associated with the Fisher-Thas-Walker flock[5] of the quadratic cone of $PG(3, q)$ (see [7]). So we have given an alternative proof of the following theorem contained in [4], [9] and [13].

Theorem 5.1 $Q(H)$ *is a generalized quadrangle with parameters* (q^2, q) *if and only if* $q \equiv -1 \pmod 3$. *Moreover, the Kantor generalized quadrangle* $Q(H)$ *is isomorphic to the generalized quadrangle associated with the Fisher-Thas-Walker flock.*

References

[1] **L. Bader.** Flocks of cones and generalized hexagons. In J. W. P. Hirschfeld, D. R. Hughes, and J. A. Thas, editors, *Advances in Finite Geometries and Designs*, Oxford, 1991. Oxford University Press.

[2] **L. Bader, G. Lunardon, and J. A. Thas.** Derivation of flocks of quadratic cones. *Forum Math.*, 2, pp. 163–174, 1990.

[3] **J. C. Fisher and J. A. Thas.** Flocks in PG(3, q). *Math. Z.*, 169, pp. 1–11, 1979.

[4] **W. M. Kantor.** Generalized quadrangles associated with $G_2(q)$. *J. Combin. Theory Ser. A*, 29, pp. 212–219, 1980.

[5] **W. M. Kantor.** Generalized polygons, SCABs and GABs. In L. A. Rosati, editor, *Buildings and the Geometry of Diagrams*, *Proceedings Como 1984*, volume 1181 of *Lecture Notes in Math.*, pages 79–158, Berlin, 1986. Springer Verlag.

[6] **N. Knarr.** A geometric construction of generalized quadrangles from polar spaces of rank three. *Resultate Math.*, 21, pp. 332–344, 1992.

---

5. The Fisher-Thas-Walker flock is defined and investigated in [3] and [14]

[7] G. Lunardon. A remark on the derivation of flocks. In J. W. P. Hirschfeld, D. R. Hughes, and J. A. Thas, editors, *Advances in Finite Geometries and Designs*. Oxford University Press, 1991.

[8] H. Lüneburg. *Translation Planes*. Springer Verlag, Berlin, 1980.

[9] S. E. Payne. Generalized quadrangles as group coset geometries. *Congr. Numer.*, 29, pp. 717–734, 1980.

[10] S. E. Payne and J. A. Thas. *Finite Generalized Quadrangles*, volume 104 of *Research Notes in Mathematics*. Pitman, Boston, 1984.

[11] S. E. Payne and J. A. Thas. Generalized quadrangles, BLT-sets and Fisher flocks. *Congr. Numer.*, 84, pp. 161–192, 1991.

[12] S. E. Payne and M. G. Tinsley. On $v_1 \times v_2(n, s, t)$ configurations. *J. Combin. Theory*, 7, pp. 1–14, 1969.

[13] J. A. Thas. Generalized quadrangles and flocks of cones. *Europ. J. Combinatorics*, 8, pp. 441–452, 1987.

[14] M. Walker. A class of translation planes. *Geom. Dedicata*, 5, pp. 135–146, 1976.

L. Bader, Dipartimento di Matematica - II Università di Roma, Via della Ricerca Scientifica, I-00133 Roma, Italy. e-mail: Bader@mat.utovrm.it

G. Lunardon Dipartimento di Matematica e Applicazioni - Università di Napoli, Complesso di Monte S. Angelo, Edificio T, Via Cinthia, I-80126 Napoli, Italy. e-mail: Lunardon@napoli.infn.it

Orthogonally divergent spreads of Hermitian curves

R. D. Baker   **G. L. Ebert**
G. Korchmáros   **T. Szőnyi**

Abstract

By a spread of an Hermitian curve $\mathcal{H}$ embedded in the Desarguesian plane $\pi = PG(2, q^2)$ we mean any collection of $q^2 - q + 1$ nonabsolute lines that partition the $q^3 + 1$ points of $\mathcal{H}$. We call such a spread orthogonally divergent (o. d. for short) if no line contains the pole of any line in the spread. The search for orthogonally divergent spreads of $\mathcal{H}$ is a reformulation of a problem posed by A. A. Bruen at Combinatorics '90, which in turn is related to the geometric construction of large sets of independent codewords in the natural linear code associated with π. In this paper we show o.d. spreads exists for any even prime power q with $q \equiv 1 \pmod 3$, and find o.d. partial spreads of small deficiency in all other cases. The inherited automorphism groups are also determined.

1. Introduction

It is easy to construct a spread of the Hermitian curve $\mathcal{H}$ embedded in the Desarguesian plane $\pi = PG(2, q^2)$. Namely, if P is any point of $\pi \setminus \mathcal{H}$, the $q^2 - q$ chords of $\mathcal{H}$ through P together with $P^\perp$ intersect $\mathcal{H}$ in a collection of $q^2 - q + 1$ blocks of $\mathcal{H}$ that partition its $q^3 + 1$ points. We thus call these $q^2 - q + 1$ lines of π a "spread" of $\mathcal{H}$. In this paper we impose the additional restriction that no line of the spread should contain the pole of any line in the spread. We think of such a spread, if it exists, as being "orthogonally divergent." In general, the problem is to find the largest possible orthogonally divergent (o.d. for short) partial spread of $\mathcal{H}$.

We first show that an o.d. spread of $\mathcal{H}$ exists for any even prime power q with $q \equiv 1 \pmod 3$. This spread is cyclically generated. We also construct maximal o.d. partial spreads of $\mathcal{H}$ with deficiency 3 for any prime power q with $q \not\equiv 0 \pmod 3$. For $q \equiv 0 \pmod 3$ maximal o.d. partial spreads of deficiency $\frac{1}{2}(q + 7)$ are constructed. The inherited automorphism groups of these partial spreads are then determined.

2. Preliminary results

Let $\pi = PG(2, q^2)$ denote the Desarguesian projective plane over the Galois field $GF(q^2)$, where q is any prime power. An *Hermitian curve* $\mathcal{H}$ in π is the set of absolute points of an Hermitian (or unitary) polarity of π. It is well known that any two Hermitian curves are projectively equivalent. If P is any point of π, then $P^\perp$ denotes the polar line of P with respect to the Hermitian polarity associated with $\mathcal{H}$. Similarly, if ℓ is any line of π, then $\ell^\perp$ denotes the pole of ℓ with respect to this polarity. If $P \in \mathcal{H}$, then $P^\perp$ meets $\mathcal{H}$ only in P; if $P \notin \mathcal{H}$, then $P^\perp$ meets $\mathcal{H}$ in $q+1$ points. We will call lines of the first type *tangents* and those of the second type *chords* (or *secants*). If $P \notin \mathcal{H}$, there are $q+1$ tangents through P (meeting $\mathcal{H}$ in the points of $P^\perp \cap \mathcal{H}$) and $q^2 - q$ chords through P.

At Combinatorics '90 in Gaeta A.A. Bruen posed the following problem concerning Hermitian curves. Let $\mathcal{B} = \{P_1, P_2, \ldots, P_t\}$ be a set of t points in π satisfying the two properties:

(a) the line $P_i P_j$ is a chord of $\mathcal{H}$ for all $i \neq j$,

(b) $P_i^\perp \cap \mathcal{B} = \emptyset$ for all i.

What is the maximum cardinality of such a set $\mathcal{B}$, hereafter referred to as a *B–set* for short, and how can we construct such sets?

The motivation for studying this problem is the following. Let p denote the characteristic of the field $GF(q^2)$, and let $C_p(\pi)$ denote the linear code over $\mathbb{Z}_p$ spanned by the rows of the incidence matrix of π, where we assume the rows are indexed by the lines of π. Then it is easy to see that the tangent lines to $\mathcal{H}$, treated as characteristic vectors in $C_p(\pi)$, are linearly independent codewords. Now for each point P of a B–set $\mathcal{B}$, construct a codeword w_P by summing the $q+1$ tangents to $\mathcal{H}$ through P and subtracting $P^\perp$. Then the support of w_P contains P but no other point of $\mathcal{B} \cup \mathcal{H}$. An easy argument now shows that $\{w_P : P \in \mathcal{B}\}$ may be appended to the tangents of $\mathcal{H}$ to create an independent set of codewords of size $|\mathcal{B}| + |\mathcal{H}|$. Thus B–sets may be used to construct "large" independent sets of codewords in a natural geometric way.

Bruen observed that if ℓ is any chord of $\mathcal{H}$ and we define $P^* = P^\perp \cap \ell$ for any point $P \in \ell \setminus \mathcal{H}$, then the $q^2 - q$ points of $\ell \setminus \mathcal{H}$ are partitioned into pairs $\{P, P^*\}$. Choosing one point from each pair yields a set $\mathcal{B}$ of cardinality $\frac{1}{2}(q^2 - q)$ that obviously satisfies properties (a) and (b) above. Such a B–set will be called *linear* since all of its points are collinear.

If $\mathcal{B} = \{P_1, P_2, \ldots, P_t\}$ is a B–set and we work dually, then $\mathcal{B}^\perp = \{P_1^\perp, P_2^\perp, \ldots, P_t^\perp\}$ is a collection of chords of $\mathcal{H}$, any two of which meet in a point off $\mathcal{H}$ by property (a). The intersections of these lines with $\mathcal{H}$ thus form a collection of pairwise disjoint blocks of $\mathcal{H}$ (treating $\mathcal{H}$ as a 2–design), and we call $\mathcal{B}^\perp$ a *partial spread* of $\mathcal{H}$. As $\mathcal{H}$ has $q^3 + 1$ points, clearly $t(q+1) \leq q^3 + 1$ and thus $t \leq q^2 - q + 1$. If this upper bound for $|\mathcal{B}|$ is achieved, then $\mathcal{B}^\perp$

becomes a (full) spread of $\mathcal{H}$ which in addition satisfies the property that no line of the spread contains the pole of any spread line (including itself). Such a spread of $\mathcal{H}$ we call *orthogonally divergent*.

3. Construction of a maximum B–set

In this section we are going to construct cyclic B-sets of maximum possible cardinality (or dually, o.d. spreads) in $PG(2, q^2)$. Our construction depends on the existence of a cyclic linear collineation group G of order $q^2 - q + 1$ that leaves $\mathcal{H}$ invariant. This group G is a subgroup of a Singer group Σ of $PG(2, q^2)$ and each point orbit under G is a complete $(q^2 - q + 1)$-arc. Moreover, $\mathcal{H}$ may be obtained as the union of $q + 1$ suitably chosen orbits under G. Finally, each point orbit under a subgroup of order $q^2 + q + 1$ in Σ is a Baer–subplane.

Following [1] let us represent $PG(2, q^2)$ by considering $GF(q^6)$ as a vector space over $GF(q^2)$. This means that the points are represented by the elements of $GF(q^6) \setminus \{0\} = GF(q^6)^*$, and $\alpha, \beta \in GF(q^6)$ represent the same point if and only if $\alpha/\beta \in GF(q^2)$. The point represented by $\alpha \in GF(q^6)$ will be denoted by (α). The lines in this representation are defined by an equation $\text{Tr}(ax) = 0$, where $a \neq 0$ is fixed and Tr stands for the trace function from $GF(q^6)$ onto $GF(q^2)$.

If ω is a generating element of $GF(q^6)$, then the group G, the Baer-subplanes and the complete arcs mentioned above are the following:

$$G = \{ \, \varphi : GF(q^6) \to GF(q^6) \; : \; \varphi(x) = x\omega^{i(q^2+q+1)}, \; i = 0, 1, \dots, q^2 - q \, \},$$

$$\text{Baer}(u) = \{ \, (\omega^{i(q^2-q+1)+u}) \; : \; i = 0, 1, \dots q^2 + q \, \} \text{ for } u = 0, 1, \dots, q^2 - q,$$

$$\text{Arc}(t) = \{ \, (\omega^{i(q^2+q+1)+t}) \; : \; i = 0, 1, \dots q^2 - q \, \} \text{ for } t = 0, 1, \dots, q^2 + q.$$

Finally, we can represent an Hermitian curve for each $a \in GF(q^3)^*$ by putting

$$\mathcal{H}(a) = \{ \, (x) \; : \; \text{Tr}(ax^{q^3+1}) = 0 \, \}.$$

One can easily see that $\mathcal{H}(a)$ is left invariant by G. In this section we will be working with the Hermitian curve $\mathcal{H} = \mathcal{H}(1)$.

First we investigate the existence of cyclic spreads of $\mathcal{H}$. One may assume that the cyclic group is G above, since any two cyclic subgroups of order $q^2 - q + 1$ are conjugate in the linear collineation group $PGU(3, q^2)$ leaving $\mathcal{H}$ invariant. Let us recall that $\mathcal{H} \cap \text{Baer}(0)$ is a line or an oval accordingly as q is even or odd. Indeed, if we consider the Baer-involution $x \to x^{q^3}$, then it fixes $\text{Baer}(0)$ pointwise and maps $\mathcal{H}$ into itself, so the assertion follows from [2] .

19

Theorem 3.1 *There exists a cyclic spread of* $\mathcal{H}$ *if and only if* q *is even. Moreover, the spread is unique if we fix the cyclic group of order* $q^2 - q + 1$.

Proof.   We can suppose that the cyclic group is G above. In [1, Cor. 2.4] it was proved that $\mathcal{H}$ is the union of certain arcs Arc(t). It is easily seen that the orbit of a line under G is a spread of $\mathcal{H}$ if and only if each Arc(t) contained in $\mathcal{H}$ meets each line of the orbit in exactly one point. Assume first that a cyclic spread of $\mathcal{H}$ exists. For a point $x \in$ Baer(0)$\cap\mathcal{H}$ let ℓ denote the line of the orbit passing through x, and let Arc(t) denote the arc containing x. As the tangents of Arc(t) at x are precisely the lines of Baer(0) ([1, Lemma 1.3]), ℓ must be a line of Baer(0). If q is odd or if q is even and ℓ is different from Baer(0)$\cap\mathcal{H}$, then we have a point $y \in \mathcal{H}\cap\ell$ not in Baer(0). Take $u(\neq 0)$ and s in such a way that $y \in$ Baer(u) and $y \in$ Arc(s). Since ℓ is a tangent of Arc(s), we see, as before, that ℓ must be a line of Baer(u). Since Baer(0) and Baer(u) have no common line, we have a contradiction. This shows that q must be even and $\ell =$ Baer(0)$\cap\mathcal{H}$.

On the other hand, if q is even the orbit of the line $\ell =$ Baer(0)$\cap\mathcal{H}$ under G turns out to be a spread of $\mathcal{H}$. Indeed, ℓ meets each Arc(t) contained in $\mathcal{H}$ in exactly one point by Theorem 2.5 of [1] . $\square$

Now we are going to decide whether our spread is o.d. or not. We make use of a particular collineation, namely $\mu : x \rightarrow x^{q^2}$, which leaves both $\mathcal{H}$ and Baer(0) invariant. Obviously, μ has order 3 and permutes the arcs of type Arc(t) among themselves. One can easily compute the fixed points of μ. Namely, let $x^\mu = x$ with $x = \omega^i$. Then $iq^2 \equiv i \pmod{q^4 + q^2 + 1}$, and hence $i(q^2 - 1) \equiv 0 \pmod{q^4 + q^2 + 1}$. Here the greatest common divisor of $q^2 - 1$ and $q^4 + q^2 + 1$ is 3, so the solutions of this congruence mod $q^4 + q^2 + 1$ are exactly $i = 0, \frac{1}{3}(q^4 + q^2 + 1)$, and $\frac{2}{3}(q^4 + q^2 + 1)$. Thus μ has exactly three fixed points.

Theorem 3.2 *There is a cyclic o.d. spread of* $\mathcal{H}$ *if and only if* q *is even and* $q \equiv 1 \pmod 3$.

Proof.   By Theorem 3.1 we may assume that q is even and the cyclic spread is the orbit of $\ell = \mathcal{H}\cap$ Baer(0) under G. Let L be the line of $PG(q^2)$ containing the Baer subline ℓ. Let Arc(s) be the orbit of $L^\perp$ under G, so Arc(s)$\cap\mathcal{H} = \emptyset$. Observe that $\ell^\mu = \ell$, and this implies that $(L^\perp)^\mu = L^\perp$, hence μ leaves Arc(s) invariant. As G is transitive on the lines of our spread, it is sufficient to check the condition of orthogonal divergence (Condition (b) in Section 2.) for L; that is, L should be disjoint from Arc(s).

Suppose first that our spread is not o.d., and hence Arc(s)$\cap L \neq \emptyset$. Now

$\text{Arc}(s) \cap L$ consists of one or two points. Since μ leaves $\text{Arc}(s) \cap L$ invariant and μ has order three, μ must fix the points of $\text{Arc}(s) \cap L$. But μ also fixes $L^{\perp} \in \text{Arc}(s) \backslash L$, and thus μ fixes at least two points of $\text{Arc}(s)$. Since $|\text{Arc}(s)| = q^2 - q + 1$, the comments above imply that μ fixes exactly three points of $\text{Arc}(s)$ and hence a cardinality argument shows $q \equiv 2 \pmod 3$.

On the other hand, if $q \equiv 2 \pmod 3$, then $|\text{Arc}(s)|$ is divisible by 3. Since μ leaves $\text{Arc}(s)$ invariant and fixes the point $L^{\perp}$ of $\text{Arc}(s)$, μ must fix two other points of $\text{Arc}(s)$. As L is left invariant by μ and $|L| = q^2 + 1 \equiv 2 \pmod 3$, μ must fix two points of L. But μ has exactly three fixed points, and thus $L \cap \text{Arc}(s) \neq \emptyset$, implying our spread is not o.d.

Since q is even, either $q \equiv 1 \pmod 3$ or $q \equiv 2 \pmod 3$, and the result now follows. $\qquad\square$

It should be remarked that the construction given in Theorems 3.1 and 3.2 yields an o.d. partial spread of size $\frac{1}{3}(q^2 - q + 1)$ when q is even and $q \equiv 2 \pmod 3$. In the next section we discuss a different construction technique which will generate a larger o.d. partial spread in this case (and other cases).

4.   Maximal B–sets of small deficiency

In this section we assume without loss of generality that the Hermitian curve $\mathcal{H}$ has equation $x^{q+1} + y^{q+1} + z^{q+1} = 0$, where (x, y, z) are homogeneous coordinates for $\pi = PG(2, q^2)$. We begin by partitioning the points of $\mathcal{H}$ in a very natural way. We left normalize our point coordinates to get a unique representation for each point of π. Letting $GF(q)$ denote the unique subfield of index 2 in $GF(q^2)$, we define $R_a = \{(1, y, z) : y^{q+1} = -(1 + a), z^{q+1} = a\}$ for each element $a \in F \equiv GF(q) \backslash \{0, 1\}$. We also define $U_1 = \{(0, 1, z) : z^{q+1} = -1\}, U_2 = \{(1, 0, z) : z^{q+1} = -1\}$, and $U_3 = \{(1, y, 0) : y^{q+1} = -1\}$. Then it is trivial to see that $\left(\bigcup_{a \in F} R_a \right) \cup U_1 \cup U_2 \cup U_3$ is a partition of the points of $\mathcal{H}$. The key ingredient is that the R_a's are "triply ruled," as we now describe.

We use $[\text{-},\text{-},\text{-}]$ to denote line coordinates in π, and we normalize line coordinates from the right. One easily checks that $(x, y, z)^{\perp} = [x^q, y^q, z^q]$ because of our choice for $\mathcal{H}$. For each $a \in F$, let $H_a = \{[u, 0, 1] : u^{q+1} = a\}, V_a = \{[u, 1, 0] : u^{q+1} = -(1+a)\}$, and $D_a = \{[0, v, 1] : v^{q+1} = -a/(1+a)\}$. Thus H_a is a partial pencil of $q + 1$ lines of π passing through the point $Q_1 = (0, 1, 0) \notin \mathcal{H}$. In fact each line of H_a meets $\mathcal{H}$ in $q + 1$ points of R_a, and thus H_a is a ruling class of R_a (which necessarily makes it a partial spread of $\mathcal{H}$). Similarly, V_a and D_a are ruling classes of R_a which consist of partial pencils of $q + 1$ chords of $\mathcal{H}$ passing through the nonabsolute points $(0, 0, 1)$ and $(1, 0, 0)$, respectively. To construct a large partial spread of $\mathcal{H}$ (which is hopefully orthogonally divergent), we select exactly one of the above three ruling classes for each $a \in F$. Let $AH = \{a \in F : H_a \text{ is chosen}\}, AV = \{a \in F : V_a \text{ is chosen}\}$, and $AD = \{a \in F : D_a \text{ is chosen}\}$. We think of the H_a's, V_a's and D_a's as being the "horizontal," "vertical," and "diagonal" ruling classes, respectively.

Since the R_a's are pairwise disjoint, the union of our partial pencils will clearly form a partial spread of $\mathcal{H}$. To check for orthogonal divergence, we check that the pole of each line in the partial spread does not lie on any line of the partial spread. Let $[u_1, 0, 1] \in H_{a_1}$ and $[u_2, 0, 1] \in H_{a_2}$ be (not necessarily distinct) lines of our chosen partial spread for some $a_1, a_2 \in AH$. Then $[u_1, 0, 1]^\perp = (u_1^q, 0, 1)$ and $(u_1^q, 0, 1) \cdot [u_2, 0, 1] = u_1^q u_2 + 1 \neq 0$ provided $u_2 \neq -1/u_1^q$. (Note that $u_1 \neq 0$ as $a_1 \neq 0$.) But

$$u_2 = -1/u_1^q \;\Rightarrow\; u_2^{q+1} = 1/u_1^{q+1}$$
$$\Rightarrow\; a_2 \;= 1/a_1.$$

Thus, if we restrict our selection process so that $a_2 \neq 1/a_1$ for any $a_1, a_2 \in AH$ (in particular, $1 \notin AH$), then no two lines (distinct or not) from the horizontal ruling classes chosen will be "orthogonal mates." In general, we say that two lines of π are *orthogonal mates with* respect to $\mathcal{H}$ if the pole of one line lies on the other. Similarly, we see that no two lines of the vertical ruling classes chosen will be orthogonal mates if $a_2 \neq -a_1/(1 + a_1)$ for any $a_1, a_2 \in AV$ (in particular, $-2 \notin AV$), and no two lines of the diagonal ruling classes chosen will be orthogonal mates if $a_2 \neq -(1 + a_1)$ for any $a_1, a_2 \in AD$ (in particular, $-\frac{1}{2} \notin AD$). Analogous computations also show that no two lines from ruling classes of different types can be orthogonal mates.

The problem now becomes one of partitioning the elements of F among the sets AH, AV and AD subject to the above restrictions. Assume first that q is odd. Let $\mathcal{P}_H$ be a partitioning of the elements of $F \setminus \{1\} = GF(q) \setminus \{0, -1, 1\}$ into unordered pairs $\{a, 1/a\}$. Let $\mathcal{P}_V$ be a partitioning of $F \setminus \{-2\}$ into unordered pairs $\{a, -a/(1 + a)\}$, and let $\mathcal{P}_D$ be a partitioning of $F \setminus \{-\frac{1}{2}\}$ into unordered pairs $\{a, -(1 + a)\}$. We now think of listing the pairs of $\mathcal{P}_H$ in one row, say the H row, the pairs of $\mathcal{P}_V$ in a second row, say the V row, and the pairs of $\mathcal{P}_D$ in a third row, say the D row. Initialize by setting AH, AV and AD each equal to the empty set. Begin the search by picking one element, say a, from any pair in the H row, and then eliminating that element from consideration in any other row. The element a gets adjoined to the set AH. Next find the unique pair in the V row containing the element most recently added (a, in this case), and adjoin the other element of that pair (namely, $-a/(1 + a)$) to the set AV. This element now gets removed from further consideration in all rows. Move to the D row, find the unique pair containing the most recently adjoined element ($-a/(1 + a)$, in this case), pick the other element of that pair (namely, $-1/(1 + a)$), and adjoin this element to AD.

Move back to row H, and continue the process. In this way we choose six elements of F as follows:

$$(\#) \qquad
\begin{array}{llll}
& & a & -(1+a) \\
AH: & & \downarrow & \downarrow \\
AV: & & -a/(1+a) & -(1+a)/a \\
& & \downarrow & \downarrow \\
AD: & -1/(1+a) & & 1/a
\end{array}$$

If there are still pairs under consideration in the H row, pick any element a' from such a pair and repeat the process to get another collection of six elements of F. Eventually, all pairs on all rows will be removed from consideration.

In the above process it is possible to get "short orbits" where the six elements chosen above are not distinct. However, we will see that this poses no serious problem.

Theorem 4.1 *Let q be any odd prime power.*

i) *If $q \not\equiv 0 \pmod 3$, there exists an orthogonally divergent partial spread of $\mathcal{H}$ of size $q^2 - q - 2$ (and hence deficiency 3). Dually, there exists a B–set of cardinality $q^2 - q - 2$. This B–set is maximal and all its points lie on the sides of a self–polar triangle.*

ii) *If $q \equiv 0 \pmod 3$, there exists an orthogonally divergent partial spread of $\mathcal{H}$ of size $q^2 - \frac{3}{2}q - \frac{5}{2}$ (and hence deficiency $\frac{q+7}{2}$). Dually, there exists a B–set of cardinality $q^2 - \frac{3}{2}q - \frac{5}{2}$, all of whose points lie on the sides of a self–polar triangle. If $q > 3$ this B–set is maximal.*

Proof. (i) Suppose first that $q \equiv 1 \pmod 3$. Let ϵ denote a primitive cube root of unity in $GF(q)$. Then the unordered pair $\{\epsilon, \epsilon^2\}$ appears in all three partitions $\mathcal{P}_H, \mathcal{P}_V$ and $\mathcal{P}_D$. Moreover, the selection process described above generates the short orbit consisting of the two elements ϵ and ϵ^2.

Without loss of generality we place $\epsilon \in AH$ and $\epsilon^2 \in AV$. However, we could just as easily choose any two of AH, AV, AD and place ϵ in one set and ϵ^2 in the other. We also have three other special pairs; namely, $\big\{-2, -\frac{1}{2}\big\} \in \mathcal{P}_H, \big\{1, -\frac{1}{2}\big\} \in \mathcal{P}_V$, and $\{1, -2\} \in \mathcal{P}_D$. The selection process $(\#)$, if started with $a = -\frac{1}{2}$, will generate the short orbit consisting of $-\frac{1}{2}, 1$, and -2. Again, without loss of generality, we place $-\frac{1}{2} \in AH, 1 \in AV$, and $-2 \in AD$ (although we could just as well place $-2 \in AH, -\frac{1}{2} \in AV$, and $1 \in AD$). We now have the same $q - 7$ distinct elements of $GF(q)$ remaining on each of the H, V, and R rows, partitioned in different ways into $\frac{1}{2}(q - 7)$ pairs in each row. No more short orbits are possible, and the selection process $(\#)$ generates $\frac{1}{6}(q - 7)$ full orbits of size 6 each. For each such orbit two

23

elements are placed in each of AH, AV, and AD. Of course, in any given row, some pairs may have neither element selected (although those elements will be selected in other rows).

We thus are able to partition all $q - 2$ elements of F among the sets AH, AV, and AD subject to the previously mentioned restrictions. Therefore we have an orthogonally divergent partial spread of $\mathcal{H}$ of size $(q-2)(q+1) = q^2 - q - 2$. Taking the pole of each such spread line, we get a B–set of the same size. Since every line in our partial spread passes through $(0,1,0), (0,0,1)$, or $(1,0,0)$ as previously discussed, every point of the B–set lies on one of the lines $(0,1,0)^{\perp} = [0,1,0], (0,0,1)^{\perp} = [0,0,1]$, or $(1,0,0)^{\perp} = [1,0,0]$. These three lines clearly form the sides of a self–polar triangle. It should also be noted that the subsets of size $q + 1$ in the B–set which are the poles of any ruling class of some R_a are Baer sublines of one side of this triangle. Finally, this B–set is clearly maximal since, dually, the only points of $\mathcal{H}$ uncovered by the partial spread are $U_1 \cup U_2 \cup U_3$. The only chords of $\mathcal{H}$ meeting $\mathcal{H}$ in a subset of these uncovered points are $[1,0,0], [0,1,0]$, and $[0,0,1]$, none of which may be added to the partial spread if we want to maintain orthogonal divergence.

Now suppose that $q \equiv 2 \pmod{3}$. The argument proceeds exactly as above except that now there are no primitive cube roots of unity in $GF(q)$. Thus the only short orbit under the selection process ($\#$) is $-\frac{1}{2} \to 1 \to -2$. The $q - 5$ points of $F \setminus \left\{ -\frac{1}{2}, 1, -2 \right\}$ are partitioned into $\frac{1}{6}(q - 5)$ orbits of size 6 by ($\#$), and we are able to construct an orthogonally divergent partial spread of $\mathcal{H}$ if size $q^2 - q - 2$ in a completely analogous fashion.

(ii) Finally suppose that $q \equiv 0 \pmod{3}$. Since $1 = -2 = -\frac{1}{2}$ in this case, we cannot use a full ruling class (of any one of the three types) for R_1. Without loss of generality we try to use as many lines as possible in the ruling class D_1. Let $[0, v_1, 1], [0, v_2, 1] \in D_1$, where $v_1^{q+1} = 1 = v_2^{q+1}$. Then $[0, v_1, 1]^{\perp} = (0, v_1^q, 1)$ and $(0, v_1^q, 1) \cdot [0, v_2, 1] = v_1^q v_2 + 1 \neq 0$ provided $v_2 \neq -1/v_1^q$. Thus we can choose $\frac{1}{2}(q + 1)$ lines from D_1 so that no two are orthogonal mates. Namely, we may as well choose $[0, v, 1]$, where $v = 1, \beta^{q-1}, \beta^{2(q-1)}, \ldots, \beta^{(q-1)^2/2}$ for some primitive element β of $GF(q^2)$. There are no primitive cube roots of unity in $GF(q)$ when $q \equiv 0 \pmod{3}$, and the $q - 3$ points of $F \setminus \{1\}$ may be partitioned into $\frac{1}{6}(q - 3)$ orbits of size 6 by ($\#$). Thus we are able to construct an orthogonally divergent partial spread of $\mathcal{H}$ of size $q^2 - \frac{3}{2}q - \frac{5}{2}$ in this case. If $q = 3$, we may add either $[0,1,0]$ or $[0,0,1]$ (but not both) to obtain a larger orthogonally divergent partial spread. For $q > 3$ a trivial counting argument shows that the only lines which could be added to the partial spread are $[0,0,1], [0,1,0], [1,0,0]$ and the remaining lines of D_1, but none of these could be added if we want to retain orthogonal divergence. $\square$

It should be noted that for $q = 3$ the above construction generates a B–set of size 2. As indicated in the proof above, this B–set can be extended by adding either $(0, 1, 0)$ or $(0, 0, 1)$, but not both. The resulting (linear) B–set of size 3 is still not maximal, and a fourth point off the line $[1, 0, 0]$ may be added. An exhaustive search shows that no B–set of size greater than 4 is possible for $q = 3$, although the general upper bound is 7 when $q = 3$.

When q is even, similar constructions are possible. We still have a short orbit when $q \equiv 1 \pmod 3$ because of the existence of primitive cube roots, but the elements $1, -2$, and $-1/2$ are no longer in $F = GF(q) \setminus \{0, -1\}$. We state the following theorem without proof.

Theorem 4.2 *Let $q > 2$ be any even prime power. Then there exists an orthogonally divergent partial spread of $\mathcal{H}$ of size $q^2 - q - 2$. Dually, there exists a B–set of cardinality $q^2 - q - 2$, which is maximal and all of whose points lie on the sides of a self–polar triangle.*

Example 1: Let $q = 9$ and $\pi = PG(2, 81)$. We choose $f(x) = x^4 + x^3 + 2$ as our primitive polynomial for $GF(81)$, and let β denote a (primitive) root of f.

As a convenient shorthand notation, we let i represent the field element β^i and we let $*$ denote the zero element of $GF(81)$. Using the selection process (#) we obtain $AH = \{10, 60\}, AV = \{30, 50\}$, and $AD = \{20, 70\}$. Since $q \equiv 0 \pmod 3$, we also choose as part of our partial spread the lines of D_1 whose line coordinates are $[*, 0, 0], [*, 8, 0], [*, 16, 0], [*, 24, 0]$ and $[*, 32, 0]$. The resulting maximal B–set, constructed as described above, of size 65 is

$\{(0, *, 7), (0, *, 15), (0, *, 23), (0, *, 31), (0, *, 39), (0, *, 47), (0, *, 55), (0, *, 63),$
$(0, *, 71), (0, *, 79), (0, *, 2), (0, *, 10), (0, *, 18), (0, *, 26), (0, *, 34), (0, *, 42),$
$(0, *, 50), (0, *, 58), (0, *, 66), (0, *, 74), (0, 6, *), (0, 14, *), (0, 22, *), (0, 30, *),$
$(0, 38, *), (0, 46, *), (0, 54, *), (0, 62, *), (0, 70, *), (0, 78, *), (0, 1, *), (0, 9, *),$
$(0, 17, *), (0, 25, *), (0, 33, *), (0, 41, *), (0, 49, *), (0, 57, *), (0, 65, *), (0, 73, *),$
$(*, 0, 1), (*, 0, 9), (*, 0, 17), (*, 0, 25), (*, 0, 33), (*, 0, 41), (*, 0, 49), (*, 0, 57),$
$(*, 0, 65), (*, 0, 73), (*, 0, 6), (*, 0, 14), (*, 0, 22), (*, 0, 30), (*, 0, 38), (*, 0, 46),$
$(*, 0, 54), (*, 0, 62), (*, 0, 70), (*, 0, 78), (*, 0, 0), (*, 0, 8), (*, 0, 16), (*, 0, 24),$
$(*, 0, 32) \}.$

Example 2: Let $q = 11$ and $\pi = PG(2, 121)$. We choose $f(x) = x^2 + x + 7$ as our primitive polynomial for $GF(121)$, and let β denote a (primitive) root of f. Using the same shorthand notation as in the above example, we see that the selection process (#) generates the sets $AH = \{24, 12, 48\}, AV = \{0, 84, 36\}$ and $AD = \{96, 72, 108\}$. Note that a "short orbit" in this case is $24 \to 0 \to 96$, as $-\frac{1}{2} = \beta^{24}$ and $-2 = \beta^{96}$. The resulting maximal B–set of size 108 is

$$\{(0,*,i) : i = 8, 18, 28, \ldots, 118\} \ \cup \ \{(0,*,i) : i = 9, 19, 29, \ldots, 119\}$$
$$\cup \ \{(0,*,i) : i = 6, 16, 26, \ldots, 116\} \ \cup \ \{(0,i,*) : i = 2, 12, 22, \ldots, 112\}$$
$$\cup \ \{(0,i,*) : i = 4, 14, 24, \ldots, 114\} \ \cup \ \{(0,i,*) : i = 1, 11, 21, \ldots, 111\}$$
$$\cup \ \{(*,0,i) : i = 2, 12, 22, \ldots, 112\} \ \cup \ \{(*,0,i) : i = 1, 11, 21, \ldots, 111\}$$
$$\cup \ \{(*,0,i) : i = 4, 14, 24, \ldots, 114\}.$$

Example 3: Let $q = 13$ and $\pi = PG(2, 169)$. We choose $f(x) = x^2 + x + 2$ as our primitive polynomial for $GF(169)$, and let β denote a primitive element of the field. The process (#) generates the sets $AH = \{70, 56, 42, 28\}$, $AV = \{0, 112, 14, 154\}$ and $AD = \{98, 140, 126\}$. The short orbits this time are $70 \to 0 \to 98$, where $-\frac{1}{2} = \beta^{70}$, and $56 \to 112$, where β^{56} and β^{112} are primitive cube roots of unity. The maximal B–set of size 154 thus constructed is $\{(0,*,i) : i \equiv 7, 8, 9, 10 \pmod{12}\} \cup \{(0,i,*) : i \equiv 5, 8, 2, 3 \pmod{12}\} \cup \{*, 0, i) : i \equiv 5, 3, 2 \pmod{12}\}$.

5. Automorphism groups of the maximal B–sets

In this section we compute the inherited automorphism groups of the maximal B–sets constructed above. Once again we assume that $\mathcal{H}$ has equation $x^{q+1} + y^{q+1} + z^{q+1} = 0$, and we let $P\Gamma U(3, q^2)$ denote the full (semilinear) unitary group. That is, $P\Gamma U(3, q^2)$ is the stabilizer of $\mathcal{H}$ in $P\Gamma L(3, q^2)$. By the inherited automorphism group of a B–set $\mathcal{B}$ we mean $H = \{\alpha \in P\Gamma U(3, q^2) : \mathcal{B}^\alpha = \mathcal{B}\}$. Throughout this section H will denote the inherited automorphism group of some maximal B–set $\mathcal{B}$ as constructed above. We represent elements of $P\Gamma L(3, q^2)$ by a field automorphism σ of $GF(q^2)$ together with a normalized nonsingular 3×3 matrix M over $GF(q^2)$. We think of an element (σ, M) of $P\Gamma L(3, q^2)$ as acting on normalized row vectors by first applying σ to each component and then post multiplying by M. The resulting row vector is then normalized. The group operation in $P\Gamma L(3, q^2)$ is given by $(\sigma_1, M_1) \cdot (\sigma_2, M_2) = (\sigma_1 \sigma_2, M_1^{\sigma_2} M_2)$, where M^σ denotes the matrix obtained by applying σ to each entry of M.

Let Δ be the self-polar triangle whose sides contain all the points of the B–sets constructed in the previous section. Thus the vertices of Δ are $\{(1, 0, 0), (0, 1, 0), (0, 0, 1)\}$.

Theorem 5.1 *Let G be the pointwise stabilizer of the vertices of Δ in $P\Gamma U(3, q^2)$. Then $G = \left\{ (\sigma, M) : \sigma \in Aut(GF(q^2)), M = \begin{pmatrix} 1 & 0 & 0 \\ 0 & \lambda & 0 \\ 0 & 0 & \delta \end{pmatrix} \right.$ for some $\lambda, \delta \in N \right\}$, where $N = \{x \in GF(q^2) : x^{q+1} = 1\}$. In particular,*

$$G \cong (\mathbb{Z}_{q+1} \times \mathbb{Z}_{q+1}) \rtimes Aut(GF(q^2)).$$

Proof. It is easy to see that the only elements of $P\Gamma L(3,q^2)$ fixing each vertex of Δ are of the form (σ, M), where $\sigma \in \mathrm{Aut}(GF(q^2))$ and $M = \begin{pmatrix} 1 & 0 & 0 \\ 0 & \lambda & 0 \\ 0 & 0 & \delta \end{pmatrix}$ with $\lambda, \delta \in GF(q^2)^*$. Consider the point $(1,y,0)$ of $\mathcal{H}$, where $y^{q+1} = -1$. Then for $(1,y,0) \cdot (\sigma, M)$ to be in $\mathcal{H}$ we must have $(\lambda y^\sigma)^{q+1} = -1$ and hence $\lambda \in N$. Using the point $(1,0,z)$ of $\mathcal{H}$, where $z^{q+1} = -1$, we similarly show $\delta \in N$. Finally, if (x,y,z) is any point of $\mathcal{H}$, it is now easy to show that $(x,y,z) \cdot (\sigma, M) \in \mathcal{H}$ and the result follows. $\qquad\square$

Corollary 5.2 *Let G_1 be the setwise stabilizer of the vertices of Δ in $P\Gamma U(3,q^2)$. Then $G_1 \cong G \rtimes S_3$. In particular, $o(G_1) = 12e(q+1)^2$, where $q = p^e$ for some prime p and some positive integer e.*

Proof. Let A be a permutation matrix corresponding to some permutation of the vertices of Δ. Let $(\sigma, M) \in G$ as described in the above theorem. Then

$$(id, A^{-1}) \cdot (\sigma, M) \cdot (id, A) = (\sigma, A^{-1}MA) \in G$$

since M is diagonal. The corollary now follows easily. $\qquad\square$

Theorem 5.3 *Let $q > 3$ be any prime power, and let $\mathcal{B}$ be a B–set as constructed in the previous section. Let α be an element of $P\Gamma U(3,q^2)$ that leaves $\mathcal{B}$ invariant. Then α permutes the sides (and hence the vertices) of Δ.*

Proof. Let $\ell_0 = [1,0,0], \ell_1 = [0,1,0]$ and $\ell_2 = [0,0,1]$ denote the three sides of Δ. Since $q > 3$, the construction of the last section guarantees that $\mathcal{B}$ contains at least four points on ℓ_1. Since the points of $\mathcal{B}$ all lie on the sides of Δ, we must have $\ell_1^\alpha = \ell_i$ for some $i, 0 \le i \le 2$. Similarly, the construction guarantees that $\mathcal{B}$ contains at least four points on ℓ_2, and hence $\ell_2^\alpha = \ell_j$ for some $j \ne i, 0 \le j \le 2$. Since $P^{(\alpha)\perp} = (P^\perp)^\alpha$ for any point P and since Δ is self–polar, the result now follows. $\qquad\square$

Corollary 5.4 *If H is the inherited automorphism group of $\mathcal{B}$, then $H \le G_1$ provided $q > 3$.*

Theorem 5.5 *Let $q > 2$ be a prime power with $q \not\equiv 0 \pmod 3$. Let $G_0 = \{(\sigma, M) \in G : \sigma = id \text{ or } \sigma = \sigma_q\}$, where G is the pointwise stabilizer of the vertices of Δ as described above and σ_q is the automorphism of $GF(q^2)$ that maps $x \to x^q$. Then $G_0 \le H$.*

Proof. Let $(\sigma, M) \in G_0$. Write $M = \begin{pmatrix} 1 & 0 & 0 \\ 0 & \lambda & 0 \\ 0 & 0 & \delta \end{pmatrix}$ for some $\lambda, \delta \in N$. The
points of $\mathcal{B}$ are of three types. Suppose first that $a \in AH$. Then $\mathcal{B}$ will contain
$q + 1$ points corresponding to this element a, namely the poles of the lines
of $H_a = \{[u, 0, 1] : u^{q+1} = a\}$. Since $[u, 0, 1]^{\perp} = (u^q, 0, 1) \sim (1, 0, 1/u^q)$,
these $q+1$ points will be $\{(1, 0, z) : z \in GF(q^2), z^{q+1} = 1/a\}$. Recall that $a \in$
$GF(q)$ with $a \neq 0, -1$. Now $(1, 0, z) \cdot (\sigma, M) = (1, 0, z^{\sigma})M = (1, 0, \delta z^{\sigma})$, where
$(\delta z^{\sigma})^{q+1} = \delta^{q+1}(z^{q+1})^{\sigma} = 1/a$ as $\sigma = id$ or $\sigma = \sigma_q$. Hence $(1, 0, z) \cdot (\sigma, M) \in \mathcal{B}$.

Similarly, if $a \in AV$, the $q + 1$ points of $\mathcal{B}$ corresponding to a are the
points $\{(1, y, 0) : y^{q+1} = -1/(1 + a)\}$. If $a \in AD$, the corresponding $q + 1$
points of $\mathcal{B}$ are $\{(0, 1, z) : z^{q+1} = -(1 + a)/a\}$. In both cases it is easily seen
that G_0 leaves each such set of points invariant. As $q \not\equiv 0 \pmod{3}$, $\mathcal{B}$ is
the union of full Baer sublines, each of which is one of the above three types.
Hence $G_0 \leq H$. $\square$

Theorem 5.6 *Let $q = 3^e$ for some integer $e \geq 2$. Then H contains an isomorphic copy of the dihedral group of order $2(q + 1)$.*

Proof. Since $q \equiv 0 \pmod{3}$, $\mathcal{B}$ contains the poles of the "half ruling class"
$\frac{1}{2}D_1 = \{[0, v, 1,] : v = 1, \beta^{q-1}, \beta^{2(q-1)}, \ldots, \beta^{(q-1)^2/2}\}$, where β is a primitive
element of $GF(q^2)$. This set of poles is $\mathcal{B}_0 = \{(0, 1, z) := 1, \beta^{q-1}, \beta^{2(q-1)}, \ldots,$
$\beta^{(q-1)^2/2}\}$. It is easy to see that (σ_q, M) leaves $\mathcal{B}_0$ invariant for any normalized
matrix of the form $M = \begin{pmatrix} 1 & 0 & 0 \\ 0 & \lambda & 0 \\ 0 & 0 & \lambda\beta^{(q-1)^2/2} \end{pmatrix}$, where $\lambda \in N$. The proof of
the previous theorem shows that each such map, being an element of G_0,
also leaves $\mathcal{B} \setminus \mathcal{B}_0$ invariant. Similarly, every collineation of the form (id, M'),
where $M' = \begin{pmatrix} 1 & 0 & 0 \\ 0 & \lambda & 0 \\ 0 & 0 & \lambda \end{pmatrix}$ for some $\lambda \in N$, leaves $\mathcal{B}_0$ and $\mathcal{B} \setminus \mathcal{B}_0$ invariant. It
is a straightforward exercise to show that these $2(q+1)$ maps form a dihedral
group. $\square$

We now address the question of determining which permutations of the
vertices of Δ are elements of H. First consider the permutation $\tau = (1\,2\,3)$.
Recall that for each $a \in AH$, the corresponding $q+1$ points of $\mathcal{B}$ are $\{(1, 0, z) :$
$z \in GF(q^2), z^{q+1} = 1/a\}$. Similarly, the $q + 1$ points of $\mathcal{B}$ corresponding to
some $a \in AV$ are $\{(1, y, 0) : y^{q+1} = -1/(1 + a)\}$, and the $q + 1$ points
of $\mathcal{B}$ corresponding to some $a \in AD$ are $\{(0, 1, z) : z^{q+1} = -(1 + a)/a\}$.
Apply τ to each type of point. We obtain $(1, 0, z)^{\tau} = (z, 1, 0) \sim (1, 1/z, 0)$,
where $(1/z)^{q+1} = a$; $(1, y, 0)^{\tau} = (0, 1, y)$, where $y^{q+1} = -1/(1 + a)$; and

28

$(0, 1, z)^\tau = (z, 0, 1) \sim (1, 0, 1/z)$, where $(1/z)^{q+1} = -a/(1+a)$. Hence $\tau \in H$ if and only if

$$\begin{cases} a \in AH \Rightarrow & -(1+a)/a \in AV \\ a \in AV \Rightarrow & -(1+a)/a \in AD \\ a \in AD \Rightarrow & -(1+a)/a \in AH \end{cases}.$$

But as one can easily deduce from the selection process $(\#)$, these conditions are all satisfied precisely when $q \equiv 2 \pmod 3$.

A completely analogous argument shows that in general no other permutation on the vertices of Δ is in H, independent of q, although for certain small values of q (such as $q = 4$) H may contain another permutation. Thus, combining all the results of this section, we have the following.

Theorem 5.7 *Let $q > 3$ by any prime power, and let $\mathcal{B}$ be a maximal $\mathcal{B}$–set as constructed in the previous section.*

(i) *If $q \equiv 0 \pmod 3$, $\mathcal{B}$ admits an automorphism group isomorphic to* $\mathbb{Z}_{q+1} \rtimes \mathbb{Z}_2$.

(ii) *If $q \equiv 1 \pmod 3$, $\mathcal{B}$ admits an automorphism group isomorphic to* $(\mathbb{Z}_{q+1} \times \mathbb{Z}_{q+1}) \rtimes \mathbb{Z}_2$.

(iii) *If $q \equiv 2 \pmod 3$, $\mathcal{B}$ admits an automorphism group isomorphic to* $(\mathbb{Z}_{q+1} \times \mathbb{Z}_{q+1}) \rtimes \mathbb{Z}_6$.

Proof. We only need concern ourselves with part (iii). Let $A = \begin{pmatrix} 0 & 0 & 1 \\ 1 & 0 & 0 \\ 0 & 1 & 0 \end{pmatrix}$ be the permutation matrix corresponding to the permutation $(1\,2\,3)$. Let $\alpha_1 = (id, A)$ and $\alpha_2 = (\sigma_q, I)$, both of which are elements of H in this case. Thus H contains the element $\alpha = (\sigma_q, A)$ of order 6. Since α normalizes the subgroup

$$\left\{ \left(id, \begin{pmatrix} 1 & 0 & 0 \\ 0 & \lambda & 0 \\ 0 & 0 & \delta \end{pmatrix} \right) : \lambda, \delta \in N \right\}$$ of H, the result follows. $\square$

Since $H \leq G_1$ for all $q > 3$, where $o(G_1) = 12e(q+1)^2$ with $q = p^e$, there is not much room for H to grow. In fact, except for some small values of q as previously indicated, the groups discussed in the above theorem are the full inherited automorphism groups.

6. Concluding remarks

We have shown that maximum $\mathcal{B}$–sets (of size $q^2 - q + 1$) exist for q even, $q \equiv 1 \pmod 3$. The construction is based on the existence of a cyclic o.d. spread of the Hermitian curve $\mathcal{H}$ in this case, and the resulting $\mathcal{B}$-set admits a linear cyclic automorphism group acting transitively on its points. Such

B–sets are $(q^2 - q + 1)$–arcs. Moreover, cyclic o.d. spreads of $\mathcal{H}$ exist only for q even, $q \equiv 1$ (mod 3).

For all other values of q we have not been able to construct maximum B–sets. It is unknown if such exist. However, we have constructed maximal B–sets of deficiency 3 for q even or odd, provided $q \not\equiv 0$ (mod 3). The points of these B–sets lie on the sides of a self-polar triangle, and they admit linear automorphism groups isomorphic to $\mathbb{Z}_{q+1} \times \mathbb{Z}_{q+1}$. For $q > 3$ and $q \equiv 0$ (mod 3), maximal B–sets of deficiency $(q + 7)/2$ have been constructed, all of whose points again lie on the sides of a self–polar triangle. This time the inherited linear collineation group is only $\mathbb{Z}_{q+1}$.

It should be mentioned that in this paper the maximal B–sets of positive deficiency were constructed so that the points were distributed as evenly as possible among the three sides of a self–polar triangle. One could easily modify the selection process (#) to produce "unbalanced" B–sets. For instance, we could construct maximal B–sets (of the same cardinalities as discussed above) in which all points lie on two sides of the self–polar triangle for q even and for $q \equiv 0$ (mod 3). For q odd with $q \not\equiv 0$ (mod 3) the remaining side would be used for only one Baer subline. Such B–sets would admit the same linear collineation groups as described in the above paragraph.

References

[1] **E. Boros and T. Szőnyi**. On the sharpness of a theorem of B. Segre. *Combinatorica*, 6, pp. 261–268, 1986.

[2] **G. Seib**. Unitäre Polaritäten endlicher projektiver Ebenen. *Arch. Math. (Basel)*, 21, pp. 103–112, 1970.

R. D. Baker, Department of Mathematical Sciences, University of Delaware, Newark, DE 19716 USA. e-mail: baker@math.udel.edu

G. Korchmáros, Istituto di Matematica, Università della Basilicata, I-85100 Potenza, Italy. e-mail: korchmaros@pzvx85.cineca.it

G. L. Ebert, Department of Mathematical Sciences University of Delaware Newark, DE 19716 USA. e-mail: ebert@math.udel.edu

T. Szőnyi, Department of Computer Science, Eötvös Loránd University, H-1088 Budapest, Múzeum krt. 6-8, Hungary. e-mail: sztomi@ludens.elte.hu

Lifts of nuclei in finite projective spaces

A. Blokhuis   F. Mazzocca *

Abstract

We present a synthetical construction of the lifting process introduced in [1] and apply this process to obtain a new result on the structure of sets in the plane admitting the maximal number of nuclei.

1. Introduction

Let B_n be a set of $q^{n-1} + q^{n-2} + \ldots + q + 1$ points, not all on a hyperplane in the n-dimensional projective space $PG(n, q)$ over the Galois field $GF(q)$, $n \geq 2$. A point not in B_n is called a nucleus of B_n if every line through it meets B_n (exactly once, of course). The set of all nuclei of B_n is denoted by $N(B_n)$. The following two fundamental results are well known.

Result 1.1 (Segre-Korchmáros, [7]) *If a, b, c are three non-collinear nuclei of B_n, then the points of B_n on the lines ab, bc, ca are collinear.*

Result 1.2 (Blokhuis-Wilbrink, [3]) *If B_n is an affine set (i.e. it is contained in the complement of a hyperplane), then $|N(B_n)| \leq q - 1$.*

The proofs of both the previous results have been given by the authors in the two dimensional case, but it is straightforward to see that they work in arbitrary dimensions. The original proof of Result 1.2 surprisingly does not use Result 1.1.

In the plane case, an elementary derivation of Result 1.2 from Result 1.1 has been obtained in [1] by using a process called "lifting". In this note we present in the general case a synthetical construction for this lifting process, only using Result 1.1 and Desargues' Theorem. The same construction is shown to be possible in a more general context. Further, we apply this lifting procedure, and the Result 1.3, in order to obtain a new result on the structure

---
*. This author gratefully acknowledges for their support the Italian M.U.R.S.T. and C.N.R.(G.N.S.A.G.A.).

of sets B_2 admitting the maximal number of nuclei.

We recall that it is conjectured [1] that the only examples of pairs $(B_2, N(B_2))$ with $|N(B_2)| = q - 1$ are the following two.

Example 1 $B_2 = (M \setminus \{x\}) \cup \{y\}$, M a line in $PG(2, q)$ and $x \in M$, $y \in PG(2, q) \setminus M$.

In this case $N(B_2)$ consists of the $q - 1$ points other than x, y on the line xy.

Example 2 $B_2 \cup N(B_2)$ consists of the ten points of a Desargues configuration in $PG(2, 5)$, where $N(B_2)$ is one of the triangles plus the center of the perspectivity.

The characterization of all sets B_2 with the maximal number of nuclei seems to be a very hard problem. A useful result is the following.

Result 1.3 (Blokhuis-Mazzocca, [1]) *If* $|N(B_2)| = q - 1$, *then no line in* $PG(2, q)$ *intersects* $B_2 \cup N(B_2)$ *in* 1 mod p *points, where* p *is the characteristic of* $GF(q)$. *In particular* $B_2 \cup N(B_2)$ *has no one-secant lines. Also if* $p = 2$, $N(B_2) = q - 1$, *then* $N(B_2)$ *is a maximal quasi-odd set (i.e. an affine set whose line intersection numbers are zero or odd) and* $B_2 \cup N(B_2)$ *is an even set.*

For more results and generalizations of the previous setting we refer the reader to [1] [5] and [6].

2.   Main Results

In this section we use the notation introduced in Section 1 and denote by ab the line through two points a and b. We set $\alpha = PG(n, q)$ and consider α as a hyperplane of $PG(n + 1, q)$.

Proposition 2.1 *Let* B_n *be such that* $N(B_n) \neq \emptyset$. *Let* v *be a point in* $PG(n + 1, q) \setminus \alpha$ *and denote by* Γ_v *the cone projecting* $N(B_n)$ *from* v. *The relation* $\sim$ *defined in* $C_v = \Gamma_v \setminus (\alpha \cup \{v\})$ *by*

(1) $x \sim y \Leftrightarrow$ *either* $x = y$ *or the line through* x, y *meets* α *in a point of* B_n,

is an equivalence relation.

Proof.   As the relation $\sim$ is obviously reflexive and symmetric, we only have to show its transitivity. Assume x, y, z are three points in C_v such that $x \sim y \sim z$. We can suppose x, y, z non-collinear, otherwise $x \sim z$ trivially. If $x' = xv \cap \alpha$, $y' = yv \cap \alpha$, $z' = zv \cap \alpha$, then the points of α given by $z'' = xy \cap x'y'$,

$x'' = yz \cap y'z'$, $y'' = xz \cap x'y'$ are collinear by Desargues' Theorem. Now as z'' and x'' belong to B_n, also y'' is on B_n, by Result 1.1 applied to the three non-collinear nuclei x', y', z'. It follows that $x \sim z$, finishing the proof. $\square$

The next proposition is an easy consequence of the previous one and the definition of Γ_v.

Proposition 2.2 *The set C_v can be partitioned into equivalence classes with respect to relation $\sim$ and the following properties hold:*

(2) every two points in C_v collinear with v are in different equivalence classes;

(3) every equivalence class has the same size as $N(B_n)$.

In the following a point set D in $PG(n+1,q) \setminus \alpha$ is called a v-lift of $N(B_n)$ if D is an equivalence class with respect to the relation $\sim$, for some point $v \in PG(n+1,q) \setminus \alpha$.

We remark that Result 1.2 is now an immediate consequence of Propositions 2.1 and 2.2. Actually, if we assume B_n affine and consider an $(n-1)$-dimensional subspace S_{n-1} in α missing B_n, then every n-dimensional subspace S_n through S_{n-1} intersecting a fixed v-lift D of $N(B_n)$ meets D exactly once. As there are at most $q-1$ of such subspaces S_n (cf. Proposition 2.1), Result 1.2 follows from (3) of Proposition 2.2.

Proposition 2.3 *Assume $|N(B_2)| = q-1$ and $B_2(\subset PG(2,q))$ not of type $(M \setminus \{x\}) \cup \{y\}$, M a line and $x \in M$, $y \notin M$ (as in Example 1). Then B_2 has the following property:*

(4) every point of B_2 is on at least one line meeting $N(B_2)$ in more than one point.

Proof. If B_2 contains a point on $q-1$ tangent lines (i.e. 1-secants) to $N(B_2)$, then B_2 is contained in the union of two lines, say L and M. Let a_1, a_2 be two points in $N(B_2)$ and assume that the line a_1a_2 meets B_2 in a point c_3 of L. If a_3 is a nucleus not collinear with a_1, a_2, then the points $c_1 = a_2a_3 \cap B_2$, $c_2 = a_3a_1 \cap B_2$ are forced to belong to L. Actually, c_1, c_2, c_3 are contained in $L \cup M$ and are collinear by Result 1.1. It follows that the set $\bar{B}_2$ of points of B_2 which are on at least one line containing more than one nucleus is contained in L.

Now, let b be a point of B_2 on M other than $L \cap B_2$. As there are $q-1$ lines through b containing a nucleus of B_2, we have $|\bar{B}_2| = 1$. It follows that B_2 is of type $(M \setminus \{x\}) \cup \{y\}$, a contradiction. $\square$

Proposition 2.4 *Assume B_2 is affine with property (4) and $|N(B_2)| = q-1$. Then no v-lift of $N(B_2)$ is contained in a plane.*

Proof. Assume a v-lift D of $N(B_2)$ is contained in a plane β and set $M = \alpha \cap \beta$. Then every point in $B_2 \setminus M$ is on $q - 1$ tangent lines to $N(B_2)$, a contradiction. $\square$

Proposition 2.5 *Assume $|N(B_2)| = q - 1$, B_2 affine and q even. If $N(B_2)$ contains more than $q/4 - 3$ collinear points, then all points in $N(B_2)$ are collinear, that is B_2 is as in Example 1.*

Proof. We can assume that B_2 has property (4) and let L be a line containing the maximum number h of collinear points in a v-lift D of $N(B_2)$. Note that h is also the maximum number of collinear points in $N(B_2)$. By Prop. 2.4 we can find a line M skew to L and intersecting D in at least two points. As D is a quasi-odd set, by Result 1.3, we have $|D \cap M| \geq 3$.

Now, a plane through M and a point in $L \cap D$ contains at least three points of $D \setminus (L \cap M)$ and so

$$4h + 3 \leq q - 1;$$

that is,

(5) $$h \leq \frac{q}{4} - 1.$$

It follows that, if $h = q/4 - 1$, then $q \geq 16$. Now, under the assumption $h = q/4 - 1$, we have that every plane through M and a point in $L \cap D$ must meet D in a projective subplane of order 2 and every line of such a subplane meets B_2 in a point. Moreover, all points of B_2, other than $M \cap \alpha$, obtained in this way are distinct. It follows that

$$2 + 6(\frac{q}{4} - 1) \leq |B_2| = q - 1,$$

i.e. $q \leq 10$, a contradiction. So we can conclude that $h < q/4 - 1$ and, because h must be odd, we have $h \leq q/4 - 3$, finishing the proof. $\square$

We remark that Prop. 2.5 improves the following result of A.A. Bruen, in case $|N(B_2)| = q - 1$ and q even.

Result 2.6 (Bruen, [4]) *If $N(B_2)$ contains more than $q/2$ collinear points, then all nuclei of B_2 are collinear.*

34

We conclude this note observing that our above defined lifting process may be useful in a more general setting, as specified in the following.

Recently, J.W.P. Hirschfeld and G. Kiss introduced the notion of a tangent set of a fixed point set B in $PG(n,q)$. It is defined as a point set $T(B)$ in $PG(n,q)$ such that the line joining any two points of $T(B)$ is a tangent to B. Note that, for a set B_n, the set $N(B_n)$ of its nuclei is a tangent set to B_n. Also, in order to define a v-lift of $N(B_n)$, we did not use the fact that every tangent to $N(B_n)$ meets B_n exactly once. So if we assume that a tangent set $T(B)$ satisfies the following property

(6) if a_1, a_2 and a_3 are non-collinear points in $T(B)$ and $b_3 \in a_1a_2 \cap B$, $b_1 \in a_2a_3 \cap B$, then $b_2 = a_3a_1 \cap b_3b_1$ is a point on B,

then a lifting procedure for $T(B)$ can be done in the same way as for a set of nuclei. For example, using the lifting process, if we assume that B is affine and $T(B)$ satisfies property (6), then it can be easily shown that $|T(B)| \leq q - 1$, as in the Blokhuis-Wilbrink result.

References

[1] A. Blokhuis and F. Mazzocca. On maximal sets of nuclei in $PG(2,q)$ and quasi-odd sets in $AG(2,q)$. In J. W. P. Hirschfeld, D . R. Hughes, and J. A. Thas, editors, *Advances in Finite Geometries and Designs*, pages 35–46, Oxford, New York, Tokyo, 1991. Oxford University Press.

[2] A. Blokhuis and F. Mazzocca. Special point sets in $PG(n,q)$ and the structure of sets with the maximal number of nuclei. *J. Geom.*, 41, pp. 33–41, 1991.

[3] A. Blokhuis and H. A. Wilbrink. A characterization of exterior lines of certain sets of points in $PG(2,q)$. *Geom. Dedicata*, 23, pp. 253–254, 1987.

[4] A. A. Bruen. Nuclei of sets of $q+1$ points in $PG(2,q)$ and blocking sets of Rédei type. *J. Combin. Theory Ser. A*, 55, pp. 130–132, 1990.

[5] G. Korchmáros and F. Mazzocca. Nuclei of point sets of size $q+1$ contained in the union of two lines in $PG(2,q)$. To appear in *Combinatorica*.

[6] F. Mazzocca. Blocking sets with respect to special families of lines and nuclei of θ_n-sets in finite n-dimensional projective and affine spaces. *Mitt. Math. Sem. Giessen*, 201, pp. 109–117, 1991.

[7] B. Segre and G. Korchmáros. Una proprietà degli insiemi di punti di un piano di Galois caratterizzante quelli formati dai punti delle singole rette esterna ad una conica. *Accad. Naz. dei Lincei, Rend. Sc. fis. Mat. Nat.*, 62, 1977.

Aart Blokhuis, Department of Mathematics and Computer Science, Eindhoven University of Technology, PO Box 513, 5600 MB Eindhoven, The Netherlands. e-mail: aartb@win.tue.nl

Francesco Mazzocca, Universitá degli Studi di Napoli, Dipartimento di Mathematica e Applicazioni "Renato Caccioppoli", Complesso Universitario Monte S. Angelo, Via Cinthia 80126 Napoli, Italy.
e-mail: geomcomb@icnucevm.cnuce.cnr.it

Large minimal blocking sets, strong representative systems, and partial unitals

A. Blokhuis   K. Metsch

Abstract

The main result is that the size of a maximal strong representative system (a set of flags with the property that every flag point only occurs on its own flag line) in $PG(2,q)$, is not between $q + 1$ and $q + \frac{1}{2}\sqrt{q}$ and is not equal to $q\sqrt{q}$ ($q \geq 49$). Hence, for $q \geq 25$, there are no minimal blocking sets of size $q\sqrt{q}$.

1. Introduction

Let Π be a projective plane of order q. A subset $\mathcal{B}$ of Π is called a *blocking set* if $\mathcal{B}$ meets every line but contains no line. A blocking set is called *minimal* if it does not contain a smaller blocking set. This implies that through every point of a minimal blocking set there is a tangent. For a minimal blocking set $\mathcal{B}$ we have the following upper and lower bounds due to Bruen [2] and Bruen and Thas [3].

$$q + \sqrt{q} + 1 \leq |\mathcal{B}| \leq q\sqrt{q} + 1.$$

Here we concentrate on the upper bound, so we study large minimal blocking sets.

A *flag* of Π is an incident point-line pair. A set

$$S = \{(P_1, l_1), ..., (P_s, l_s)\}$$

of flags is called a *strong representative system*, if

$$P_i \in l_j \iff i = j.$$

For the size s of a strong representative system S, it was shown by Illés, Szőnyi, Wettl [4] that $s \leq q\sqrt{q} + 1$, with equality iff S consists of the incident point-tangent pairs of a unital. We denote the set of points occurring in a flag of S by $P(S)$ or simply S and the set of lines by $\ell(S)$. Points in $P(S)$ and lines in $\ell(S)$ are called *special* and other points and lines *ordinary*. A strong representative system is called *maximal* if it is not part of a larger strong representative system. A minimal blocking set $\mathcal{B}$ gives rise to a maximal strong representative system if we associate to each point $P \in \mathcal{B}$ the flag consisting of P and a tangent through P. It was shown by Illés, Szőnyi, Wettl [4] in the desarguesian and by Wettl [5] in the general case that a maximal strong representative system of size $q + 1$ either has all points collinear or all lines concurrent. Notice that a maximal strong representative system cannot have size less than $q + 1$. In this note, we obtain the following results.

Theorem 1.1 In $PG(2, q)$, there are no maximal strong representative systems S with $q + 1 < |S| < q + \frac{1}{2}\sqrt{q}$.

Theorem 1.2 If $q \geq 49$ is a square, then a strong representative system of size $q\sqrt{q}$ is part of a unital. Consequently, there is no minimal blocking set of size $q\sqrt{q}$.

A *partial unital* is a set $\mathcal{U}$ of points such that every point of $\mathcal{U}$ lies on a tangent and no line contains more than $\sqrt{q} + 1$ points of $\mathcal{U}$. Again we can get a strong representative system from a partial unital. We say that a strong representative system is (part of) a unital if the flags are (a subset of) the incident point, tangent pairs of a unital.

Theorem 1.3 If $q \geq 25$, then a partial unital of size $q\sqrt{q} - 1$ is either a minimal blocking set or part or a unital.

In the desarguesian case we can do better if q is odd.

Theorem 1.4 For odd $q \geq 25$, a partial unital of size $q\sqrt{q} - 1$ in $PG(2, q)$ is part of a unital.

2.   Small strong representative systems

In this section, we shall obtain a lower bound on the size of a maximal strong representative system using the following recent result on nuclei. A *nucleus* of a set S of points in a projective (or affine) plane, is a point $P \notin S$ such that every line on P meets S.

Result 2.1 (A. Blokhuis [1]) A set S of size $q + a$ in $AG(2, q)$ has at most $a(q - 1)$ nuclei.

Remark. Note that the same bound holds for a set S in $PG(2,q)$ if there is a line missing S.

Theorem 2.2 *Suppose that S is a maximal strong representative system of $PG(2,q)$ with $q+a$ elements. Then either $a = 1$ or $q \leq 2a(2a-1)$.*

Proof. Assume that $a > 1$ and $q > 2a(2a-1)$. Recall that a point or a line is called *special* if it occurs in some flag of S and *ordinary* otherwise. Note that $a > 1$ implies that not all points on a line, or all lines on a point can be special. It is also not possible that all lines have a special point, since in this case $P(S)$ would be a blocking set and we would have $a \geq 1 + \sqrt{q}$ by [2]. It follows from the above remark that $P(S)$ has at most $a(q-1)$ nuclei. Since the strong representative system is maximal, every point that is not on a special line must be a nucleus of $P(S)$.

We call a point *covered*, if it lies on a special line. Let b be the maximum number of special lines on a point, and choose a point P that is on b special lines. Then each ordinary line on P has at least $q - (q+a-b) = b-a$ points that are not covered. Each such point must be a nucleus of $P(S)$. Hence $(q+1-b)(b-a) \leq a(q-1)$, so

$$f(b) := (q+1-b)(b-a) - a(q-1) \leq 0.$$

Since

$$f(2a+1) = f(q-a) = (a+1)(q-2a) - a(q-1) = q - a(2a+1) > 0,$$

it follows that $b \leq 2a$ or $b \geq q-a+1$.

Assume that $b \leq 2a$. Let M be the set of covered points, and for $X \in M$, denote by r_X the number of special lines on X. Since there are $q+a$ special lines, we have

$$\sum_{X \in M} r_X = (q+a)(q+1)$$

and

$$\sum_{X \in M} r_X(r_X - 1) = (q+a)(q+a-1).$$

In view of $r_X \leq b$ for $X \in M$, it follows that

$$|M| = \sum_{X \in M} r_X - \sum_{X \in M} (r_X - 1)$$
$$\leq (q+a)(q+1) - \frac{1}{b}(q+a)(q+a-1).$$

Since the points not in M are nuclei and since the number of nuclei is at most $a(q-1)$, we have $|M| + a(q-1) \geq q^2 + q + 1$. Hence

$$(q+a)(q+1) - \frac{1}{b}(q+a)(q+a-1) + a(q-1) \geq q^2 + q + 1.$$

It follows that $q^2 - q(2ab - 2a + 1) + a(a - 1) + b \leq 0$. Hence $q < 2ab - 2a + 1$. But $b \leq 2a$ and $q > 2a(2a - 1)$, so $q \geq 2a(2a - 1) + 1 \geq 2ab - 2a + 1$, a contradiction.

Hence $b \geq q + 1 - a$. Dually, there is a line l with at least $q + 1 - a$ special points. Let P be a point on b special lines $l_1, \ldots, l_b$. Then $P \notin l$ (indeed, if P were in l, then there would be at least $2b$ special points, b on l and b on the special lines on P). Since $a > 1$, there must be an ordinary line g on P.

Assume that l has an ordinary point G not on g. Choose a point Q on g that is not covered (there are at least $q + 1 - 2a$ choices for Q). Since S is a maximal strong representative system, the line QG must have a special point. This point is not on l. Hence there are at least $q + 1 - 2a$ special points not on l, and at least $q + 1 - a$ special points on l. This is too much.

Hence every point of l other than $l \cap g$ is special, so $g \cap l$ must be an ordinary point. Since this holds for every ordinary line on P, it follows that g is the only ordinary line on P. Hence P lies on n special lines $l_1, \ldots, l_n$ with special points $l \cap l_i$. Every other special pair must be of the form (X, x) with $X = l \cap g$ and $x \neq l$. But there can be at most one such pair in S. This contradicts $a > 1$. $\qquad\square$

3. Large strong representative systems

Throughout this section S denotes a strong representative system in Π, a projective plane of order q, with $s = q\sqrt{q} + 1 - d$ elements, where d is called the *deficiency* of S. For every line l, we set $k_l = |l \cap P(S)|$ and $d_l = \sqrt{q} + 1 - k_l$. We shall show first that $d \geq 0$ with equality if and only if q is a square and S is a unital.

Lemma 3.1 *Let N be the set of ordinary lines. Then*

$$\sum_{l \in N} 1 = q^2 + q + 1 - s, \tag{1}$$

$$\sum_{l \in N} k_l = sq, \tag{2}$$

$$\sum_{l \in N} k_l(k_l - 1) = s(s - 1), \tag{3}$$

$$\sum_{l \in N} d_l = d(q + \sqrt{q} + 1), \text{ and} \tag{4}$$

$$\sum_{l \in N} d_l(d_l - 1) = d(q + \sqrt{q} + d - 1). \tag{5}$$

Proof. Equations (2) and (3) hold, since every point of S lies on q ordinary lines and since any two points of S are joined by an ordinary line. Equations (4) and (5) are consequences of (1), (2) and (3), since $k_l = \sqrt{q} + 1 - d_l$. $\qquad\square$

Theorem 3.2 *We have $d \geq 0$ with equality if and only if S is a unital.*

Proof. From equations (4) and (5) in Lemma 3.1, we obtain $\sum_{l \in N} d_l^2 = d(2q + 2\sqrt{q} + d)$. Since the left hand side is non-negative, we see that $d \geq 0$ or $d \leq -(2q + 2\sqrt{q})$ that is $|S| \leq q\sqrt{q} + 1$ or $|S| \geq q\sqrt{q} + 2q + 2\sqrt{q} + 1$. Since this holds for every strong representative system of Π and because every subset of S is also a strong representative system, we see that we must have $|S| \leq q\sqrt{q} + 1$ or equivalently $d \geq 0$.

Now assume that $d = 0$. Then equations (4) and (5) in Lemma 3.1 imply that every ordinary line meets S in $\sqrt{q} + 1$ points so that S is a unital. $\square$

From now on we assume that q is a square.

Lemma 3.3 *a) Suppose that every line meets S in at most $\sqrt{q} + 1$ points, so $P(S)$ is a partial unital. Then every ordinary line meets S in at least $\sqrt{q} + 1 - d$ points.*

b) Suppose that there is no line missing S and that every line meets S in at most $\sqrt{q} + 1$ points, so $P(S)$ is a blocking set as well as a partial unital. Denote by m be the number of ordinary lines that satisfy $d_l \geq 1$. Then $m \leq (d - 1)(q + \sqrt{q} + 1) + 2 - d$.

Proof. a) Consider a point P of S and an ordinary line l through P. Since P lies on a special line and because every ordinary line through P meets S in at most $\sqrt{q} + 1$ points, we see that $q\sqrt{q} + 1 - d = |S| \leq |l \cap S| + (q - 1)\sqrt{q}$ so that $|l \cap S| \geq \sqrt{q} + 1 - d$.

b) By part a) and the hypotheses, we have $0 \leq d_l \leq d$ for every ordinary line. Let M be the set of ordinary lines l that satisfy $d_l \geq 1$. Then Lemma 3.1 implies that

$$d(q + \sqrt{q} + d - 1) = \sum_{l \in M} d_l(d_l - 1) \leq d \sum_{l \in M} (d_l - 1) = d^2(q + \sqrt{q} + 1) - dm.$$

Hence $m \leq (d - 1)(q + \sqrt{q} + 1) + 2 - d$. $\square$

Lemma 3.4 *a) Suppose that $2d \leq \sqrt{q}$. If there is a line with more than $\sqrt{q} + 1$ special points, then there are at most $d - 1$ lines missing S.*

b) Suppose that $d < \sqrt{q}$. There is a line with more than $\sqrt{q} + 1$ special points if and only if there is a point that lies on more than $\sqrt{q} + 1$ special lines.

Proof. a) Suppose that a line l_0 meets S in $\sqrt{q} + 1 + e$ points with $e \geq 1$. Then $d_{l_0}(d_{l_0} - 1) = e(e + 1)$. Assume that there are d (or more) lines missing

S. Then Lemma 3.1 (5) implies that $d(q + \sqrt{q}) + e(e+1) \le d(q + \sqrt{q} + d - 1)$ so that $e(e+1) \le d(d-1)$, that is

$$e \le d - 1.$$

Let $P_1, \ldots, P_u$ be the points of l_0 that lie on more than $\sqrt{q} + 1$ special lines and let $\sqrt{q} + 1 + t_i$ be the number of special lines through P_i. Since the points of $S \cap l_0$ lie on a unique special line, we see that l_0 meets at most

$$\sqrt{q} + 1 + e + (q - \sqrt{q} - e)(\sqrt{q} + 1) + \sum_{i=1}^{u} t_i = q\sqrt{q} + 1 - e\sqrt{q} + \sum_{i=1}^{u} t_i$$

special lines. Since the number of special lines is $s = q\sqrt{q} + 1 - d$, we conclude that

$$\sum_{i=1}^{u} t_i \ge e\sqrt{q} - d.$$

Consider a point P_i. Let M_i be the set of ordinary lines through P_i different from l_0. Then $|M_i| = q - \sqrt{q} - 1 - t_i$ and

$$q\sqrt{q} + 1 - d = |S| = (\sqrt{q} + 1 + e) + (\sqrt{q} + 1 + t_i) + \sum_{l \in M_i} k_l$$

$$= (\sqrt{q} + 1 + e) + (\sqrt{q} + 1 + t_i) + (q - \sqrt{q} - 1 - t_i)(\sqrt{q} + 1) - \sum_{l \in M_i} d_l.$$

Hence

$$- \sum_{l \in M_i} d_l = t_i\sqrt{q} - e - d \ge t_i(\sqrt{q} - e - d).$$

If M_i' is the set of lines l of M_i with $d_l < 0$, then we conclude

$$\sum_{l \in M_i'} d_l(d_l - 1) \ge -2 \sum_{l \in M_i} d_l \ge 2t_i(\sqrt{q} - e - d).$$

If M' is the union of the sets M_i', $i = 1, \ldots, u$, we obtain

$$\sum_{l \in M'} d_l(d_l - 1) \ge \sum_{i=1}^{u} 2t_i(\sqrt{q} - e - d) \ge 2(e\sqrt{q} - d)(\sqrt{q} - e - d).$$

Since the d lines l missing S satisfy $d_l(d_l - 1) = q + \sqrt{q}$, we have

$$d(q + \sqrt{q}) + 2(e\sqrt{q} - d)(\sqrt{q} - e - d) \le d(q + \sqrt{q} + d - 1)$$

by Lemma 3.1 (5). Hence

$$f(e) := 2(e\sqrt{q} - d)(\sqrt{q} - e - d) \le d(d-1).$$

Since $1 \le e \le d - 1$ and because f has degree 2 in e, it follows that $f(1) \le d(d-1)$ or $f(d-1) \le d(d-1)$. However, since $2d \le \sqrt{q}$, we have

$$f(1) = 2(\sqrt{q} - d)(\sqrt{q} - d - 1) > d(d-1)$$

and

$$f(d-1) = 2((d-1)\sqrt{q} - d)(\sqrt{q} - 2d + 1) \ge 2((d-1)\sqrt{q} - d) \ge 2d(2d-3).$$

In view of $1 \le e \le d - 1$, we have $d \ge 2$. Hence

$$f(d-1) \ge 2d(2d-3) > d(d-1).$$

This contradiction proves part a).

b) Suppose a point P lies on $t \ge \sqrt{q} + 2$ special lines and assume that every line through P has at most $\sqrt{q} + 1$ special points. Then

$$\begin{aligned} |S| &\le t + (q + 1 - t)(\sqrt{q} + 1) \\ &\le \sqrt{q} + 2 + (q - \sqrt{q} - 1)(\sqrt{q} + 1) = q\sqrt{q} - \sqrt{q} + 1, \end{aligned}$$

contradicting the hypothesis. Thus a point on at least $\sqrt{q} + 2$ special lines lies on a line with at least $\sqrt{q} + 2$ special points. By duality, a line with at least $\sqrt{q} + 2$ special points has a point lying on at least $\sqrt{q} + 2$ special lines. $\square$

Proposition 3.5 If $2d \le \sqrt{q}$ then there exist at most d lines that miss S and equality holds if and only if S can be obtained from a unital by removing d points.

Proof. In view of $d(q + \sqrt{q} + d - 1) < (d+1)(q + \sqrt{q})$, equation (5) in Lemma 3.1 implies that there are at most d lines that miss S. Now assume that there are d lines that miss S. We have to show that S is a unital from which d points have been removed. We proceed by induction on d. If $d = 0$ then Lemma 3.2 shows that S is a unital. Now assume that $d \ge 1$. In view of Lemma 3.4, every line meets S in at most $\sqrt{q} + 1$ points, and every point lies on at most $\sqrt{q} + 1$ special lines.

Let P be a point not in S, denote by α the number of passants through P, and by t the number of special lines through P. Then P lies on $q + 1 - \alpha - t$ lines that are not special lines and not passants and these lines meet S in at most $\sqrt{q} + 1$ points. Hence

$$\begin{aligned} |S| &\le t + (q + 1 - \alpha - t)(\sqrt{q} + 1) \le \\ &\le (q+1)(\sqrt{q} + 1) - (\alpha + t)\sqrt{q} = q\sqrt{q} + q + \sqrt{q} + 1 - (\alpha + t)\sqrt{q}. \end{aligned}$$

Since $|S| > q\sqrt{q} - \sqrt{q} + 1$, it follows that $t + \alpha \le \sqrt{q} + 1$. In particular, a point lying on a passant of S lies on at most $\sqrt{q}$ special lines. Dually, if a point P lies on no special line, then every line through P meets S in at most $\sqrt{q}$ points.

Let m be the number of points lying on one of the d lines that miss S. Then

$$m \geq d(q+1) - \frac{1}{2}d(d-1) > (d-1)(q+\sqrt{q}+1) + 2 - d.$$

The statement dual to the statement in Lemma 3.3 b) shows therefore that there exists a point P that lies on no special line. Since every line through P meets S in at most $\sqrt{q}$ points and in view of $|S| > (q+1)(\sqrt{q}-1)$, we conclude that P lies on a line with precisely $\sqrt{q}$ special points. Let P_i, $i = 1, \ldots, u := q - \sqrt{q}$, be the ordinary points of l other than P. We denote by α_i the number of passants through P_i, and by t_i the number of special lines through P_i, $i = 1, \ldots, u$. Then $t_i + \alpha_i \leq \sqrt{q} + 1$. We have $\sum_{i=1}^{u} t_i = |S| - \sqrt{q}$, since every special line that meets l not in S contains a point P_i. It follows that

$$\sum_{i=1}^{u} \alpha_i \leq \sum_{i=1}^{u}(\sqrt{q}+1-t_i) = (q-\sqrt{q})(\sqrt{q}+1) - \sum_{i=1}^{u} t_i$$
$$= (q-\sqrt{q})(\sqrt{q}+1) - |S| + \sqrt{q} = q\sqrt{q} - |S| = d - 1.$$

Since there are d passants, it follows that some passant l' must pass through P. Furthermore, l' is the only passant through P, since the lines through P meet S in at most $\sqrt{q}$ points and since $|S| > (q-1)\sqrt{q}$. Since P does not lie on a special line it follows that $S' := S \cup \{P, l'\}$ is a strong representative system with deficiency $d - 1$. Furthermore, there are $d - 1$ lines missing S'. Now apply the induction hypothesis to complete the proof. $\qquad\square$

4. Strong representative systems with deficiency 1

In this section, we shall show that every partial unital with deficiency 1 is a unital from which one point has been removed.

Let S be a strong representative system with deficiency 1. Proposition 3.5 says that there is at most one line missing S with equality only if S is obtained from a unital by removing one point. We therefore only have to show that there is a line missing S. We shall succeed to do so for $q \neq 4, 9, 16, 25$.

Lemma 4.1 *If $q \geq 49$ and if S is a strong representative system with deficiency 1 in $PG(2, q)$ then there is a line missing S.*

Proof. Assume that every line meets S. If every line meets S in at most $\sqrt{q}+1$ points, then, by Lemma 3.3 a), every ordinary line meets S in at least $\sqrt{q}$ points. In this case, $d_l(d_l - 1) = 0$ for every line that is not a tangent. But equation (5) in Lemma 3.1 shows that this is not possible.

Hence, there is a line l_0 that meets S in $\sqrt{q}+1+e$ points with $e \geq 1$. As in the proof of Lemma 3.4, we denote by $P_1, \ldots, P_u$ the points of l_0 that lie on more than $\sqrt{q}+1$ tangents, by $\sqrt{q}+1+t_i$ the number of tangents through P_i, and by M' the set of lines l through one of the points P_i such that $d_l < 0$ and $l \neq l_0$. As in the proof of Lemma 3.4, we have

$$\sum_{i=1}^{u} t_i \geq e\sqrt{q} - 1$$

and

$$\sum_{l \in M'} d_l(d_l - 1) \geq 2(e\sqrt{q} - 1)(\sqrt{q} - e - 1) =: f(e).$$

Lemma 3.1 (5) shows that $f(e) \leq q + \sqrt{q}$. In view of $q \geq 49$, we have

$$f(1) = 2(\sqrt{q} - 1)(\sqrt{q} - 2) > q + \sqrt{q}$$

and

$$f(\sqrt{q} - 2) = 2(q - 2\sqrt{q} - 1) > q + \sqrt{q}.$$

Hence $e \neq 1, \sqrt{q} - 2$. Since f has degree 2 in e, it follows that $1 \leq e \leq \sqrt{q} - 2$ is not possible. Hence $e \geq \sqrt{q} - 1$ and $d_{l_0}(d_{l_0} - 1) \geq (\sqrt{q} - 1)(\sqrt{q} - 2)$.

In the same way it follows that every line l that meets S in more than $\sqrt{q}+1$ points satisfies $d_l(d_l - 1) \geq (\sqrt{q} - 1)(\sqrt{q} - 2)$. But $2(\sqrt{q} - 1)(\sqrt{q} - 2) > q + \sqrt{q}$, so equation (5) in Lemma 3.1 shows that every line other than l_0 meets S in at most $\sqrt{q}+1$ points.

Consider a point $P \in l_0 \cap S$. Since S has deficiency 1 and since l_0 has deficiency $-e$, the sum of the deficiencies of the ordinary lines other than l_0 on P must be $e + 1$. Since l_0 is the only line with negative deficiency, it follows from equation (4) in Lemma 3.1 that

$$(\sqrt{q} + 1 + e)(e + 1) - e = |l_0 \cap S|(e + 1) + d_{l_0} \leq q + \sqrt{q} + 1.$$

But $e \geq \sqrt{q} + 1$, a contradiction. □

5.   Strong representative systems with deficiency 2

In this section we consider a strong representative system with $q\sqrt{q} - 1$ points. Now two difficulties occur. First of all, we are no longer able to prove that every line meets S in at most $\sqrt{q}+1$ points. Secondly we shall find parameters for a very regular hypothetical partial unital with deficiency two that is not part of a unital.

In this section we assume that every line meets S in at most $\sqrt{q}+1$ points (so S is a partial unital) and that $q \geq 16$. Lemma 3.4 shows that every point lies on at most $\sqrt{q}+1$ tangents.

An ordinary line which meets S in precisely i points will be called an i-line, and by b_i we denote the number of i- lines.

45

Lemma 5.1 *A point of S either lies on a unique $(\sqrt{q}-1)$-line and on no $\sqrt{q}$-line, or it lies on no $(\sqrt{q}-1)$-line and on two $\sqrt{q}$-lines. In particular, every ordinary line meets S in 0, $\sqrt{q}-1$, $\sqrt{q}$ or $\sqrt{q}+1$ points.*

Proof. Obvious. □

Lemma 5.2 *a) If there is no line missing S then every ordinary line meets S in $\sqrt{q}-1$ or $\sqrt{q}+1$ points. Furthermore the number of lines that meet S in $\sqrt{q}-1$ points is equal to $q+\sqrt{q}+1$.*

b) If l is a line missing S, then every point of l lies on at most $\sqrt{q}$ tangents. Dually, if P is a point lying on no tangent, then every line through P meets S in at most $\sqrt{q}$ points.

c) There is a line missing S if and only if there is a point lying on no tangents.

Proof. a) Since there is no line missing S, every ordinary line meets S in $\sqrt{q}-1$, $\sqrt{q}$ or $\sqrt{q}+1$ points. Since the right hand sides of equations (4) and (5) in Lemma 3.1 coincide, it follows that there do not exist lines meeting S in precisely $\sqrt{q}$ points. Now equation (2) in Lemma 3.1 shows that the number of $(\sqrt{q}-1)$- lines is equal to $q+\sqrt{q}+1$.

b) Suppose that l is a line missing S and denote by t the number of tangents that pass through a point P of l. Since every line meets S in at most $\sqrt{q}+1$ points, we see that the lines through P cover at most $t+(q-t)(\sqrt{q}+1) = q\sqrt{q}+q-t\sqrt{q}$ points of S. Since this number must be at least $|S| = q\sqrt{q}-1$, we see that $t \le \sqrt{q}$.

c) It suffices to verify one direction, the other implication follows then by duality. Assume that there is no line missing S and that there exists a point P that lies on no tangent. Then part a) shows that every ordinary line meets S in $\sqrt{q}-1$ or $\sqrt{q}+1$ points so that b) implies that every line through P meets S in $\sqrt{q}-1$ points. Hence $|S| = (q+1)(\sqrt{q}-1)$, which is a contradiction. □

Lemma 5.3 *It is not possible that there is exactly one line that misses S.*

Proof. Assume that there is a unique line l_0 missing S. By Lemma 5.1, every ordinary line other than l_0 meets S in at least $\sqrt{q}-1$ points. Lemma 3.1 implies that the number of $(\sqrt{q}-1)$-lines is $b_{\sqrt{q}-1} := \frac{1}{2}(q+\sqrt{q}+2)$ and that the number of $\sqrt{q}$-lines is $b_{\sqrt{q}} = q-1$.

By Lemma 5.2, there exists a point not lying on a tangent. Assume that there are two points not lying on tangents. Then the statement dual to Proposition 3.5 shows that the set of tangents of S are all but two tangents from a unital so that S consists of all but two points of a unital. But then there are two lines missing S, a contradiction. Hence there is a unique point P_0 not lying on tangents.

First consider the case in which $P_0 \in l_0$. By 5.2 b), every point of l_0 lies on at most $\sqrt{q}$ tangents. Since there are $q\sqrt{q} - 1$ tangents, it follows that l_0 has a unique point Q_0 that lies on $\sqrt{q} - 1$ tangents while the $q - 1$ points $\neq P_0, Q_0$ of l_0 lie on $\sqrt{q}$ tangents. Dualizing, we see that P_0 must lie on $q - 1$ lines that meet S in $\sqrt{q}$ points. Since $b_{\sqrt{q}} = q - 1$, this means that every $\sqrt{q}$-line passes through P_0. Let P be a point $\neq P_0, Q_0$ of l_0. Then P lies on $\sqrt{q}$ tangents, which together cover $\sqrt{q}$ points of S, and on $q - \sqrt{q}$ lines that cover the remaining $|S| - \sqrt{q} = q\sqrt{q} - \sqrt{q} - 1 = (q - \sqrt{q})(\sqrt{q} + 1) - 1$ points of S. But this is only possible if P lies on a unique $\sqrt{q}$-line, a contradiction.

Now consider the case that P_0 is not on l_0. Since the lines through P_0 meet S in at most $\sqrt{q}$ points (Lemma 5.2), every line through P_0 meets S in $\sqrt{q} - 1$ or $\sqrt{q}$ points. In view of $|S| = (q - \sqrt{q})\sqrt{q} + (\sqrt{q} + 1)(\sqrt{q} - 1)$, we see that P_0 lies on precisely $q - \sqrt{q}$ lines that meet S in $\sqrt{q}$ points. Let l be one of these lines. Then each of the $\sqrt{q}$ points of $l \cap S$ lies on a second $\sqrt{q}$-line. Hence there are at least $\sqrt{q}$ lines that meet S in $\sqrt{q}$ points and do not pass through P_0. But there are only $q - 1$ lines that meet S in $\sqrt{q}$ points and P_0 lies already on $q - \sqrt{q}$ of them. This contradiction completes the proof of the lemma. $\qquad\square$

Lemma 5.4 *If there does not exist a line missing S, then q is even.*

Proof.   By Lemma 5.2, every ordinary line meets S in $\sqrt{q} - 1$ or $\sqrt{q} + 1$ points. Furthermore, every point of S lies on a unique tangent and a unique line that meets S in $\sqrt{q} - 1$ points. If q is odd, then $\gcd(\sqrt{q} - 2, \sqrt{q}) = 1$. However, the main result in the next section shows that this situation cannot occur. $\qquad\square$

Remark.   If q is even, then we do not know if there exists a partial unital with deficiency two which meets every line. The first difficult case is $q = 16$. Is it possible to find a set S of 63 points in $PG(2, 16)$ with intersection numbers 1, 3 and 5 and such that every point lies on a unique tangent?

6. $\{1, m, n\}$-sets in $PG(2, q)$

In this section we consider a set S of points in $PG(2, q)$, q a square, with intersection numbers $1, m, n$ satisfying $1 < m < n$. We call a line that meets S in 1, m or n points respectively a *tangent*, a *short* line or a *long* line. We assume that every point of S lies on a unique tangent and on a unique short line. These conditions are very restrictive and we expect strong results; in fact, we do not believe that such sets exist. However, we shall only be able to prove the following proposition.

Proposition 6.1 *Suppose S is a set of points in $PG(2, q)$, q a square, having the properties described above. Then $(m - 1, n - 1) \neq 1$.*

Before we start with the proof of this proposition, we prove a lemma needed later on in the proof.

Lemma 6.2 *Let S be a set of points in the affine plane $AG(2, q)$, where $q = p^s$ for some prime number p. Suppose Π_1, Π_2 and Π_3 are three parallel classes with the following property: Exactly one line l_i of Π_i meets S in a number of points which is not zero modulo p. Then l_1, l_2 and l_3 are concurrent.*

Proof. We may assume that Π_1 consists of the vertical lines, that Π_2 consists of the horizontal lines, and that Π_3 consists of the lines with equations $x + y = c$, $c \in F_q$. Furthermore, we may assume that l_1 and l_2 meet in $(0, 0)$. Let (x_i, y_i), $i = 1, \ldots, s := |S|$, be the points of S. We are interested in the sums $X := \sum_{i=1}^{s} x_i$ and $Y := \sum_{i=1}^{s} y_i$. If $l \in \Pi_1 \setminus \{l_1\}$, then the points of $l \cap S$ together will contribute nothing to the sum X, because all points of $l \cap S$ have the same x-coordinate and because of $|l \cap S| = 0$ (in F_q). Also the points of the exceptional line l_1 will contribute nothing to the sum X, because l_1 has equation $x = 0$. Hence $X = 0$. The same argument used for the lines of Π_2 shows that $Y = 0$. Hence $X + Y = 0$. Now we use the lines of Π_3 to determine $X + Y$ again.

Since the lines $l \in \Pi_3$ have equations $x + y = c$, $c \in F_q$, we see as before that the points of $l \cap S$ together will contribute nothing to the sum $X + Y$, if $l \in \Pi_3 \setminus \{l_3\}$. Furthermore, if l_3 has equation $x + y = c_0$, then the points of $l_3 \cap S$ contribute $|l \cap S| c_0$ to the sum $X + Y$. Thus $X + Y = |l \cap S| c_0$. As $X + Y = 0$ and $|l \cap S| \neq 0$ (in F_q), it follows that $c = 0$, that is, l_3 passes through $(0, 0)$. □

Lemma 6.3 a) *The number of points of S is $s = m + (q - 1)(n - 1)$.*
b) $s(1 + \frac{1}{m} + \frac{q-1}{n}) = q^2 + q + 1$.
c) $(n - 1)(m + 1) = q(n - 1 - m - \frac{q - (n-1)^2}{n-m} m)$ and $n < 2 + \sqrt{q}$.

Proof. a) Since every point of S lies on a unique tangent and a unique short line, we have $s = m + (q - 1)(n - 1)$.

b) Since every every point of S lies on a unique tangent, a unique short line and $q - 1$ long lines, the number of tangents is s, the number of short lines is $\frac{s}{m}$ and the number of long lines is $\frac{s(q-1)}{n}$. Thus $s + \frac{s}{m} + \frac{s(q-1)}{n} = q^2 + q + 1$.

c) The equation in c) follows from a) and b). Assume that $n \geq 2 + \sqrt{q}$.
Then $\frac{(n-1)^2-q}{n-m} \geq \frac{(n-1)^2-q}{n-2} \geq 2$ so that $(n-1)(m+1) \geq q(n-1-m+2m) = q(n-1+m) > q(m+1)$ which is absurd, since $n \leq q+1$. $\qquad\square$

Lemma 6.4 *For a point P not in the set S, let t_P be the number of tangents of S through P. Then $\sum_{P \notin S} t_P = sq$ and $\sum_{P \notin S}(t_P - n)^2 = \frac{s}{m}(n-m)^2$.*

Proof.   Since S has s tangents, we have

$$\sum_{P \notin S} 1 = q^2 + q + 1 - s, \qquad \sum_{P \notin S} t_P = sq \qquad \text{and} \qquad \sum_{P \notin S} t_P(t_P - 1) = s(s-1).$$

Thus

$$A := \sum_{P \notin S}(t_P - n)^2 = s(s-1) - (2n-1)sq + n^2(q^2 + q + 1 - s).$$

Using $q^2 + q + 1 - s = s(\frac{1}{m} + \frac{q-1}{n})$ (see Lemma 6.3), we obtain

$$
\begin{aligned}
\frac{A}{s} - \frac{(n-m)^2}{m} &= (s-1) - (2n-1)q + n^2(\frac{1}{m} + \frac{q-1}{n}) - \frac{1}{m}(n-m)^2 \\
&= (s-1) - (2n-1)q + n(q-1) - m + 2n = 0.
\end{aligned}
$$

$\qquad\square$

Lemma 6.5 *Set $\hat{s} = q^2 + q + 1 - s$. Suppose that $0 \leq k < n - m$ and that $t_i, i = 1, \ldots, \hat{s}$, are integers satisfying $\sum_{i=1}^{\hat{s}} t_i = sq$ and $t_i \equiv k \pmod{n-m}$. If $q \geq 16$, then $\sum_{i=1}^{\hat{s}}(t_i - n)^2 \geq \frac{s}{m}(n-m)^2$ with equality if and only if $k = 0$ and $t_i \in \{m, n\}$ for all $i = 1, \ldots, \hat{s}$.*

Proof.   By Lemma 6.3 b), we have $\hat{s} = s(\frac{1}{m} + \frac{q-1}{n})$. It follows that

$$\sum_{i=1}^{\hat{s}} t_i = sq = \hat{s}n - \frac{s}{m}(n-m).$$

Hence the average value for t_i is $n - \frac{s(n-m)}{m\hat{s}}$. Using $s = m + (q-1)(n-1)$ and $2 \leq m < n < 2 + \sqrt{q}$, we see that the average value for t_i is between $n - 1$ and n. Thus $\sum(t_i - n)^2$ obtains its minimum value, if $t_i \in \{m+k, n+k\}$ for all i. Hence

$$\sum_{i=1}^{\hat{s}}(t_i - n)^2 \geq x(n - m - k)^2 + yk^2$$

where x and y are defined by $x + y = \hat{s}$ and

$$x(m+k) + y(n+k) = \sum_{i=1}^{\hat{s}} t_i = sq,$$

49

that is,

$$x = \frac{\hat{s}k}{n-m} + \frac{s}{m}. \quad \text{and} \quad y = \hat{s} - \frac{\hat{s}k}{n-m} - \frac{s}{m}$$

Now assume that $\sum_{i=1}^{\hat{s}}(t_i - n)^2 \leq \frac{s}{m}(n-m)^2$. Then

$$\begin{aligned}
\frac{s}{m}(n-m)^2 &\geq x(n-m-k)^2 + yk^2 \\
&= (x+y)k^2 - 2x(n-m)k + x(n-m)^2 \\
&= \hat{s}k^2 - 2x(n-m)k + (\frac{\hat{s}k}{n-m} + \frac{s}{m})(n-m)^2.
\end{aligned}$$

If $k = 0$ the we obtain equality so that $t_i \in \{m+k, n+k\}$ for all i. Assume that $k > 0$. Then it follows that

$$\begin{aligned}
0 &\geq \hat{s}k - 2x(n-m) + \hat{s}(n-m) \\
&= -\hat{s}k - \frac{2s}{m}(n-m) + \hat{s}(n-m) \\
&\geq \hat{s} - \frac{2s}{m}(n-m).
\end{aligned}$$

Since $\hat{s} = s(\frac{1}{m} + \frac{q-1}{n})$, we conclude that $2s(n-m) \geq \hat{s}m = s + s\frac{q-1}{n}m$ so that $(2n - 1 - 2m)n \geq (q-1)m$. Since $q - 1 \geq 2n$ (this follows from $q \geq 16$ and $n < 2 + \sqrt{q}$, cf. 6.3 c), and $m \geq 2$, it follows that $(2n-5)n \geq 2(q-1)$ or $2((n-1)^2 - q) \geq n - 2$. Since $(n-1)^2 \leq q$ by the preceding lemma, this is a contradiction. $\qquad\square$

Lemma 6.6 *If $(n-1, m-1) = 1$ and $q \geq 16$ then every point outside the set S lies on m or n tangents. Furthermore there are exactly $\frac{s}{m}$ points that lie on m tangents. We have $m = \sqrt{q} - 1$ and $n = 1 + \sqrt{q}$.*

Proof. For every point P not in S, we denote by t_P the number of tangents and by m_P the number of m-lines through P. Then

$$m + (q-1)(n-1) = s = t_P + m_P m + (q + 1 - t_P - m_P)n$$

so that

$$t_P(n-1) + m_P(n-m) = q + 2n - 1 - m$$

Since $(n-1, m-1) = 1$ and hence $(n-1, n-m) = 1$, we see that this equation determines t_P uniquely modulo $n - m$. The preceding two lemmas show that $t_P \in \{m, n\}$ for every point P not in S. Lemma 6.3 b) shows that

$$\sum_{P \notin S} t_P = sq = \frac{s}{m} \cdot m + (q^2 + q + 1 - s - \frac{s}{m})n.$$

It follows that there are exactly $\frac{s}{m}$ points that lie on m tangents. Let P be a point satisfying $t_P = n$. Then $t_P(n-1) + m_P(n-m) = q + 2n - 1 - m$ implies that $n - m$ divides $q - (n-1)^2$. It follows therefore from Lemma 6.3 c) that q divides $(n-1)(m+1)$. Since $m+1 \leq n \leq 1 + \sqrt{q}$, this implies that $m = \sqrt{q} - 1$ and $n = \sqrt{q} + 1$. $\qquad\square$

Lemma 6.7 *For $q \geq 16$ the case $(n-1, m-1) = 1$ does not occur.*

Proof. Assume that $(n-1, m-1) = 1$. Then $m = \sqrt{q} - 1$ and $n = \sqrt{q} + 1$ by the preceding lemma. Since $(n-1, m-1) = 1$, it follows that q is odd.

Let l be a long line. Then the short lines that meet l in a point of S are concurrent. This can be seen as follows. Let P_i be a point of $l \cap S$ and let l_i be the short line through P_i, $i = 1, 2, 3$. Consider the affine plane $\mathbf{A}$ that has l as line at infinity, and the parallel-classes Π_i, $i = 1, 2, 3$, that contains l_i. Then Lemma 6.2 shows that the lines l_1, l_2 and l_3 are concurrent. Therefore all short lines that meet the long line l in a point of S are concurrent.

Let P be the point in which these $\sqrt{q} + 1$ short lines meet. Since every point lies on at most $\sqrt{q} + 1$ short lines, every short line through P must meet l in a point of S. Let l_1 and l_2 be two short lines through P. Set $P_1 := l_1 \cap l$ and let Q be a point $\neq P_1$ of $l_1 \cap S$. Then each line which joins Q to one of the $\sqrt{q}$ points of $l \backslash \{P_1\}$ is a long line. Since l_2 is short, we see that Q lies on a long line l' that meets l in a point Q' of S and that meets l_2 in a point outside S. Hence l' meets two of the short lines through P and therefore every short line that meets l' in a point of S will pass through P (use the same argument we used for the line l). However, since P lies on $\sqrt{q} + 1$ short lines and because l' and l_2 meet in a point outside S, there must be a point on $l' \cap S$ which does not lie on one of the short lines through P. This contradiction completes the proof. $\qquad\square$

References

[1] **A. Blokhuis**. On nuclei and affine blocking sets. Preprint, 1993.

[2] **A. A. Bruen**. Blocking sets in finite projective planes. *SIAM J. Appl. Math.*, 21, pp. 380–392, 1971.

[3] **A. A. Bruen and J. A. Thas**. Blocking sets. *Geom. Dedicata*, 6, pp. 193–203, 1977.

[4] **T. Illés, T. Szőnyi, and F. Wettl**. Blocking sets and maximal strong representative systems in finite projective planes. *Mitt. Math. Sem. Giessen*, 201, pp. 97–107, 1991.

[5] **F. Wettl**. Nuclei in finite non-desarguesian planes. In A. Beutelspacher, F. Buekenhout, F. De Clerck, J. Doyen, J. W. P. Hirschfeld, and J. A. Thas, editors, *Finite Geometry and Combinatorics*, pages 405–412, Cambridge, 1993. Cambridge University Press.

Aart Blokhuis and Klaus Metsch, Department of Mathematics and Computer Science, Eindhoven University of Technology, PO Box 513, 5600 MB Eindhoven, The Netherlands. e-mail: aartb@win.tue.nl — kmetsch@win.tue.nl

The complement of a geometric hyperplane in a generalized polygon is usually connected

A. E. Brouwer

Abstract

We show that with few exceptions the subgeometry of a finite generalized polygon induced by the objects far away from (i.e., in general position w.r.t.) a given flag, is connected.

1. Generalized polygons

Let $(X, \mathcal{L})$ be a nondegenerate generalized n-gon, and let $x \in X$, $L \in \mathcal{L}$, $x \in L$. Let $(Y, \mathcal{M})$ be the subgeometry of $(X, \mathcal{L})$ induced by the points and lines in general position w.r.t. x, and let $(Z, \mathcal{N})$ be the subgeometry of $(Y, \mathcal{M})$ induced by the points and lines in general position w.r.t. (x, L). (Here two objects are said to be in general position when they are contained in opposite chambers. In particular, for two points x and y of a generalized $2m$-gon, this means $d(x, y) = m$ in the collinearity graph, or $d(x, y) = 2m$ in the point-line incidence graph.)

We want to show that $(Y, \mathcal{M})$ and $(Z, \mathcal{N})$ are connected. If $n = 2$, then $Z = Y = X \backslash \{x\}$ and $\mathcal{M} = \mathcal{L}$, $\mathcal{N} = \mathcal{L} \backslash \{L\}$, so that indeed both geometries are connected (and are generalized 2-gons again).

If $n = 3$, then $(X, \mathcal{L})$ is a projective plane, $(Y, \mathcal{M})$ a dual affine plane, and $(Z, \mathcal{N})$ what is sometimes called a biaffine plane: all points not on a given line, and all lines not on a given point. Again, clearly both geometries are connected.

Now let us assume that $n > 3$ and $(X, \mathcal{L})$ is finite of order (s, t), so that $n \in \{4, 6, 8, 12\}$. Each line of $\mathcal{M}$ has s points in Y, so if $s = 1$ then $(Y, \mathcal{M})$ will be disconnected (unless $|Y| = 1$). Similarly, if $t = 1$ then $(Z, \mathcal{N})$ will be disconnected (unless $|\mathcal{N}| = 1$). Moreover, if $t = 1$ and $n = 2m$, then $(X, \mathcal{L})$ is the flag graph of a generalized m-gon of order (s, s), and considering $(Y, \mathcal{M})$ in the former setting is equivalent to considering $(Z, \mathcal{N})$ in the latter setting. Thus, we may suppose $s, t > 1$ and $n \in \{4, 6, 8\}$.

The main purpose of the present note is to prove the following theorem.

Theorem 1.1 *Let $(X, \mathcal{L})$ be a thick finite generalized n-gon of order (s, t). Then*

(i) The subgeometry $(Y, \mathcal{M})$ (or, indeed, the complement of an arbitrary geometric hyperplane in $(X, \mathcal{L})$) is connected, except possibly in the cases $(n, s, t) = (6, 2, 2), (8, 2, 4)$.

(ii) The subgeometry $(Z, \mathcal{N})$ is connected, except possibly in the cases $(n, s, t) = (4, 2, 2), (6, 2, 2), (6, 3, 3), (8, 2, 4), (8, 4, 2)$.

(iii) For the stated possibly exceptional parameter sets actual exceptions do occur.

2. Examples

Let us first discuss the exceptions. There is a unique generalized quadrangle of order $(2,2)$, and in it $(Y, \mathcal{M})$ is the 1-skeleton of the cube, and $(Z, \mathcal{N})$ is the union of two quadrangles.

The only known generalized hexagons of order (q, q) are those gotten from $G_2(q)$ - these are not self-dual, except when q is a power of three. There is a unique generalized hexagon of order $(2,2)$ (up to duality) (Cohen & Tits [2]). Let $P = G_r$ be the maximal parabolic of $G = G_2(q)$ corresponding to a short root r and let points be the cosets of P, and lines the cosets of $Q = G_s$, where s is the other fundamental root. (The corresponding generalized hexagon is called the dual of the classical $G_2(q)$ generalized hexagon.) Then the q^5 points at distance 3 from x are parametrized by $E := U_{srsrs}^-$, the product of the five root groups X_α for positive α distinct from r. Now $[E, E] = X_{3r+2s}$, so $E/[E, E]$ is elementary abelian of order q^4. The group E fixes the $q + 1$ lines on x, and hence a system of $q + 1$ parallel classes of lines in $(Y, \mathcal{M})$. We see that the geometry on Y given by the union of a parallel classes has connected components of size (at most) q^{a+1}. In particular, for $q = 2$ we find that $(Y, \mathcal{M})$ has two connected components of 16 points each, and $(Z, \mathcal{N})$ is the union of four 8-gons. Similarly, for $q = 3$ we find that $(Z, \mathcal{N})$ has three connected components, each of size 81.

If we consider the classical $G_2(q)$ generalized hexagon, i.e., take for the points the cosets of G_s, then for $q = 3$ nothing changes since G has an automorphism interchanging long and short roots. But for $q = 2$ we now find that $(Y, \mathcal{M})$ is connected (Cohen & Tits [2]; see also [1], p. 384). However, there are other geometric hyperplanes with disconnected complement: $(X, \mathcal{L})$ contains a generalized hexagon $(W, \mathcal{K})$ of order $(1,2)$; if H is the set of 21 points on the lines of $\mathcal{K}$ but not in W, then each line meets H in precisely 1 point, and $X \backslash H$ has two connected components (of sizes 14 and 28). A complete description of the geometric hyperplanes in the two generalized hexagons of order $(2,2)$ can be found in [3].

Finally, the only known generalized octagons are those gotten from $^2F_4(q)$. In the particular case of the generalized octagon of order $(2,4)$ belonging to $^2F_4(2)$, explicit inspection reveals that both $(Y, \mathcal{M})$ and $(Z, \mathcal{N})$ have two connected components with 512 points each. The automorphism group of each component has order 10240 and acts sharply 2-arc transitively.

Thus, we verified part (iii) of the theorem.

Considering the case $s > 1$, $t = 1$ for a moment, this means that we saw that for $(n, s, t) = (8, 2, 1)$, $(12, 2, 1)$ and $(12, 3, 1)$ there exist generalized polygons such that $(Y, \mathcal{M})$ is disconnected; for each of these parameter sets a unique generalized polygon is known.

3. An eigenvalue argument

Connectedness of various graphs will be proved here using an eigenvalue argument I learned from Willem Haemers. Suppose Γ is a connected graph with second largest eigenvalue θ, and Δ is a regular subgraph of valency $r > \theta$. Then Δ is connected.

[Indeed, the multiplicity of the valency as eigenvalue of a regular graph Δ equals the number of connected components of Δ. But the largest eigenvalue of Γ has multiplicity 1 (by Perron-Frobenius), so by interlacing it follows that Δ is connected.]

So, we need the second largest eigenvalue of the collinearity graph of a generalized n-gon of order (s, t). As is well known (cf. [1], p. 203) this is

$$\theta = s - 1 + \sqrt{ast}$$

with $a = 0, 1, 2, 3$ for $n = 4, 6, 8, 12$, respectively. The geometry $(Y, \mathcal{M})$ (or, indeed, the complement of an arbitrary geometric hyperplane in $(X, \mathcal{L})$) has $t + 1$ lines on each point, and s points on each line, and hence its collinearity graph is regular of valency $r = (s - 1)(t + 1)$, and our sufficient condition for connectedness becomes $(s - 1)(t + 1) > s - 1 + \sqrt{ast}$, i.e.,

$$s - 1 > \sqrt{as/t}.$$

If $s > 1$, then this is automatically fulfilled for generalized quadrangles ($n = 4$). For generalized hexagons we find $s = t = 2$ as only possible exception. Finally, for generalized octagons we find $s = 2$, $t \le 4$. But since $2st$ is a square this means that $t = 4$.

This proves part (i) of the theorem.

4. Far away from a flag

In the cases we are considering ($n = 4, 6, 8$), the geometry $(Z, \mathcal{N})$ has the same point set as $(Y, \mathcal{M})$, but has lost a parallel class $\mathcal{C}$ of lines. Suppose that it is disconnected, so that Z is the disjoint union of two nonempty sets U, V where no line of $\mathcal{N}$ meets both U and V. Consider the corresponding partition of the adjacency matrix A of the collinearity graph of $(Y, \mathcal{M})$, and its condensed form, the matrix B of average row sums of the blocks of A. With $r = (s-1)(t+1)$, $u = |U|$, $v = |V|$, we find

$$B = \begin{pmatrix} r-\epsilon & \epsilon \\ \epsilon u/v & r-\epsilon u/v \end{pmatrix}$$

where ϵ is the average number of neighbours a point of U has in V. The eigenvalues of B are r and $r - \epsilon - \epsilon u/v$, and these interlace the eigenvalues of A, so we must have

$$(s-1)(t+1) - \epsilon(1 + u/v) \leq s - 1 + \sqrt{ast}.$$

Let us look more closely at ϵ. If n_i lines of C have i points in U and $s-i$ points in V, then $u = \sum i n_i$, $v = \sum (s-i) n_i$, $\epsilon = (1/u) \sum i(s-i) n_i$, so that

$$\epsilon(1 + \frac{u}{v}) = (\frac{1}{u} + \frac{1}{v}) \sum i(s-i) n_i.$$

This expression is maximized for given u and v by having all lines meet U in the same number of points, and then it equals s. Thus,

$$(s-1)(t+1) - s \leq s - 1 + \sqrt{ast},$$

i.e., $(s-1)(t-1) \leq 1 + \sqrt{ast}$. For $n = 4$ $(a = 0)$ this implies $s = t = 2$. For $n = 6$ $(a = 1)$ this implies $s = t \in \{2, 3\}$ (since st is a square). Finally, for $n = 8$ $(a = 2)$ this implies $\{s, t\} = \{2, 4\}$ (since $2st$ is a square).

This proves part (ii) of the theorem.

5. The infinite case

As Hans Cuypers remarked, it is easy to prove that $(Y, \mathcal{M})$ and $(Z, \mathcal{N})$ are connected in the case of not necessarily finite thick generalized quadrangles.

Theorem 5.1 *Let $(X, \mathcal{L})$ be a thick generalized quadrangle. Then the subgeometries $(Y, \mathcal{M})$ (or, indeed, the complement of an arbitrary geometric hyperplane in $(X, \mathcal{L})$) and $(Z, \mathcal{N})$ are connected, except that in case $(X, \mathcal{L})$ is the unique generalized quadrangle of order $(2, 2)$, the geometry $(Z, \mathcal{N})$ is the union of two quadrangles.*

Proof.   The part about the complement of a geometric hyperplane is well-known and easy to prove, so let us consider $(Z, \mathcal{N})$. Since $(Y, \mathcal{M})$ is connected (and $Y = Z$) it suffices to show that two points $y, z \in Z$ that are joined by a line K concurrent with L are joined by a chain in $(Z, \mathcal{N})$. By duality we may assume that each line has at least four points. Let us denote incidence by $*$ and collinearity (concurrency) by $\sim$. Let M, N be lines distinct from K on y, z, respectively. Then $M, N \in \mathcal{N}$. A chain $y * M * p * H * q * N * z$ will join y and z in $(Z, \mathcal{N})$ unless $H \sim L$ or $p \sim x$ or $q \sim x$. Let $A = \{a | x \sim a \sim y\} \backslash L$ and $B = \{b | x \sim b \sim z\} \backslash L$. If $q \in N \backslash B$, then q is collinear with all except at most one element of A, for if $q \not\sim a, a'$ then at most one of the two lines joining q to ya or ya' meets L and the other will achieve the connection. But now choose two points $q, q' \in N \backslash B$. Since they cannot have a common neighbour in A, it follows that $|A| = 2$ so that $t = 2$. But in that case $s \neq 3$ and we can find a third point $q'' \in N \backslash B$, a contradiction. $\qquad \square$

For $n > 4$, nothing is known.

References

[1] **A. E. Brouwer, A. M. Cohen, and A. Neumaier.** *Distance-regular Graphs.* Springer Verlag, Berlin, 1989.

[2] **A. M. Cohen and J. Tits.** On generalized hexagons and a near octagon whose lines have three points. *European J. Combin.*, 6, pp. 13–27, 1985.

[3] **D. Frohardt and P. Johnson.** Geometric hyperplanes in generalized hexagons of order $(2, 2)$. In preparation.

A. E. Brouwer, Dept. of Math., Eindhoven Univ. of Technology, P. O. Box 513, 5600 MB Eindhoven, the Netherlands. e-mail: aeb@win.tue.nl

Locally co-Heawood graphs

A. E. Brouwer   D. G. Fon-der-Flaass
S. V. Shpectorov

Abstract

Let Δ be the incidence graph of the unique biplane on 7 points, that is, the bipartite complement of the Heawood graph. We find that there are precisely three connected graphs that are locally Δ, on 36, 48 and 108 vertices, where the last graph is an antipodal 3-cover of the first one.

1. Introduction

Let notation be as in [1]. (In particular, $\sim$ denotes adjacency, $\Gamma_i(\gamma)$ is the collection of vertices at distance i from γ in Γ, $\Gamma(\gamma) := \Gamma_1(\gamma)$, and $\gamma^\perp :=$ $\{\gamma\} \cup \Gamma(\gamma)$.) The Heawood graph H is the smallest cubic graph of girth 6; it is bipartite, the incidence graph of the Fano plane. The co-Heawood graph Δ is its bipartite complement, the nonincidence graph of the Fano plane, i.e., the incidence graph of the unique biplane on 7 points. (Thus, $\Delta = H_3$.) The graph Δ has 14 vertices, valency 4, is bipartite, is distance-regular of diameter 3 and has distance distribution diagram

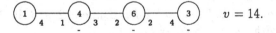

$$v = 14.$$

Its automorphism group is $G \simeq PGL(2,7)$ of order 336 acting distance transitively.

The graph Δ occurs in the Suzuki chain $S_0 = 4K_1$, $S_1 = \Delta$, S_2, S_3, S_4, S_5 of graphs on 4, 14, 36, 100, 416, 1782 vertices, respectively. Each graph S_{i+1} of this chain is locally S_i. In particular, the graph $\Sigma := S_2$ is locally Δ, it is strongly regular with parameters $(v, k, \lambda, \mu) = (36, 14, 4, 6)$.

In this note we determine all connected graphs that are locally Δ, i.e., all connected graphs Γ such that for each vertex $\gamma \in \Gamma$ its neighbourhood $\Gamma(\gamma)$ induces a subgraph isomorphic to Δ. There turn out to be three such graphs. The smallest one is the graph Σ on 36 vertices. The largest one, T, on 108 vertices, is an antipodal 3-cover of Σ, but is not distance-regular. Its

distance-distribution diagram is

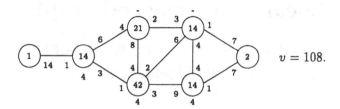

$v = 108.$

This graph was described explicitly in Neumaier [2]; the above distance-distribution diagram was given by Soicher [pers. comm.]. Soicher [3] proved that Σ and T are the unique connected locally Δ graphs with an automorphism group transitive on ordered triangles, and asked whether this transitivity assumption could be removed. In this note this question is answered in the negative. Indeed, there is a third locally Δ graph Φ_{12} on 48 vertices, where Φ_m is defined as follows: Take for the vertex set the set $Z_m \times I_4$ (where I_4 is an arbitrary set of size 4), and let (a, b) be adjacent to the 14 vertices $(a \pm 4, b)$, $(a \pm 2, b')$, $(a \pm 1, b')$ for $b' \neq b$.

For any two vertices x, y of a graph Γ at mutual distance 2, let $\mu(x, y) :=$ $\Gamma(x) \cap \Gamma(y)$ denote the (graph induced on the) set of common neighbours of x and y. We shall call such a subgraph of Γ a μ-graph.

Theorem 1.1 *Let Γ be a connected locally Δ graph. Then one of the following three cases occurs:*

(a) $\Gamma \simeq \Sigma$. Thus, Γ has 36 vertices, and $\mu(x, y) \simeq C_4 + 2K_1$ for any two vertices x, y at distance 2. The graph Σ has diameter 2, and $Aut\,\Sigma \simeq U_3(3) : 2$ acting rank 3 with vertex stabilizer $L_2(7) : 2$.

(b) $\Gamma \simeq T$. Thus, Γ has 108 vertices, and for each vertex x of Γ we have $\mu(x, y) \simeq C_4$ for 21 vertices y, and $\mu(x, y) \simeq K_1$ for 42 vertices y. The graph T has diameter 4, and $Aut\,T \simeq (3 \times U_3(3)) : 2$ acting rank 7 with vertex stabilizer $L_2(7) : 2$.

(c) $\Gamma \simeq \Phi_{12}$. Thus, Γ has 48 vertices, and for each vertex x of Γ we have $\mu(x, y) \simeq C_8$, C_6, C_4, $6K_1$, $3K_1$ and $2K_1$ for 3, 8, 3, 2, 2, 12 vertices y, respectively. The graph Φ_{12} has diameter 3, and $Aut\,\Phi_{12} \simeq D_{24} \times Sym(4)$, acting transitively on the vertices (but not on the edges).

Note that there are infinitely many graphs that are locally E, if E is the graph obtained from Δ by deleting one edge. Examples are given by the graphs Φ_m ($m = \infty$ or $m \geq 9$, $m \neq 12$).

2.   Generalities

Let Γ be a connected locally Δ graph.

Lemma 2.1 *For any two vertices x, y at distance 2 in Γ, the subgraph $\mu(x, y)$ of Γ is one of the following:*
(i) mK_1 *(a coclique of size m), with $1 \le m \le 7$,*
(ii) $C_4 + mK_1$ *(a quadrangle and m isolated points), with $0 \le m \le 2$,*
(iii) $C_6 + mK_1$ *(a hexagon and m isolated points), with $0 \le m \le 1$, or*
(iv) C_8 *(an octagon).*

Proof.   For $z \in \mu(x, y)$, we see in $\Gamma(z) \simeq \Delta$ the vertices x, y without common neighbour, so in $\Gamma(z)$ these vertices have distance 2 or 3. In the former case x, y, z have 2 common neighbours, in the latter case none. Thus, $\mu(x, y)$ is a union of polygons and isolated points. Since Δ is bipartite, the polygons have an even length. If C is an induced $2g$-gon in Δ, then the $4g$ edges meeting C in one vertex have their other end point in $\Delta \setminus C$, so that $4g \le 4(14 - 2g)$, i.e., $g \le 4$. Inspection of Δ learns that no two polygon components can occur, and at most the indicated number of isolated points.   □

We can describe the polygons in Δ: there are 21 quadrangles (all equivalent under G), and any two vertices of Δ at distance 2 determine a unique quadrangle. The distance distribution diagram around a quadrangle (or around an octagon, or around a pair of vertices at distance 3) is

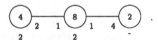

There are 21 octagons, all equivalent under G. They arise as the sets of vertices at distance 1 to a quadrangle (or to a pair of vertices at mutual distance 3). There are 84 hexagons, falling into two G-orbits.

If $a \sim z \sim b$ in $\Gamma(x)$, then locally in z we see that a and b have a common neighbour x, and hence there is a unique other point y adjacent to a, b, z, and $\mu(x, y)$ contains a polygon on the 2-claw $a \sim z \sim b$. Thus: any 2-claw in $\Gamma(x)$ determines a unique polygon. Let us first determine such polygon systems in Δ.

Lemma 2.2 *Let $\mathcal{P}$ be a system of polygons in Δ such that any 2-claw of Δ is in a unique polygon from $\mathcal{P}$. Then we have one of the following seven possibilities:*
(i) $\mathcal{P}$ *consists of all 21 quadrangles;*
(ii) $\mathcal{P}$ *consists of 14 hexagons;*
(iii) $\mathcal{P}$ *consists of 3 octagons and 10 hexagons;*

(iv) $\mathcal{P}$ consists of 1 quadrangle, 1 octagon and 12 hexagons;
(v) $\mathcal{P}$ consists of 3 quadrangles, 3 octagons and 8 hexagons;
(vi) $\mathcal{P}$ consists of 1 quadrangle, 4 octagons and 8 hexagons;
(vii) $\mathcal{P}$ consists of 3 quadrangles, 6 octagons and 4 hexagons.

Proof.   A small computer program quickly finds all possibilities (since Δ is a very small graph).   □

    The systems of hexagons found in case (ii) are described in detail in Section 4.

    Our classification will be in three parts corresponding to the possibilities for the non-singleton components of the μ-graphs: (i) always a quadrangle, (ii) always a hexagon, (iii) somewhere an octagon occurs. [Note that if no octagon occurs, and for some point x we have case (i) for the system of polygons on $\Gamma(x)$, then the same holds for the neighbours of x, so that case (i) holds everywhere.]

3.   Quadrangles only

In this section we assume that Γ is a locally Δ graph in which all non-singleton components of μ-graphs are quadrangles, and conclude that Γ must be isomorphic to one of the graphs Σ (on 36 vertices) and T (on 108 vertices) mentioned in the introduction.

    Since any $K_{1,1,2}$ is in a unique octahedron, we find two (flag transitive) Buekenhout-Tits geometries of type

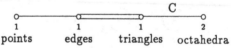

It is not difficult to find these graphs explicitly (essentially by reconstructing the distance-distribution diagram given earlier), but using the abovementioned result by Soicher we get the classification almost for free. Let $\tilde{\Gamma}$ be the universal cover of Γ modulo triangles. Then $\tilde{\Gamma}$ is locally Δ and is triangulable. Starting with Σ we find $\tilde{\Sigma}$, and by Soicher [3] we have $\tilde{\Sigma} \simeq T$.

Lemma 3.1 $\tilde{\Gamma} \simeq T$.

Proof.   Given an isomorphism (really, isomorphic embedding) $\psi : x^{\perp} \to T$ (with $x \in \tilde{\Gamma}$) and a neighbour y of x, there is a unique isomorphism $\psi^{y} : y^{\perp} \to T$ such that $\psi^{y}(z) = \psi(z)$ for $z \in x^{\perp} \cap y^{\perp}$ (because $Aut\,\Delta$ is vertex transitive and its vertex stabilizer is $Sym(4)$ with natural action on the neighbours of the fixed vertex). Fix $a \in \tilde{\Gamma}$ and an isomorphism $\varphi_a : a^{\perp} \to T$. For each path $\pi = a \sim ... \sim y \sim z$ we find a unique isomorphism $\varphi_{\pi} : z^{\perp} \to T$, using the above recipe: $\varphi_{\pi} = (\varphi_{\pi_0})^z$, where π_0 is the path $a \sim ... \sim y$. Define a map

$\varphi : \tilde{\Gamma} \to T$ by $\varphi(z) = \varphi_\pi(z)$ for any path π from a to z. As soon as we have shown that φ is well-defined, it is clear that it is the required isomorphism. (Indeed, it is a covering map, and T has no proper covers for which triangles lift to triangles.) Since $\tilde{\Gamma}$ is triangulable, it suffices to show that for any triangle $\pi = a \sim b \sim c \sim a$ we have $\varphi_{abca} = \varphi_a$. (Note that by definition $\varphi_{aba} = \varphi_a$.) Or, equivalently, it suffices to show that $\varphi_{ab}(z) = \varphi_{ac}(z)$ for $z \in \Gamma(b) \cap \Gamma(c)$, $z \neq a$. But for such a z the graph $\mu(a, z)$ contains a quadrangle $bcde$ (i.e., we have an octahedron $\{a, b, c, d, e, z\}$), and z is uniquely determined by both $\{a, b, c, e\}$ in $b^\perp$ and $\{a, b, c, d\}$ in $c^\perp$ (and the same holds for the images of these points in T), so indeed $\varphi_{ab}(z) = \varphi_{ac}(z)$. □

Thus, Γ is a quotient of T. Since T and Γ are locally isomorphic, the vertices of a fiber have mutual distance at least 4, and from the distance-distribution diagram of T we see that $\Gamma \simeq T$ or $\Gamma \simeq \Sigma$.

4. Hexagons only

Considering Δ as the nonincidence graph of the Fano plane $(X, \mathcal{L})$, we can describe the 84 hexagons as follows: given an edge of Δ, i.e., an antiflag (x, L), we have the hexagons:

type I: 3 Lines on x and 3 Points on L

type IIa: 3 Lines not on x and distinct from L and 3 Points on L

type IIb: 3 Lines on x and 3 Points not on L and distinct from x.

An *antipolarity* σ of $(X, \mathcal{L})$ is a permutation of $X \cup \mathcal{L}$ interchanging X and $\mathcal{L}$ and such that $x \in \sigma(y)$ if and only if $x \neq y$ and $y \notin \sigma(x)$ (for $x, y \in X$). The Fano plane has 8 antipolarities (all conjugate under $L_3(2)$); if we represent the Fano plane by Points i and Lines $\bar{i}$ ($i \in \mathbb{Z}_7$) with incidence $i + 1, i + 2, i + 4 \in \bar{i}$, then the map interchanging i and $\bar{i}$ is an antipolarity.

Now we can describe the systems of 14 hexagons in Δ covering all 2-claws. There are precisely 16 such systems (all conjugate under $PGL(2, 7)$). Given an antipolarity σ we have the system $[\sigma]_a$ consisting of all hexagons of types I and IIa for the antiflags $(x, \sigma(x))$, and the system $[\sigma]_b$ consisting of all hexagons of types I and IIb for these antiflags.

Now let Γ be a locally Δ graph in which all non-singleton components of μ-graphs are hexagons. Let us label each $\Gamma(x)$ with Δ in such a way that the system of hexagons is of type $[\sigma]_a$ for some σ. This then determines which half of $\Gamma(x)$ is called the set of Points and which half the set of Lines. We shall prove below that if y is a Point in $\Gamma(x)$, then x is a Point in $\Gamma(y)$. Color an edge xy red if y is a Point in $\Gamma(x)$ and green otherwise. Since Δ is bipartite, no two adjacent edges have the same color. But then any triangle xyz yields a contradiction.

Remains to prove our claim. Let y be a Point in $\Gamma(x)$. There are three type I hexagons on y, belonging to the antiflags $(a, \sigma(a))$ with $y \in \sigma(a)$. Let us call the vertices in $\Gamma_2(x)$ corresponding to these a'. There are three type IIa hexagons on y, belonging to the same antiflags. Let us call the corresponding vertices a''. Now in $\Gamma(y)$ we see six hexagons on x: for each Point z distinct from y in $\Gamma(x)$ the graph $\mu(y, z)$ contains the path $k' \sim L \sim x \sim M \sim k''$ where $K = \sigma(k)$ is the Line on y, z, and p is the third Point on K, and L, M are the other two Lines on p, with $k \in L$. In $\Gamma(y)$ the three vertices x, k', k'' do not have a common neighbour (since by definition the hexagons k' and k'' do not have common Lines), so they are collinear Points or concurrent Lines. But since for three choices of z the hexagon $\mu(y, z)$ is a hexagon of type IIa and consequently x, k', k'' are Points, it follows that x is a Point in $\Gamma(y)$, as desired.

This shows that no such graph Γ exists.

5. At least one octagon

Let Γ be a connected locally Δ graph and assume that at least one μ-graph in Γ is an octagon. We shall prove that Γ is uniquely determined. This is the most difficult case, because the vertex stabilizer in $Aut\,\Gamma$ is (much) smaller than $Aut\,\Delta$.

The induced octagons of Δ form one orbit under $Aut\,\Delta$. (Viewing Δ as the nonincidence graph of the Fano plane $(X, \mathcal{L})$, the octagons are the sets of 4 Points not on L and 4 Lines not on x for the flags (x, L).) For any octagon O in Δ, let O_i be the set of vertices outside O adjacent to i vertices in O. We have $|O_2| = 4$ and O_2 induces a quadrangle, and $|O_4| = 2$. We call the two vertices of O_4 the *centres* of O (in Δ). Each vertex of O is adjacent to a unique centre of O.

Lemma 5.1 *Let a, b be vertices of Γ such that $O = \mu(a, b)$ is an octagon, and let x, y be opposite vertices of O. Then $\mu(x, y)$ is an octagon, with a, b as a pair of opposite vertices. Precisely one neighbour of a in $\mu(x, y)$ is a centre of O in $\Gamma(a)$.*

Proof. Let x and y have common neighbours p, q in $\Gamma(a)$ and r, s in $\Gamma(b)$. Then $\mu(x, y)$ contains the paths $p \sim a \sim q$ and $r \sim b \sim s$. But p, q, r, s belong to the same bipartite half of $\Gamma(x)$ so that p, q are nonadjacent to r, s. $\quad\square$

From this lemma, it follows that every vertex x of Γ occurs in some octagon $\mu(a, b)$: if a is in some octagon $\mu(x, y)$, with opposite point b, then $O = \mu(a, b)$ is an octagon, and all vertices of $\Gamma(a) \setminus O$ occur in some octagon $\mu(u, v)$ for opposite vertices u, v of O. Since Γ is connected, the claim follows.

Lemma 5.2 *Let a, b be vertices of Γ such that $O = \mu(a, b)$ is an octagon. Let x, x' be opposite vertices of O, and let c, d be the centres of O in $\Gamma(a)$*

and $\Gamma(b)$, respectively, that are adjacent to x, x'. Then c and d are opposite vertices in the octagon $\mu(x, x')$.

Proof.   Suppose not, and let u, u' be the other two vertices of O adjacent to c and d. For $z = x, x', u, u'$ we see in $\Gamma(z)$ that a and b belong to the same bipartite half, and hence, since $a \sim c$ and $b \sim d$, the vertices c and d belong to the same bipartite half, so that z is not isolated in $\mu(c, d)$. Therefore $\mu(c, d)$ is an octagon. Let u be the vertex opposite x in this octagon, and consider the octagon $\mu(x, u)$. If r denotes the common neighbour of x and u in O, then $\mu(x, u)$ contains the path $c \sim a \sim r \sim b \sim d$. Let s be the vertex opposite a in $\mu(x, u)$, so that $s \sim d$. Then $\mu(a, s)$ is an octagon with x and u opposite, and either c or r is a centre of it in $\Gamma(a)$. If c is centre of $\mu(a, s)$, then $\mu(a, s)$ has a path $x \sim t \sim u'$ in common with O, impossible. Thus r is centre of $\mu(a, s)$, and if O_4 is the quadrangle $e \sim f \sim g \sim h \sim e$ with $e \sim x$, $f \sim r$, then $\mu(a, s)$ contains the path $x \sim e \sim f \sim g \sim u$. Now interchange the rôles of x, u and x', u'. We find for some s' that $\mu(a, s')$ contains the path $x' \sim e \sim f \sim g \sim u'$. But then $e \sim f \sim g$ is covered twice. Contradiction. $\square$

Let $O = \mu(x, y) = abcda'b'c'd'$ be an octagon. Call the edge ab of O thick if b is a centre of the octagon $\mu(a, a')$ in $\Gamma(a)$, and thin otherwise. The previous lemma implies that if b is a centre of $\mu(a, a')$, then b' also is, and a is a centre of $\mu(b, b')$, so that the concept thick is well-defined. Each octagon $\mu(x, y)$ has four thin and four thick edges (and these alternate).

Lemma 5.3 Let a, c be vertices at distance 2 on an octagon $O = \mu(x, y)$ (say, $a \sim b \sim c$ in O). Then $\mu(a, c)$ is a hexagon $xby\bar{y}\bar{b}\bar{x}$ and $\mu(\bar{x}, \bar{y})$ is an octagon $\bar{O}$. If ab is thick in O, then $\bar{c}\bar{b}$ is thick in $\bar{O}$.

Proof.   Let $\bar{x}$ and $\bar{y}$ be the centres of O adjacent to a, c in $\Gamma(x)$ and $\Gamma(y)$, respectively, and let $\bar{b}$ be the common neighbour of a and c in the octagon $\bar{O}$. Then $\mu(a, c)$ is as claimed. Let O contain the path $abcda'$ and let $\bar{O}$ contain the path $a\bar{b}cd\bar{a}'$. Then similarly $\mu(a', c)$ is the hexagon $xdy\bar{y}\bar{d}\bar{x}$. Let $O' = \mu(c, c')$. Looking at $\Gamma(c)$ we see that the common neighbour of x and y in O'_2 is adjacent to the common neighbour of $\bar{x}$ and $\bar{y}$ in O'_2. Thus, both pairs $\{b, \bar{b}\}$ and $\{d, \bar{d}\}$ have one point in O'_2 and one point in O'_4. Thus, $\bar{b}c$ is thick in $\bar{O}$ if and only if bc is thin in O. $\square$

Lemma 5.4 Let $O' = \mu(a, a')$ be an octagon with opposite vertices x, y and let v be the centre of O' in $\Gamma(a)$ nonadjacent to x, y. Let $O = \mu(x, y)$ contain the path $a \sim p \sim q$, where ap is thin and pq thick. Then $v \sim q$.

Proof.   Let $O = abcda'b'c'd'$ with ab thick, so that $p = d'$ and $q = c'$.

Following the notation of the previous lemma, we see in $\bar{O} = \mu(\bar{x}, \bar{y})$ the path $a \sim \bar{b} \sim c$ with $c\bar{b}$ thick, and vertices a', c', with c' opposite c. In O' we have thick edges $x\bar{x}$ and $y\bar{y}$, so that $v \in \bar{O}$. Now $v \sim a$, and av is a thick edge of $\bar{O}$, so $v \neq \bar{b}$ and hence $v \sim c' = q$. $\qquad \square$

Let $O = \mu(x,y) = a'b'c'd'a''b''c''d''$ be an octagon in $\Gamma(x)$. Let $O_4 = \{A, B\}$ and $O_2 = \{a, b, c, d\}$ with $A \sim a', c', a'', c''$ and $a \sim b \sim c \sim d \sim a$ and $a \sim a', a''$; $b \sim b', b''$, etc. Let moreover the edge $a'b'$ be thick in O. We can now find all polygons in $\Gamma(x)$. First of all we have the octagon O:

$$a'b'c'd'a''b''c''d''.$$

Next we have the four hexagons $\mu(x,m)$ for m at distance 2 from x in the octagons $\mu(\xi', \xi'')$, $\xi \in \{a, b, c, d\}$:

$$Aa'b'bb''a''$$
$$aa'b'Bb''a''$$
$$Ac'd'dd''c''$$
$$cc'd'Bd''c''.$$

Let for $\xi \in \{a, b, c, d\}$ the vertices ξ''' and ξ'''' be the centres of $\mu(\xi', \xi'')$ adjacent to ξ' and ξ'', respectively, but not adjacent to x, y. By our last lemma we have $a''' \sim c''$, $a'''' \sim c'$, $b''' \sim d'$, $b'''' \sim d''$, $c''' \sim a'$, $c'''' \sim a''$, $d''' \sim b''$, $d'''' \sim b'$. This forces the following paths:

$$
\begin{array}{llll}
\mu(x, a''') : & aa'Ac''b'' & \text{and} & \mu(x, d''') : & dd'Bb''c'' \\
\mu(x, a'''') : & aa''Ac'b' & \text{and} & \mu(x, d'''') : & dd''Bb'c' \\
\mu(x, b''') : & bb'Bd'a'' & \text{and} & \mu(x, c'''') : & cc''Aa''d' \\
\mu(x, b'''') : & bb''Bd'a' & \text{and} & \mu(x, c''') : & cc'Aa'd''.
\end{array}
$$

Now we have seen all 2-claws on A and B, and on the vertices of O only the 2-claws $aa'd''$, $dd''a'$; $bb'c'$, $cc'b'$; $cc''b''$, $bb''c''$; $dd'a''$, $aa''d'$ are not yet covered. Concerning each of these pairs of 2-claws, there are two possibilities: either they form a quadrangle together, when two of the above μ-graphs coincide to form an octagon, or they are covered by hexagons. In no case is a 2-claw in O_2 covered, so we also need the quadrangle

$$abcd.$$

Since the edges of O_2 must be covered exactly twice by other polygons, we have either $b''' = c''''$, $b'''' = c'''$ or $a''' = d'''$, $a'''' = d''''$. By symmetry we may assume that we are in the first case. Thus, we have octagons:

$$bb'Bd'a''Ac''c$$
$$bb''Bd'a'Ac'c.$$

We have now seen that $\Gamma(x)$ contains 3 octagons, namely $\mu(x, y)$, $\mu(x, b''')$ and $\mu(x, b'''')$. Moreover, $\mu(b''', b'''')$ is again an octagon (since b''' and b'''' are centres of $\mu(b', b'')$ adjacent to the same vertices of that octagon). Since y was arbitrary, any pair in $\{x, y, b''', b''''\}$ determines an octagon. In this way we find a partition of Γ into groups of size 4, let us call them *tetrads*, such that any two vertices from the same tetrad determine an octagon (and all octagons occur in this way).

Next, consider the tetrads $\{x, y, b''', b''''\}$ and $\{\cdot, A, b'', b'\}$ or $\{\cdot, b, a', a''\}$. We see that if $\{p_1, p_2, p_3, p_4\}$ and $\{q_1, q_2, q_3, q_4\}$ are tetrads, with q_3, q_4 opposite in $\mu(p_1, p_2)$, then (for some appropriate ordering of $\{q_1, q_2, q_3, q_4\}$) q_i, q_j are opposite in $\mu(p_k, p_l)$ for any choice of distinct i, j, k, l. Let us call two such tetrads *adjacent* when this situation occurs, i.e., when each point of one is adjacent to three points of the other.

The graph T on the tetrads is regular of valency 4, and is locally a 3-path, and hence is a quotient of the graph on $\mathbb{Z}$ defined by $m \sim m \pm 1, m \pm 2$. Call two tetrads *close neighbours* when they have two common neighbours in T (i.e., are represented by m and $m \pm 1$).

Consider the tetrads $\{x, y, b''', b''''\}$ and $\{\cdot, A, b'', b'\}$ and $\{\cdot, B, c', c''\}$. We see that if $P = \{p_1, p_2, p_3, p_4\}$ and $Q = \{q_1, q_2, q_3, q_4\}$ and $R = \{r_1, r_2, r_3, r_4\}$ are tetrads where Q and R are close neighbours of P, and $p_1 \not\sim q_1, r_1$, then $q_1 \not\sim r_1$. Thus, the graph is a direct product $T \times I_4$ (possibly with a $Sym(4)$-twist in case $|T| < \infty$), and $(t, i) \sim (t \pm 1, j), (t \pm 2, j)$ for $i \neq j$. This accounts for 12 neighbours of every point. For x the remaining two neighbours are a and d, so that $(t, i) \sim (t \pm 4, i)$. Finally, $a \sim d$, so that $|T| = 12$ and no twist occurs.

This completes the determination of Γ.

References

[1] A. E. Brouwer, A. M. Cohen, and A. Neumaier. *Distance-regular Graphs*. Springer Verlag, Berlin, 1989.

[2] A. Neumaier. Rectagraphs, diagrams and Suzuki's sporadic simple group. In E. Mendelsohn, editor, *Algebraic and Geometric Combinatorics*, volume 15 of *Ann. Discrete Math.*, pages 305–318, 1982.

[3] L. Soicher. On simplicial complexes related to Suzuki sequence graphs. In M. Liebeck and J. Saxl, editors, *Groups, Combinatorics and Geometry, Proceedings of the L.M.S. Durham Symposium on Groups and Combinatorics, 1990*, volume 165 of *London Math. Soc. Lecture Note Series*, pages 240–248. Cambridge University Press, 1992.

A. E. Brouwer, Dept. of Math., Eindhoven Univ. of Technology, P. O. Box 513, 5600 MB Eindhoven, the Netherlands. e-mail: aeb@win.tue.nl

D. G. Fon-der-Flaass, Institute of Math., 630090 Novosibirsk-90, Russia. e-mail: flaass@math.nsk.su

S. V. Shpectorov, Institute for System Analysis, 9 Pr. 60 Let Oktyabrya, 117312 Moscow, Russia. e-mail: ssh@cs.vniisi.msk.su

A theorem of Parmentier characterizing projective spaces by polarities

F. Buekenhout

Abstract

We present a remarkable result obtained in 1974 by Anne Parmentier
with some help of the author. This result has remained unpublished
and unknown since then. The idea is to axiomatize and characterize
the structure consisting of a pair (P, π) where P is a projective space
and π is a polarity of P in such a way that the system of axioms
weakens the usual requirements on P.

1. Introduction

We start with a *linear space* $P = (\mathcal{P}, \mathcal{L})$ namely a set of *points* $\mathcal{P}$, equipped
with a family $\mathcal{L}$ of subsets of $\mathcal{P}$, called lines, such that every pair of points is
contained in a unique line and every line has at least two points. A *subspace*
of P is a set of points X such that for any distinct points p, q in X, the line
pq containing p and q is contained in X.

We now consider a symmetric relation π on the set $\mathcal{P}$. If $p \in \mathcal{P}$, p^π denotes
the set of all $x \in \mathcal{P}$ such that $p\pi x$. We call p^π, the "polar hyperplane" of p.
If $X \subseteq \mathcal{P}$, X^π denotes the set $\bigcap_{p \in X} p^\pi$.

We shall say that the pair (P, π) is a *linear space* with *polarity* if the following
hold:

(1) for every line ℓ of P and point p of P, either $\ell \subseteq p^\pi$ or $\ell \cap p^\pi$ is a point.
   With other words, each p^π is a "projective hyperplane" or "geometric
   hyperplane" of P.

(2) for each line ℓ, $\ell = (\ell^\pi)^\pi$

(3) for each point p, $p^\pi \neq \mathcal{P}$.

Theorem. *([3]) If (P, π) is a linear space with polarity, then P satisfies the
Pasch-Veblen axiom i.e. P is a projective space (with possibly lines of two
points). Consequently, if all lines of P have at least three points, P is a
projective space in the classical sense. If moreover, P is of finite dimension,*

69

then π is a polarity in the classical sense.

The proof given by Parmentier was rather long. In [1], there is a shorter proof. It will be given here. Recently ([2]), the result appears as a consequence of a more general theory.

2. Proof of the theorem

1) For each point p, p^π is a maximal proper subspace of the linear space P, in view of (3) and (1).
2) We shall construct a dual space P^*.
3)

Lemma. *If a, b are distinct points and if $x \in P \backslash (a^\pi \cap b^\pi)$ then there is a unique point $y \in P$ such that y^π contains x and $a^\pi \cap b^\pi$. Moreover y is on the line ab.*

Proof. The hyperplane x^π intersects the line ab in a unique point y because $x \notin a^\pi \cap b^\pi$. Then $(a^\pi \cap b^\pi)^\pi$ contains a, b hence y and so y^π contains $a^\pi \cap b^\pi$ as well as x. Now, assume that z were a point other than y such that z^π contains $a^\pi \cap b^\pi$ and x.

Put $\ell = ab$. Then $a^\pi \cap b^\pi = \ell^\pi$ and by (2), $(\ell^\pi)^\pi = \ell$, hence $z \in \ell$ and moreover $z \in x^\pi$, thus $z = y$.

4) If a, b are distinct points, then $a^\pi \neq b^\pi$. Otherwise, let $c \in P \backslash a^\pi$. Then c^π intersects the line ab in some point p and p^π contains a^π and c, hence by 1), $p^\pi = P$, contradicting (3).

5) Let the *dual* P^* be defined as a pair $(P^*, \mathcal{L}^*)$ where P^* is the set of all p^π, $p \in P$ and a member of $\mathcal{L}^*$ is any set $\ell^* = \{p^\pi | p \in \ell\}$ where $\ell \in \mathcal{L}$.

6) Now P^* is a linear space.
7)

Lemma. *P^* is a projective space.*

Proof. It suffices to check the Pasch-Veblen axiom.

Let a, b, c, d, e be distinct points of P such that a^π, b^π, c^π and a^π, d^π, e^π are collinear in P^* on distinct lines. By 4), $a^\pi, b^\pi, c^\pi, d^\pi, e^\pi$ are distinct as well. By the definition of $\mathcal{L}^*$, a, b, c and a, d, e are collinear in P.

Here $b^\pi \cap d^\pi$ is not contained in $c^\pi \cap e^\pi$ for otherwise $(bd)^\pi \subseteq (ce)^\pi$, hence $(bd)^{\pi\pi} \supseteq (ce)^{\pi\pi}$ which means $bd = ce$ by (2), but then $a^\pi, b^\pi, \ldots, e^\pi$ would be collinear.

Then, let x be a point of $b^\pi \cap d^\pi$, not in $c^\pi \cap e^\pi$. Lemma 3 provides a unique point y in P, such that y^π contains x and $c^\pi \cap e^\pi$. Moreover $y \in ce$, hence y^π is collinear with c^π and e^π.

We claim that y^π is also collinear with b^π and d^π. Let $z \in b^\pi \cap d^\pi$, $z \neq x$. The line xz is in $b^\pi \cap d^\pi$ and it intersects a^π in some point u. Therefore $u \in a^\pi \cap b^\pi$

and to $a^\pi \cap d^\pi$ and so $u \in c^\pi \cap e^\pi$. Thus $x \neq u$. Consequently $xz = xu$ and since y^π contains x and $c^\pi \cap e^\pi$ we see that y^π contains z, hence it contains $b^\pi \cap d^\pi$.

8) P is a projective space. Indeed, if a, b, c, d, e are distinct points, such that a, b, c and a, d, e are collinear on distinct lines, then Lemma 7) provides some y^π collinear with b^π and d^π and with c^π and e^π in P^*, hence y is collinear with b and with c and e, in P. □

References

[1] **F. Buekenhout**. *Géométrie Projective*. Université Libre de Bruxelles, Bruxelles, 1974. 177 pages.

[2] **F. Buekenhout and A. M. Cohen**. Diagram geometry. Book manuscript.

[3] **A. Parmentier**. Caractérisations des polarités dans les espaces projectifs et linéaires. Master's thesis, Université Libre de Bruxelles, 1974.

F. Buekenhout, Université Libre de Bruxelles, Campus Plaine C.P.216, Bd du Triomphe, B-1050 Bruxelles. e-mail: fbueken@ulb.ac.be

Geometries with diagram $\underset{}{\circ}\overset{L}{\rule{1.2em}{0.4pt}}\underset{}{\circ}\overset{P^*}{\rule{1.2em}{0.4pt}}\underset{}{\circ}$

F. Buekenhout   O.H. King

Abstract

We study flag-transitive geometries with diagram

$$\underset{0}{\circ}\overset{L}{\rule{2em}{0.4pt}}\underset{1}{\circ}\overset{P^*}{\rule{2em}{0.4pt}}\underset{2}{\circ}$$

The residues of 0-, 1- and 2- elements are respectively dual Petersen graphs, generalised digons and finite linear spaces. The finite linear space must be either a projective plane, in which case there are known to be exactly two geometries, or a complete graph on 4 vertices. We show that in the latter case there are no geometries with a flag-transitive automorphism group acting primitively on the 2-elements. We give examples where an automorphism group acts imprimitively.

1.   Introduction

Let Γ be a flag-transitive, residually connected geometry with diagram

$$\underset{0}{\circ}\overset{L}{\rule{2em}{0.4pt}}\underset{1}{\circ}\overset{P^*}{\rule{2em}{0.4pt}}\underset{2}{\circ}$$

and let G be a flag-transitive subgroup of $Aut(\Gamma)$. We call the 0-, 1- and 2- elements respectively points, lines and planes (or circles). The diagram indicates that the residue of a plane is a finite linear space L, the residue of a point is a Petersen graph with planes as vertices and lines as edges, and the residue of a line is a generalised digon. We assume that Γ satisfies a natural condition: that a pair of points is incident with at most one line.

Given any point x, each line in $Res(x)$ is incident with 2 planes and each plane with 3 lines; and given any plane π, each point in $Res(\pi)$ is incident with 3 lines and each line is incident with $n+1$ points for some number n. It follows that each line is incident with two points and 2 planes. The diagram may be written

$$\underset{n}{\circ}\overset{L}{\rule{2em}{0.4pt}}\underset{2}{\circ}\overset{P^*}{\rule{2em}{0.4pt}}\underset{1}{\circ}$$

with the types 0,1,2 implicit from left to right. The finite linear space

$$\underset{n}{\circ}\overset{L}{\rule{2em}{0.4pt}}\underset{2}{\circ}$$

is a design with parameters $2 - (v, n+1, 1)$ for some v, and the design parameter r is given as 3. Fisher's Inequality applies to yield $r \geq n + 1$, i.e. $n = 1$ or 2.

If $n = 2$ then L is the projective plane of order 2 and our diagram is equivalent to o—P—o——o (i.e. with the types written in the order 2,1,0 from left to right). This is the diagram of a P-geometry. S. Shpectorov ([4]) has shown that there are exactly two such geometries, and they are associated with the groups M_{22} and its triple cover $\hat{M}_{22}$. In the first geometry the automorphism group M_{22} acts primitively on 2-elements, but in the second case the group $\hat{M}_{22}$ acts imprimitively. We should mention that S. Shpectorov and A.A. Ivanov have completely determined P-geometries of arbitrary rank, although not all their work is yet published (see for example [2] , [3]).

If $n = 1$ then L is the complete graph on 4 points. We write L as c in this case and refer to the planes of Γ as circles. Henceforth we assume that $n = 1$ and Γ has diagram o—c—o—P^*—o . We prove that there are no geometries for which a flag-transitive automorphism group G acts primitively on circles. Our approach uses results of Wong ([5]) and Gorenstein and Gilman ([1]) to reduce to the case where $G \cong PSL(2, q)$. We then show that $PSL(2, q)$ admits no geometries with this diagram.

Now assume that G is a flag-transitive subgroup of $Aut(\Gamma)$. For the present we do not assume that G acts primitively on circles.

Given Γ, the point-line graph $\bar{\Gamma}$ is a graph whose vertices are the points of Γ with distinct points x, y adjacent in $\bar{\Gamma}$ if they are incident to a line of Γ. The action of G on $\bar{\Gamma}$ is (vertex) transitive and since a point is incident with 15 lines and a line with 2 points, $\bar{\Gamma}$ is regular of valency 15. The residual connectedness of Γ ensures that $\bar{\Gamma}$ is a connected graph. The following is the Petersen graph with its edges labelled $y_1, ..., y_{15}$ and with alternative labels y, b, c, d, e for $y_1, ..., y_5$. We shall use this graph and refer to the labels in making observations on the line graph of the Petersen graph.

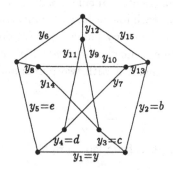

Diagram 1

2.   Preliminary Results

Lemma 2.1 *Let $\bar{\Gamma}_0$ be the neighbourhood graph of a vertex x of $\bar{\Gamma}$, ie. $\bar{\Gamma}_0$ is the full subgraph of $\bar{\Gamma}$ on the fifteen neighbours of x, and let G_0 be the stabilizer in G of x. Then $\bar{\Gamma}_0$ is regular and contains a subgraph isomorphic to $\mathcal{L}$ (the line graph of the Petersen graph) and preserved by G_0.*

Proof.   For any distinct vertices y_i, y_j of $\bar{\Gamma}_0$, the flag-transitivity of G implies the existence of $g \in G$ such that $g(x, xy_i) = (x, xy_j)$ (where xy_i, xy_j are the lines of Γ incident with x, y_i and x, y_j respectively), ie. $g \in G_0$ and $g(y_i) = y_j$. Thus G_0 is transitive on the vertices of $\bar{\Gamma}_0$ and so $\bar{\Gamma}_0$ is regular.

In the residue of x in Γ, if two lines, ℓ_i and ℓ_j, are incident with a common circle C then the second points, y_i and y_j say, of ℓ_i and ℓ_j are incident with C and y_i, y_j are incident with a line ℓ. Now let $\ell_1, ..., \ell_{15}$ be the lines of Γ incident with x and let $y_1, ..., y_{15}$ be the corresponding second points. Then y_i, y_j are adjacent in $\bar{\Gamma}_0$ whenever ℓ_i, ℓ_j are incident with a common circle in the residue of x. But $Res(x)$ is a Petersen graph P with circles as vertices and lines as edges, so the vertices of $\bar{\Gamma}_0$ correspond to the edges of P and the adjacency just described corresponds to adjacency of edges in P. Thus we have a copy of $\mathcal{L}$ in $\bar{\Gamma}_0$ preserved by G_0. Of course $\bar{\Gamma}_0$ may contain further edges, but these do not concern us here. $\square$

Lemma 2.2 *If G_0 represents the stabilizer in G of a point x then $G_0 \cong A_5$ or S_5.*

Proof.   Given x, its stabilizer G_0 acts flag-transitively on the $Res(x)$. The lines and circles of Γ incident with x form a Petersen graph with the lines as edges and the circles as vertices. The only groups acting faithfully and flag-transitively on the Petersen graph are A_5 and S_5, so it remains to show that G_0 acts faithfully.

Let $\mathcal{L}$ be the copy of the line graph of the Petersen graph described in Lemma 2.1, lying inside $\bar{\Gamma}_0$. The vertices of $\mathcal{L}$ are points $y_1, ... y_{15}$ of Γ but are also edges of a Petersen graph, hence the labelling in Diagram 1. It is clear that any automorphism of $\mathcal{L}$ that fixes a vertex and its four neighbours fixes every vertex of $\mathcal{L}$. Let $g \in G_0$ fix every vertex of $\bar{\Gamma}_0$ and let y be any vertex of $\bar{\Gamma}_0$. Then the neighbourhood graph $\bar{\Gamma}_y$ of y in $\bar{\Gamma}$ contains a subgraph $\mathcal{L}_1$ isomorphic to $\mathcal{L}$, and $g \in G_y$. If y corresponds to the edge y_1 in Diagram 1, then $\mathcal{L}_1$ contains x, and the four neighbours of x in $\mathcal{L}_1$ are b, c, d and e. Since all five vertices are fixed by g, all vertices of $\bar{\Gamma}_y$ are fixed by g. Thus g fixes all vertices of $\bar{\Gamma}_y$ for all y at distance 1 from x in $\bar{\Gamma}$. Given that $\bar{\Gamma}$ is connected, an inductive argument shows that g fixes every vertex of $\bar{\Gamma}$. Hence $g = 1$ and

G_0 is faithful on $\bar{\Gamma}_0$. □

Lemma 2.3 *Let (x, ℓ, C) be a flag in Γ with G_0, G_1, G_2 the stabilizers in G of x, ℓ and C respectively.*

(a) *If $G_0 \cong S_5$ then*
  (i) $G_1 \cong 2 \times D_8$
  (ii) $G_2 \cong 2.S_4$
  (iii) $G_0 \cap G_1 \cong D_8$
  (iv) $G_0 \cap G_2 \cong 2 \times S_3$
  (v) $G_1 \cap G_2 \cong 2^3$
  (vi) $G_0 \cap G_1 \cap G_2 \cong 2^2$.

(b) *If $G_0 \cong A_5$ then*
  (i) $G_1 \cong D_8$ or 2^3
  (ii) $G_2 \cong S_4$
  (iii) $G_0 \cap G_1 \cong 2^2$
  (iv) $G_0 \cap G_2 \cong S_3$
  (v) $G_1 \cap G_2 \cong 2^2$
  (v) $G_0 \cap G_1 \cap G_2 \cong 2$.

Proof.

(a) Let y be the second point incident with ℓ (x being the first), then for any $g_1, g_2 \in G_1 \setminus G_0$, $g_1 g_2 \in G_0 \cap G_1$.

  If y is as represented in Diagram 1 then $G_0 \cap G_1$ permutes b, c, d, e, and in view of the discussion in the proof of Lemma 2.2, $G_0 \cap G_1$ acts faithfully on $\{b, c, d, e\}$. One observes that the possible permutations in Aut $\mathcal{L}$ are $(bc), (de), (bd)(ce)$ and products, and that all occur. Thus $G_0 \cap G_1$ has order 8, and the only possibility in S_5 is $G_0 \cap G_1 \cong D_8$. This proves (iii).

  Since G is flag transitive it contains elements mapping (x, ℓ, C) to (y, ℓ, C), ie. $G_1 \not\leq G_0$. Hence $G_0 \cap G_1$ is of index 2 in G_1, and $G_1 \cong D_8.2$. We may take C to be incident to b, c and let C_1 be the circle incident to ℓ, x, d and e. The following diagram illustrates the situation:

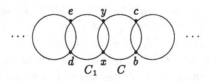

Diagram 2

We know that $\ell = xy$ is incident with just two circles, C and C_1, so any element of G_1 either fixes C and C_1 or interchanges them. In any event, the vertices x, y, b, c, d, e are permuted. Let $g \in G_1$ switch x, y, then we may multiply by (some of) $(bd)(ce), (bc), (de)$ to get an element of G_1 switching x and y but fixing b, c, d and e, so G_1 contains the permutation (xy). This permutation commutes with $G_0 \cap G_1$ so $G_1 \cong 2 \times D_8$, proving (i). In $G_1 \cap G_2$ we have commuting permutations $(xy), (bc)$ and (de), and we observe that $(bd)(ce) \notin G_2$, so $G_1 \cap G_2 \cong 2^3$, proving (v).

Now let $h \in G_2$ fix x, y, b and c then either h fixes d and e or it switches them. If h fixes d and e then $h = 1$; if $h(d) = e$ then $h^2 = 1$. Therefore the kernel of the action of G_2 on $\{x, y, b, c\}$ has order 2. Acting as a subgroup of $Sym\{x, y, b, c\}$ we know that $G_0 \cap G_2$ contains the permutation (bc). If we consider the flag (x, xb, C) then $G_0 \cap G_2$ also contains (yc). Since $G_1 \cap G_2$ contains (xy) we conclude that G_2 acts as S_4 on $\{x, y, b, c\}$ ie. $G_2 \cong 2.S_4$. This proves (ii).

To get the structure of $G_0 \cap G_2$ we need to observe that G_0 acts imprimitively on the vertices of $\mathcal{L}$. In fact there is only one imprimitive action and the blocks of imprimitivity are $\{y_1, y_{10}, y_{12}\}$, $\{y_2, y_8, y_{11}\}$, $\{y_3, y_6, y_7\}$, $(y_4, y_{14}, y_{15}\}$, $\{y_5, y_9, y_{13}\}$ where $y_1, ..., y_{15}$ are the edges in the Diagram 1. The action of G_0 on the blocks is transitive and faithful so that S_5 is realised as the full symmetric group on these five blocks. If the blocks are labelled 1,2,3,4,5 respectively then the permutations $(de), (bc)$ and (cy) correspond to (45), (23) and (13). Thus $G_0 \cap G_2$ contains a subgroup isomorphic to $2 \times S_3$. On the other hand $G_0 \cap G_2$ contains the kernel of G_2 on $\{x, y, b, c\}$ and acts as S_3 (at most) on $\{y, b, c\}$. Hence $G_0 \cap G_2$ has order 12 and $G_0 \cap G_2 \cong 2 \times S_3$, proving (iv).

Finally $G_0 \cap G_1 \cap G_2$ is a proper subgroup of $G_0 \cap G_1$ but contains (bc) and (de), so $G_0 \cap G_1 \cap G_2 \cong 2^2$.

(b) Let us start this time with G_2 and consider the kernel of G_2 on $\{x, y, b, c\}$. An element of this kernel acts on $\{x, y, b, c, d, e\}$ as the identity or (de), but as noted above, (de) corresponds to the permutation (45) on the blocks of imprimitivity for G_0 on $\mathcal{L}$. Since $(45) \notin A_5$ we conclude that the kernel of G_2 on $\{x, y, b, c\}$ fixes d and e as well, so is reduced to the identity in G. We find that G_0 contains all permutations of $\{y, b, c\}$ and that G_y contains all permutations of $\{x, b, c\}$. Thus $G_2 \cong S_4$ and $G_0 \cap G_2 \cong S_3$, proving (ii) and (iv).

If we look at $G_0 \cap G_1$ we find that the only permutations of $\{b, c, d, e\}$ lying in G_0 are now the identity, $(bc)(de), (be)(cd)$ and $(bd)(ce)$, so $G_0 \cap G_1 \cong 2^2$, proving (iii).

Now G_1 again contains $G_0 \cap G_1$ as a subgroup of index 2 so that G_1 has order 8. If $g \in G_1$ switches x and y then, treating elements of G_1 as permutations of $\{x, y, b, c, d, e\}$, we can multiply g by an element of $G_0 \cap G_1$ to get either (xy) or $(xy)(bc)$. If we get (xy) then G_1 is generated by (xy), $(bc)(de)$ and $(bd)(ce)$, so $G_1 \cong 2^3$. If we get $(xy)(bc)$ instead G_1 has five involutions and two elements of order 4 (of which one is $(xy)(becd)$) so that $G_1 \cong D_8$. In either case $G_1 \cap G_2 \cong 2^2$ and $G_0 \cap G_1 \cap G_2 \cong 2$. $\qquad\qquad\qquad\qquad\qquad\qquad\qquad\qquad\qquad\qquad\qquad\qquad$ □

Lemma 2.4 *Let (x, ℓ, C) be a flag of Γ with G_0 and G_2 being the stabilizers in G of x and C. Then G is generated by G_0 and G_2.*

Proof. Let $F = <G_0, G_2>$. We show that F is transitive on the points of Γ, from which it follows that $F = G$.

We consider the action of F on $\bar{\Gamma}$ and recall that $\bar{\Gamma}$ is connected. Let Δ be the (vertex) orbit of F containing x. We suppose that there is a point not in Δ and arrive at a contradiction: we choose a point v not in Δ whose distance from x is minimal amongst the points not in Δ. First observe that G_2 acts transitively on the four points: x, y, b and c incident to C. Thus Δ contains a point of $\bar{\Gamma}_0$, but G_0 acts transitively on the points of $\bar{\Gamma}_0$ so all the points of $\bar{\Gamma}_0$ lie in Δ. Therefore the distance from x to v is at least 2.

Let u be a point adjacent to v such that $dist(u, x) = dist(v, x) - 1$, and let w be a point adjacent to u such that $dist(w, x) = dist(u, x) - 1$. Then $u, w \in \Delta$, so $G_u \leq F$ (where G_u is the stabilizer of u in G). However G_u is transitive on the points adjacent to u, so v and w lie in the same orbit of F, a contradiction to $v \notin \Delta$.

Hence F is transitive on the points of Γ as claimed, and $F = G$. $\qquad$ □

3.   Reduction to PSL$(2, q)$

Result 1. (W.J. Wong [5]) *Let G be a primitive permutation group on a finite set Ω, such that the stabilizer G_α in G of an element α of Ω has an orbit of length 3. Then G is isomorphic to one of the following groups with G_α as shown:*

G	G_α
$G(3,p)$ (prime $p \neq 3$)	3
$G(6,p)$ (prime $p \neq 3$)	S_3
A_5	S_3
S_5	D_{12}
$PGL(2,7)$	D_{12}
$PSL(2,11)$	D_{12}
$PSL(2,13)$	D_{12}
$PSL(2,q)$ (prime $q \equiv \pm 1 \, mod \, 16$)	S_4
$SL(3,3)$	S_4
$AutSL(3,3)$	$S_4 \times 2$

Theorem 3.1 *If G acts primitively on the circles in Γ and if Γ has an even number of circles then $G \cong PSL(2,q)$ for some prime $q \equiv \pm 1 \mod 16$, and $G_0 \cong A_5$.*

Proof. Let Ω be the set of circles and let $C \in \Omega$. Then by Lemma 2.3, $G_C \cong 2.S_4$ or S_4. Since G is primitive on Ω, $\{C\}$ is the only trivial orbit of G_C, and since $|\Omega|$ is even there must be another orbit of odd length. But whether G_C be $2.S_4$ or S_4 the only possibility for an odd orbit length is 3. Thus G_C has an orbit of length 3 and Wong's Theorem (Result 1) applies. Note that $SL(3,3)$ and $AutSL(3,3)$ have orders 5616 and 11232 respectively, so neither contain A_5 and neither are candidates for G. The only possibility that remains is $PSL(2,q)$ with $G_C \cong S_4$, ie. $G_0 \cong A_5$.   □

Result 2. (Gorenstein and Gilman [1]) *If G is a finite simple group and if a Sylow 2-subgroup of G has nilpotency class 2 then G is isomorphic to one of: $L_2(q)$, $q \equiv \pm 7 \pmod{16}$; $U_3(2^n)$, $n \geq 2$; $Sz(2^n)$, n odd, $n \geq 2$; $L_3(2^n)$, $n \geq 2$; $PSp_4(2^n)$, $n \geq 2$; A_7.*

The following table gives the orders of the groups listed above together with the order of a Sylow 2-subgroup.

G	Order of G	Order of $S \in S_2(G)$
$L_2(q)$, $q \equiv \pm 7 \pmod{16}$	$\frac{q(q-1)(q+1)}{2}$	8
$U_3(2^n)$, $(n \geq 2)$	$\frac{2^{3n}(2^{2n}-1)(2^{3n}+1)}{(3,2^n+1)}$	2^{3n}
$Sz(2^{2m+1})$, $(m \geq 1)$,	$2^{4m+2}(2^{4m+2}+1)(2^{2m+1}-1)$	2^{4m+2}
$L_3(2^n)$, $(n \geq 2)$	$\frac{2^{3n}(2^{2n}-1)(2^{3n}-1)}{(3,2^n-1)}$	2^{3n}
$PSp_4(2^n)$, $(n \geq 2)$	$2^{4n}(2^{2n}-1)(2^{4n}-1)$	2^{4n}
A_7	2520	8

Lemma 3.2 *If G acts primitively on the circles in Γ and if Γ has an odd number of circles then either G is simple or $G_0 \cong S_5$ and G has a simple subgroup of index 2.*

Proof.   Suppose that G has a proper, non-trivial normal subgroup H. If $H \leq G_2$ then since G acts transitively on circles H fixes each circle. For each point x, we can find two circles in $Res(x)$ that have no other point in common. Clearly H then fixes x. Thus H fixes all points so $H = 1$, contradiction.

Since G acts primitively on circles, G_2 is a maximal subgroup of G. Hence $HG_2 = G$ and $|G/H| \leq 48$. But now consider G_0 : since $G_0 \cap H$ is normal in G_0, $|G_0/G_0 \cap H| = 1, 2$ or ≥ 60. If $|G_0/G_0 \cap H| \geq 60$ then $|G_0H| = |G_0/G_0 \cap H| \cdot |H| \geq 60|H|$ which is impossible. Therefore either $G_0 \cong A_5$ and $G_0 \leq H$ or $G_0 \cong S_5$ and $G_0 \cap H \cong S_5$ or A_5. If $G_0 \leq H$ then H contains every point stabilizer (G being transitive on points), and in particular (with reference to Lemma 2.3) H contains G_0 and G_y; clearly $G_0 \cap G_2$ and $G_y \cap G_2$ generate G_2 so $G_2 \leq H$, contradiction. If $G_0 \cong S_5$ and $G_0 \cap H \cong A_5$ then also $G_y \cap H \cong A_5$ and $G_0 \cap H \cap G_2$ and $G_y \cap H \cap G_2$ generate a subgroup of G_2 which acts as S_4 on $\{x, y, b, c\}$ (reference to Lemma 2.3); either we have the whole of G_2, in which case $G_2 \leq H$ and again we have a contradiction or $|G_2 \cap H| = 24$, ie. H has a subgroup isomorphic to S_4, and $|G/H| = |G_2/G_2 \cap H| = 2$.

Thus if G is non-simple then it has a normal subgroup H of index 2. Moreover $G_0 \cap H$ is transitive on the neighbours of x, so the group generated by $G_0 \cap H$ and its conjugates in G is transitive on the points of Γ; this latter group must be the whole of H. In other words H is generated by the unique subgroups of index 2 of point stabilizers in G. It follows that H is the only proper non-trivial normal subgroup of G. Let Q be a minimal normal subgroup of H then the G-conjugates of Q generate a normal subgroup of G, ie. generate H. Thus H is a direct product of isomorphic minimal normal subgroups. However Γ has an odd number of circles so $|G|/|G_2|$ is odd, ie. a Sylow 2-subgroup of G has order 16 and a Sylow 2-subgroup of H has order 8; moreover H has a subgroup isomorphic to S_4 so a Sylow 2-subgroup of H is isomorphic to D_8. This situation cannot occur if H is a direct product of two or more isomorphic subgroups, so $H = Q$ and H is simple.   □

Theorem 3.3 *If G acts primitively on the circles in Γ and if Γ has an odd number of circles then either $G \cong PSL(2, q)$ or G has a subgroup H of index 2 isomorphic to $PSL(2, q)$ acting flag-transitively on Γ. In each case q is a prime power $\equiv \pm 7 \bmod 16$.*

Proof. $|G|/|G_2|$ is odd so a Sylow 2-subgroup of G has order 16 (resp. 8) if $G_0 \cong S_5$ (resp. $G_0 \cong A_5$). Therefore G_1 is a Sylow 2-subgroup of G. We know that $G_1 \cong 2 \times D_8$ when $G_0 \cong S_5$ and $G_1 \cong D_8$ or 2^3 when $G_0 \cong A_5$, but in the latter situation $G_2 \cong S_4$ and so G has a subgroup isomorphic to D_8. Hence $G_1 \cong 2 \times D_8$ or D_8, ie. a Sylow 2-subgroup of G has nilpotency class 2.

By Lemma 3.2, either G is simple or $G_0 \cong S_5$ and G has a simple subgroup of index 2.

Suppose that G is simple. From the list following Result 2 we observe that the only possibilities for G are $PSL(2,q)$ or A_7 with G_1 being D_8 in each case. In the case of A_7, however, S_4 is not a maximal subgroup (being contained in a subgroup isomorphic to A_6). There remains only $PSL(2,q)$ with $q \equiv \pm 7 \bmod 16$.

Now suppose that $G_0 \cong S_5$ and that G has a simple subgroup H of index 2. We saw in the proof of Lemma 3.2 that H must be transitive on the points of Γ. We know that $H \cap G_0 \cong A_5$ and that this group is flag-transitive on $Res(x)$, so H is flag-transitive on Γ. Now by the first part of this Theorem, if H is a simple flag-transitive group on Γ then $H \cong PSL(2,q)$ with $q \equiv \pm 7 \bmod 16$. $\square$

4. $PSL(2,q)$

Theorem 4.1 *There are no geometries with diagram* $\circ \!\!-\!\!\overset{c}{-}\!\!-\!\! \circ \!\!-\!\!\overset{P^*}{-}\!\!-\!\! \circ$ *on which $PSL(2,q)$ acts flag-transitively.*

Proof. We suppose that there does exist such a geometry Γ and reach a contradiction. The Sylow 2-subgroups of $PSL(2,q)$ are dihedral so $G_1 \cong D_8$, $G_0 \cong A_5$ and $G_2 \cong S_4$ (alternatively, $PSL(2,q)$ doesn't have subgroups isomorphic to S_5 or $2.S_4$). Moreover, for $PSL(2,q)$ to contain a subgroup isomorphic to S_4 we require $q \equiv \pm 1 \bmod 8$.

The groups G_0, G_1 and G_2 satisfy the conditions (i)-(vi) set out in Lemma 2.3(b). If $q \equiv -1 \bmod 8$ then $q^2 \equiv 1 \bmod 8$ and G_0, G_1 and G_2 may be considered as subgroups of $PSL(2,q^2)$. As our objective henceforth is to show that the conditions of Lemma 2.3(b) cannot be satisfied, we may assume that $q \equiv 1 \bmod 8$.

Recall from the proof of Lemma 2.3 that G_1 may be represented as a permutation group on the letters $\{x,y,b,c,d,e\}$. Let $\alpha = (xy)(becd)$ and $\beta = (xy)(bc)$; then G_1 has a presentation $< \alpha, \beta : \alpha^4 = \beta^2 = 1, \beta\alpha = \alpha^3\beta >$ and $G_1 \cap G_0$ and $G_1 \cap G_2$ are given by $\{1, \alpha\beta, \alpha^2, \alpha^3\beta\}$ and $\{1, \beta, \alpha^2, \alpha^2\beta\}$ respectively, with $G_0 \cap G_1 \cap G_2 = \{1,\alpha^2\}$ being the centre of G_1.

We set up the elements of G as 2×2 matrices in the following way: each element of G may be represented by an element of $SL(2,q)$ acting naturally on a 2-dimensional vector space V, and a basis for V may be chosen with respect to which

$$\alpha^2 = \begin{bmatrix} i & 0 \\ 0 & -i \end{bmatrix},$$

where $i \in GF(q)$ is chosen such that $i^2 = -1$.

Now G_1 is a subgroup of $C_G(\alpha^2)$, the elements of which are represented by

$$\begin{bmatrix} \lambda & 0 \\ 0 & \lambda^{-1} \end{bmatrix} \text{ and } \begin{bmatrix} 0 & \lambda \\ -\lambda^{-1} & 0 \end{bmatrix}$$

where λ ranges over the non-zero values of $GF(q)$. Clearly $C_G(\alpha^2)$ is dihedral of order $q - 1$, with two elements of order 4, namely

$$\begin{bmatrix} \tau & 0 \\ 0 & \tau^{-1} \end{bmatrix} \text{ and } \begin{bmatrix} \tau^{-1} & 0 \\ 0 & \tau \end{bmatrix}$$

where $\tau = \rho^{(q-1)/8}$ for some primitive element ρ of $GF(q)$. We may choose ρ so that

$$\alpha = \begin{bmatrix} \tau & 0 \\ 0 & \tau^{-1} \end{bmatrix}$$

and i such that $\tau^2 = i$. At present

$$\beta = \begin{bmatrix} 0 & \lambda \\ -\lambda^{-1} & 0 \end{bmatrix}$$

for some $\lambda \in GF(q)\backslash\{0\}$, but we may refine the choice of basis for V such that

$$\beta = \begin{bmatrix} 0 & 1 \\ -1 & 0 \end{bmatrix}$$

and this may be done without changing the representation of α (the refinement may be simply a scaling of the second basis vector).

We now turn to G_2 which acts as the full symmetric group on $\{x, y, b, c\}$ and $\beta = (xy)(bc)$ is in the normal subgroup of G_2 of order 4. We can begin to realize the elements of G_2 as 2×2 matrices by noting that $C_{G_2}(\beta)$ is isomorphic to D_8 and that $C_G(\beta)$ consists of the matrices

$$\begin{bmatrix} \lambda & \mu \\ -\mu & \lambda \end{bmatrix} \text{ and } \begin{bmatrix} \theta & \varphi \\ \varphi & -\theta \end{bmatrix}$$

where $\lambda, \mu, \theta, \varphi \in GF(q)$ satisfy $\lambda^2 + \mu^2 = 1 = -\theta^2 - \varphi^2$. Of course $C_G(\beta)$ is dihedral of order $q - 1$ and has two elements of order 4, namely

$$\begin{bmatrix} j & j \\ -j & j \end{bmatrix} \text{ and } \begin{bmatrix} j & -j \\ j & j \end{bmatrix}$$

82

where $j \in GF(q)$ satisfies $j^2 = 1/2$ (such j always exists for $q \equiv 1 \bmod 8$).
Let

$$\gamma = \begin{bmatrix} j & j \\ -j & j \end{bmatrix}$$

and note that α^2 is a non-central involution in $C_{G_2}(\beta)$, then $\gamma^2 = \beta$ and $C_{G_2}(\beta)$ is generated by γ and α^2. Moreover $C_{G_2}(\beta)$ has two subgroups isomorphic to 2^2, namely $\{1, \gamma^2, \alpha^2, \gamma^2\alpha^2\}$ and $\{1, \gamma^2, \gamma\alpha^2, \gamma^3\alpha^2\}$; the first of these is just $\{1, \beta, \alpha^2, \alpha^2\beta\}$, ie. is $G_1 \cap G_2$. On $\{x, y, b, c\}$, $G_1 \cap G_2$ is given by $\{id, (xy)(bc), (bc), (xy)\}$ and so is not normal in G_2. It follows that the normal subgroup of G_2 of order 4 is given by $\{1, \gamma^2, \gamma\alpha^2, \gamma^3\alpha^2\}$, indeed G_2 will be the normalizer in G of this subgroup.

Recall that $G_0 \cap G_1 = \{1, \alpha\beta, \alpha^2, \alpha^3\beta\}$ and that $G_0 \cap G_2 \cong S_3$, so that if δ is one of the two involutions of $G_0 \cap G_2$ not in G_1 then $\alpha\beta\delta \in G_0$. If we represent the blocks of imprimitivity of G_0 on $\bar{\Gamma}_0$ by 1,2,3,4 and 5 (containing y, b, c, d and e respectively) then $\alpha\beta = (24)(35)$ or $(25)(34)$ and $\delta = (12)(45)$ or $(13)(45)$; any combination gives the order of $\alpha\beta\delta$ as 5. We have a matrix representative for $\alpha\beta$, and δ is an involution normalizing $\{1, \gamma^2, \gamma\alpha^2, \gamma^3\alpha^2\}$. We know that G_2 has nine involutions of which five centralize β; by considering G_2 as $Sym\{x, y, b, c\}$ where $\beta = (xy)(bc)$ we see that the only involution of G_2 centralizing β that also lies in G_0 is the one given by the permutation (bc) and this we may identify as α^2. Since $\delta \notin G_1$ we conclude that δ does not centralize β, ie. $\delta^{-1}\beta\delta = \gamma\alpha^2$ or $\gamma^3\alpha^2$. As an involution, δ is represented by a matrix of the form

$$\begin{bmatrix} \eta & \xi \\ \epsilon & -\eta \end{bmatrix}$$

for some $\eta, \xi, \epsilon \in GF(q)$ satisfying $-\eta^2 - \epsilon\xi = 1$. The possibility $\beta\delta = \delta\gamma\alpha^2$ gives

$$\begin{bmatrix} 0 & 1 \\ -1 & 0 \end{bmatrix} \begin{bmatrix} \eta & \xi \\ \epsilon & -\eta \end{bmatrix} = \pm \begin{bmatrix} \eta & \xi \\ \epsilon & -\eta \end{bmatrix} \begin{bmatrix} j & j \\ -j & j \end{bmatrix} \begin{bmatrix} i & 0 \\ 0 & -i \end{bmatrix}$$

and the possibility $\beta\delta = \delta\gamma^3\alpha^2$ gives

$$\begin{bmatrix} 0 & 1 \\ -1 & 0 \end{bmatrix} \begin{bmatrix} \eta & \xi \\ \epsilon & -\eta \end{bmatrix} = \pm \begin{bmatrix} \eta & \xi \\ \epsilon & -\eta \end{bmatrix} \begin{bmatrix} j & -j \\ j & j \end{bmatrix} \begin{bmatrix} i & 0 \\ 0 & -i \end{bmatrix}.$$

Write k for ij (so that $k^2 = -1/2$) and consider the $+$ option in the first equation:

$$\begin{bmatrix} \epsilon & -\eta \\ -\eta & -\xi \end{bmatrix} = k \begin{bmatrix} \eta - \xi & -(\eta + \xi) \\ \eta + \epsilon & \eta - \epsilon \end{bmatrix}$$

from which $\epsilon = -\eta(k+1)/k = \eta(2k-1)$, $\xi = -\eta(k-1)/k = -\eta(2k+1)$ and $\eta^2 = -1/4$, ie. $\eta = \pm i/2$. Thus

83

$$\delta = \eta \begin{bmatrix} 1 & -(1+2k) \\ 2k-1 & -1 \end{bmatrix}. \tag{1}$$

Similarly the $-$ option in the first equation and the $+$ and $-$ options in the second equation yield:

$$\delta = \eta \begin{bmatrix} 1 & 2k-1 \\ -(1+2k) & -1 \end{bmatrix} \tag{2}$$

$$\delta = \eta \begin{bmatrix} 1 & 1-2k \\ 2k+1 & -1 \end{bmatrix} \tag{3}$$

$$\delta = \eta \begin{bmatrix} 1 & 1+2k \\ 1-2k & -1 \end{bmatrix}. \tag{4}$$

The four cases gives $\alpha\beta\delta$ as follows:

$$\alpha\beta\delta = \pm\tau/2 \begin{bmatrix} i(2k-1) & -i \\ -1 & 2k+1 \end{bmatrix} \tag{1}$$

$$\alpha\beta\delta = \pm\tau/2 \begin{bmatrix} -i(2k+1) & -i \\ -1 & 1-2k \end{bmatrix} \tag{2}$$

$$\alpha\beta\delta = \pm\tau/2 \begin{bmatrix} i(2k+1) & -i \\ -1 & 2k-1 \end{bmatrix} \tag{3}$$

$$\alpha\beta\delta = \pm\tau/2 \begin{bmatrix} i(1-2k) & -i \\ -1 & -(2k+1) \end{bmatrix}. \tag{4}$$

We calculate:

$$(\alpha\beta\delta)^5 = \pm\tau^5/8 \begin{bmatrix} (2k+3)(2-i) & 3i \\ 3 & (2k-3)(2i-1) \end{bmatrix} \tag{1}$$

$$(\alpha\beta\delta)^5 = \pm\tau^5/8 \begin{bmatrix} (3-2k)(2-i) & 3i \\ 3 & -(3+2k)(2i-1) \end{bmatrix} \tag{2}$$

$$(\alpha\beta\delta)^5 = \pm\tau^5/8 \begin{bmatrix} (2k-3)(2-i) & 3i \\ 3 & (3+2k)(2i-1) \end{bmatrix} \tag{3}$$

$$(\alpha\beta\delta)^5 = \pm\tau^5/8 \begin{bmatrix} -(2k+3)(2-i) & 3i \\ 3 & (3-2k)(2i-1) \end{bmatrix} \tag{4}$$

In each case we reach the same conclusion: q is a power of 3.

Hence we are reduced to the case $q = 3^a$ where a is even (since $q \equiv 1$ mod 8). Observe that $-1/2 = 1$ so $k^2 = 1$, ie. $k = \pm 1$. Therefore in this particular setting α, β, γ and δ all have coefficients in GF(9). Now $\alpha\beta$, α^2, δ must generate G_0, and γ, α^2, δ must generate G_2, so G_0 and G_2 both lie inside PSL(2,9). By Lemma 2.4, $< G_0, G_2 > = G$ so $G \leq PSL(2,9)$. However the index of G_0 in G is then at most 6, ie. the number of points of Γ is a most 6. This is impossible because $\bar{\Gamma}$ has valency 15.

This concludes the proof of Theorem 4.1.   □

We get:

Theorem. *There are no flag-transitive residually connected geometries with diagram* ○—^{c}—○—$^{P^*}$—○ *for which the automorphism group acts primitively on circles.*

Corollary. *There is exactly one flag-transitive residually connected geometry with diagram* ○—^{L}—○—$^{P^*}$—○ *for which the automorphism group acts primitively on circles.*

5.   Circle-imprimitive Geometries

We now turn to some examples of geometries for which the group G acts imprimitively on circles. We should like to thank D. V. Pasechnik, A. Pasini and S. Shpectorov for helpful discussion which confirmed one example and generated others.

We give four examples, three being quotients of the first.

Example 1

Let V be the vector space 2^6 over GF (2). We embed a Petersen graph P in V with the edges and vertices of P being (some of the) 1- and 2- dimensional subspaces of V.

Choose a basis $v_1, ..., v_6$ for V. The vertices of P will be 2-dimensional subspaces $A - J$ as given below and as marked on the graph, and edges are appropriate intersections, some of which are marked on the graph by their non-zero vector. Incidence is inclusion.

$A = < v_1, v_2 >$ $B = < v_2, v_3 >$ $C = < v_3, v_4 >$
$D = < v_4, v_5 >$ $E = < v_5, v_1 >$ $F = < v_1 + v_2, v_6 >$
$G = < v_2 + v_3, v_1 + v_3 + v_4 + v_5 + v_6 >$ $H = < v_3 + v_4, v_6 >$
$I = < v_4 + v_5, v_1 + v_2 + v_6 >$ $J = < v_1 + v_5, v_3 + v_4 + v_6 >$

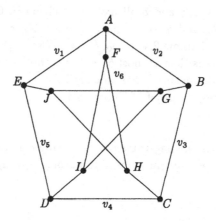

Diagram 3

It is not difficult to check that the stabilizer of P in $SL(6,2)$ acts faithfully on P and is isomorphic to S_5.

We now construct a geometry Γ as follows:

Let AG(6,2) be the affine space on the vertices of V. The points of Γ are the 64 points of AG(6,2). The lines of Γ are the 15 affine lines through the origin that appear as edges of P together with all affine lines parallel to these 15, so there are 480 lines in 15 parallel classes. The circles of Γ are the 10 affine planes through the origin that appear as vertices of P together with all affine planes parallel to these 10, so there are 160 circles in 10 parallel classes. Incidence is inclusion. The group $G = Aut\Gamma$ is $V.S_5$. It is then clear that G is transitive on points of Γ and that the stabilizer of the origin is transitive on flags of P, so G is flag-transitive. Also G is faithful on Γ and Γ is (residually) connected. The residue of the origin is P. The residue of the circle A in Diagram 3 is a complete graph because it contains the 3 marked lines through the origin together with 3 parallel lines. Finally the residue of the line $\{0, v_1\}$ contains the points $0, v_1$ and the circles A, E.

Example 2

Consider further the action of S_5 on V in Example 1. It has one fixed non-zero vector: $v_2 + v_3 + v_5 + v_6$. Let V' be the quotient of V by $\{0, v_2 + v_3 + v_5 + v_6\}$. In effect we are setting $v_6 = v_2 + v_3 + v_5$ and taking $v_1, v_2, ...v_5$ as a basis for V'. We get an embedding of the Petersen graph in V'. The stabilizer in $SL(5,2)$ is again S_5 and a construction analogous to Example 1 produces a geometry with 32 points, 240 lines and 80 circles. The automorphism group of this geometry is $2^5.S_5$.

Example 3

In Example 2 we find that S_5 has one non-zero fixed vector in $V' : v_1 + v_2 + v_3 + v_4 + v_5$. Let V'' be the quotient of V' by $\{0, v_1 + v_2 + v_3 + v_4 + v_5\}$. In effect we are setting $v_5 = v_1 + v_2 + v_3 + v_4$ and taking $v_1, ..., v_4$ as a basis for V''. We get an embedding of the Petersen graph in V''. The stabilizer in $SL(4, 2)$ is once more S_5 and we arrive at a geometry with 16 points, 120 lines and 40 circles (and now the 120 lines are all the affine lines).

Example 4

Returning to Example 1 we can consider the action of A_5 on V. It has 3 fixed non-zero vectors: $v_2 + v_3 + v_5 + v_6$, $v_1 + v_2 + v_3 + v_4 + v_5$ and $v_1 + v_4 + v_6$. Let W be the quotient of V by $\{0, v_1 + v_4 + v_6\}$ then the Petersen graph may be embedded in W but its stabilizer in $SL(5, 2)$ is A_5 rather than S_5. We get a geometry on 32 points whose automorphism group is $2^5 . A_5$. Thus this geometry is not isomorphic to that in Example 2.

References

[1] **R. Gilman and D. Gorenstein**. Finite groups with Sylow 2-subgroups of class two: I, II. *Trans. Amer. Math. Soc.*, 207, pp. 1–101; 103–125, 1975.

[2] **A. A. Ivanov and S. V. Shpectorov**. Geometries for sporadic groups related to the Petersen graph: I. *Comm. Algebra*, 16, pp. 925–954, 1988.

[3] **A. A. Ivanov and S. V. Shpectorov**. Geometries for sporadic groups related to the Petersen graph: II. *European J. Combin*, 10, pp. 347–361, 1989.

[4] **S. V. Shpectorov**. A geometric characterization of the group M_{22}. In *Investigations in the Algebraic Theory of Combinatorial Objects*, pages 112–123, Moscow, 1985. VNIISI.

[5] **W. J. Wong**. Determination of a class of primitive permutation groups. *Math. Z.*, 99, pp. 235–246, 1967.

F. Buekenhout, Université Libre de Bruxelles, Département de Mathématique, Campus Plaine, C. P. 216, B-1050 Bruxelles,Belgium. e-mail: fbueken@ulb.ac.be

O.H. King, Department of Mathematics and Statistics, The University of Newcastle upon Tyne, Newcastle upon Tyne,NE1 7RU,U.K.
e-mail: O.H.King@newcastle.ac.uk

Remarks on finite generalized hexagons and octagons with a point-transitive automorphism group

F. Buekenhout   H. Van Maldeghem *

Abstract

We show that the only point-transitive representations of the groups displayed in the ATLAS [2] on a finite generalized hexagon or octagon are the natural ones.

1. Introduction.

Let Γ be a thick, finite generalized hexagon (resp. octagon) of order (s,t) and G a group of automorphisms of Γ acting transitively on the points. Assume furthermore that G is almost simple, so there is a nonabelian simple group S with

$$S \trianglelefteq G \leq \mathrm{Aut}S.$$

We want to show that "small" G are ruled out, in particular that S cannot be a sporadic group. As a matter of fact, we consider all groups displayed in the main section of the ATLAS [2]. Let us call these groups "ATLAS-groups", then we can formulate our main results as follows:

Theorem 1.1 *If an* ATLAS-*group acts transitively on the points of a generalized hexagon, then it is has socle $G_2(q)$ ($q = 2,3,4,5$) or $^3D_4(2)$ and it acts in the natural way on a 'classical' generalized hexagon or its dual.*

Theorem 1.2 *If an* ATLAS-*group acts transitively on the points of a generalized octagon, then it is has socle $^2F_4(2)'$ and it acts in the natural way on the 'classical' generalized octagon of order $(2,4)$ or its dual.*

Theorem 1.1 will be proved in section 3 and theorem 1.2 will be proved in section 4.

---

*. This author is supported by the National Fund for Scientific Research (Belgium).

2. Some known facts.

2.1. Generalized Hexagons.

Let Γ be a generalized hexagon of v points and order (s,t). Then $v = (s+1)(1+st+s^2t^2)$, st is a perfect square ([3]) and $s \leq t^3 \leq s^9$ ([4]). Also, the rational number

$$\frac{st(s+1)(t+1)(1+st+s^2t^2)}{2[s^2t+t^2s-st+s+t\pm(s-1)(t-1)\sqrt{st}]} \tag{1}$$

is an integer ([5]).

2.2. Generalized Octagons.

Let Γ be a generalized octagon of v points and order (s,t). Then $v = (s+1)(1+st)(1+s^2t^2)$, $2st$ is a perfect square ([3]) and $s \leq t^2 \leq s^4$ ([5]). Also, the rational number

$$\frac{st(s+1)(t+1)(1+st)(1+s^2t^2)}{4[s^2t+t^2s-2st+s+t\pm(s-1)(t-1)\sqrt{2st}]} \tag{2}$$

is an integer ([5]).

3. Generalized hexagons.

In this section, we prove theorem 1.1.

We use the notation above and put $u = \sqrt{st}$ and $w = s+t$. Rewriting condition (1) we have that

$$\frac{u^2(1+w+u^2)(1\pm u+u^2)}{2(w-u)} \tag{3}$$

must be an integer for both choices of signs.

Suppose that G acts transitively on the v points of a thick generalized hexagon Γ of order (s,t) and G is a one of the simple groups listed in the ATLAS [2], see also tables 2 and 3 below. Since $v = (1+s)(1+st+s^2t^2)$, the latter expression divides $|G|$. Let p be a prime dividing $1+st+s^2t^2$. Then $1+st+s^2t^2 \equiv 0 \pmod{p}$, hence $s^3t^3 \equiv 1 \pmod{p}$. If $st \equiv 1 \pmod{p}$, then clearly $1+st+s^2t^2 \equiv 3 \pmod{p}$ and so $p=3$. In the other case, 1 must have three distinct third roots in $GF(p)$, so $p-1$ is divisible by 3 or in other words, $p \equiv 1 \pmod 3$. Note that for any integer n, $1+n+n^2$ is never divisible by 9. Put $a(G)$, or simply a, for the largest integer divisible by 3, but not by 9, all of whose other prime divisors are congruent to 1 (mod 3) and such that $a(G)$ divides $|G|$. We now distinguish between "small" groups and "larger" ones, the larger ones being $E_7(2)$, M and $E_8(2)$.

90

3.1. Small Groups.

Given G, it turns out that $a(G)$ only depends on its "socle" S except for $S \cong Sz(8)$ in which case we consider $a(\mathrm{Aut}\,Sz(8))$. Obviously, un upper bound U for u is given by the fourth root of $a(G)$. We can then look at table 1; it contains all values for $(1 + st + s^2t^2)$ for given u, $2 \leq u \leq 136$. We consider the largest number $U^* \leq U$ such that $1 + (U^*)^2 + (U^*)^4$ divides $|G|$. This is clearly a new upper bound for u. Hence $st \leq (U^*)^2$ and since $s \leq t^3$, this implies

$$s \leq \sqrt[4]{(st)^3} \leq \sqrt{(U^*)^3}.$$

So

$$v \leq \lfloor (\sqrt{(U^*)^3} + 1) \rfloor (1 + (U^*)^2 + (U^*)^4)$$

and we denote the latter by $h(G)$. If $a(G) > 3$ (that means, if $U > 1$), then we list the values for $a(G)$, U, U^* and $h(G)$ (if $U^* \geq 2$) in table 2, in which we also include the number $P(G)$ defined as the smallest permutation degree of S. The value for $P(G)$ follows from [6] for $^2E_6(2)$ and $E_6(2)$; from [7] for the sporadic groups and from the ATLAS [2] for the other groups. The "ATLAS-groups" with $a(G) \leq 3$ are A_5, A_6, $L_2(11)$, $L_2(17)$, $L_2(16)$, $L_2(23)$, M_{11}, $U_4(2)$, M_{12}, $S_4(4)$ and $U_5(2)$.

In a lot of cases, we have $h(G) < P(G)$ which is a contradiction. If $U^* = 2$, then $s = t = 2$ and by [8], Γ is the unique classical generalized hexagon $H(2)$ arising from the classical group $U_3(3) \cong G_2(2)'$. Only one simple group is a proper subgroup of $U_3(3)$, namely $L_3(2)$. But this group does not act transitively on the 63 points of $H(2)$ because 63 does not divide $|L_3(2)| = 168$. Of course if $S \cong U_3(3)$, then G acts transitively on exactly two generalized hexagons, namely $H(2)$ and its dual. The only remaining sporadic group is Suz. The largest possible value for s or t is 8 (when u=4; in general the largest value for s or t is $\sqrt{u^3}$, see above). Now Suz contains an element θ of order 11. Since $11 \equiv 2 \pmod 3$ and $11 > s + 1, t + 1$, θ fixes at least one point x, all lines through x, all points on all lines through x, etc. So θ fixes everything, a contradiction. In the sequel, we shall refer to this argument by the expression: *a group element of order 11 cannot live in Γ*. We consider the other groups in turn. Note that $u > 2$ (by the argument above), so $(s,t) \neq (2,2)$. A similar argument kills $(s,t) = (2,8)$ and $(s,t) = (8,2)$. Indeed, the generalized hexagons with these orders are unique by [1] and the related simple group is $^3D_4(2)$. Its proper simple subgroups are $L_3(2)$, $L_2(8)$ and $U_3(3)$ ([2]). None of these groups has a divisible by 13, which shows our assertion.

$\boxed{L_2(13)}$ Here $u \leq 4$ and so $(s,t) = (3,3)$ or $(s,t) = (4,4)$. In the latter case, $v = 1365$ and this does not divide $|\mathrm{Aut}(L_2(13)| = 2184$. In the former

u	$1 + u^2 + u^4$	u	$1 + u^2 + u^4$	u	$1 + u^2 + u^4$	u	$1 + u^2 + u^4$
2	3.7	36	13.31.43.97	70	3.1657.4831	104	3.67.163.3571
3	7.13	37	3.7.31.43.67	71	3.1657.5113	105	67.163.11131
4	3.7.13	38	3.7.67.1483	72	7.751.5113	106	3.19.199.11131
5	3.7.31	39	7.223.1483	73	3.7.751.1801	107	3.7.13.19.127.199
6	31.43	40	3.7.223.547	74	3.7.13.61.1801	108	7.13.61.127.193
7	3.19.43	41	3.547.1723	75	7.13.61.5701	109	3.7.61.193.571
8	3.19.73	42	13.139.1723	76	3.1951.5701	110	3.7.571.12211
9	7.13.73	43	3.13.139.631	77	3.1951.6007	111	12211.12433
10	3.7.13.37	44	3.7.283.631	78	6007.6163	112	3.4219.12433
11	3.7.19.37	45	7.19.109.283	79	$3.7^2.43.6163$	113	3.13.991.4219
12	7.19.157	46	3.7.19.103.109	80	$3.7^2.43.6481$	114	7.13.991.1873
13	3.61.157	47	3.7.37.61.103	81	7.13.73.6481	115	3.7.1873.4447
14	3.61.211	48	13.37.61.181	82	3.7.13.73.2269	116	$3.7^2.277.4447$
15	211.241	49	3.13.19.43.181	83	3.19.367.2269	117	$7^2.277.13807$
16	3.7.13.241	50	3.19.43.2551	84	19.37.193.367	118	3.31.151.13807
17	3.7.13.307	51	7.379.2551	85	3.37.193.2437	119	3.31.151.14281
18	$7^3.307$	52	3.7.379.919	86	3.7.1069.2437	120	13.1117.14281
19	$3.7^3.127$	53	3.7.409.919	87	7.13.19.31.1069	121	3.7.13.19.37.1117
20	3.127.421	54	7.409.2971	88	3.7.13.19.31.373	122	3.7.19.37.43.349
21	421.463	55	3.13.79.2971	89	3.7.373.8011	123	7.43.349.2179
22	$3.13^2.463$	56	3.13.31.79.103	90	8011.8191	124	3.7.2179.5167
23	$3.7.13^2.79$	57	31.103.3307	91	3.2791.8191	125	3.19.829.5167
24	7.79.601	58	3.7.163.3307	92	3.43.199.2791	126	13.19.829.1231
25	3.7.31.601	59	3.7.163.3541	93	7.43.199.1249	127	3.13.1231.5419
26	3.7.19.31.37	60	7.523.3541	94	3.7.13.229.1249	128	$3.7^2.337.5419$
27	19.37.757	61	3.7.13.97.523	95	3.7.13.229.1303	129	$7^2.31.337.541$
28	3.271.757	62	3.13.97.3907	96	7.67.139.1303	130	3.7.31.541.811
29	3.13.67.271	63	37.109.3907	97	3.67.139.3169	131	3.7.811.17293
30	$7^2.13.19.67$	64	3.19.37.73.109	98	3.31.313.3169	132	97.181.17293
31	$3.7^2.19.331$	65	3.7.19.73.613	99	31.313.9901	133	3.13.97.181.457
32	3.7.151.331	66	7.613.4423	100	3.7.13.37.9901	134	3.13.79.229.457
33	7.151.1123	67	$3.7^2.31.4423$	101	3.7.13.37.10303	135	7.43.61.79.229
34	3.397.1123	68	$3.7^2.13.19^2.31$	102	7.19.79.10303	136	3.7.43.61.6211
35	3.13.97.397	69	$13.19^2.4831$	103	3.7.19.79.3571	137	3.7.37.73.6211

Table 1.

S	a	U	U^*	$h(S)$	$P(G)$	S	a	U	U^*	$h(G)$	$P(G)$
$L_3(2)$	3.7	2	2	63	7	HS	3.7	2	2	63	100
$L_2(8)$	3.7	2	2	63	9	J_3	3.19	2	1		6156
$L_2(13)$	3.7.13	4	4	2457	14	$U_3(11)$	3.37	3	1		1332
A_7	3.7	2	2	63	7	$O_8^+(2)$	3.7	2	2	63	120
$L_2(19)$	3.19	2	1		20	$O_8^+(2)$	3.7	2	2	63	119
$L_3(3)$	3.13	2	1		13	$^3D_4(2)$	$3.7^2.13$	6	4	2457	819
$U_3(3)$	3.7	2	2	63	28	$L_3(11)$	3.7.19	4	2	63	133
$L_2(25)$	3.13	2	1		26	A_{12}	3.7	2	2	63	12
$L_2(27)$	3.7.13	4	4	2457	28	M_{24}	3.7	2	2	63	24
$L_2(29)$	3.7	2	2	63	30	$G_2(4)$	3.7.13	4	4	2457	416
$L_2(31)$	3.31	3	1		32	McL	3.7	2	2	63	275
A_8	3.7	2	2	63	8	A_{13}	3.7.13	4	4	2457	13
$L_3(4)$	3.7	2	2	63	21	He	3.7^2	3	2	63	2058
$Sz(8)$	3.7.13	4	4	2457	65	$O_7(3)$	3.7.13	4	4	2457	351
$L_2(32)$	3.31	3	1		33	$S_6(2)$	3.7.13	4	4	2457	364
$U_3(4)$	3.13	2	1		65	$G_2(5)$	3.7.31	5	5	7929	3906
$U_3(5)$	3.7	2	2	63	50	$U_6(2)$	3.7	2	2	63	672
J_1	3.7.19	4	2	63	266	$R(27)$	3.7.13.19.37	20	11	54.10^4	2.10^4
A_9	3.7	2	2	63	9	$S_8(2)$	3.7	2	2	63	120
$L_3(5)$	3.31	3	1		31	Ru	3.7	2	2	63	4060
M_{22}	3.7	2	2	63	22	Suz	3.7.13	4	4	2457	1782
J_2	3.7	2	2	63	100	$O'N$	$3.7^3.19.31$	27	5	7929	122760
$S_6(2)$	3.7	2	2	63	28	Co_3	3.7	2	2	63	276
A_{10}	3.7	2	2	63	10	$O_8^+(3)$	3.7.13	4	4	2457	1080
$L_3(7)$	$3.7^3.19$	11	2	63	57	$O_8^-(3)$	3.7.13	4	4	2457	1066
$U_4(3)$	3.7	2	2	63	112	$O_{10}^+(2)$	3.7.31	5	5	7929	496
$G_2(3)$	3.7.13	4	4	2457	351	$O_{10}^-(2)$	3.7	2	2	63	495
$S_4(5)$	3.13	2	1		156	Co_2	3.7	2	2	63	2300
$U_3(8)$	3.7.19	4	2	63	513	Fi_{22}	3.7.13	4	4	2457	3510
$U_3(7)$	$3.7^3.43$	14	2	63	344	HN	3.7.19	4	2	63	1140000
$L_4(3)$	3.13	3	1		40	$F_4(2)$	$3.7^3.13$	6	4	2457	69615
$L_5(2)$	3.7.31	5	5	7929	31	Ly	3.7.31.37.67	136	5	7929	8835156
M_{23}	3.7	2	2	63	23	Th	$3.7^2.13.19.31$	32	5	7929	$> 10^8$
$L_3(8)$	$3.7^2.73$	10	2	63	73	Fi_{23}	3.7.13	4	4	2457	31671
$^2F_4(2)'$	3.13	2	1		1600	Co_1	$3.7^2.13$	6	4	2357	98280
A_{11}	3.7	2	2	63	11	J_4	3.7.31.37.43	31	6	19995	$> 10^8$
$Sz(32)$	31	2	0		1025	$^2E_6(2)$	$3.7^2.13.19$	13	4	2457	3968055
$L_3(9)$	3.7.13	4	4	2457	91	$E_6(2)$	$3.7^3.13.31.73$	74	9	186004	139503
$U_3(9)$	3.73	3	1		730	Fi_{24}'	$3.7^3.13$	10	4	2457	306936
						B	$3.7^2.13.19.31$	32	5	7929	$> 10^{10}$

Table 2.

case, the stabilizer G_x of a point x contains an element θ of order 3. In G, there are exactly 91 Sylow 3-subgroups. Hence θ must fix exactly 4 points of Γ. These 4 points form a set of imprimitivity. Since there are 4 lines through x, θ must fix one of these lines, say l. If y is another fixed point for θ, then θ also fixes the point on l nearest to y. Consequently, θ fixes l pointwise. The 91 lines thus obtained form a partition of the point set and G acts transitively on the set $\mathcal{L}$ of such lines. Hence G acts primitively on that set and since the stabilizer G_l of l normalizes θ, G_l is isomorphic to D_{12} or D_{24} (see [2]). So we can identify the 91 lines in $\mathcal{L}$ with the 91 pairs of points of the projective line over $GF(13)$.

Suppose first $G \cong L_2(13)$. Then $|G_x| = 3$, so no involution can fix a point in Γ. Every involution fixes exactly 7 lines of $\mathcal{L}$ (that is the number of pairs it stabilizes on the projective line $PG(1,13)$). These seven lines are mutually *opposite* (on maximal distance) since otherwise a point is fixed. Identify an arbitrary line $l \in \mathcal{L}$ with the pair $\{(0),(\infty)\}$. All pairs $\{(r),(s)\}$, $r, s \in GF(13)^t$, with r/s a square in $GF(13)$ can be stabilized under a certain involution also stabilizing $\{(0),(\infty)\}$, and the others cannot. There are 30 such pairs and by the preceding argument they are all opposite l. Left are five orbits of length 12 under the stabilizer of $\{(0),(\infty)\}$. Two of these orbits contain all pairs of the form $\{(0),(r)\}$ and $\{(\infty),(r)\}$ with r a square, resp. a non-square, denote them by $O_\square$, resp. $O_{\boxtimes}$. The other three orbits contain pairs $\{(s),(2s)\}$, resp. $\{(s),(5s)\}$, $\{(s),(6s)\}$ and we denote them by O_i, $i = 2, 5, 6$ respectively. Since there are 36 elements of $\mathcal{L}$ at distance 4 from l, at least one of the sets O_i, $i = 2, 5, 6$, must have all its elements at distance 4 from l. Suppose $\{(r),(s)\} \in O_2 \cup O_5$ is a line at distance 4 from l. Let $y_{r,s}$ (resp. $l'_{r,s}$) be the unique point (resp. line) of Γ at distance 1 (resp. 2) from l and at distance 3 (resp. 2) from $\{(r),(s)\}$. Applying θ, we see that also $\{(3r),(3s)\}$ is at distance 4 from l. Define $y_{3r,3s}$ and $l'_{3r,3s}$ as above. Obviously $y_{r,s} = y_{3r,3s}$. One can verify that $\{(r),(s)\}$ and $\{(3r),(3s)\}$ lie at distance 6 from each other, so $l'_{r,s} \neq l'_{3r,3s}$. Hence θ acts transitively on the lines distinct from l through each point of l. This implies that G acts transitively on the pairs of lines (m, m'), with $m \in \mathcal{L}$ and $m' \notin \mathcal{L}$. So G acts transitively on the lines not in $\mathcal{L}$, but this action is imprimitive with sets of imprimitivity of order 3. The stabilizer of such a set is A_4 (see [2]). So A_4 acts regularly on the 12 points of the three lines of a set of imprimitivity. This implies the existence of an involution swapping any two points on a line not in $\mathcal{L}$, or in other words, swapping two lines of $\mathcal{L}$ at distance 4. But the pairs $\{(0),(\infty)\}$ and $\{(r),(s)\}$ are swapped by an element of $PGL_2(13) \setminus L_2(13)$. Hence the lines of $\mathcal{L}$ at distance 4 from l are precisely all elements of O_6, $O_\square$ and $O_{\boxtimes}$ and θ fixes all lines meeting l. Therefore, the sets $L = \{\{(\infty),(r)\} \| r \in \{0,1,3,9\}\}$ and $L' = \{\{(\infty),(r)\} \| r \in \{0,2,5,6\}\}$ consist of 4 lines meeting a common line.

But the automorphism determined by adding 12 to each coordinate maps $\{(\infty), (1)\}$ to l and $\{(\infty), (3)\}$ to $\{(\infty), (2)\}$, hence L should be mapped to L', but it is not, as one can verify immediately.

Next suppose $G \cong PGL_2(13)$. Then $|G_x| = 6$ and so there is an involution θ fixing x. Also, θ fixes exactly 7 lines of $\mathcal{L}$, among them the unique line l of $\mathcal{L}$ incident with x. If all other fixed lines of $\mathcal{L}$ have distance 4 from l, then θ fixes at least two points on each of these lines and so θ fixes points at distance 5 from l and from other fixed lines, hence θ fixes an ordinary hexagon. In the other case, this is trivially true. Hence θ fixes a subhexagon of order $(1,3)$ or $(3,1)$ (since θ fixes at least 7 lines from $\mathcal{L}$, any other possible configuration of fixed structure contains at most 5 elements of $\mathcal{L}$). So *theta* fixes either 26 or 52 points. Since there are no involutions in G_x other than θ, this implies that θ has exactly 14, resp. 7 conjugates in G, which means that θ is normalized by a group of order at least 2.3.13, contradicting the information on $L_2(13)$ given in the ATLAS [2].

$\boxed{L_2(27)}$ As in the previous case, $(s,t) = (3,3)$ or $(s,t) = (4,4)$. Also $(s,t) = (4,4)$ is eliminated the same way. If $(s,t) = (3,3)$, then the stabilizer of a point x has order $3^3.[G : L_2(27)]$ and it follows from the ATLAS [2] that it normalizes an elementary abelian subgroup of order 3^3. There are 28 such groups (two by two disjoint) and therefore any element of such group fixes exactly 13 points of Γ. As before, these 13 points form a set Ω of imprimitivity containing the points of some line l through x (considering an element of order 3 in G_x). Any other point of Ω defines a point nearest to l which must also be in Ω. Hence we can assume that there is another line l' through x all of whose points are in Ω. By transitivity, every point of Ω is incident with two lines all of whose points are in Ω. This is impossible in view of $|\Omega| = 13$.

$\boxed{Sz(8)}$ Again $(s,t) = (3,3)$ or $(s,t) = (4,4)$. In the former case, a group element of order 5 cannot live in Γ. In the latter case, we deduce as above that every of the 65 Sylow 2-groups fixes 21 points, which form a set of imprimitivity. As before, such a set cannot exist.

$\boxed{G_2(3)}$ Again $(s,t) = (3,3)$ or $(s,t) = (4,4)$. In the latter case, v does not divide $|G|$. In the former case, Γ must be the classical generalized hexagon since G essentially has only one transitive representation on 364 points.

$\boxed{L_5(2)}$ In view of $a = 3.7.31$, only $(s,t) = 5$ is possible. Consider a point x. The stabilizer G_x contains a group H of order 2^9. A line through x is in an orbit of length 1,2 or 4 under H. If it is in an orbit of length 4, then some other line through x is in an orbit of length 2. Hence a group H^* of order 2^8 fixes a line l through x. Since $s = t = 5$, this group must fix another line through x and another point on l, etc. So at least an ordinary hexagon is fixed by H^*. A point y on one of the sides has an orbit of size at most 4, the stabilizer of y fixes also another point on the same side and there are only 2 points left on that side. Hence a group of order at least 2^5 fixes a hexagon and all points on one of its sides. Similarly, a group of order at least 2^2 fixes a hexagon, all points on one of its sides and all lines through one of its vertices. But then all elements of Γ are fixed (since this generates a subhexagon of order $(5,5)$), a contradiction.

$\boxed{L_3(9)}$ Here $(s,t) = (3,3)$ or $(s,t) = (4,4)$. In the former case, a group element of order 5 cannot live in G. In the latter case, the stabilizer G_x of a point x contains a group of order 3^5. As above, this group fixes a hexagon and a subgroup of order 3^3 fixes everything, a contradiction.

$\boxed{{}^3D_4(2)}$ We have to rule out $(s,t) = (3,3)$ and $(s,t) = (4,4)$. In both cases, the stabilizer G_x of a point x contains an element of order 7 and such an element cannot live in Γ.

$\boxed{G_2(4)}$ Order $(3,3)$ is ruled out by the presence of an element of order 5 in G. The representation on 1365 points is essentially unique, hence the "classical" generalized hexagon of order $(4,4)$ and its dual arise.

$\boxed{A_{13}}$ Every order is ruled out by the presence of an element of order 13, which cannot live in Γ.

$\boxed{O_7(3)}$ If $(s,t) = (3,3)$, then a group element of order 5 cannot live in G. If $(s,t) = (4,4)$, then, as in the case of $L_3(9) \trianglelefteq G$, the presence of a group of order 3^8 in the stabilizer of a point leads to a contradiction.

$\boxed{S_6(3)}$ See $O_7(3) \trianglelefteq G$; both groups have the same order.

$\boxed{G_2(5)}$ Here, only $(s,t) = (5,5)$ is possible, it occurs and it is unique up to duality (by the information in the ATLAS [2]).

$\boxed{R(27)}$ Here, $u = 11, 10, 4, 3, 2$. But $u \leq 4$ is impossible in view of $P(G)$. If $u = 11$, then $s = t = 11$ and a group element of order 13 cannot live in Γ. If $u = 10$, then $s = 4, 5, 10, 20$ or 25. But $s = 4, 10, 20$ or 25 implies v does not divide $|G|$ ($s = 4$ and $s = 25$ are also ruled out by the fact that (3) is not an integer in these cases). Hence $(s,t) = (5,20)$. The order of the stabilizer of a point is $2^2.3^b.19$, with $b = 8$ or 9 depending on $G \cong R(27)$ or $G \cong R(27) : 3$. So G_x properly contains the normalizer of a Sylow 19-subgroup, which is a maximal subgroup, a contradiction.

$\boxed{O_8^+(3)}$ In view of $P(S) = 1080$, the only possibility here is $s = t = 4$. The order of the stabilizer of a point of Γ is divisible by $2^{12}.3^{11}.5$. The presence of a Sylow 3-subgroup of G_x of order at least 3^{11} leads to a contradiction as in the case $S \cong L_3(9)$ or $S \cong O_7(3)$.

$\boxed{O_8^-(3)}$ Whatever the order of Γ, a group element of order 41 cannot live in G.

$\boxed{O_{10}^+(2)}$ Whatever the order of Γ, a group element of order 17 cannot live in Γ.

$\boxed{E_6(2)}$ In view of table 1, $u = 9$ or $u \leq 5$. In the latter case, v would be smaller than $P(G)$; in the former case, a group element of order 31 cannot live in Γ (because $s, t \leq 27$).

This completes the case of small groups.

3.2. Larger Groups.

In this case, we can still compute U as above, but it is too large to use table 1 to find U^*. But we can use U to compute $h(G)$ as in the previous paragraph. For $G \cong M$, we obtain

$$
\begin{aligned}
a(G) &= 3.7^6.13^3.19.31 \\
h(G) &\approx 107.10^{14} \\
P(G) &\approx 927.10^{17}
\end{aligned}
$$

This rules out $G \cong M$. In general, there is obviously a minimal value U_* for u such that the derived number $h(G)$ is larger then $P(G)$. So $U_* \leq u \leq U$. We now develop a method to reduce the bound U until it is below U_* without having to calculate all values for $1 + u^2 + u^4$. Consider a divisor d of $a(G)$, preferably larger than or just a little bit smaller than U. The number of u such that d divides $1 + u^2 + u^4$ is limited and usually none of these values for u (except maybe very small ones which are in conflict with U_* anyway) give a $1 + u^2 + u^4$ dividing $a(G)$. Hence we can recalculate U starting from $a(G)/p$, where P is the smallest prime dividing d. We refer to this procedure as "reduction modulo d". We usually take for d a prime or a prime power, so that we can do this reduction a few times and end up with no value of u left. Let us illustrate this in the case of $G \cong E_7(2)$. We have here

$$
\begin{aligned}
a(G) &= 3.7^3.13.19.31.43.73.127 \\
P(G) &= 277347807 \\
U &= 1331 \\
U_* &= 35
\end{aligned}
$$

It is easy to calculate by hand that only $u = 1250$ gives rise to $1 + u^2 + u^4$ divisible by 73.127. Indeed, every u giving $1 + u^2 + u^4$ divisible by 73 is congruent to 8, 9, 64 or 65 modulo 73, similarly, every suitable u must also be congruent to 19, 20, 107 or 108 modulo 127. Only 1250 satisfies these conditions and is smaller 1331. But in the same way, we see that 43 does not divide $1 + u^2 + u^4$ in that case, so $d = 43.73.127$ gives us no solutions and the new upper bound is $U' = 519$. We now see that $1250 > 519$, so the new upper bound becomes 455. Putting $d = 127$, the possible values for u are 19, 107, 146, 234, 273, 361 and 400. This gives us:

u	$1 + u^2 + u^4$	u	$1 + u^2 + u^4$
19	$3.7^3.127$	20	$3.127.421$
107	$3.7.13.19.127.199$	108	$7.13.61.127.193$
146	$3.13^2.127.7057$	147	$13^2.127.21757$
234	$7.127.433.7789$	235	$3.7.19.127.139.433$
273	$19.31.127.74257$	274	$3.19.31.127.25117$
361	$3.7^3.13^2.127.769$	362	$3.7^3.127.331.397$

Hence only $u = 19$ would do, but it is smaller than U_*. The new upper bound now becomes 396. A reduction modulo 43.73 (with no possible u smaller than 396) gives the new upper bound 154. Reduction modulo 7^3 yields $u = 19$ or 20 (too small) and the new upper bound 95. Table 1 now shows that $u < 35$, ruling out $G \cong E_7(2)$.

The group $E_8(2)$ is much harder to handle because it is much larger. We have:

$$\begin{aligned} a(G) &= 3.7^4.13^2.19.31^2.43.73.127.151.241.331 \\ P(G) &\approx 293.10^{15} \\ U &= 571575 \\ U_* &= 1500 \end{aligned}$$

Using reductions again, we have ruled out $E_8(2)$ by computer using CAYLEY.

We will apply this method of reduction again in the next section. We will not have to use the computer again.

This completes the proof of theorem 1.1.

4. Generalized octagons.

In this section, we prove theorem 1.2.

We use the notation of subsection 2.2. Put $u = \sqrt{\frac{st}{2}}$ and $w = s + t$. We rewrite the rational number (2) of section 2.2 as

$$\frac{u^2(1 + w + 2u^2)(1 + 2u^2)(1 \pm 2u + 2u^2)}{2(w \pm 2u)} \tag{4}$$

This must be an integer for both choices of signs.

Suppose that G acts transitively on the v points of a thick generalized octagon Γ of order (s, t) and G is again one of the simple groups listed in the

u	$1+4u^4$	u	$1+4u^4$	u	$1+4u^4$
2	5.13	12	5.53.313	22	5^2.37.1013
3	5^2.13	13	5.73.313	23	5.13.17.1013
4	5^2.41	14	5.73.421	24	5.13.17.1201
5	41.61	15	13.37.421	25	1201.1301
6	5.17.61	16	5.13.37.109	26	5.281.1301
7	5.17.113	17	5.109.613	27	5.17.89.281
8	5.29.113	18	5.137.613	28	5^3.13.17.89
9	5.29.181	19	5.137.761	29	5^3.13.1741
10	13.17.181	20	29^2.761	30	1741.1861
11	5.13.17.53	21	$5^2.29^2$.37	31	5.397.1861

Table 3.

ATLAS [2]. Obviously $v = (1+s)(1+st)(1+s^2t^2)$ divides $|G|$. Let p be a prime dividing $1+s^2t^2$. Then $1+s^2t^2 \equiv 0 \pmod{p}$, hence $s^2t^2 \equiv -1 \pmod{p}$ and -1 is a square in $GF(p)$ which implies $p = 2$ or $p \equiv 1 \pmod{4}$. Since st is even, $p \neq 2$. So we put $a(G)$, or simply a, for the largest integer all of whose other prime divisors are congruent to 1 (mod 4) and such that $a(G)$ divides $|G|$. We now again distinguish between "small" groups and "larger" ones, this time, the larger ones being only M and $E_8(2)$.

4.1. Small Groups.

As before, $a(G)$ only depends on the socle S of G except in the cases $S \cong Sz(32)$ and $S \cong L_2(32)$ in which case we consider the respective automorphism groups. We can copy the arguments of subsection 3.1 almost word by word. An upper bound U for u is given by the fourth root of $a(G)/2$. We can then look at table 4; it contains all values for $(1 + s^2t^2)$ for given u, $2 \leq u \leq 31$. We consider the largest number $U^* \leq U$ such that $1 + 4(U^*)^4$ divides $|G|$. This is clearly a new upper bound for u. By inspection of the orders of the small ATLAS-groups, it turns out that only for 24 among them $U^* > 1$. We list them is table 5 together with their order, the order d of their outer automorphism group, a, U and U^*.

Note that, if $u = 3$, then $1 + st = 19$, hence $|G|$ must be divisible by 19. For the groups in table 5 with $U^* \geq 3$, this is only true for Th, $^2E_6(2)$, B and $E_7(2)$. In this case however, $\{s,t\} = \{3,6\}$ and no group element of prime 31 nor 17 can live in Γ contradicting the fact that one of these primes divides the order of the four groups mentioned. So we may assume $u \neq 3$. There is only one case where $U^* > 3$ and that is if $S \cong Sz(32)$. Here $u = 4$ and

S	$|S|$	d	a	U	U^*
$L_2(25)$	$2^3.3.5^2.13$	2^2	$5^2.13$	3	3
$U_3(4)$	$2^6.3.5^2.13$	2^2	$5^2.13$	3	3
$S_4(5)$	$2^6.3^2.5^4.13$	2	$5^4.13$	6	3
$L_4(3)$	$2^7.3^6.5.13$	2^2	5.13	2	2
$^2F_4(2)'$	$2^{11}.3^3.5^2.13$	2	$5^2.13$	3	3
$Sz(32)$	$2^{10}.5^2.31.41$	5	$5^3.41$	5	4
$L_3(9)$	$2^7.3^6.5.7.13$	2^2	5.13	2	2
$G_2(4)$	$2^{12}.3^3.5^2.7.13$	2	$5^2.13$	3	3
A_{13}	$2^9.3^5.5^2.7.11.13$	2	$5^2.13$	3	3
$O_7(3)$	$2^9.3^9.5.7.13$	2	5.13	2	2
$S_6(3)$	$2^9.3^9.5.7.13$	2	5.13	2	2
Ru	$2^{14}.3^3.5^3.7.13.29$	1	$5^3.13.29$	10	3
Suz	$2^{13}.3^7.5^2.7.11.13$	2	$5^2.13$	3	3
$O_8^+(3)$	$2^{12}.3^{12}.5^2.7.13$	$2^3.3$	$5^2.13$	3	3
$O_8^-(3)$	$2^{10}.3^{12}.5.7.13.41$	2^2	$5.13.41$	5	2
Fi_{22}	$2^{17}.3^9.5^2.7.11.13$	2	$5^2.13$	3	3
$F_4(2)$	$2^{24}.3^6.5^2.7^2.13.17$	2	$5^2.13.17$	6	3
Th	$2^{15}.3^{10}.5^3.7^2.13.19.31$	1	$5^3.13$	4	3
Fi_{23}	$2^{18}.3^{13}.5^2.7.11.13.17.23$	1	$5^2.13.17$	6	3
Co_1	$2^{21}.3^9.5^4.7^2.11.13.23$	1	$5^4.13$	6	3
$^2E_6(2)$	$2^{36}.3^9.5^2.7^2.11.13.17.19$	2.3	$5^2.13.17$	6	3
$E_6(2)$	$2^{36}.3^6.5^2.7^3.13.17.31.73$	2	$5^2.13.17.73$	17	3
Fi'_{24}	$2^{21}.3^{16}.5^2.7^3.11.13.17.23.29$	2	$5^2.13.17.29$	14	3
B	$2^{41}.3^{13}.5^6.7^2.11.13.17.19.23.31.47$	1	$5^6.13.17$	30	3
$E_7(2)$	$2^{63}.3^{11}.5^2.7^3.11.13.17.19.31.43.73.127$	1	$5^2.13.17.73$	17	3

Table 4.

$\{s, t\} = \{4, 8\}$, so a group element of order 31 cannot live in Γ. Hence, for the rest of the proof, we have $u = 2$ and hence $(s, t) = (2, 4)$ or $(s, t) = (4, 2)$. So if $|G|$ contains a prime p distinct from 13 and greater then 6, then we obtain a contradiction since a group element of order p could not live in Γ. Only for the first five groups of table 5 we have that $p = 7$ does not divide $|G|$. Moreover, the orders of the first two groups are not divisible by $3^2 = 1 + st$, a contradiction. We consider the other groups in turn.

$\boxed{S_4(5)}$ Note that necessarily $s = 4$ because otherwise $3^3 = (1+s)(1+st)$ and this does not divide $|G|$. The order of the stabilizer G_x of a point x of Γ is divisible by 5, so consider an element θ of order 5 in G_x. It has to fix all three lines through x and all points other than x on these lines, etc. So θ is the identity, a contradiction.

$\boxed{L_4(3)}$ Here, $s = 2$ since 5^2 does not divide $|G|$. The stabilizer G_x of a point x contains a Sylow 3-subgroup H of order 3^3. The latter fixes at least two lines through x, all points on these lines, at least one other line through those points, etc. This is enough to conclude that H fixes a suboctagon Γ' of order $(2, 1)$ and H has all orbits of length 27 on the points and lines not in Γ' (otherwise a group element fixes the "geometric closure" of Γ' and a fixed element not in Γ', which is Γ itself). But the number of points outside Γ' in Γ is 1710 and this is not divisible by 27.

$\boxed{^2F_4(2)'}$ Every transitive action on 1755 or 2925 points is primitive by the information in the ATLAS [2], hence Γ is the usual generalized octagon or its dual.

This completes the case of small groups.

4.2. Larger Groups.

We first deal with $G \cong M$. We have

$$
\begin{aligned}
a(G) &= 5^9 . 13^3 . 17 . 29 . 41 \\
P(G) &\approx 927 . 10^{17} \\
U &= 2158 \\
U_* &= 373
\end{aligned}
$$

The lower bound U_* is achieved as in subsection 3.2. We do a reduction modulo 5^5 (cp. subsection 3.2). Suppose $5^5 | 1 + s^2 t^2$. This means that $s^2 t^2 = 4u^4 \equiv -1$ (mod 5^5) or $u^4 \equiv 781$ (mod 5^5). This has only four solutions not larger than 2158, namely 1028, 1029, 2096 and 2097. But for none of these values $(1 + st)(1 + s^2 t^2)$ is a divisor of $|M|$. Hence

$$1 + s^2 t^2 | 5^4 . 13^3 . 17 . 29 . 41 \approx 2775 . 10^7,$$

101

giving us the new upper bound $U' = 298 < 375 = U^*$.

Similarly, we deal with $G \cong E_8(2)$. Here

$$
\begin{aligned}
a(G) &= 5^5.13^2.17^2.41.73.241 \\
P(G) &\approx 293.10^{15} \\
U &= 2290 \\
U_* &= 170
\end{aligned}
$$

Reduction modulo 5^5 gives the new bound 1531. Reduction modulo 5^4 (possibilities for u are 221, 222, 403, 404, 846, 847, 1471, 1472) gives the new bound 1024. Reduction modulo 73.241 ($u = 570$) gives the new bound 350. Reduction modulo 241 ($u = 88, 89, 152, 153, 329, 330$) gives the new bound 259. Reduction modulo 17^2 ($u = 125, 126, 163, 164$) gives finally the upper bound 128, contradicting $U_* = 170$.

This completes the proof of theorem 1.2.

References

[1] **A. M. Cohen and J. Tits**. On generalized hexagons and a near octagon whose lines have three points. *European J. Combin.*, 6, pp. 13–27, 1985.

[2] **J. H. Conway, R. T. Curtis, S. P. Norton, R. A. Parker, and R. A. Wilson**. *Atlas of Finite Groups*. Clarendon Press, Oxford, 1985.

[3] **W. Feit and G. Higman**. The non-existence of certain generalized polygons. *J. Algebra*, 1, pp. 114–131, 1964.

[4] **W. Haemers and C. Roos**. An inequality for generalized hexagons. *Geom. Dedicata*, 10, pp. 219–222, 1981.

[5] **D. G. Higman**. Invariant relations, coherent configurations and generalized polygons. In M. Hall Jr. and J. H. van Lint, editors, *Proceedings of the Advanced Study Institute on Combinatorics, Breukelen*, volume 55 of *Mathematical Centre Tracts*, pages 247–363, Dordrecht, 1975. Reidel.

[6] **M. W. Liebeck and J. Saxl**. On the orders of maximal subgroups of the finite exceptional groups of Lie type. *Proc. London Math. Soc.*, 55, pp. 299–330, 1987.

[7] **V. D. Mazurov**. The minimal permutation representation of the Thompson group. *Algebra and Logic*, 27, pp. 350–365 (supplement), 1988.

[8] **J. Tits**. Sur la trialité et certains groupes qui s'en déduisent. *Publ. Math. IHES*, 2, pp. 14–60, 1959.

F. **Buekenhout**, Université Libre de Bruxelles, Campus Plaine 216, B – 1050 Bruxelles. e-mail: fbueken@ulb.ac.be

H. **Van Maldeghem**, Universiteit Gent, Galglaan 2, B – 9000 Gent. e-mail: hvm@cage.rug.ac.be

Block-transitive t-designs, II: large t

P. J. Cameron   C. E. Praeger

Abstract

We study block-transitive t-(v, k, λ) designs for large t. We show that there are no nontrivial block-transitive 8-designs, and no nontrivial flag-transitive 7-designs. There are no known nontrivial block-transitive 6-designs; we show that the automorphism group of such a design, or of a flag-transitive 5-design with more than 24 points, must be either an affine group over $GF(2)$ or a 2-dimensional projective linear group. We begin the investigation of these two cases, and construct a flag-transitive 5-(256, 24, λ) design for a suitable value of λ.

1.   Introduction

A t-(v, k, λ) *design* is a pair $\mathcal{D} = (X, B)$, where X is a set of v points, B a set of k-element subsets of X called blocks, such that any t points are contained in exactly λ blocks, for some $t \leq k$ and $\lambda > 0$. Such a design (X, B) is said to be *trivial* if B consists of all the k-element subsets of X. A *flag* in a design $\mathcal{D}$ is an incident point-block pair. A subgroup G of the automorphism group of $\mathcal{D}$ is said to be *block-transitive* if G is transitive on B ; $\mathcal{D}$ is *block-transitive* if Aut($\mathcal{D}$) is. *Point-* and *flag-transitivity* are defined similarly. For information about t-designs, see Hughes and Piper [11].

In this paper we consider nontrivial block-transitive t-designs with t large. We use a result of Ray-Chaudhuri and Wilson [16] together with the finite simple group classification to show in Section 2 that $t \leq 7$.

Theorem 1.1 *Let* $\mathcal{D} = (X, B)$ *be a nontrivial t-design admitting a group* G *of automorphisms. If* G *is block-transitive then* $t \leq 7$, *while if* G *is flag-transitive then* $t \leq 6$.

Thus there are no nontrivial block-transitive 8-designs, and no non-trivial flag-transitive 7-designs. We conjecture that more is true (see also Kourovka Notebook, [14], Problem 11.45).

Conjecture 1.2 *There are no nontrivial block-transitive 6-designs.*

In Section 2 we identify (Proposition 2.3) a short list of groups as

103

the only possible candidates for automorphism groups of block-transitive 7-designs. We examine several of the groups on this list and conclude (Proposition 2.4) that such a design would have as automorphism group either an affine group over $GF(2)$ or a 2-dimensional projective linear group. In Section 3 we extend Alltop's investigation [1] of block-transitive 5-designs admitting affine groups. In particular we give a construction of a flag-transitive 5-$(256, 24, \lambda)$ design with automorphism group $AGL(8, 2)$ and with $\lambda = 2^{21}.3^2.5^2.7.31$. Finally in Section 4 we discuss flag-transitive 5-designs admitting $PSL(2, q)$.

Note that for positive integers $s < t$, any block-transitive t-design $\mathcal{D} = (X, \beta^G)$ is also a block-transitive s-design. Also the *complementary* design $\mathcal{D}^c = (X, (X \backslash \beta)^G)$ is a block-transitive t-design (if $k+t \leq v$, see [11], Exercise 1.39), but if $\mathcal{D}$ is flag-transitive then $\mathcal{D}^c$ is not necessarily flag-transitive.

2. Automorphism groups of block-transitive t-designs, t large

Let $\mathcal{D} = (X, B)$ be a t-(v, k, λ) design, where $t \geq 2$, and suppose that a subgroup G of $\mathrm{Aut}(\mathcal{D})$ is block-transitive on $\mathcal{D}$. The starting point for our investigation is the following theorem which can be deduced from a result of Ray-Chaudhuri and Wilson [16].

Theorem 2.1 (Ray-Chaudhuri and Wilson.) *Let $\mathcal{D} = (X, B)$ be a t-(v, k, λ) design with $t \geq 2$, and let G be a subgroup of $\mathrm{Aut}(\mathcal{D})$.*

(a) If G is block-transitive on $\mathcal{D}$ then G is $[t/2]$-homogeneous on points.

(b) If G is flag-transitive on $\mathcal{D}$ then G is $[(t + 1)/2]$-homogeneous on points.

Proof. (a) The theorem of Ray-Chaudhuri and Wilson [16] shows that the incidence matrix of $[t/2]$-sets of points against blocks has full rank. Now the argument of Block's Lemma gives the result of case (a).

(b) The derived design of $\mathcal{D}$ with point set $X \backslash \{x\}$ and block set $\{\beta \backslash \{x\} \mid \beta \in B, x \in \beta\}$ is a block-transitive $(t - 1)$-design. By part (a), G is transitive on X and, if $t \geq 3$, then G_x is $[(t - 1)/2]$-homogeneous on $X \backslash \{x\}$. Hence G is $[(t + 1)/2]$-homogeneous on X. $\square$

One consequence of the finite simple group classification is that the only 6-homogeneous permutation groups on X are the alternating and symmetric groups on X (see [5]), and as $\mathrm{Alt}(X)$ and $\mathrm{Sym}(X)$ are k-homogeneous for all $k \leq v - 2$, only trivial designs admit these groups. Thus an immediate corollary to Theorem 2.1 is that there are no nontrivial block-transitive 12-designs and no nontrivial flag-transitive 11-designs. By being more careful we can improve this observation considerably. The classification of all finite 3-homogeneous subgroups G of $\mathrm{Sym}(X)$ also follows from the finite simple group classification (see Livingstone and Wagner [13], Cameron [5] and Liebeck [12]); such a group G satisfies one of the following.

List 2.2 *(a) G is $Alt(X)$ or $Sym(X)$,*
(b) G is M_v with $|X| = v$ where $v \in \{11, 12, 22, 23, 24\}$, or $Aut M_{22}$ with $|X| = 22$,
(c) $G = M_{11}$ with $v = 12$,
(d) $PSL(2, q) \leq G \leq P\Gamma L(2, q)$ with $|X| = q + 1, q$ a prime power, $q \geq 5$.
(e) $G = AGL(d, 2)$ with $|X| = 2^d \geq 8$.

We note that not every group satisfying (d) is 3-homogeneous on X. First we prove a straightforward consequence of Theorem 2.1 using this classification.

Proposition 2.3 *Let $\mathcal{D} = (X, B)$ be a nontrivial t-(v, k, λ) design, and let G be a group of automorphisms of $\mathcal{D}$.*
(a) Suppose that G is block-transitive and $t \geq 6$. Then one of the following holds.
   (i) t is 10 or 11, and G is M_{12} or M_{24} with $v = 12$ or 24 respectively.
   (ii) t is 8 or 9, and G is one of the Mathieu groups in (b) of List 2.2 (with $v \neq 22$), or $G = P\Gamma L(2, 32)$ with $v = 33$.
   (iii) t is 6 or 7, and G is one of the groups in List 2.2.
(b) Suppose that G is flag-transitive and $t \geq 5$. Then one of the following holds.
   (i) t is 9 or 10, and G is M_{12} or M_{24} with $v = 12$ or 24 respectively.
   (ii) t is 7 or 8, and G is one of the Mathieu groups in (b) of List 2.2 (with $v \neq 22$), or
   (iii) t is 5 or 6, and G is one of the groups in List 2.2.

Proof. Since $\mathcal{D}$ is nontrivial, G is not $Alt(X)$ or $Sym(X)$ and $k \geq t + 1$. By Theorem 2.1 if G is block-transitive then G is $[t/2]$-transitive on X. If $t \geq 10$ then t is 10 or 11 and G is 5-homogeneous whence by Livingstone and Wagner [13] G is 5-transitive and so G is M_{24} with $v = |X| = 24$, or $G = M_{12}$ with $v = 12$. If t is 8 or 9 then G is 4-homogeneous and it follows, since $v > k \geq t + 1 \geq 9$, that either G is one of the Mathieu groups in (b) ($v \neq 22$) of List 2.2 or G is $P\Gamma L(2, 32)$ with $v = 33$. If t is 6 or 7 then G is 3-homogeneous and so is one of the groups in List 2.2. If G is flag-transitive then, by Theorem 2.1, G is $[(t+1)/2]$-transitive on X, and the result follows as above. (Note that when $t = 7, G$ is not $PSL(2, 8)$ or $P\Gamma L(2, 8)$ since these groups would be k-homogeneous.) $\square$

105

Now we shall examine the possible block-transitive actions of the Mathieu groups on nontrivial t-designs for large t.

Proposition 2.4 (a) Let $v = 22, 23,$ or 24. If the Mathieu group M_v or $\operatorname{Aut} M_v$ is block-transitive on a t-(v, k, λ) design $\mathcal{D} = (X, B)$ for some t, k, λ, then $t \leq v - 19 \leq 5$.

(b) Let $v = 11$ or 12. If the Mathieu group M_v is block-transitive on a t-(v, k, λ) design $\mathcal{D} = (X, B)$ for some t, k, λ, then $t \leq v - 7 \leq 5$.

(c) If $G = M_{11}$ is block-transitive on a t-$(12, k, \lambda)$ design $\mathcal{D} = (X, B)$ for some t, k, λ, then $t \leq 5$.

Proof. Let G be the Mathieu group M_v or $\operatorname{Aut} M_{22}$ and assume that G is block-transitive on a t-(v, k, λ) design $\mathcal{D} = (X, B)$. By replacing $\mathcal{D}$ by the complementary design if necessary, we may assume that $k \leq v/2$. The number $b = |B|$ of blocks satisfies $bk(k-1)\ldots(k-t+1) = v(v-1)\ldots(v-t+1)\lambda$, and b divides $|G|$ since G is block-transitive. If v is 22, 23 or 24 and $t \geq v - 18$, then 19 would divide b, and hence 19 would divide $|G|$ which is not the case. Similarly, if v is 11 or 12 and $t \geq v - 6$, then we deduce that 7 divides $|G|$ which again is not the case. Finally if $G = M_{11}, v = 12$ and G is block-transitive, then as above we may assume that $k \leq 6$. Then if $t \geq 6$ we deduce that 7 divides $|G|$ which is not the case. Hence $t \leq 5$. $\quad\square$

It is now a simple matter to complete the proof of Theorem 1.1. If $\mathcal{D} = (X, B)$ is a nontrivial t-(v, k, λ) design admitting G acting block-transitively with $t \geq 8$, or flag-transitively with $t \geq 7$, then by Propositions 2.3 and 2.4, $v = 33$ and $G = P\Gamma L(2, 32)$. As in the proof of Proposition 2.4, G is block-transitive on $\mathcal{D}$ and on its complementary design, both of which are t-designs and one of which has block size at most 16; we then deduce that 29 divides the number of blocks which leads to a contradiction since 29 does not divide $|G|$. Thus Theorem 1.1 is proved. $\quad\square$

Moreover it follows from Propositions 2.3 and 2.4 that a group G of automorphisms which is block-transitive on a nontrivial 6-design must satisfy (d) or (e) of List 2.2.

3. Designs in Affine Spaces

In this section we discuss t-designs $\mathcal{D} = (X, B)$ admitting $G = AGL(d, 2) < \operatorname{Sym}(X)$ acting block-transitively, where $v = |X| = 2^d \geq 8$, $t \geq 4$. Such designs have been studied by W.O. Alltop [1], see also [11]. We summarise some of his work below.

The group G is transitive on 3-element subsets and has two orbits on 4-element subsets of $X = AG(d, 2)$, a d-dimensional affine space over $GF(2)$. The orbits of G on 4-element subsets are the collection S_0 of all affine planes, and the collection S_1 of all 4-element subsets which generate a 3-dimensional affine space. (An s-element subset which generates an $(s-1)$-dimensional affine space is called an *independent* s-element subset.) Also G has two orbits on 5-element subsets of X, namely the set T_0 of all 5-element subsets which contain an affine plane, and the set T_1 of independent 5-element subsets. Moreover,

$$\frac{|S_1|}{|S_0|} = 2^d - 4, \quad \text{and} \quad \frac{|T_1|}{|T_0|} = \frac{2^d - 8}{5}.$$

Suppose now that $\mathcal{D} = (X, B)$ is a nontrivial t-$(2^d, k, \lambda)$ design admitting G, with $t = 4$ or $t = 5$. Then $B = \beta^G = \{\beta^g | g \in G\}$ for some k-element subset β of X where $t < k < 2^d - t$, and by replacing $\mathcal{D}$ by its complementary design if necessary we may assume that $k \leq 2^{d-1}$. Conversely if β is any k-element subset of X (where $k \leq 2^{d-1}$) and if σ_i, τ_i are the number of 4-element subsets, 5-element subsets of β which belong to S_i, T_i respectively, for $i = 0, 1$, then Alltop [1] showed that (X, β^G) is a 4-design if and only if

$$\sigma_1 = \sigma_0(2^d - 4),$$

a 5-design if and only if

$$\tau_1 = \tau_0(2^d - 8)/5,$$

and, surprisingly, (X, β^G) is a 5-design if and only if it is a 4-design, and the two equations above are each equivalent to

$$\sigma_0 = \binom{k}{4}/(2^d - 3). \tag{1}$$

One simple consequence of the fact that $2^d - 3$ must divide $\binom{k}{4}$ (with $k \leq 2^{d-1}$) is that $2^d - 3$ must be divisible by at least two distinct primes, and the smallest value of d for which this is true is $d = 8$. The divisibility condition for $d = 8$ implies that $k \in \{23, 24, 25, 46, 47, 69\}$, and Alltop [1] constructed an example of a 5-$(2^8, 24, \lambda)$ design admitting $AGL(8, 2)$ by this method, that is by finding a 24-element subset of $AG(8, 2)$ which contained exactly 42 affine planes.

It is possible to determine precisely the values of k and d for which a divisibility condition of the type above holds. In general there will be several infinite families of possibilities. A discussion of this in a similar context is given in Gamble [10] using the Chinese Remainder Theorem. However it is not at all clear how to construct k-element subsets containing a given number $\binom{k}{4}/(2^d - 3)$ of affine planes in general. Alltop [1] gives a heuristic argument to suggest that such sets may exist often.

We extend Alltop's work by constructing a 5-$(2^8, 24, \lambda)$ design on which $AGL(8, 2)$ is flag-transitive, not merely block-transitive. First we observe that flag-transitivity imposes extra restrictions on the parameters of a design; of course the complementary design of a flag-transitive design need not be flag-transitive so we can no longer assume that k is at most $v/2$.

Lemma 3.1 Suppose that $\mathcal{D} = (X, B)$ is a nontrivial 5-$(2^d, k, \lambda)$ design admitting a flag-transitive group of automorphisms $G = AGL(d, 2)$, where $X = AG(d, 2)$, and $d \geq 3$, for some k, λ.

(a) Then each block contains $\sigma_0 = \binom{k}{4}/(2^d - 3)$ affine planes (whence $2^d - 3$ divides $\binom{k}{4}$), and k divides $|G|$. In particular $d \geq 8$.

(b) For $d = 8$, k is 24 or 210.

Proof. Part (a) follows from our discussion above of Alltop's results together with the observations that the block stabilizer of a block β is transitive on β. Now suppose $d = 8$. The condition that $2^8 - 3$ divides $\binom{k}{4}$ implies that k or $2^8 - k$ lies in $\{23, 24, 25, 46, 47, 69\}$, and the condition that k divides $|AGL(8, 2)|$ further restricts k to be one of 24, 25, 210. Suppose that $k = 25$. Since the order of a Sylow 5-subgroup of G is 25 and since the stabilizer G_β of a block β is transitive on β, G_β must contain a Sylow 5-subgroup P of G. Now P has 9 orbits of length 25 in X and $N_G(P)$ is transitive on them. Thus each of these 9 orbits of P of length 25 is a block of $\mathcal{D}$. Each of these orbits is of the form $\beta = \{v + w | v \in O_1, w \in O_2\}$ where O_1 and O_2 are P-orbits of length 5, O_1 and O_2 span disjoint 4-dimensional vector subspaces, and any four points of O_i are independent, for $i = 1, 2$. The affine planes contained in β are the sets $\{v + w, v' + w, v + w', v' + w'\}$ where $v, v' \in O_1, w, w' \in O_2$, and β contains $\binom{5}{2}^2 = 100$ such planes. However, by part (a), if β is to be a block of a 5-design then β must contain $\sigma_0 = \binom{24}{4}/(2^8 - 3) = 50$ affine planes. Thus there is no such design with $k = 25$.

In a flag-transitive 5-design admitting $AGL(8, 2)$ each block must contain 42 affine planes if $k = 24$, and 321,220 planes if $k = 210$. We shall give a construction of such a design with block size 24. We doubt whether an example exists with block size 210. There were two "candidates" for such a 5-$(2^8, 210, \lambda)$ design, neither of which worked. These were the subgroups S_{10} and $GL(4, 2) \times GL(2, 2)$ of $GL(8, 2)$, each having orbit lengths 1, 45, 210 on V. (The first is contained in $Sp(8, 2)$. The second acts on V via its identification with $V(4, 2) \otimes V(2, 2)$, its orbits being the sets of tensors of ranks 0, 1, 2. In each case, it is not difficult to count the affine planes contained in the orbit of length 210.)

Question. Is there a flag-transitive 5-$(256, 210, \lambda)$ design with automorphism group $AGL(8, 2)$, for some λ?

As preparation for our construction of a flag-transitive 5-$(2^8, 24, \lambda)$ de-

sign with automorphism group $AGL(8,2)$, we present the design criterion of
Alltop [1], Lemma 3.1(a), in a different form.

In any 5-$(2^d, k, \lambda)$ design $\mathcal{D} = (X, B)$ with $X = AG(d, 2)$ there will
be many blocks containing the zero vector 0. For a k-element subset β of X
containing $0, \beta = \{x_1, \ldots, x_k = 0\}$, define the following subset of $V = \mathbb{Z}_2^k$:

$$C_0(\beta) = \{(f(x_1), \ldots, f(x_k)) | f \in X^*\}$$

where X^* is the dual of X (here X is viewed as a d-dimensional vector space
over GF(2)). Then $C_0(\beta)$ is a subspace of V of dimension d_0 equal to the
dimension of the linear span of β in X. Let $C(\beta)$ be the subspace of all
even weight vectors in $C_0(\beta)^\perp$, so $C_0(\beta)^\perp = C(\beta) \oplus \langle(0, \ldots, 0, 1)\rangle$. Then dim
$C(\beta) = k - d_0 - 1$. A justification for this seemingly artificial construction is
provided by the following lemma, in which, for a subset S of β, we denote by
χ_S the vector $(v_1, \ldots, v_k)$ of V with $v_i = 1$ if and only if $x_i \in S$. This lemma
provides a coding theory setting for deciding whether a given k-element subset
of X generates a 5-design. After proving it we give a partial converse which
may be used for construction of such designs.

Lemma 3.2 *(a) Let S be a subset of β. If $\chi_S \in C(\beta)$ then $\sum_{x \in S} x = 0$ in X.
Conversely if $\sum_{x \in S} x = 0$ in X then $\chi_S \in C(\beta)$ if $|S|$ is even, $\chi_{S\setminus\{0\}} \in C(\beta)$
if $|S|$ is odd and $0 \in S$, and $\chi_{S \cup \{0\}} \in C(\beta)$ if $|S|$ is odd and $0 \notin S$. Further
if $\chi_S \in C(\beta)$ and S is non-empty then $|S| \geq 4$, that is $C(\beta)$ has minimum
weight at least 4.*

*(b) The affine planes in β are the 4-element subsets S of β for which
$\chi_S \in C(\beta)$ (that is the supports of weight 4 vectors of $C(\beta)$). Moreover
(X, β^G) is a 5-design if and only if $C(\beta)$ contains exactly $\binom{k}{4}/(2^d - 3)$ weight
4 vectors.*

*(c) If (X, β^G) is a 5-design and G is flag-transitive then G_β induces a
transitive group of automorphisms of the code $C(\beta)$.*

Proof. (a) If $\chi_S \in C(\beta) \subseteq C_0(\beta)^\perp$, then for every $f \in X^*$ we have

$$0 = \sum_{x \in \beta} f(x)\chi_S(x) = \sum_{x \in S} f(x) = f(\sum_{x \in S} x),$$

whence $\sum_{x \in S} x = 0$ in X. Suppose then that $\sum_{x \in S} x = 0$. The computation
above shows that $\chi_S \in C_0(\beta)^\perp$. Thus, if $|S|$ is even, then $\chi_S \in C(\beta)$. Suppose
that $|S|$ is odd. If $0 \in S$ then $\sum_{x \in S\setminus\{0\}} x = 0$ and $|S\setminus\{0\}|$ is even, and we have
just shown that $\chi_{S\setminus\{0\}} \in C(\beta)$. Similarly, if $0 \notin S$ then $\sum_{x \in S \cup \{0\}} x = 0$ and
$|S \cup \{0\}|$ is even, and again $\chi_{S \cup \{0\}} \in C(\beta)$. If $\chi_S \in C(\beta), S$ nonempty, then
as $C(\beta)$ consists of even weight vectors, $|S| \geq 2$ is even. If $S = \{x, y\}$, then

109

as we have just shown, $x + y = 0$, that is, $x = y$, which is a contradiction. Hence $|S| \geq 4$.

(b) The affine planes in β are the 4-element subsets S of β with zero sum, $\sum_{x \in S} x = 0$. By part (a) these are precisely the supports of weight 4 vectors in $C(\beta)$. Finally, by Lemma 3.1(a), (X, β^G) is a 5-design if and only if there are $\binom{k}{4}/(2^d - 3)$ weight 4 vectors in $C(\beta)$.

(c) Suppose (X, β^G) is a 5-design and G is flag-transitive on it. Since $C(\beta) = C_0(\beta)^\perp \cap 1^\perp$, where 1 is the all-1 vector, we have $C(\beta)^\perp = \langle C_0(\beta), 1 \rangle$, and it suffices to show that this space is G-invariant. Clearly G fixes 1. Suppose that the element $g : x \mapsto xA + c$ of G induces the permutation π of $\{x_1, \ldots, x_k\}$. Then, for any $f \in X^*$, we have

$$(f(x_1\pi), \ldots, f(x_k\pi)) = (f(x_1 A + c), \ldots, f(x_k A + c))$$
$$= (f'(x_1), \ldots, f'(x_k)) + f(c)1,$$

where $f' \in X^*$ is defined by $f'(x) = f(xA)$. This vector belongs to $\langle C_0(\beta), 1 \rangle$, as required. $\square$

Now we prove a partial converse of Lemma 3.2. Lemma 3.2 proved that if (X, β^G) is a 5-design then $C(\beta)$ is a code consisting of even weight vectors with minimum weight 4, dimension $k - d_0 + 1$, and containing $\binom{k}{4}/(2^d - 3)$ weight 4 vectors, where d_0 is the dimension of the span of β in X. We prove the converse of this in the case where $d_0 = d$.

Lemma 3.3 *Let $V = \mathbb{Z}_2^k$, let T be the standard basis of V, and let W be the codimension 1 subspace of V consisting of all even weight vectors in V. Suppose that C is a subspace of W of dimension $k - d - 1$ having minimum weight 4 and containing exactly $\binom{k}{4}/(2^d - 3)$ weight 4 vectors. Then, denoting by $\bar{D}$ the image $(D + C)/C$ in V/C of a subset D of V, the set $\bar{V} \backslash \bar{W}$ has the structure of an affine space $AG(d, 2)$ and $\bar{T}$ is a k-element subset of $\bar{V} \backslash \bar{W}$ containing $\binom{k}{4}/(2^d - 3)$ affine planes, whence $\mathcal{D} = (\bar{V} \backslash \bar{W}, \bar{T}^{AGL(d,2)})$ is a 5-$(2^d, k, \lambda)$ design admitting $AGL(d, 2)$ as a block-transitive automorphism group for some λ. Further, the automorphism group Aut C of C induces a group of affine transformations of $\bar{V} \backslash \bar{W}$ which fixes $\bar{T}$ setwise. In particular if Aut C is transitive on components of vectors in V then $\mathcal{D}$ is a flag-transitive 5-design.*

Proof. Now $\bar{V}$ is a $(d+1)$-dimensional vector space with $\bar{W}$ a codimension 1 subspace. Hence $\bar{V} \backslash \bar{W}$ is an affine space $AG(d, 2)$. Moreover $\bar{T}$ is a subset of $\bar{V} \backslash \bar{W}$ of cardinality k, since any identification of elements of T modulo C would correspond to vectors of weight 2 in C. A 4-element subset of $\bar{T}$ is an affine plane in $\bar{V} \backslash \bar{W}$ if and only if it has zero sum modulo C, that is, if and only if it corresponds to a 4-element subset of T which is the support of a vector in C. Hence $\bar{T}$ contains $\binom{k}{4}/(2^d - 3)$ affine planes whence, by Lemma 3.1, $\mathcal{D}$ is a block-transitive 5-design admitting $G = AGL(d, 2)$.

The automorphism group A of the code C is a group of permutations of the components of vectors in V which leaves C invariant. Thus A acts as a group of linear transformations of V leaving W, C and T invariant. It follows that A induces a group of affine transformations of $\bar{V}\backslash\bar{W}$ which leaves $\bar{T}$ invariant. In particular if A is transitive on components then A is transitive on $\bar{T}$ and hence $\mathcal{D}$ is flag-transitive. □

We show that there is an example of a flag-transitive 5-design (X, β^G) when $d = 8$ and $k = 24$ by constructing a binary linear code of length 24, dimension 15, minimum weight 4, containing exactly 42 vectors of weight 4, and admitting a transitive automorphism group.

Construction 3.4 *Let C_0 be the extended binary Golay code of length 24, dimension 12 and minimum weight 8 (see [7], Chapter 11). It is easily checked that any coset of C_0 in $V = \mathbb{Z}_2^{24}$ which contains a weight 4 vector contains exactly six weight 4 vectors (forming a sextet). Let S be an octad, that is the support of some weight 8 vector in C_0; and let H be an extended Hamming code (see [7]) of length 8, dimension 4 and minimum weight 4 based on S, and embedded in V (by replacing each vector $h \in H$ by the vector in V which is zero off S and agrees with h on S). We claim that $C = H + C_0$ has dimension 15, minimum weight 4 and contains exactly 42 vectors of weight 4. Further H can be chosen in such a way that C has a transitive automorphism group.*

Proof. Since $\chi_S \in H \cap C_0$ it follows that $\dim(H + C_0)/C_0 = 3$, whence $C = H + C_0$ has dimension 15. Also C has minimum weight 4 and, since each nontrivial coset of $\langle \chi_S \rangle$ in H consists of two weight 4 vectors, each nontrivial coset of C_0 in C contains at least 1, and hence exactly 6, weight 4 vectors. Thus C contains exactly 42 weight 4 vectors. The automorphism group $A = AutC_0$ of C_0 is M_{24} and the setwise stabilizer A_S of S in M_{24} is $2^4.A_8$ (see [8]). Further, the stabilizer A_H of H in A_S is $2^4.AGL(3, 2)$. There are two conjugacy classes of subgroups $AGL(3, 2)$ in A_8 (corresponding to the stabilizer of a point or a hyperplane in $PG(3, 2)$) and hence essentially two different ways to choose the Hamming code H. If we choose a copy of H corresponding to the stabilizer of a hyperplane, then A_H stabilizes a partition P of the 24 components into 3 octads and an affine space (or Hamming code) on each of these octads. Moreover (see [8]) A_H is contained in the subgroup $A_C = 2^6 : (PSL(3, 2) \times S_3)$ permuting the 3 octads (and their Hamming codes) transitively; it is the stabilizer of an octad in A_C. The subgroup A_C preserves C and is transitive on the 24 components. □

Thus putting together this construction and the result of Lemma 3.3 we have as an immediate corollary:

Corollary 3.5 *There is a flag-transitive 5-$(256, 24, \lambda)$ design with automorphism group $AGL(8, 2)$.*

Since $G = AGL(8, 2)$ is a maximal subgroup of GA_{256} by [15], $AGL(8, 2)$ is the full automorphism group of the design. Moreover, since there are $|AGL(8, 2) : (2^6{:}(PSL(3, 2) \times S_3))|$ blocks in the design we constructed, it follows that the parameter $\lambda = 2^{21}.3^2.5^2.7.31$.

Finally we record the necessary and sufficient conditions for (X, β^G) to be a block-transitive 6-design admitting $G = AGL(d, 2)$ in terms of the 6-element subsets of β.

Proposition 3.6 *Let $G = AGL(d, 2)$ and $X = AG(d, 2)$ for $d \geq 5$.*

(a) The group G has four orbits on the set $\binom{X}{6}$ of 6-element subsets of X. Two of these are the sets U_0 and U_3 of all 6-element subsets γ which generate an affine space of dimension 3 and 5 respectively. The other two orbits are the sets U_1 and U_2 of all 6-element subsets which generate a 4-dimensional affine space and which contain no affine plane or a unique affine plane respectively. Further,

$$
\begin{aligned}
|U_0| &= 2^d(2^d - 1)(2^d - 2)(2^d - 4)/2^4.3, \\
|U_1| &= |U_0|(2^d - 8)/3.5, \\
|U_2| &= |U_0|(2^d - 8), \\
|U_3| &= |U_0|(2^d - 8)(2^d - 16)/2^{11}.3.7.31.
\end{aligned}
$$

(b) For a k-element subset β of X, (X, β^G) is a block-transitive 6-design admitting G if and only if the numbers u_i of 6-element subsets of β lying in U_i, for $i = 0, 1, 2, 3$, satisfy

$$
u_2 = 15u_1 = u_0(2^d - 8), u_3 = u_0(2^d - 8)(2^d - 16)/2^{11}.3.7.31.
$$

Equivalently (X, β^G) is a block-transitive 6-design admitting G if and only if (X, β^G) is a 4-design (that is the number σ_0 of affine planes in β is $\binom{k}{4}/(2^d - 3)$),

$$
u_0 = 15\binom{k}{6}/(2^d - 3)(2^d - 5) \tag{2}
$$

and

$$u_1 = u_0(2^d - 8)/15. \tag{3}$$

Proof. (a) Since G is transitive on independent ordered 6-tuples of points of X, U_3 is an orbit. Similarly since G is transitive on the set of 3-dimensional affine subspaces of X, and $AGL(3,2)$ is transitive on 6-element subsets of $AG(3,2)$, U_0 is an orbit. So consider the set of 6-element subsets γ which generate a 4-dimensional affine space. Consider γ as a subset of $AG(4,2) = \mathbb{Z}_2^4$. Since γ generates $AG(4,2)$ and since $AGL(4,2)$ is transitive on independent ordered 5-tuples from $AG(4,2)$ we may assume that γ contains $0, e_1 = (1,0,0,0), e_2 = (0,1,0,0), e_3 = (0,0,1,0)$ and $e_4 = (0,0,0,1)$. If γ contains no affine planes then $\gamma \cap \langle e_i, e_j, e_l \rangle = \{0, e_i, e_j, e_l\}$ for distinct i, j, l, whence $\gamma = \{0, e_1, e_2, e_3, e_4, (1,1,1,1)\}$. Hence U_1 is an orbit. Suppose then that γ contains an affine plane. Without loss of generality we may assume that γ contains $\delta = \{0, e_1, e_2, e_1 + e_2\}$, and as the stabilizer in $AGL(4,2)$ of δ is transitive on independent pairs from $AG(4,2) \backslash \delta$, U_2 is also an orbit. It is straightforward to compute the sizes of these orbits.

(b) By a result in the folklore of this subject (see [11] or [6], Proposition 1.3) (X, β^G) will be a 6-design if and only if the ratio of the numbers u_i of 6-element subsets of β lying in U_i to the cardinality of U_i is independent of i, that is if and only if $u_0/|U_0| = u_1/|U_1| = u_2/|U_2| = u_3/|U_3|$ which yields

$$u_1 = u_0(2^d - 8)/15, u_2 = u_0(2^d - 8) \text{ and } u_3 = u_0(2^d - 8)(2^d - 16)/2^{11}.3.7.31.$$

We obtain some equivalent necessary and sufficient conditions as follows. By Lemma 3.1, (X, β^G) is a 4-design if and only if the number σ_0 of affine planes in β is $\sigma_0 = \binom{k}{4}/(2^d - 3)$. Counting the number of pairs (γ, ν) where $\nu \subseteq \gamma \subseteq \beta$, ν is an affine plane, and $|\gamma| = 6$ we find that

$$\sigma_0 \binom{k-4}{2} = 3u_0 + u_2,$$

since each $\gamma \in U_2$ contains a unique affine plane and each 6-element subset of $AG(3,2)$ contains 3 affine planes. It follows that, if (X, β^G) is a 4-design, then $u_2 = u_0(2^d - 8)$ if and only if $u_0 = \sigma_0 \binom{k-4}{2}/(2^d - 5)$, that is,

$$u_0 = 15 \binom{k}{6}/(2^d - 3)(2^d - 5). \tag{4}$$

Further, by [1], (X, β^G) is a 4-design if and only if the number τ_1 of independent 5-element subsets of β is

$$\tau_1 = \sigma_0(k-4)(2^d - 8)/5 = \binom{k}{4}(k-4)(2^d - 8)/5(2^d - 3).$$

Counting the number of pairs (γ, ν) where $\nu \subseteq \gamma \subseteq \beta$, ν is an independent 5-element subset of γ, and $|\gamma| = 6$, we find that

$$\tau_1(k-5) = 6u_1 + 4u_2 + 6u_3 \tag{5}$$

since all 5-element subsets of 6-sets γ in U_1 or U_3 are independent, and for $\gamma \in U_2$ an independent 5-element subset of γ contains 3 of the 4 points of the unique affine plane in γ. Thus again, if (X, β^G) is a 4-design, and equation (2) holds then (using equation (5))

$$
\begin{aligned}
u_1 + u_3 &= \sigma_0 \binom{k-4}{2}(2^d - 8)(2^d - 15)/15(2^d - 5) \\
&= \binom{k}{6}(2^d - 8)(2^d - 15)/(2^d - 3)(2^d - 5). \tag{6}
\end{aligned}
$$

(We note that equations (2), (6), $u_2 = u_0(2^d - 8)$, and $\sigma_0 = \binom{k}{4}/(2^d - 3)$ imply that $u_0 + u_1 + u_2 + u_3 = \binom{k}{6}$, so there is no extra information there.) It follows that (X, β^G) is a 6-design if and only if it is a 4-design, equation (2) holds, and $u_1 = u_0(2^d - 8)/15 = \binom{k}{6}(2^d - 8)/(2^d - 3)(2^d - 5)$. $\square$

4. Designs in projective lines

The only groups from List 2.2 not yet discussed in detail are the projective groups G satisfying $PSL(2,q) \leq G \leq P\Gamma L(2,q)$ with G acting 3-homogeneously on the $q + 1$ points of the projective line $X = PG(1,q) = GF(q) \cup \{\infty\}$. Designs with point set X admitting such groups G have been investigated by Assmus and Mattson [3] and Alltop [2], and more recently by Laue and Schmalz (see [17]). Alltop constructed an infinite family of block-transitive 4-$(q+1, q/2, \lambda)$ designs admitting $G = PGL(2,q)$ in the case where $q = 2^{2a+1} \geq 32$, where $\lambda = (2^{2a} - 3)(2^{2a-1} - 1)$, namely (X, β^G) where β is a $2a$-dimensional subspace of $X \backslash \{\infty\}$ considered as a $(2a + 1)$-dimensional vector space over $GF(2)$.

We have not made much progress in this case in showing that no block-transitive 6-designs or flag-transitive 5-designs admitting G exist. We note that the cross-ratio gives a complete invariant for the set of orbits of $PGL(2,q)$ on ordered 4-tuples of points from X, that is (a, b, c, d) and (a', b', c', d') lie in the same orbit under $PGL(2,q)$ if and only if

$$\frac{(a-c)(b-d)}{(a-d)(b-c)} = \frac{(a'-c')(b'-d')}{(a'-d')(b'-c')}.$$

The cross-ratio takes all values in $GF(q)$ except 0 and 1. In general, the cross-ratio of four given points takes six different values depending on the order of the points. Such a 4-set lies in an orbit of size $(q + 1)q(q - 1)/4$. However, harmonic quadruples (cross-ratio -1, 2 or $\frac{1}{2}$), or equianharmonic quadruples (cross-ratio $-\omega$ or $-\omega^2$, where ω is a primitive cube root of unity), form orbits of size $(q+1)q(q-1)/8$ or $(q+1)q(q-1)/12$ respectively. Special care is needed in characteristic 2 (no harmonic quadruples) and 3 (harmonic quadruples form an orbit of length $(q + 1)q(q - 1)/24$; no equianharmonic quadruples). For example, suppose that $q \equiv -1 \pmod 6$ and β is a k-subset of $X = PG(1, q)$, $G = PGL(2, q)$. Then (X, β^G) is a 4-design if and only if β contains $3\binom{k}{4}/(q - 2)$ harmonic quadruples and $6\binom{k}{4}/(q - 2)$ quadruples from each "ordinary" orbit. The case $q = 11$, $\beta = \{0, 1, 3, 4, 5, 9\}$ gives a well-known example. However, the conditions for a 5-design or 6-design will be much more complicated.

In the other direction, we have the following result.

Proposition 4.1 *Let $\mathcal{D} = (X, \beta^G)$ be a block-transitive 6-$(q+1, k, \lambda)$ design admitting a group G where $PSL(2, q) \leq G \leq P\Gamma L(2, q)$ with $q = p^a \geq p, p$ a prime, and G is 3-homogeneous on $X = PG(1, q)$, for some k and λ. Then the following all hold, where $a_0 = |G|/(q + 1)q(q - 1)$ (so that $\frac{1}{2} \leq a_0 \leq a$):*

(a) The number b of blocks divides $(q+1)q(q-1)a_0$ and $b \geq (q+1)q(q-1)/6$.

(b) $q - 2$ divides $k(k - 1)(k - 2)(k - 3)a_0$.

(c) $(q - 2)(q - 3)$ divides $k(k - 1)(k - 2)(k - 3)(k - 4)a_0$.

(d) $(q - 2)(q - 3)(q - 4)$ divides $k(k - 1)(k - 2)(k - 3)(k - 4)(k - 5)a_0$.

Proof. The inequality for b comes from the theorem of Ray-Chaudhuri and Wilson [16], the divisibility from the block-transitivity. The remaining parts of the Proposition come from the familiar divisibility conditions for t-designs (see [11]). □

These divisibility conditions appear to have few solutions. The first two are $(q, k) = (59, 22)$ and $(67, 16)$.

Similar results hold for flag-transitive 5-designs:

Proposition 4.2 *Let $\mathcal{D} = (X, \beta^G)$ be a flag-transitive 5-$(q + 1, k, \lambda)$ design admitting a group G where $PSL(2, q) \leq G \leq P\Gamma L(2, q)$ with $q = p^a \geq p, p$ a prime, and G is 3-homogeneous on $X = PG(1, q)$, for some k and λ. Then the following all hold, where $a_0 = |G|/(q + 1)q(q - 1)$ (so that $\frac{1}{2} \leq a_0 \leq a$):*

(a) The number r of blocks containing a point divides $q(q - 1)a_0$ and $r > q(q - 1)/2$.

(b) k divides $q(q^2 - 1)a_0$.

(c) $q - 2$ divides $(k - 1)(k - 2)(k - 3)a_0$.

(d) $(q - 2)(q - 3)$ divides $(k - 1)(k - 2)(k - 3)(k - 4)a_0$.

Proof. The set of blocks containing a point $x \in X$ (with the point x removed) form the blocks of a 4-$(q, k - 1, \lambda')$ design $\mathcal{D}_x$ for some λ', and hence by the result of Ray-Chaudhuri and Wilson [16], the number of these blocks $r \geq q(q-1)/2$. This is the Petrenjuk bound, and by [4], [9], the only nontrivial 4-designs attaining this bound are the 4-$(23, 7, 1)$ design with automorphism group M_{23} and its complement. Suppose that $q = 23$ and $r = 23.11$. Then k must be 8 (rather than 17), $\mathcal{D}$ must be the 5-$(24, 8, 1)$ design with $\text{Aut}\mathcal{D} = M_{24}$, and $G = \text{Aut}\mathcal{D} \cap PGL(2, 23) = PSL(2, 23)$. However, see [8], $|G| = bk$, while an involution in G fixes a flag, and hence G is not flag-transitive. Hence $r > q(q - 1)/2$. Moreover since G is flag-transitive, $vr = bk$ divides $|G|$ which divides $(q+1)q(q-1)a_0$, where $b = |\beta^G|$ and $v = q+1$. This proves (a) and (b). Moreover, since $\mathcal{D}$ is both a 4-design and a 5-design, $\binom{v}{4}$ divides $b\binom{k}{4}$, and $\binom{v}{5}$ divides $b\binom{k}{5}$; using the fact that bk divides $(q+1)q(q-1)a_0$ we obtain (c) and (d). □

Again, the conditions of the Proposition have comparatively few solutions. The first few for which q is prime are $(q, k) = (17, 8)$, $(23, 8)$, $(23, 16)$, $(47, 12)$, and $(233, 24)$. (Note that, if q is prime, then $a = 1$ and the design is sharply flag-transitive.)

From the last two results, we can deduce the following:

Corollary 4.3 *No 6-design admits a group G satisfying $PSL(2, q) \leq G \leq P\Gamma L(2, q)$ acting flag-transitively.*

Proof. Such a design satisfies both $b \geq (q + 1)q(q - 1)/6$ and $b \leq (q + 1)q(q - 1)a/k$, where $q = p^a$, p prime. So $k \leq 6a \leq 6\log_2 q$. The divisibility conditions imply that $k \geq c.q^{1/2}$. The finite number of possibilities allowed by these conditions are easily checked by hand. □

As we noted in our previous paper [6], if (X, β^G) is a flag-transitive 5-design, then so also is $(X, \beta^{P\Gamma L(2,q)})$, so we assume first in our search for such designs, that $G = P\Gamma L(2, q)$, with $q = p^a$ as in Proposition 4.2. Let $x \in \beta$, let F be the flag (x, β), and let $H = PGL(2, q)$.

Lemma 4.4 *Let $G, H, \mathcal{D}$ be as above, and suppose that $H_F \neq 1$. Then one of the following holds.*

(a) $H_\beta \leq D_{2(q\pm1)}$, $H_F = Z_2$, and H has two orbits on flags. Moreover, either $H_\beta = D_{2k}$ is transitive on β, or $H_\beta = D_k$ has two orbits of length $k/2$ in β.

(b) $H_\beta = Z_p^c.Z_u \leq Z_p^a.Z_{q-1}$, $k = p^c$, H_β is transitive on β, $G_\beta = N_G(H_\beta)$, and H has $a/|G_\beta : H_\beta| = a/|HG_\beta : H|$ orbits on blocks. Also $H_F = Z_u, u$ divides $q - 1$, and $u < 2a/|G_\beta : H_\beta|$.

Proof. Suppose that p divides $|H_F|$, and let $g \in H_F$ have order p. Then x is the unique point of X fixed by g. Since G_β is transitive on β and H_β is normal in G_β, each characteristic subgroup of H_β has equal length orbits in β. It follows that $O_p(H_\beta) = 1$, and hence either H_β is contained in $D_{2(q\pm1)}$ with $p = 2$ or H_β is A_4 or S_4 with $p = 3$, or $H_\beta = A_5$ with $p = 2$ or 3, or $PSL(2, p^{a_0}) \leq H_\beta \leq PGL(2, p^{a_0})$ for some divisor a_0 of a. In all cases all subgroups of H_β of order p are conjugate in H_β, so each subgroup of H_β of order p fixes a unique point of β. Also, since H_β is normal in G_β and G_β is transitive on β, each point of β is fixed by some subgroup of H_β of order p. It follows that H_β is transitive on β, that is, H is flag-transitive. Hence vr divides $|H| = q(q^2 - 1) = vq(q - 1)$. By Proposition 4.2(a), $r > q(q-1)/2$ and it follows that $H_F = 1$, which is a contradiction. It follows both that p does not divide $|H_F|$ and that H is not flag-transitive. Suppose next that $PSL(2, p^{a_0})$ is normal in H_β for some divisor a_0 of a. Since p does not divide $|H_F|$, x does not lie in the orbit of H_β of length $p^{a_0} + 1$. It follows that $H_F = 1$ which contradicts our assumptions. Hence H_β is soluble.

Note that $H_F \leq H_x = Z_p^a.Z_{q-1}$, and as p does not divide $|H_F|$, $H_F \leq Z_{q-1}$, say $H_F = \langle h \rangle \simeq Z_u$, and h fixes x and one other point, say x' of X. Since H_β is normal in G_β and G_β is transitive on β, each point of β is fixed by some subgroup Z_u of H_β, conjugate to H_F by an element of G_β. Note that $\beta = x^{G_\beta}$. Further, H_F fixes δ points of x^{H_β}, where $\delta = |N_{H_\beta}(H_F) : H_F|$, and, as h fixes just two points of X, δ is 1 or 2. Suppose that H_F is normal in H_β. Then $|x^{H_\beta}| = |H_\beta : H_F| = \delta = 1$ or 2. If H_F is normal in G_β then $k = |\beta| = |x^{G_\beta}| \leq 2$, which is a contradiction. Hence H_F is not normal in G_β. Then $u = 2$ and the normal closure of H_F in G_β has order 4, and is contained in H_β; hence $H_\beta = Z_2 \times Z_2$, and as $k \geq 5, \beta$ contains x and x' and $k = 6$. But then $N_H(H_\beta) = A_4$ or S_4 fixes β setwise (as β is the set of points of X fixed by elements of H_β of order 2) which is a contradiction. Hence H_F is not normal in H_β.

Suppose next that $H_\beta \leq D_{2(q\pm1)}$. Then $u = |H_F| = 2$ and $H_\beta \simeq D_{2w}$ for some $w > 2, w$ dividing $q \pm 1$. Suppose first that $\delta = 2$. Then w is even, $x' \in x^{H_\beta} \subseteq \beta, w = |x^{H_\beta}|$, and there are two conjugacy classes in H_β of non-central involutions, each of size $w/2$. Since G_β is transitive on β, β consists of all the fixed points of just one or of both of these classes, that is either $k = w$ ($\beta = x^{H_\beta}$) or $k = 2w$ respectively. In the former case, since H is not flag-transitive, H has 2 orbits on blocks and $H_\beta = D_{2k}$. Then $G_\beta = N_G(H_\beta), H_\beta = N_H(H_\beta), H$ is block-transitive and $H_\beta = D_k$ has two orbits of length $w = k/2$ in β. Now suppose that $\delta = 1$. Then w is odd,

$w = |x^{H_\beta}|, x' \notin x^{H_\beta}$, and $N_H(H_\beta) = D_{4w}$ interchanges x^{H_β} and $(x')^{H_\beta}$. It follows that $x' \notin \beta$. Since $N_G(H_\beta)$ fixes $x^{H_\beta} \cup (x')^{H_\beta}$ setwise (the set of points fixed by involutions in H_β) it follows that G_β has index 2 in $N_G(H_\beta)$ and $\beta = x^{H_\beta}$. Also, there are two equal sized conjugacy classes in H of subgroups D_{2w}, w odd, and hence, since H is not flag-transitive, H has two equal sized orbits on blocks (and hence on flags).

If $H_\beta \leq Z_p^a.Z_{q-1} = H_{x'}$ then, since H_F is not normal in H_β, $O_p(H_\beta) \neq 1$ and $x' \notin \beta$. Since p does not divide $|H_F|, O_p(H_\beta)$ is semiregular on β and each subgroup Z_u of H_β fixes a unique point of β. It follows that k is equal to the number of subgroups of H_β of order u, that is $k = |H_\beta : H_F|$. So $\beta = x^{H_\beta}$. Moreover β is fixed setwise by $N_G(H_\beta)$, so $G_\beta = N_G(H_\beta)$ and H has $a/|G_\beta : H_\beta| = a/|HG_\beta : H|$ orbits on blocks. By Proposition 4.2(a), $u < 2a/|G_\beta : H_\beta|$.

The remaining possibilities for H_β are $H_\beta = A_4, S_4$ and A_5. Here u is 2,3,4 or 5, and in all cases if two cyclic subgroups of H_β are conjugate in G_β then they are already conjugate in H_β. Also in all cases $N_{H_\beta}(H_F) = D_{2u}$ so $\delta = 2, x' \in \beta$, and β is the set of fixed points of all the H_β conjugates of H_F. Thus $\beta = x^{H_\beta}$ and β is fixed setwise by $N_G(H_\beta)$, whence $G_\beta = N_G(H_\beta)$ and $H_\beta = N_H(H_\beta)$. It follows that H is flag-transitive, which is a contradiction.$\square$

References

[1] W. O. Alltop. 5-designs in affine spaces. *Pacific J. Math.*, 39, pp. 547–551, 1971.

[2] W. O. Alltop. An infinite class of 5-designs. *J. Combin. Theory*, 12, pp. 390–395, 1972.

[3] E. F. Assmus Jr. and H. F. Mattson Jr. New 5-designs. *J. Combin. Theory*, 6, pp. 122–151, 1969.

[4] A. Bremner. A diophantine equation arising from tight 4-designs. *Osaka J. Math.*, 16, pp. 353–356, 1979.

[5] P. J. Cameron. Finite permutation groups and finite simple groups. *Bull. London Math. Soc.*, 13, pp. 1–22, 1981.

[6] P. J. Cameron and C. E. Praeger. Block-transitive t-designs I: point-imprimitive designs. To appear in *Discrete Math.*

[7] P. J. Cameron and J. H. van Lint. *Designs, Graphs, Codes and their Links*, volume 22 of *London Math. Soc. Stud. Texts*. Cambridge Univ. Press, Cambridge, 1991.

[8] J. H. Conway, R. T. Curtis, S. P. Norton, R. A. Parker, and R. A. Wilson. *Atlas of Finite Groups*. Clarendon Press, Oxford, 1985.

[9] H. Enomoto, N. Ito, and R. Noda. Tight 4-designs. *Osaka J. Math.*, 16, pp. 39–43, 1979.

[10] **G. Gamble**. Block-transitive 3-designs with affine automorphism group. University of Western Australia Research Report, University of Western Australia, 1992.

[11] **D. R. Hughes and F. C. Piper**. *Design Theory*. Cambridge University Press, Cambridge, 1985.

[12] **M. W. Liebeck**. The affine permutation groups of rank 3. *Proc. London Math. Soc.*, 54, pp. 477–516, 1987.

[13] **D. Livingstone and A. Wagner**. Transitivity of finite permutation groups on unordered sets. *Math. Z.*, 90, pp. 393–403, 1965.

[14] **V. D. Mazurov and E. I. Khukro**. *Unsolved Problems in Group Theory. The Kourovka Notebook*. Russian Academy of Sciences, 1992.

[15] **B. Mortimer**. Permutation groups containing affine groups of the same degree. *J. London Math. Soc*, 15, pp. 445–455, 1977.

[16] **D. K. Ray-Chaudhuri and R. M. Wilson**. On t-designs. *Osaka J. Math.*, 12, pp. 737–744, 1975.

[17] **B. Schmalz**. t-Designs zu vorgegebener Automorphismengruppe. *Bayreuth. Math. Schr.*, 41, pp. 1–164, 1992.

P. J. Cameron, School of Mathematical Sciences, Queen Mary and Westfield College, Mile End Road, London E1 4NS, U.K.
e-mail: P.J.Cameron@qmw.ac.uk

C. E. Praeger, Department of Mathematics, University of Western Australia, Nedlands, W.A. 6009, Australia.
e-mail: praeger@madvax.maths.uwa.edu.au

Generalized Fischer spaces

H. Cuypers

Abstract

A generalized Fischer space is a partial linear space in which any two
intersecting lines generate a subspace isomorphic to an affine plane or
the dual of an affine plane. We give a classification of all finite and infinite generalized Fischer spaces under some nondegeneracy conditions.

1. Introduction

Let $\Pi = (P, L)$ be a *partial linear space,* that is a set of *points* P together
with a set L of subsets of P of cardinality at least 2 called *lines,* such that
each pair of points is in at most one line. A subset X of P is called a *subspace*
of Π if it has the property that any line meeting X in at least two points
is contained in X. A subspace X together with the lines contained in it is a
partial linear space. Subspaces are usualy identified with these partial linear
spaces. As the intersection of any collection of subspaces is again a subspace,
we can define for each subset X of P the subspace *generated* by X to be the
smallest subspace containing X. This subspace will be denoted by $\langle X \rangle$. A
plane is a subspace generated by two intersecting lines.

In [3], [6] partial linear spaces are considered in which all planes are
either isomorphic to an affine plane or the dual of an affine plane. Such spaces
are called *generalized Fischer spaces.* Since the affine plane of order 2 is not
generated by two intersecting lines, all lines of a generalized Fischer space
contained in some plane have at least 3 points. A generalized Fischer space
all of whose lines contain 3 points is called a *Fischer space.* Fischer spaces are
closely related to 3-transpositions. See [1], [3], [4], [7] and Section 3.

Let Π be a generalized Fischer space and suppose x and y are points
of Π. We say that x and y are *collinear* , notation $x \perp y$, if there is a line
containing them. If x and y are distinct collinear points, then the unique line
containing them will be denoted by xy. By $x^\perp$ we denote the set of all points
y with $x \perp y$ or $y = x$. The space Π is called $\perp$-*reduced* if $x^\perp = y^\perp$ implies
$x = y$. It is called $\not\perp$-*reduced* if $x^\perp \setminus \{x\} = y^\perp \setminus \{y\}$ implies $x = y$ and *reduced*
if it is both $\perp$-and $\not\perp$- reduced. We call Π *irreducible* if it is reduced and both
the graph $(P, \perp)$ and its complement are connected.

In [6] a complete classification of all finite irreducible generalized Fischer

121

spaces is given. Recently some new results on geometries and groups related to generalized Fischer spaces are obtained, without any finiteness assumptions. See [2], [4], [5], [9] and [10]. These results make it possible to classify all irreducible generalized Fischer spaces, finite and infinite:

Theorem 1.1 *Let* Π *be an irreducible generalized Fischer space, then* Π *is isomorphic to one of the following:*

1. *the geometry of hyperbolic lines in a nondegenerate symplectic space;*

2. *the geometry of 2-sets in a set* Ω;

3. *the geometry of elliptic lines in a nondegenerate orthogonal* $GF(2)$-*space;*

4. *the geometry of hyperbolic lines in a nondegenerate unitary* $GF(4)$-*space;*

5. *the geometry of tangent lines in a nondegenerate orthogonal* $GF(3)$-*space;*

6. *the geometry of tangent lines in a nondegenerate unitary* $GF(4)$-*space;*

7. *one of the Fischer spaces obtained from the sporadic Fischer groups* Fi_{22}, Fi_{23} *and* Fi_{24}.

In the next section we will give a brief description of the geometries occurring in the above theorem; a proof of that result is given in Section 3.

2. Examples

We give a brief description of the geometries appearing in our main result Theorem 1.1.

2.1 *The geometry of hyperbolic lines in a symplectic space.*

Let V be a vector space equipped with a symplectic form f. The 1-spaces of V are the points and the sets of points contained in a 2-space on which the form is nontrivial are the lines of our space.

2.2 *The geometry of 2-sets in some set* Ω.

Let Ω be a set, then the points of our space are the subsets of size 2 in Ω. A line is the triple of points contained in a subset of size 3 of Ω.

2.3 *The geometry of elliptic lines in a nondegenerate orthogonal* $GF(2)$-*space.*

Let V be a vector space over $GF(2)$ and suppose Q is a nondegenerate quadratic form on Q. Points are the 1-spaces of V on which Q is nontrivial, lines are the sets of points in a 2-space on which Q induces an elliptic form.

2.4 *The geometry of tangent lines in a nondegenerate orthogonal $GF(3)$-space.*

Suppose V is a $GF(3)$-space and Q a nondegenerate quadratic form on V. The points are the 1-spaces $\langle v \rangle$ of V with $Q(v) = 1$, the lines are the sets of 3 points in the 2-spaces of V on which Q has a unique 1-dimensional radical.

2.5 *The geometry of hyperbolic lines in a nondegenerate unitary $GF(4)$-space.*

Suppose V is a vector space over $GF(4)$, and h is a nondegenerate hermitian form on V. The points are the 1-spaces of V on which h vanishes, a line is the set of points in a 2-space on which h is nondegenerate.

2.6 *The geometry of tangent lines in a nondegenerate unitary $GF(4)$-space.*

Let V and h as above. Now we take as points the 1-spaces of V on which h does not vanish, and as lines the sets of points contained in a 2-space on which h has a 1-dimensional radical.

2.7 *The Fischer spaces related to the sporadic Fischer groups Fi_i, where $i = 22, 23, or 24$.*

The sporadic Fischer groups contain a unique conjugacy class D of 3-transpositions. Take the elements of D as points and the subsets of D of size 3 contained in a subgroup $\langle d, e \rangle$, where d and e are noncommuting elements in D, as lines.

It is not hard to check that, under some mild conditions on the dimension, the spaces described above are indeed irreducible generalized Fischer spaces.

3. Proof of the Theorem

In this section we give a proof of the main theorem of this paper.

Suppose Π is a connected generalized Fischer space. It follows from [3] that all lines in Π have the same number of points. First we consider the case where all lines contain 3 points. As already observed in the introduction, there is a correspondence between Fischer spaces and classes of 3-transpositions. We explain this connection.

Let $\Pi = (P, L)$ be a connected Fischer space, then to every point p of Π we can attach an involution t_p in the automorphism group of Π that fixes p, interchanges the points on every line through p that are different from p and fixes all the other points. Then the set $\{t_p \mid p \in P\}$ is a set of 3-transpositions in the subgroup of the automorphism group of Π generated by these involutions. Since Π is connected, these involutions form a conjugacy class of 3-transpositions. The subgroup of $\mathrm{Aut}(\Pi)$ generated by these involutions will be denoted by $G(\Pi)$.

On the other hand, if D is a conjugacy class of 3-transpositions, then let L be the set of all triples $D \cap \langle d, e \rangle$ where d, e are noncommuting elements of D. Then (D, L) is a connected generalized Fischer space. Up to a center the groups $\langle D \rangle$ and $G(D, L)$ are isomorphic.

Proposition 3.1 *Let Π be an irreducible Fischer space, then $G(\Pi)$ contains no nontrivial normal solvable subgroup.*

Proof. This is 2.2 of [4]. □

Proposition 3.2 *Let Π be an irreducible Fischer space, then Π is one of the spaces of the conclusion of Theorem 1.1.*

Proof. By the above, Π is one of the spaces obtained from a conjugacy class of 3-transpositions generating a group with no nontrivial normal solvable subgroup. Hence we can apply Theorem 1.1 of [4], and we find that Π is one of the spaces appearing in the conclusion of Theorem 1.1, or $G(\Pi)$ is one of the two triality groups $P\Omega_8^+(2) : \Sigma_3$ or $P\Omega_8^+(3) : \Sigma_3$. However, in these last two cases the complement of the graph $(P, \perp)$ is not connected. □

This handles the case where all lines contain 3 points, and from now on we can assume that all lines contain at least 4 points. So suppose that Π is an irreducible generalized Fischer space with all lines containing at least 4 points.

Proposition 3.3 *Suppose Π contains only dual affine planes. Then Π is isomorphic to the geometry of hyperbolic lines in a nondegenerate symplectic space.*

Proof. Since all planes of Π are dual affine, we can apply the results of [2], see also [8]. Hence it follows that Π is the geometry of hyperbolic lines in

CUYPERS: GENERALIZED FISCHER SPACES

some symplectic space. Since Π is $\not\!\perp$-reduced this space is nondegenerate. $\square$

It remains to consider the case where there are both affine and dual affine planes in Π. (If Π contains only affine planes, it is not $\perp$-reduced.)

Thus assume that both types of planes occur in Π. Then the main result of [6] reads as follows.

Lemma 3.4 *[6] Let π be an affine plane in Π, and x a point not in π. If x is not collinear to some point in π, then there is a line in π all of whose points are not collinear with x.*

This lemma has the following important consequence.

Proposition 3.5 *[6] Let x be a point in Π. Then the lines and affine planes on x form a nondegenerate polar space.*

The following lemma is concerned with dual affine planes.

Lemma 3.6 *Let π be a dual affine plane and x a point not in π. Then x is at least not collinear with all the points of some line of π or with all but one of the points on some transversal coclique of π.*

Proof. Suppose x is collinear with some point y in the plane π. If there is at most one line on y in π completely contained in $x^{\perp}$ then x is noncollinear with all or all but one of the points on a transversal coclique in π. Thus assume that there are at least two lines l and m of π on y in $x^{\perp}$. Then $\langle x, l \rangle$ is an affine plane and all lines on y in π are in $x^{\perp}$, see [3]. Let z be a point on m different from y. Then by the above lemma there is a line in $\langle x, l \rangle$ consisting of points not collinear to z. This line has to be parallel to the line through x and y. Now suppose t is a point in the transversal coclique of π on y and let n be a line on t in π. Then n meets l and m in points u and v respectively, different from y. As the line through x and u meets every line parallel to xy, it contains a point not collinear to v. Hence $\langle x, n \rangle$ is dual affine and t is the unique point not in $x^{\perp}$. Thus x is not collinear to all points of the transversal coclique of π on y except for y. $\square$

Lemma 3.7 *If two dual affine planes π_1 and π_2 meet in two noncollinear points, they meet in at least all but two points of a transversal coclique. Moreover, if π_1 contains a point that is not collinear with all points of a transversal coclique of π_2 then they meet in all points of a transversal coclique.*

Proof. Suppose π_1 and π_2 are two dual affine planes meeting in two non-

125

collinear points x and y. Choose points $z_i \in \pi_i$, $i = 1, 2$, such that $x \perp z_i \perp y$ and $z_1 \perp z_2$. By the above lemma it is possible to choose z_1 and z_2 in such a way that both planes $\langle z_1, x, z_2 \rangle$ and $\langle z_1, y, z_2 \rangle$ are dual affine. Now let u be the point on xz_1 not collinear with z_2 and w the point on z_1z_2 not collinear to y. Let v be the intersection point of uw with xz_2. By the previous lemma all but one of the lines in π_2 through v contain a point not collinear to u and thus generate together with u a dual affine plane. Let l be such a line not on y. Then l meets yz_2 in a point t say, and $t \perp w$. The line tw meets yz_1 in some point s. If $s \perp u$ then l and us meet in a point of $\pi_1 \cap \pi_2$. Thus at least all but one of the lines l give rise to an intersection point of π_1 and π_2, different lines giving different points. This implies that $\pi_1 \cap \pi_2$ contains at least all but two points of the transversal coclique of π_i on x, $i = 1, 2$.

If there is a point in π_1 that is not collinear with all points of a transversal coclique of π_2, then the proof of Lemma 3.5 and 3.6 of [2] applies. This proves the lemma. □

Lemma 3.8 *Lines contain at most 5 points.*

Proof. Suppose lines contain at least 6 points. Let π be an affine plane and x a point collinear with some but not all points of π. Fix a line l in π containing a unique point y not collinear with x. Let m be the line in π containing the points not collinear with x. Now let z be a point collinear with all points on a line of π through y but different from l. This is possible since the lines and affine planes on y form a nondegenerate polar space. Then z is not collinear with a point on l different from y and a point u on m different from y. Thus z is collinear with at least all but one of the points of the transversal coclique on x in the plane $\langle x, l \rangle$. Let v be a point in that coclique collinear with z. Then the plane $\langle z, y, v \rangle$ meets the transversal coclique in at least 3 points, two of them not collinear with u. But z and u are also not collinear, which implies that u is noncollinear with all points in the plane $\langle z, y, v \rangle$ contradicting $u \perp y$. □

Proposition 3.9 *The space Π is isomorphic to the geometry of tangent lines in some nondegenerate unitary $GF(4)$-space.*

Proof. The space Π satisfies the conditions of Theorem 1.2 of [5]. (Notice that the collinearity graph of Π has diameter 2, cf. [3]) If there is a point x in Π such that the lines and affine planes on x form a polar space of rank at least 3, then the proposition follows easily from Theorem 1.3 of [5] and the fact that lines contain 4 or 5 points.

Thus assume that the polar space of lines and planes on a given point has rank 2. In view of the results of [3] we only have to show that Π is finite, and since the diameter of the collinearity graph of Π is 2 it suffices to prove that the generalized quadrangle of lines and planes on a point of Π is finite.

Fix a point x and suppose the generalized quadrangle Q_x of lines and affine planes on x is infinite. Since all lines in Q_x contain $s+1$ points with $s = 4$ or 5, the number of lines through a point of the generalized quadrangle Q_x is constant and thus infinite.

We obtain a contradiction in a number of steps.

Let D be the set of lines on x in some fixed dual affine plane π on x.

Step 1. If a point of Q_x is collinear with 2 points of D it is collinear with all points in D.

Proof. See [3] $\qquad\qquad\qquad\qquad\qquad\qquad\qquad\qquad\qquad\qquad\qquad\qquad$ $\square$

Step 2. Let K be a line in Q_x missing D, then K contains at least 2 points not collinear with any point in D.

Proof. Every point in D is collinear with one point on K, every point on K with 0,1 or all points of D. Hence there are at least $|K| - |D| = 2$ points on K not collinear with any point of D. $\qquad\qquad\qquad\qquad\qquad\qquad$ $\square$

Let p be a point in Q_x collinear with all points of D.

Step 3. There is a dual affine plane π' on x and a point x' in Π, which is not collinear with precisely all the points of a transversal coclique in π', such that the lines in π' on x and xx' are collinear with p in Q_x.

Proof. Since there are infinitely many lines on p not meeting D, there are infinitely many pairwise noncollinear points in Q_x collinear with p that are not collinear with any point of D. This implies that there is a point x' in Π collinear with x such that xx' and p are collinear and that is not collinear with all the points of a transversal coclique in π and we are done, or there are two noncollinear points y and z in Π both collinear with x that are not collinear with some fixed point v on a line in π through x, but still xz and xy being collinear in Q_x with p. Then take $x' = v$ and $\pi' = \langle x, y, z \rangle$. $\qquad$ $\square$

Without loss we can now assume that there is a point y collinear with x that is not collinear with all the points of a transversal coclique in π. Let π' be the plane generated by y and two points of a transversal T of π contained in $y^\perp$. Then by Lemma 3.7 the planes π and π' meet in all points of the transversal. But that means that the lines and dual affine planes on x meeting π' in a point respectively a transversal coclique or line but not in the transversal of π' having a point not in $x^\perp$ form an affine plane. Denote this plane by A.

Step 4. Lines in Q_x contain 5 points.

Proof. Suppose lines in Q_x contain 6 points. Let K be a line through p in Q_x meeting A in a point and fix a point p' on K different from p and not in A. Then p' is collinear with a unique point in A. Suppose M is a line on p' not meeting A. Since there are infinitely many lines on p', M exists. Each point of A is collinear with one point on M. Each point on M with 0, 1, 4 or 16 points of A. But then 4 points on M are collinear with 4 points in A and 2 points in M with no point in A. Hence p' is collinear with 4 points in A. A contradiction.   □

Step 5. A Contradiction.

Proof. Let M be a line in Q_x not meeting A. Then every point of A is collinear with one point in M and every point in M with 0, 1, 3 or 9 in A. Thus if M contains a point collinear with only one point in A then the other points in M are also collinear to some point in A, and there is a second point on M collinear with just one point in A.

As in the previous step let p' be a point on a line through p meeting A, which is different from p, and not in A. So p' is collinear with just one point in A. Let M be a line through p' not meeting A and suppose q is a second point on M collinear with just one point in A. Let N be a line on p not meeting A. Then all points on N except p are not collinear with any point in A. But q is collinear with a point r in N different from p. Now the line qr contains a point with no neighbors in A and a point with just one neighbor in A, which contradicts the above.   □

Since Π is finite we can apply the results from [3], [6] and the proposition is proved.   □

The main theorem follows now from the Propositions 3.2, 3.3 and 3.9.

References

[1] **F. Buekenhout**. La géométrie des groupes de Fischer, 1974. Unpublished notes.

[2] **H. Cuypers**. Symplectic geometries, transvection groups and modules. To appear in *J. Combin. Theory Ser. A*.

[3] **H. Cuypers**. On a generalization of Fischer spaces. *Geom. Dedicata*, 34, pp. 67–87, 1990.

[4] **H. Cuypers and J. I. Hall**. The classification of 3-transposition groups with trivial center. In M. Liebeck and J. Saxl, editors, *Groups, Combinatorics and Geometry, Proceedings of the L.M.S. Durham Symposium on Groups and Combinatorics, 1990*, volume 165 of *London Math. Soc. Lecture Note Series*, pages 121–138. Cambridge University Press, 1992.

[5] **H. Cuypers and A. Pasini**. Locally polar geometries with affine planes. *European J. Combin.*, 13, pp. 39–57, 1992.

[6] **H. Cuypers and E. E. Shult**. On the classification of generalized Fischer spaces. *Geom. Dedicata*, 34, pp. 89–99, 1990.

[7] **B. Fischer**. Groups generated by 3-transpositions I. *Invent. Math.*, 13, pp. 232–246, 1971. See also University of Warwick Lecture Notes.

[8] **J. I. Hall**. Hyperbolic lines in symplectic spaces. *J. Combin. Theory Ser. A*, 47, pp. 284–298, 1988.

[9] **J. I. Hall**. Graphs, geometry, 3-transpositions, and symplectic F_2-transvection groups. *Proc. London Math. Soc. Ser. 3*, 58, pp. 112–136, 1989.

[10] **P. Johnson**. Polar spaces of arbitrary rank. *Geom. Dedicata*, 35, pp. 229–250, 1990.

H. Cuypers, Department of Mathematics, Eindhoven University of Technology, P.O.Box 513, 5600 MB, Eindhoven, The Netherlands.
e-mail: hansc@win.tue.nl

Ovoids and windows in finite generalized hexagons

V. De Smet   H. Van Maldeghem [*]

Abstract

We characterize some finite Moufang hexagons as the only generalized hexagons containing "a lot of" thick ideal subhexagons or as the only hexagons containing ovoids all of whose points are regular.

1. Introduction

A generalized hexagon of order $(s,t), s,t \geq 1$ is a $1 - (v, s+1, t+1)$ design $S = (\mathcal{P}, \mathcal{B}, I)$ whose incidence graph has girth 12 and diameter 6, also denoted by $S(s,t)$. If $s = t$, S is said to have order s. The only known finite generalized hexagons with $s, t > 1$ arise from the Chevalley groups $G_2(q)$ and $^3D_4(q)$ and have respective order (q,q) and (q, q^3), q power of a prime. We denote the $^3D_4(q)$-hexagon by $H(q, q^3)$, its dual by $H(q^3, q)$ and we denote the $G_2(q)$-hexagon by $H(q)$, its dual by $H^*(q)$. An explicit description of these is given in Kantor [2].

Note that $H^*(q)$ is always a subhexagon of $H(q, q^3)$; dually $H(q)$ is a subhexagon of $H(q^3, q)$. A subhexagon S' of order (s', t') is called *ideal* if $t = t'$ (see Ronan [4]). Furthermore, S is called *thick* if $s, t > 1$. Note that $s = 1$ or $t = 1$ corresponds to the incidence graph of a projective plane. With these definitions, $H(q)$ is a thick ideal subhexagon of $H(q^3, q)$. Now consider the following configuration in a generalized hexagon S. Let L_1 and L_2 be two lines at distance 6 (in the incidence graph) from each other and let p_1, p_2, p_3 be three distinct points on L_1. There are points p'_1, p'_2, p'_3 on L_2 at distance 4 from resp. p_1, p_2, p_3 and there are unique chains $p_i I M_i I p''_i I M'_i I p'_i$, $i = 1, 2, 3$. The configuration consisting of the lines L_1, L_2, M_i, M'_i, $i = 1, 2, 3$ and the points p_i, p'_i, p''_i, $i = 1, 2, 3$, is called a *window* of S. By the transitivity of the collineation group of $H(q^3, q)$ there is a subhexagon isomorphic to $H(q)$ containing any given window. It is our aim to show the contrary, namely that if every window of a thick generalized hexagon S is contained in an ideal subhexagon, then S is isomorphic to $H(q^3, q)$.

Let $d(x, y)$ denote the distance between x and y in the incidence graph

---

[*]. Research Associate at the National Fund for Scientific Research (Belgium)

of a generalized hexagon $S = (\mathcal{P}, \mathcal{B}, I)$, $x, y \in \mathcal{P} \cup \mathcal{B}$. Denote by $\Gamma_i(x)$, with $x \in \mathcal{P} \cup \mathcal{B}$, the set of all elements of $\mathcal{P} \cup \mathcal{B}$ at distance i from x. If $d(x, y) = 4$ for $x, y \in \mathcal{P}$, then there is a unique point z collinear to both and we denote $z = x * y$. Define $\mathcal{W}(x, y) = \Gamma_6(z) \cap \Gamma_4(x) \cap \Gamma_4(y)$. If $u \in \mathcal{W}(x, y)$ then we denote by z^u the set of points collinear with z and at distance 4 from u. If this set is independent from the choice of u, then z^u is called an *ideal line* (see Ronan [4]) and is denoted by $\langle x, y \rangle$. Now fix a point $p \in \mathcal{P}$ and suppose that $\langle x, y \rangle$ is an ideal line for every pair $(x, y) \in \mathcal{P}^2$ such that $d(x, p) = d(y, p) = 2$ and $d(x, y) = 4$, then we call p *half-regular*. If moreover $\langle p, z \rangle$ is an ideal line for every point z at distance 4 from p, then we call p *regular*. This is motivated by the facts that (1) if all points of S are regular and S has order s, then $S \cong H(q)$, see Ronan [4], (2) a derivation can be defined in a regular point of S and if $s = t$, then this is a generalized quadrangle (see Van Maldeghem - Bloemen [7]). These properties are very similar to properties of generalized quadrangles with regular points (see Payne - Thas [3]). In fact, for generalized quadrangles of order s, one can show that, if every point of an ovoid is regular, then the generalized quadrangle is classical and arises from a Chevalley group $S_4(2^e)$. In this paper, we extend this property to generalized hexagons, an ovoid of a generalized hexagon of order s being a set of $s^3 + 1$ points at distance 6 from each other. There is one difference though: the existence of ovoids in $H(q)$ is only proved for $q = 3^e$. For q even, there are no ovoids (see e.g. Thas [6]) and for other values of q, the question remains open.

Also our characterization of $H(q^3, q)$ has an analogue for generalized quadrangles, (see Payne - Thas [3], 5.3.5. ii, dual), a window in a generalized quadrangle being a quadrilateral with one more "transversal".

2. Proof of the results

2.1. Characterization by windows

Lemma 2.1 *Let $S(s, t)$ be a finite generalized hexagon which contains a proper subhexagon $S'(s', t)$, which in turn contains a proper subhexagon $S''(s'', t)$. Then $s = t^3$, $s' = t$ and $s'' = 1$.*

Proof. From Haemers and Roos [1] it follows that $s \leq t^3$ (1).
From Thas [5] we have $s \geq s'^2 t$ (2) and $s' \geq s''^2 t$ (3).
So (1) and (2) gives $s'^2 \leq t^2$ or $s' \leq t$ (4).
Now (3) and (4) gives $s'' = 1$, and so $s' = t$.
From (1) and (2) it then follows that $s = t^3$. $\qquad\square$

Theorem 2.2 *Let* $S = (\mathcal{P}, \mathcal{B}, I)$ *be a finite generalized hexagon of order* (s, t) *with* $s \geq 3$.
There exists a proper ideal subhexagon through every window of S *iff* S *is isomorphic to* $H(q^3, q)$, $s = q^3$ *and* $t = q$.

Proof.

$\Leftarrow$ See introduction.

$\Rightarrow$ In order to proof that S is Moufang we have to proof that S has ideal lines. [4] So we must proof that for all $a, b \in \mathcal{P}$ with $d(a, b) = 4$ and $a * b = c$ we have $\langle a, b \rangle = c^z$ for all $z \in \mathcal{W}(a, b)$.

Step 1: We show that $c^z = c^{z'}$ for all $z, z' \in \mathcal{W}(a, b)$ such that c, z and z' form a window with the same two lines , say L and M. Suppose z_1 and z_2 are such elements of $\mathcal{W}(a, b)$.

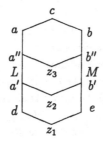

Figure 1.

Let $S_{12}(c, z_1, z_2)$ be the proper ideal subhexagon through the window c, z_1, z_2, L and M. Since $s \geq 3$ and S_{12} is proper, there exists another point b'' on M, with $b'' \notin S_{12}$. The shortest path between b'' and L gives rise to the point $z_3 \in \mathcal{W}(a, b)$ (see figure 1). Let $S_{13}(c, z_1, z_3)$ be the proper ideal subhexagon through the window c, z_1, z_3, L and M. Remark that $z_2 \notin S_{13}$. Finally, let $S_{23}(c, z_2, z_3)$ be the proper ideal subhexagon through the window c, z_2, z_3, L and M. Note that $z_1 \notin S_{23}$. We will now look at some intersections of those subhexagons. Let $\mathcal{D}_2 = S_{12}(c, z_1, z_2) \cap S_{23}(c, z_2, z_3)$, then $\mathcal{D}_2$ is a proper ($z_1 \notin S_{23}$ and $z_3 \notin S_{12}$) ideal subhexagon of S_{12} and S_{23}. Let $\mathcal{D}_3 = S_{13}(c, z_1, z_3) \cap S_{23}(c, z_2, z_3)$, then $\mathcal{D}_3$ is a proper ($z_1 \notin S_{23}$ and $z_2 \notin S_{13}$) ideal subhexagon of S_{13} and S_{23}. If we apply the lemma to

$$S(s,t) \supset S_{12} \supset \mathcal{D}_2,$$
$$S(s,t) \supset S_{13} \supset \mathcal{D}_3$$
$$\text{and } S(s,t) \supset S_{23} \supset \mathcal{D}_3$$

we have that S has order (t^3, t), S_{12}, S_{13} and S_{23} have order (t, t) and $\mathcal{D}_2$ and $\mathcal{D}_3$ are thin ideal subhexagons. Now we can apply a corollary of the theorem of Thas [5] to the following pairs of generalized hexagons:

(1) $S_{12} \supset \mathcal{D}_2$.

$z_1 \in S_{12} \backslash \mathcal{D}_2$ and not collinear with a point of $\mathcal{D}_2$,

so z_1 is at distance 3 from $1 + t$ lines of $\mathcal{D}_2 \subset S_{23}$.

(2) $S_{13} \supset \mathcal{D}_3$.

$z_1 \in S_{13} \backslash \mathcal{D}_3$ and not collinear with a point of $\mathcal{D}_3$,

so z_1 is at distance 3 from $1 + t$ lines of $\mathcal{D}_3 \subset S_{23}$.

(3) $S \supset S_{23}$.

$z_1 \in S \backslash S_{23}$ and not collinear with a point of S_{23},

so z_1 is at distance 3 from $1 + t$ lines of S_{23}.

From (1), (2) and (3) it follows that $\mathcal{D}_2$ and $\mathcal{D}_3$ have $1 + t$ lines in common which are at distance 3 from z_1.

Case 1: Suppose that all those $1 + t$ lines are at distance 3 from c. From the thinness of $\mathcal{D}_2$ and $\mathcal{D}_3$ it follows that $c^{z_1} = c^{z_2} = c^{z_3}$.

Case 2: Suppose at least one of those $1 + t$ lines is at distance 5 from c, say L_2. Let $(c, L_0, l_0, L_1, l_1, L_2)$ denote the shortest path between the line L_2 and c. Since c and $L_2 \in \mathcal{D}_2$ ($\mathcal{D}_3$) it follows that $L_0, l_0, L_1, l_1 \in \mathcal{D}_2$ ($\mathcal{D}_3$). From $\mathcal{D}_2$ it then follows that $d(z_2, l_0) = 4$ and $d(z_2, l_1) = 6$. So there is a second point on L_2 at distance 4 from z_2. We deduce $d(a', l_1) = 4$. But l_1 and L lie in $\mathcal{D}_3$ so the shortest path between them is also in $\mathcal{D}_3$, a contradiction. So case 2 cannot occur and case 1 proves step 1.

Step 2:

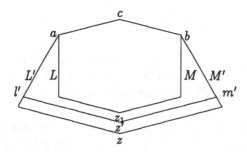

Figure 2.

Suppose $z \in \mathcal{W}(a, b)$ so that c, z_1, z, L and M do not form a window. There exists a thin ideal subhexagon $\mathcal{D}$ through c, z_1, L and M (see step 1). So L' and $M' \in \mathcal{D}$. Let l' and m' be the respective second points on L' and M' in $\mathcal{D}$ and denote $l' * m' = z'$. Then from the thinness of $\mathcal{D}$ we have $c^{z'} = c_1^z$. But applying step 1 we obtain $c^{z'} = c^z$. $\qquad \Box$

2.2. Characterization by subhexagons

Lemma 2.3 *Let $S = (\mathcal{P}, \mathcal{B}, I)$ be a finite generalized hexagon of order (s, t). Through every 2 opposite, half-regular points there exists exactly one thin ideal subhexagon.*

Proof. Let p_1 and p_2 be two opposite, half-regular points of S. Since they are opposite, we can take all lines through p_1 and consider the unique $t + 1$ shortest paths of length 4 between p_2 and those $t + 1$ lines. So we get in an unique way, $t + 1$ points collinear with p_1 and $t + 1$ points collinear with p_2. Call them respectively $x_0, \ldots, x_t$ and $y_0, \ldots, y_t$ with $x_i \sim y_i$, $i = 0, \ldots, t$.

Because $d(x_i, y_{i+1}) = 6$, $i = 0, \ldots, t$ (mod $t + 1$), we can do the same construction with each of these $t + 1$ couples of opposite points to get for each x_i, in a unique way $t - 1$ points collinear with x_i. We call them x_k^i, $k = 1, \ldots, t - 1$. Similarly we obtain for each y_{i+1}, $t - 1$ points collinear with y_{i+1} and call them y_k^{i+1}, $k = 1, \ldots, t - 1$ and we can do this in such a way that $x_k^i \sim y_k^{i+1}$.

Now we have all the $2(t^2 + t + 1)$ points we need for a thin ideal subhexagon. We still have to consider the lines through x_k^i and through y_k^i, $i = 0, \ldots, t$ and $k = 1, \ldots, t - 1$. Since p_1 and p_2 are half-regular, the hyperbolic lines $\langle x_i, x_j \rangle = \{x_0, \ldots, x_t\}$ and $\langle y_i, y_j \rangle = \{y_0, \ldots, y_t\}$ are ideal. Now each of the y_k^i belongs to $\mathcal{W}(x_i, x_{i-1})$, so $d(y_k^i, x_j) = 4$ for every $j \in \{0, \ldots, t\}$. So for each j, there is a line (x_j, x_l^j) containing a point collinear with y_k^i. But the point x_l^j on that line belongs to $\mathcal{W}(y_j, y_{j+1})$, so $d(x_l^j, y_i) = 4$. Since $y_i \sim y_k^i$ it must be that $x_l^j \sim y_k^i$. In this way we obtain all other lines of the thin ideal subhexagon.

Remark that for a fixed y_i, all y_k^i are collinear with some point x_l^j, $\forall j \in \{0, \ldots, t\} \backslash \{k\}$ and that no two of the y_k^i's can be collinear with the same x_l^j. Moreover, for two points y_k^i and y_l^j with $i \neq j$, there is exactly one x_n^m collinear with both.

Indeed, there cannot be more than one, otherwise we would have a quadrangle in S. So if we look at all $t - 1$ points $x_{nm}^m \sim y_k^i$, $m \in \{0, \ldots, t\} \backslash \{i, j\}$, there is always one and only one y_s^j, $s = 1, \ldots, t - 1$, collinear with one of those x_{nm}^m's. Since we have $t - 1$ such y_s^j's and $t - 1$ such x_{nm}^m's, there is exactly one of the x_{nm}^m collinear with y_l^j. The same arguments hold for the x_k^i.

It is now straightforward to check that there is always a path of length ≤ 6 between two of the constructed elements. So we indeed have a thin ideal subhexagon. □

Lemma 2.4 *Let $S = (\mathcal{P}, \mathcal{B}, I)$ be a finite generalized hexagon of order s which contains an ovoid $\mathcal{O}$ for which all points are half-regular. Then every thin ideal subhexagon of S contains exactly 2 points of $\mathcal{O}$.*

Proof.

(1) From lemma 2.3 it follows that through every 2 points of $\mathcal{O}$ there is exactly one thin ideal subhexagon $\mathcal{D}$. Moreover $\mathcal{D}$ cannot contain more than 2 points of $\mathcal{O}$ since every other point of $\mathcal{D}$ is at distance ≤ 4 from one of those 2 points of $\mathcal{O}$. So in total there are $\dfrac{(s^3+1)s^3}{2}$ thin ideal subhexagons which contain two points of $\mathcal{O}$.

(2) Suppose there are α thin ideal subhexagons in S. We count in two different ways the number of pairs $(x, \mathcal{D})$ with $x \in \mathcal{P}$, $\mathcal{D}$ a thin ideal subhexagon of S and $x \in \mathcal{D}$. It then follows that

$$\alpha \leq \frac{(1+s).(1+s^2+s^4).s^3}{2.(1+s+s^2)} = \frac{(s^3+1).s^3}{2}$$

The lemma follows from (1) and (2). □

Corollary 2.5 *From the equality in the proof of lemma 2.4 it follows that through every point $x \in \mathcal{P}$ there are s^3 thin ideal subhexagons. This means that through every 2 points of S there exists a thin ideal subhexagon.*

Theorem 2.6 *Let $S = (\mathcal{P}, \mathcal{B}, I)$ be a finite generalized hexagon of order s containing an ovoid. Every point of an ovoid $\mathcal{O}$ is regular iff S is isomorphic to $H(q)$, $q = s$.*

Proof.

$\Leftarrow$ This follows from Ronan [4].

$\Rightarrow$ Due to Ronan [4] we have to prove that S has ideal lines. So, for two points $x, y \in \mathcal{P}$ with $d(x, y) = 4$, $z = x * y$ we must prove that $\langle x, y \rangle = z^w$, $\forall w \in \mathcal{W}(a, b)$.

From lemma 2.4 it follows that there are s thin ideal subhexagons $\mathcal{D}_i$, $i = 1, \ldots, s$ containing x and y. They can be obtained by choosing a point y_i on a line through y at distance 5 from x and they all contain 2 points of $\mathcal{O}$. Since $z^w = z^{w'}$ $\forall w, w' \in \mathcal{W}(a, b) \cap \mathcal{D}_i$, we have to prove that $z^{w_1} = z^{w_2} = \ldots = z^{w_s}$ with $w_i \in \mathcal{D}_i$, $i = 1, \ldots, s$.

Case 1: $x \in \mathcal{O}$ or $y \in \mathcal{O}$ then it is immediate that $\langle x, y \rangle$ is ideal.

Case 2: x and z are collinear with the same unique point p_x of $\mathcal{O}$. Let p_y be the unique point of $\mathcal{O}$ collinear with y and denote the line through y and p_y by L. With every point p on $L\backslash\{y\}$ there corresponds a thin ideal subhexagon $\mathcal{D}_p$ through x, y and p.

First we look at $\mathcal{D}_{p_y}$ and the hyperbolic line $\langle x, y \rangle_{p_y}$ in $\mathcal{D}_{p_y}$. We will show that the hyperbolic lines $\langle x, y \rangle_p$ in the other $s - 1$ $\mathcal{D}_p$'s are the same. Let k be a point of $L\backslash\{y, p_y\}$ and let $\mathcal{D}_k$ be the thin ideal subhexagon through x, y and k. From lemma 2.4 we know that $\mathcal{D}_k$ contains two points of $\mathcal{O}$. Since every point of $\mathcal{D}_k$ is at distance ≤ 4 from at least one of those two points of $\mathcal{O}$, the point z is at distance 4 from one of them, say p. Since p and L are in $\mathcal{D}_k$, also the shortest path between them lies in $\mathcal{D}_k$. Denote $p * k$ by a. Also the shortest path between a and the line through x and z lies in $\mathcal{D}_k$. Denote $a * x$ by b (see figure 3).

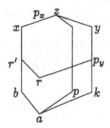

Figure 3.

Suppose that $\langle x, y \rangle_k = z^a$ is different from $\langle x, y \rangle_{p_y}$. So there is a line M through z on which the point c at distance 4 from a is different from the point d at distance 4 from r. Denote $c * a$ by e and $d * r$ by f. Since p_y and p are regular, we have ideal lines $\langle r', p_y \rangle$ and $\langle b, r \rangle = \langle b, k \rangle$ (see figure 3) . From $z \in \mathcal{W}(b, k)$ it follows that $e \in \langle b, k \rangle$, so $d(r, e)$ must be 4. Denote $r * e$ by g. From $a \in \mathcal{W}(r', p_y)$ it follows that $g \in \langle r', p_y \rangle$ and so $d(z, g)$ must be 4 which is a contradiction.

Case 3: y and z are collinear with the same unique point p_y of $\mathcal{O}$. This is similar to case 2.

Case 4: x, y and z are collinear to different points of $\mathcal{O}$, say respectively p_x, p_y and p_z.

(i) If $p_z \in z^w$ for some $w \in \mathcal{W}(x, y)$ then $\langle x, y \rangle$ is ideal since p_z is regular.

(ii) So suppose there is a point t on the line through z and p_z at distance 4 from a point $w \in \mathcal{W}(x, y)$. By case 2 we have that $\langle t, y \rangle$ is ideal , so $\langle x, y \rangle$ is ideal. $\qquad \square$

References

[1] **W. Haemers and C. Roos**. An inequality for generalized hexagons. *Geom. Dedicata*, 10, pp. 219–222, 1981.

[2] **W. M. Kantor**. Generalized polygons, SCABs and GABs. In L. A. Rosati, editor, *Buildings and the Geometry of Diagrams, Proceedings Como 1984*, volume 1181 of *Lecture Notes in Math.*, pages 79–158, Berlin, 1986. Springer Verlag.

[3] **S. E. Payne and J. A. Thas**. *Finite Generalized Quadrangles*, volume 104 of *Research Notes in Mathematics*. Pitman, Boston, 1984.

[4] **M. A. Ronan**. A geometric characterization of Moufang hexagons. *Invent. Math.*, 57, pp. 227–262, 1980.

[5] **J. A. Thas**. A restriction on the parameters of a subhexagon. *J. Combin. Theory*, 21, pp. 115–117, 1976.

[6] **J. A. Thas**. Existence of ovoids and spreads of classical finite polar spaces. Unpublished, 1989.

[7] **H. Van Maldeghem and I. Bloemen**. Generalized hexagons as amalgamations of generalized quadrangles. Preprint, 1992.

V. De Smet, H. Van Maldeghem. University of Ghent, Galglaan 2, B-9000 Gent. e-mail: vds@cage.rug.ac.be, hvm@cage.rug.ac.be

Flag-transitive $L.C_2$ geometries

D. Ghinelli

Abstract

We consider locally C_2 geometries where all planes are linear spaces of order (r, s) (i.e. 2-designs with parameters 2-$(r(s+1)+1, r+1, 1)$). We call them $L.C_2$ geometries, for short. We give a classification of flag-transitive $L.C_2$ geometries, under the hypothesis that residues of points are classical generalized quadrangles of order (s, t), $s, t > 1$.

1. Introduction

An $L.C_2$ geometry of order (r, s, t) is a Buekenhout geometry with three types of varieties (points, lines and planes) belonging to the diagram

$$
\begin{array}{ccc}
\overset{L}{\underset{r}{\circ}\!\!-\!\!\!-\!\!\!-\!\!\!-\!\!\!-\!\!\underset{s}{\circ}\!\!=\!\!=\!\!\underset{t}{\circ}}
\end{array}
$$

where the stroke

$$
\underset{r}{\circ}\!\!-\!\!\!-\!\!\!\overset{L}{-}\!\!\!-\!\!\!-\!\!\underset{s}{\circ}
$$

denotes the class of linear spaces of order (r, s), namely 2-$(r(s+1)+1, r+1, 1)$ designs, while

$$
\underset{s}{\circ}\!\!=\!\!=\!\!\underset{t}{\circ}
$$

is the class of generalized quadrangles (abbreviated GQ) of order (s, t). The reader is referred to [12] for background and notation, and obviously to [6], [7], [21] (see [33], [36] [37] and [38] for further details on GQs).

Our purpose is to classify all $L.C_2$ geometries with a flag-transitive automorphism group G (i.e. an automorphism group transitive on maximal flags). We call them *flag-transitive $L.C_2$ geometries*.

In all of this paper, we denote by Γ a flag-transitive $L.C_2$ geometry of order (r, s, t) with $t > 1$ (i.e. we exclude grids as residues of points) and $s > 1$ (i.e. only thick GQs appear), and by $G \leq \mathrm{Aut}(\Gamma)$ a flag-transitive group acting on Γ. If x is a variety of Γ, then G_x is the stabilizer of x in G, K_x is the elementwise stabilizer of the residue Γ_x of x (i.e. the kernel of the action of G_x on Γ_x) and $\overline{G}_x = G_x/K_x$ is the action of G_x on Γ_x.

Clearly the residue Γ_P of a point P of Γ must be a flag-transitive GQ, but unfortunately the classification of flag-transitive GQs is not yet completed. However (see [23] p. 98, Summary) apart from four sporadic cases (given by $T_2^*(O_q)$ for some oval O_q in $PG(2, q)$ with $q = 4$ or 16, and their duals), the only presently known flag-transitive thick GQs are the *classical* ones with $t > 1$ (see the proof of lemma 1.3 below, for a definition). Therefore it seems natural to make the further assumption that the *residues of points are classical* GQs. Then Γ is said to be a *classical* $L.C_2$ geometry.

When $r = 1$, then

$$\underset{1}{\circ} \overset{L}{\rule{2.5cm}{0.4pt}} \underset{s}{\circ} = \underset{1}{\circ} \overset{c}{\rule{2.5cm}{0.4pt}} \underset{s}{\circ}$$

is the diagram of a circle geometry (i.e. a complete graph on $s + 2$ vertices). Thus $L.C_2$ geometries of order $(1, s, t)$ belong to the diagram $c.C_2$

$$\underset{1}{\circ} \overset{c}{\rule{2.5cm}{0.4pt}} \underset{s}{\circ}\!=\!\!=\!\!\underset{t}{\circ}$$

and the classification of flag-transitive classical $c.C_2$ geometries was given in theorem 4 of [15], namely:

Theorem 1.1 *If Γ is a flag-transitive $c.C_2$ geometry with thick classical residues then*

(i) Γ *is either an affine polar space of order 2 or a standard quotient of such an affine polar space, or*

(ii) Γ *is one of the six examples $\Gamma_1, \cdots, \Gamma_6$ below, for which we recall some essential information:*

geometry	group[1]	(s, t)	residues[2]
Γ_1	$U_5(2)$	$(3, 3)$	$W(3)$
Γ_2	McL	$(3, 9)$	$Q_5^-(3)$
Γ_3	$O_6^-(3)$	$(4, 2)$	$H_3(2^2)$
Γ_4 Smith	$3 \cdot O_6^-(3)$	$(4, 2)$	$H_3(2^2)$
Γ_5 Patterson	Suz	$(9, 3)$	$H_3(3^2)$
Γ_6 Yoshiara	$Aut(HS)$	$(9, 3)$	$H_3(3^2)$

The reader is referred to 4.5 in [15] for a proof, which brings together contributions by Buekenhout and Hubaut [10], Blokhuis and Brouwer ([3] and

---

1. We only give the minimal flag-transitive automorphism group.
2. Isomorphism type of the point residues.

an unpublished paper quoted in [5], page 399), Weiss and Yoshiara ([40] and also [42]).

The proof essentially depends on three celebrated results on permutation groups, namely Seitz theorem [34] (see [23], theorem C.7.1 or [28], § 4), a theorem of Suzuki [35] and Tits [39] on 1–point extensions of $L_2(q)$, and the classification of 2–transitive permutation groups [11].

A proof of theorem 1.1 can also be achieved using coset enumeration (Yoshiara [42]). For the case in which grids appear as point-residues, see [29] which also explains the reason of our assumption $t > 1$.

The purpose of this paper is to prove a classification theorem for $r \geq 2$. Clearly, residues of planes are flag-transitive linear spaces; from the classification of flag-transitive linear spaces announced in [9] we have immediately the following result (see section 2 and 3 of [9] for a description of the examples or, in this paper, section 2; see also [8] for a discussion of the background to this problem and for further references).

Lemma 1.2 *Let Γ be a flag-transitive $L.C_2$ geometry of order (r, s, t) with classical point-residues, $r, s, t > 1$, and let $G \leq \mathrm{Aut}(\Gamma)$ be a flag-transitive group on Γ. Then either the residue Γ_π of a plane π of Γ is one of the following*

(1) a desarguesian projective space of dimension $d \geq 2$,

(2) a Hermitian unital,

(3) a Ree unital,

(4) a Witt-Bose-Shrikhande space,

(5) a desarguesian affine space of dimension $d \geq 2$ and the group $\overline{G}_\pi$ induced on Γ_π by the stabilizer G_π is not 1-dimensional,

(6) a non desarguesian translation affine plane,

(7) a Hering space,

   or

(8) Γ_π has $q = p^d$ points (p prime) and $\overline{G}_\pi$ is a subgroup of the group $A\Gamma L(1, q)$ of 1-dimensional semilinear affine transformations.

The following two lemmas will be needed in the next section:

Lemma 1.3 *Let Γ be an $L.C_2$ geometry of order (r, s, t), then $r \leq s$. If Γ has classical point-residues and $s, t > 1$, then s and t are powers of the same prime.*

Proof.   Trivially, given a maximal flag (P, ℓ, π) of Γ, the number $r + 1$ of points in ℓ is less than or equal to the number $s + 1$ of lines through P in the linear space π. Since $t > 1$, grids are excluded as residues of points. Thus the classical GQs occurring as point residues are necessarily the geometries of points and lines of a nonsingular algebraic variety of $PG(d, q)$ of one of the following types (see [33], 3.1.1, p. 36):

(i) a nonsingular quadric $Q_4(q)$ or an elliptic quadric $Q_5^-(q)$ of projective index 1 in $PG(d, q)$, $d = 4$ or 5 (here $(s, t) = (q, q)$ or (q, q^2)),

(ii) a nonsingular Hermitian variety $H_d(q^2)$ of $PG(d, q^2)$, $d = 3$ or 4 (here $(s, t) = (q^2, q)$ or (q^2, q^3)),

(iii) a symplectic variety $W(q)$ of $PG(3, q)$ (here $(s, t) = (q, q)$).

Therefore s and t are in all cases powers of the characteristic p of $GF(q)$. □

Lemma 1.4 *Let Γ be a flag-transitive classical $L.C_2$ geometry of order (r, s, t), with $r, s, t > 1$ and with flag-transitive group $G \leq \mathrm{Aut}(\Gamma)$. If (P, π) is a point-plane flag in Γ, then one of the following occurs:*

(1) *The group $\overline{G}_P$ contains a classical group in its natural action on the classical generalized quadrangle Γ_P.*

(2) $\Gamma_P = Q_4(2)$ *and* $\overline{G}_P = A_6$.

(3) $\Gamma_P = Q_4(3)$ *or* $W(3)$, *and* $\overline{G}_P = 2^4 \cdot A_5$, $2^4 \cdot S_5$ *or* $2^4 \cdot Frob(20)$.

(4) $\Gamma_P = H_3(3^2)$ *or* $Q_5^-(3)$, *and* $\overline{G}_P = L_3(4) \cdot 2$ *or* $L_3(4) \cdot 2^2$.

Furthermore, in case (1) the stabilizer $(\overline{G}_P)_\pi$ acts as a 2-transitive group on the lines through P incident with π.

Proof. The first part of the statement follows from a theorem of Seitz [34] (see [23], theorem C.7.1, and also [28], §4).

As for the last part, we note that the stabilizer of a "line" π of the $GQ(s, t)$ Γ_P (hence of a plane π on the point P of Γ) is 2-transitive on the "points" of the "line" π of Γ_P (hence on the lines of Γ through P, incident with π). □

Dually, in a classical $GQ(s, t)$ (hence in Γ_P) on which we have the natural action of a classical group, the stabilizer of a "point" is 2-transitive on the "lines" through the "point". Therefore for every point-line flag (P, ℓ) of Γ the stabilizer $(\overline{G}_P)_\ell$ is 2-transitive on the planes through P incident with ℓ.

In the rest of the paper we apply 1.2, 1.3 and 1.4 to prove (see theorem 4.1 for a precise statement) that a classical flag-transitive $L.C_2$ geometry is either one of the following known examples (see [12], [30], [32])

1. a polar space (of rank 3),

2. the A_7 geometry, that is the flat C_3 geometry for the alternating group A_7,

3. an affine polar space (of rank 3),

4. a standard quotient of an affine polar space (of rank 3),

   or

5. $r = s - 1$ with $s = 3, 9$. In these small sporadic cases, we also give information on the possible groups acting on the residues (see 4.1 (iii)).

In any case the only linear spaces which appear as residues of points are (desarguesian) projective or affine planes, for which we use the diagrams

$$\underset{s}{\circ}\!\!-\!\!-\!\!-\!\!\underset{s}{\circ} = \underset{s}{\circ}\!\!-\!\!-\!\!\overset{\displaystyle L}{-\!\!-\!\!-\!\!-}\!\!\underset{s}{\circ}$$

or

$$\underset{s-1}{\circ}\!\!-\!\!-\!\!\overset{\displaystyle Af}{-\!\!-\!\!-}\!\!\underset{s}{\circ} = \underset{s-1}{\circ}\!\!-\!\!-\!\!\overset{\displaystyle L}{-\!\!-\!\!-}\!\!\underset{s}{\circ}$$

The proof of 4.1 (ii) suggests the following conjecture.

Conjecture 1.5 In the small cases in 4.1 (iii) Γ is a (possibly improper) standard quotient of an affine polar space.

By a result of Cuypers [14], in order to prove the conjecture, it is enough to prove that in each of the above cases the following property holds

(LL) *Given any two points, there is at most one line on them.*

2. The cases where $\overline{G}_\pi$ is almost simple or affine

In this section we consider the cases (1)-(7) of lemma 1.2, which correspond to the examples of linear spaces with a group almost simple or affine in [9].

Proposition 2.1 *Let π be a plane of a flag-transitive classical $L.C_2$ geometry of order (r, s, t), with $r, s, t > 1$. Then the residue Γ_π of π cannot be*
  (i) *a desarguesian projective or affine space of dimension $d > 2$,*
 (ii) *a Ree unital,*
(iii) *a Hering space.*

Proof.
  (i) Let Γ_π be a desarguesian projective space $PG(d, q)$ of dimension $d \geq 2$ and order $q = p^h$ (p prime). Then

$$r = q, \qquad s = q^{d-1} + q^{d-2} + \cdots + q + 1 - 1 = q(q^{d-2} + \cdots + 1).$$

By lemma 1.3, s must be a power of a prime. Hence $d = 2$.
Similarly, if Γ_π is a d-dimensional desarguesian affine space $AG(d, q)$ ($d \geq 2$), then

$$r = q - 1, \qquad s = q^{d-1} + q^{d-2} + \cdots + q,$$

so that lemma 1.3 implies $d = 2$. This completes the proof of (i).

(ii) For any integer $e \geq 0$ and $q = 3^{2e+1}$, the *Ree unital* $U_R(q)$ of order q is the linear space of order $(r, s) = (q, q^2 - 1)$ whose points and lines are, respectively, the Sylow 3-subgroups and the involutions of the Ree group $^2G_2(q)$ (whose order is $q^3(q^3 + 1)(q - 1)$). A point and a line are incident, if and only if the involution normalizes the Sylow 3-subgroup (see [24]).

If $\Gamma_\pi \simeq U_R(q)$, since $s = (q-1)(q+1)$ is a power of a prime, by lemma 1.3 we get $q = 2$, a contradiction.

(iii) Hering spaces are the two nonisomorphic flag-transitive linear spaces of order (r, s) on $v = r(s+1) + 1 = 3^6$ points with line size 3^2 constructed by Hering in [20]. Therefore $r = 3^2 - 1$, so that $s = 3^4 + 3^2 = 90$ which is not a power of a prime, a contradiction. $\qquad \square$

Proposition 2.2 *Let π be a plane of a flag-transitive classical $L.C_2$ geometry of order (r, s, t), with $r, s, t > 1$. If the residue Γ_π of π is a Hermitian unital, then Γ_π is $AG(2, 3)$ and $r = 2$.*

Proof. For any prime power q, the *Hermitian unital* $U_H(q)$ of order q is the linear space on $q^3 + 1$ points with line size $q + 1$ whose points and lines are respectively the absolute points and nonabsolute lines of a unitary polarity in $PG(2, q^2)$, the incidence being the natural one. Any group $\mathcal{G}$ with $PSU(3, q) \leq \mathcal{G} \leq P\Gamma U(3, q)$ acts flag transitively on $U_H(q)$.

If $\Gamma_\pi \simeq U_H(q)$, then $r = q$ and $s = [(v-1)/r] - 1 = q^2 - 1 = (q-1)(q+1)$. Therefore, lemma 1.3 implies $q = 2$, and so $r = 2$, $s = 3$ and Γ_π is isomorphic to the affine plane $AG(2, 3)$. $\qquad \square$

Proposition 2.3 *Let π be a plane of a flag-transitive classical $L.C_2$ geometry of order (r, s, t), with $r, s, t > 1$. Then the residue Γ_π of π cannot be a Witt-Bose-Shrikhande space.*

Proof. Starting from the group $PSL(2, 2^n)$ with $n \geq 3$, Kantor (see [22]) defined a flag-transitive linear space $\mathcal{W}(2^n)$ as follows: the points are the subgroups of $PSL(2, 2^n)$ isomorphic to the dihedral group of order $2(2^n + 1)$, the lines are the involutions of $PSL(2, 2^n)$, a point being incident with a line if and only if the subgroup contains the involution. $\mathcal{W}(2^n)$ has $2^{n-1}(2^n - 1)$ points, line size $r + 1 = 2^{n-1}$ (hence $2^n + 1$ lines on every point). $\mathcal{W}(2^n)$ is called a Witt-Bose-Shrikhande space because the first description (using the Miquelian inversive plane of order 2^n) dates back to the classical paper of Witt [41], while a second geometric description, using a complete conic C (i.e. the union of an irreducible conic and its nucleus) in $PG(2, 2^n)$, was given by Bose and Shrikhande in [4]. Here the points of $\mathcal{W}(2^n)$ are the lines of $PG(2, 2^n)$ disjoint from C, the lines of $W(2^n)$ are the points of the plane outside C, the incidence being the natural one.

The group $P\Gamma L(2, 2^n)$, which is the stabilizer of C in the automorphism group of the plane, is isomorphic to $\text{Aut}(W(2^n))$ and is flag-transitive on $W(2^n)$. Any group $\mathcal{G}$ with $PSL(2, 2^n) \leq \mathcal{G} \leq P\Gamma L(2, 2^n)$ acts flag-transitively on $W(2^n)$.

If $\Gamma_\pi \simeq W(2^n)$, then

$$r = 2^{n-1} - 1 \quad \text{and} \quad s = 2^n.$$

Furthermore, the stabilizer of a point P in π induces a subgroup

$$(\overline{G}_\pi)_P \leq (P\Gamma L(2, 2^n))_P$$

hence has an order dividing $2(2^n + 1)n$.

This gives a contradiction in case (1) of lemma 1.4, since $(\overline{G}_\pi)_P$ should be 2-transitive on the $s + 1$ lines through a point P of $W(2^n)$, by lemma 1.4.

Case (2) of lemma 1.4 is impossible, since then $s = 2$ implies $n = 1$ and $r = 0$ (while $r > 1$).

In the cases (3) and (4) of lemma 1.4, we have $s = 3$ or 9, contradicting $s = 2^n$. This completes the proof. □

Corollary 2.4 *Let Γ be a flag-transitive classical $L.C_2$ geometry of order (r, s, t) with $r, s, t > 1$, and let $G \leq \text{Aut}(\Gamma)$ be flag-transitive on Γ. If $\overline{G}_\pi$ (π a plane of Γ) is almost simple (i.e. has a non abelian simple normal subgroup N such that $N \trianglelefteq \overline{G}_\pi \leq \text{Aut}(N)$) then $\Gamma_\pi \simeq PG(2, q)$ and*

$$PSL(3, q) \leq \overline{G}_\pi \leq P\Gamma L(3, q).$$

Proof. This follows immediately from propositions 2.1 (i) and (ii), 2.2, 2.3, and from the results in [9] (sections 1 and 2). □

Next we consider the case where $\overline{G}_\pi$ is affine. By 2.1 (i) and (iii) we only have to consider the cases where Γ_π is a desarguesian affine plane or a non desarguesian translation affine plane.

Proposition 2.5 *Let π be a plane of a flag-transitive classical $L.C_2 = Af.C_2$ geometry of order $(s-1, s, t)$, with $s, t > 1$. If the residue Γ_π of π is desarguesian and the group $\overline{G}_\pi$ is not 1-dimensional, then $\overline{G}_\pi$ is 2-transitive (hence given in [26]).*

Proof. If Γ_π is a desarguesian affine plane and the group $\overline{G}_\pi$ is not 1-dimensional, then one of the following holds (see 3.1 in [9])

(a) $\overline{G}_\pi$ is 2-transitive (hence given in [26]).

(b) $s = 11$ or 23 and $\overline{G}_\pi$ is one of the three soluble flag-transitive groups given in [17], table II,

(c) $s = 9, 11, 19, 29$ or 59, the last term in the derived series of $(\overline{G}_\pi)_P$ (P a point in π) is $2 \cdot A_5$ and $\overline{G}_\pi$ is given in [17], table II.

Now, it is not too difficult to rule out cases (b) and (c) using table II in [17] and lemma 1.4. In case (1) of lemma 1.4, $(\overline{G}_P)_\pi$ is 2-transitive on the $s + 1$ lines through P in π, therefore $(1 + s)s$ must divide $|(\overline{G}_P)_\pi| = |\overline{G}_\pi|/s^2$. Otherwise we have case (c) with $s = 9$, $\Gamma_P = H_3(3^2)$ and $(\overline{G}_\pi)_P = SL_2(9)$ or $SL_2(9) \cdot 2 = M_{10}$. Hence $\overline{G}_\pi \simeq ASL(2,9)$ or $9^2 \cdot M_{10}$, which are 2-transitive. This completes the proof. □

Proposition 2.6 Let π be a plane of a flag-transitive classical $L.C_2$ geometry of order (r, s, t) with $r, s, t > 1$. Then the residue Γ_π of π cannot be a non desarguesian translation affine plane.

Proof.   If Γ_π is a non desarguesian translation affine plane, then one of the following holds (see 3.2 in [9])

(a) Γ_π is one of the Lüneburg planes, constructed in [25]. These are affine planes of order $s = q^2$, where $q = 2^{2e+1} \geq 8$, and $^2B_2(q) \leq (\overline{G}_\pi)_P \leq$ Aut($^2B_2(q)$)).

(b) Γ_π is the Hering plane of order 27 constructed in [19]. Here $(\overline{G}_\pi)_P = SL(2,13)$ and $\overline{G}_\pi$ is 2-transitive on the points of Γ_π.

(c) Γ_π is the nearfield plane of order 9. Here there are seven possibilities for $\overline{G}_\pi$, given by Foulser in [18], §5.

In case (a), $(\overline{G}_\pi)_P$ contains a Suzuki group $Sz(2^{2e+1})$, $e \geq 1$. This gives a contradiction, since $s \geq 64$ implies that $(\overline{G}_\pi)_P$ is classical, and no classical group satisfies this condition. Therefore case (a) does not occur.

In cases (b) and (c) the automorphism group Aut(Γ_π) is 2-homogeneous.

The stabilizer $(\overline{G}_\pi)_P$ of a point P in Hering's plane of order 27 is isomorphic to $SL(2,13)$. Now $(\overline{G}_\pi)_P$ is 2-transitive on the 28 lines through P in π, because $\overline{G}_P$ is classical, hence $28 \cdot 27$ must divide $|SL(2,13)| = 13 \cdot 2^3 \cdot 3 \cdot 7$, a contradiction. Therefore case (b) cannot occur.

The stabilizer of a point in the nearfield plane of order 9 is isomorphic to $S_5 \cdot 2^4 \cdot 2$, preserves a pairing of the 10 points at infinity and acts on this set of 5 pairs as the symmetric group S_5 ([1], see also [18] and [25]). Thus the order 9 divides $5! \cdot 2^4 \cdot 2$, which gives a contradiction and case (c) is impossible. □

3. The 1-dimensional affine case

Let Γ be a flag-transitive classical $L.C_2$ geometry of order (r, s, t) with $r, s, t > 1$ and let $G \leq \mathrm{Aut}(\Gamma)$ be a flag-transitive group on Γ.

In this section, Γ_π (π a plane of Γ) has $v = q = p^d$ points (p prime) and $H := \overline{G}_\pi$ is a subgroup of the group

$$A\Gamma L(1, q) = \{x \longmapsto ax^\sigma + b \,|\, a, b \in GF(q), \ a \neq 0, \ \sigma \in \mathrm{Aut}(GF(q))\}$$

of 1-dimensional semilinear affine transformations.

Remark 3.1 Since $|A\Gamma L(1, q)| = q(q-1)d$, $|H| = |\overline{G}_\pi|$ divides $q(q-1)d$.

The stabilizer H_P of a point P of π is a subgroup of the stabilizer of a point (for instance the point 0) in $A\Gamma L(1, q)$. Thus

$$H_P \leq [A\Gamma L(1, q)]_0 = \{x \longmapsto ax^\sigma \,|\, a \in GF(q) - \{0\}, \ \sigma \in \mathrm{Aut}(GF(q))\},$$

which is a semidirect product of two cyclic groups: one of order $q - 1$ (the multiplicative group of $GF(q)$) and one of order d (the group $\mathrm{Aut}(GF(q))$). We write $H_P \leq (q - 1) \cdot d$.

The commutator subgroup $[H_P, H_P]$ is then a cyclic group whose order divides $q - 1$ and is trivial if $d = 1$.

In particular, case (1) of lemma 1.4 cannot occur since H_P is not 2-transitive on the $s + 1$ lines through P in π.

Remark 3.2 Γ_π is a linear space of order (r, s) with $v = r(s+1)+1 = q = p^d$ points, thus $r = (p^d - 1)/(s + 1)$ and every line of Γ_P has exactly

$$r + 1 = \frac{p^d + s}{s + 1}$$

points. Since H is flag-transitive, $r + 1$ divides $|H|$ which divides $q(q-1)d$ by the first remark. Therefore

$$\frac{p^d + s}{s + 1} \qquad \text{divides} \qquad p^d(p^d - 1)d.$$

Next we examine the cases (2), (3) and (4) of lemma 1.4.

Proposition 3.3 *There is no flag-transitive geometry Γ belonging to the diagram*

$$\begin{array}{ccccc}
\circ & \!\!\!\!\!\!\!\!\!\!\!\!\!\!\!\! & \circ & \!\!\!\!\!\!\!\!\!\!\!\!\!\!\!\! & \circ \\
2 & & 2 & & 2
\end{array}$$

with a flag-transitive group $G \leq \text{Aut}(\Gamma)$ such that for any point-line pair (P, π)

$$\overline{G}_\pi \leq A\Gamma L(1, q), \qquad \Gamma_P = Q_4(2) \quad \text{and} \quad \overline{G}_P = A_6.$$

Proof. Since $\Gamma_P = Q_4(2)$ has 15 points, the order of the stabilizer of a "line" π of Γ_P is

$$|(\overline{G}_P)_\pi| = \frac{|\overline{G}_P|}{15} = \frac{|A_6|}{15} = 24.$$

Hence $(\overline{G}_P)_\pi$ is a maximal subgroup of order 24 of A_6, therefore it is isomorphic to S_4. The commutator subgroup of S_4 is A_4 which is not cyclic, contradicting remark 3.1. $\qquad \square$

Proposition 3.4 *There is no classical flag-transitive $L.C_2$ geometry Γ belonging to the diagram*

with $r > 1$ and with a flag-transitive group $G \leq \text{Aut}(\Gamma)$ such that $\overline{G}_\pi$ is a subgroup of $A\Gamma L(1, q)$.

Proof. Here $\Gamma_P = H_3(3^2)$ and $\overline{G}_P = L_3(4) \cdot 2$ or $L_3(4) \cdot 2^2$. Thus $(\overline{G}_P)_\pi = L_2(9)$ or $L_2(9) \cdot 2$, respectively. Now $L_2(9)$ is simple, and we are very far away from the semidirect product we should have by remark 3.1. Also $L_2(9) \cdot 2$ is highly unsolvable and, again by remark 3.1, this group cannot occur. $\qquad \square$

Proposition 3.5 *There is no flag-transitive $L.C_2$ geometry Γ belonging to the diagram*

with a flag-transitive group $G \leq \text{Aut}(\Gamma)$ such that $\overline{G}_\pi$ is a subgroup of $A\Gamma L(1, q)$.

Proof. In this case $\Gamma_\pi = PG(2, 3)$ and it is well known that a 1-dimensional affine group cannot act flag-transitively on $PG(2, 3)$ (see for example Dembowski [16]). $\qquad \square$

It remains to consider the cases (see lemma 1.4) in which Γ has diagram

$$Af$$

o————o====o
2   3 t

with $t = 3$ or 9. In both cases, since $(r, s) = (2, 3)$ we have $p^d = 9$, thus $p = 3$ and $d = 2$.

4.   The main theorem

In this section we apply the previous results to prove the following classification theorem (see theorem 1.1 for the case $r = 1$).

Theorem 4.1 *Let Γ be a flag-transitive Buekenhout geometry belonging to the diagram*

$$L$$

o————o====o
r s t

with classical point-residues, $r, s, t > 1$, and let $G \leq \mathrm{Aut}(\Gamma)$ be flag-transitive on Γ. Then one of the following holds

(i) *Γ_π is a desarguesian projective plane and Γ is either a polar space or the A_7 geometry,*

(ii) *Γ_π is a desarguesian affine plane and Γ is an affine polar space or a standard quotient of an affine polar space,*

(iii) *Γ_π is a desarguesian affine plane of order 3 or 9, the group $\overline{G}_\pi$ induced by G_π is 2-transitive on points (hence is given in [26], Appendix 1), and either $(s, t) = (3, 3)$, $\Gamma_P = Q_4(3)$ or $W(3)$, and $\overline{G}_P = 2^4 \cdot A_5$, $2^4 \cdot S_5$ or $2^4 \cdot Frob(20)$, or $(s, t) = (9, 3)$ or $(3, 9)$, $\Gamma_P = H_3(3^2)$ or $Q_5^-(3)$, and $\overline{G}_P = L_3(4) \cdot 2$ or $L_3(4) \cdot 2^2$.*

Proof.   From the classification of flag-transitive linear spaces announced in [9] we have for the residue Γ_π of a plane π the 8 possibilities given in lemma 1.2.

Since Γ has classical point-residues, s and t are powers of the same prime (lemma 1.3). This implies that desarguesian projective spaces of dimension $d \geq 3$ and Ree unitals cannot occur as plane residues (prososition 2.1). Using lemmas 1.3 and 1.4 we proved that Hermitian unitals of order $q > 2$ (prososition 2.2) and Witt-Bose-Shrikhande spaces (prososition 2.3) cannot occur. Hence when the group $\overline{G}_\pi$ is almost simple (i.e. in cases (1)–(4) of lemma 1.2), we have necessarily $\Gamma_\pi \simeq PG(2, q)$ and $PSL(3, q) \leq \overline{G}_\pi \leq P\Gamma L(3, q)$.

Now Γ has diagram

o————o====o
q q t

with $t = q$ when $\Gamma_P = Q_4(q)$ or $W(q)$, $t = q^2$ when $\Gamma_P = Q_5^-(q)$, $t = \sqrt{q}$ when $\Gamma = H_3(q)$ (q a square) or $t = q\sqrt{q}$ when $\Gamma = H_4(q)$ (q a square). It follows from [2] that Γ is either a polar space or the A_7 geometry. This proves (i).

In the affine case (i.e. in cases (5)–(7) of lemma 1.2), since s and t are powers of the same prime by lemma 1.3, desarguesian affine spaces of dimension $d \geq 3$ and Hering spaces cannot appear as plane residues (prososition 2.1(i) and (iii)). We proved in prososition 2.6 that non desarguesian translation affine planes cannot occur as plane residues. Hence the only remaining case is when Γ_π is a desarguesian affine plane of order $s = q$.

Now Γ belongs to the diagram

$$Af$$

$$q - 1 \qquad q \qquad t$$

and by prososition 2.5, if the group $\overline{G}_\pi$ is not 1-dimensional, then $\overline{G}_\pi$ is 2-transitive (hence given in [26]). Since $r = q - 1 > 1$, case (2) of lemma 1.4 cannot occur. Therefore we are either in the cases (3) and (4) of lemma 1.4, and these give (iii), or the group $\overline{G}_P$ contains a classical group in its natural action on the classical generalized quadrangle Γ_P. To settle this last case, we recall that, if the following property holds

(LL) *Given any two points, there is at most one line on them*

a result of Cuypers [14] implies that Γ is either an affine polar space or a standard quotient of an affine polar space. Hence, to get result (ii), it suffices to prove that (LL) holds in Γ when the group $\overline{G}_P$ contains a classical group in its natural action on the classical generalized quadrangle Γ_P.

By contradiction, assume that there are two points P and Q joined by at least two different lines ℓ, ℓ'. Clearly, ℓ and ℓ' are not coplanar (since the residue of a plane is a linear space), and all the lines joining P and Q are pairwise non coplanar. In the $GQ(q, t)$ which is the residue of P, these lines form a set of pairwise non collinear points, hence their number is at most the number $qt + 1$ of points of an ovoid of Γ_P. This implies that there are at most qt lines different from ℓ through P and Q. Now $G_{P,\ell}$ fixes a set of at most $qt(q - 1)$ lines on P (namely the lines different from ℓ meeting ℓ in a point different from P), therefore also $(\overline{G}_P)_\ell$ fixes a set of at most $qt(q - 1)$ lines on P. This is impossible since $(\overline{G}_P)_\ell$ is a classical stabilizer of the 'point' ℓ in the classical GQ Γ_P, hence must be transitive on the $q^2 t$ 'points' of Γ_P non collinear with ℓ (i.e. on the $q^2 t$ lines through P non coplanar with ℓ), and $q^2 t > qt(q - 1)$. This contradiction proves that property (LL) holds in Γ. This completes the proof of (ii).

In the 1-dimensional affine case (i.e. in case (8) of lemma 1.2), case (1) of lemma 1.4 cannot occur (see remark 3.1).

In case (2) of lemma 1.4, $s = t = 2$. Since $1 < r \leq s$ (lemma 1.3), we have $r = s = t = 2$. We ruled out this possibility in prososition 3.3.

In case (3) of lemma 1.4, $s = 3$, thus $1 < r \leq s$ implies $r = 2$ or 3 and residues of planes are affine or projective planes of order 3. But $r = 3$ cannot occur (see prososition 3.5) and so the only possible case is an $Af.C_2$ geometry of order $(2, 3, 3)$.

In case (4) of lemma 1.4, if $s = 3$ (i.e. $\Gamma_P = Q_5^-(3)$, thus $t = 9$) we have, similarly, that the only possible case is an $Af.C_2$ geometry of order $(2, 3, 9)$.

The case $s = 9$ cannot occur in the 1-dimensional affine case, as proved in prososition 3.4. □

Remark 4.2 (added in proof) A. Del Fra has informed us that he has now proved our conjecture, showing that (LL) holds in each of the cases of 4.1 (iii) and using [14], as suggested in the introduction.

Acknowledgement

I would like to thank Antonio Pasini for all his useful comments.

References

[1] **J. André**. Projektive Ebenen über Fastkörpern. *Math. Z.*, 62, pp. 137–160, 1955.

[2] **M. Aschbacher**. Finite geometries of type C_3 with flag-transitive automorphism groups. *Geom. Dedicata*, 16, pp. 195–200, 1984.

[3] **A. Blokhuis and A. E. Brouwer**. Uniqueness of a Zara graph on 126 points and non-existence of a completely regular two-graph on 288 points. In P. J. de Doelder, J. de Graaf, and J. H. van Lint, editors, *Papers Dedicated to J. J. Seidel*, pages 6–19. Eindhoven University, Eindhoven, 1984.

[4] **R. C. Bose and S. S. Shrikhande**. On the construction of sets of mutually orthogonal Latin squares and the falsity of a conjecture of Euler. *Trans. Amer. Math. Soc.*, 95, pp. 191–209, 1960.

[5] **A. E. Brouwer, A. M. Cohen, and A. Neumaier**. *Distance-regular Graphs*. Springer Verlag, Berlin, 1989.

[6] **F. Buekenhout**. Diagrams for geometries and groups. *J. Combin. Theory Ser. A*, 27, pp. 121–151, 1979.

[7] **F. Buekenhout**. The basic diagram of a geometry. In *Geometries and Groups, Proceedings of the Conference in Honour of H. Lenz*, volume 893 of *Lecture Notes in Math.*, pages 1–29, Berlin, 1981. Springer Verlag.

[8] **F. Buekenhout, A. Delandtsheer, and J. Doyen**. Finite linear spaces with flag-transitive groups. *J. Combin. Theory Ser. A*, 49, pp. 268–293, 1988.

[9] F. Buekenhout, A. Delandtsheer, J. Doyen, P. B. Kleidman, M. W. Liebeck, and J. Saxl. Linear spaces with flag-transitive automorphism groups. *Geom. Dedicata*, 36, pp. 89–94, 1990.

[10] F. Buekenhout and X. Hubaut. Locally polar spaces and related rank 3 groups. *J. Algebra*, 45, pp. 391–434, 1977.

[11] P. J. Cameron. Finite permutation groups and finite simple groups. *Bull. London Math. Soc.*, 13, pp. 1–22, 1981.

[12] P. J. Cameron. *Projective and Polar Spaces*, volume 13 of *QMW Maths Notes*. Queen Mary and Westfield College, University of London, 1992.

[13] J. H. Conway, R. T. Curtis, S. P. Norton, R. A. Parker, and R. A. Wilson. *Atlas of Finite Groups*. Clarendon Press, Oxford, 1985.

[14] H. Cuypers. Finite locally generalized quadrangles with affine planes. *European J. Combin.*, 13, pp. 439–454, 1992.

[15] A. Del Fra, D. Ghinelli, T. Meixner, and A. Pasini. Flag-transitive extensions of C_n-geometries. *Geom. Dedicata*, 37, pp. 253–273, 1991.

[16] P. Dembowski. *Finite Geometries*. Springer Verlag, Berlin, 1968.

[17] D. A. Foulser. The flag-transitive collineation groups of the finite Desarguesian affine planes. *Canad. J. Math.*, 16, pp. 443–472, 1964.

[18] D. A. Foulser. Solvable flag-transitive affine groups. *Math. Z.*, 86, pp. 191–204, 1964.

[19] C. Hering. Eine nicht-Desarguessche zweifach transitive affine Ebene der Ordnung 27. *Abh. Math. Sem. Univ. Hamburg*, 34, pp. 203–208, 1969.

[20] C. Hering. Two new sporadic doubly transitive linear spaces. In C. A. Baker and L. M. Batten, editors, *Finite Geometries*, volume 103 of *Lecture Notes in Pure and Appl. Math.*, pages 127–129, New York, 1985. M. Dekker.

[21] D. R. Hughes and F. C. Piper. *Design Theory*. Cambridge University Press, Cambridge, 1985.

[22] W. M. Kantor. Plane geometries associated with certain 2-transitive groups. *J. Algebra*, 37, pp. 489–521, 1975.

[23] W. M. Kantor. Generalized polygons, SCABs and GABs. In L. A. Rosati, editor, *Buildings and the Geometry of Diagrams, Proceedings Como 1984*, volume 1181 of *Lecture Notes in Math.*, pages 79–158, Berlin, 1986. Springer Verlag.

[24] H. Lüneburg. Some remarks concerning the Ree groups of type (G_2). *J. Algebra*, 3, pp. 256–259, 1966.

[25] H. Lüneburg. *Translation Planes*. Springer Verlag, Berlin, 1980.

[26] M. W. Liebeck. The affine permutation groups of rank 3. *Proc. London Math. Soc.*, 54, pp. 477–516, 1987.

[27] **G. Lunardon and A. Pasini**. A result on C_3 geometries. *European J. Combin.*, 10, pp. 265–271, 1989.

[28] **T. Meixner**. *Klassische Tits Kammersysteme mit einer Transitiven Automorphismengruppe*. Number 174 in Mitt. Math. Sem. Giessen. University of Giessen, Giessen, 1986.

[29] **T. Meixner and A. Pasini**. A census of known flag-transitive extended grids. In A. Beutelspacher, F. Buekenhout, F. De Clerck, J. Doyen, J. W. P. Hirschfeld, and J. A. Thas, editors, *Finite Geometry and Combinatorics*, pages 249–268, Cambridge, 1993. Cambridge University Press.

[30] **A. Pasini**. Geometric and algebraic methods in the classification of geometries belonging to Lie diagrams. *Ann. Discrete Math.*, 37, pp. 315–356, 1988.

[31] **A. Pasini**. On locally polar geometries whose planes are affine. *Geom. Dedicata*, 34, pp. 35–56, 1990.

[32] **A. Pasini**. Quotients of affine polar spaces. *Bull. Soc. Math. Belg.*, 42, pp. 643–658, 1990.

[33] **S. E. Payne and J. A. Thas**. *Finite Generalized Quadrangles*, volume 104 of *Research Notes in Mathematics*. Pitman, Boston, 1984.

[34] **G. M. Seitz**. Flag-transitive subgroups of Chevalley groups. *Ann. of Math.*, 97, pp. 27–56, 1973.

[35] **M. Suzuki**. Transitive extension of a class of doubly transitive groups. *Nagoya Math. J.*, 27, pp. 159–169, 1966.

[36] **G. Tallini**. Ruled graphic systems. In A. Barlotti, editor, *Atti Convegno di Geom. Combin. e sue Appl.*, pages 385–393, Perugia, 1971.

[37] **G. Tallini**. Sistemi grafici rigati. Ist. Mat. Univ. Napoli Rel. n. 8, University of Napoli, 1971.

[38] **G. Tallini**. Strutture di incidenza dotate di polarità. *Rend. Sem. Mat. Fis. Milano*, 41, pp. 3–42, 1971.

[39] **J. Tits**. Non-existence de certaines extensions transitives I. Groupes projectifs à une dimension. *Bull. Soc. Math. Belg.*, 23, pp. 481–492, 1971.

[40] **R. Weiss and S. Yoshiara**. A geometric characterization of the groups HS and Suz. *J. Algebra*, 133, pp. 182–196, 1990.

[41] **E. Witt**. Uber Steinersche Systeme. *Abh. Math. Sem. Univ. Hamburg*, 12, pp. 265–275, 1938.

[42] **S. Yoshiara**. A classification of flag-transitive classical $c.C_2$ geometries by means of generators and relations. *European J. Combin.*, 12, pp. 159–181, 1991.

D. Ghinelli, Dipartimento di Matematica, Università di Roma "La Sapienza", P.le A. Moro, 2 , I-00185 Roma, Italy. e-mail: dina@itcaspur.bitnet

On nonics, ovals and codes in Desarguesian planes of even order

D. G. Glynn*

Abstract

A nonic in $PG(2,q)$, $q = 2^h$, is defined to be either a non-degenerate conic plus its nucleus, or a degenerate conic, different from a repeated line, minus its nucleus. It is noted that every nonic is in the dual-line code of the plane, and so several questions arise. Can we have collections of nonics generating this code? Each oval (or $(q+2)$-arc) is the sum (mod 2) of various numbers of nonics – what is the minimum number?

1. Introduction

There are four kinds of *conics* in a projective plane $\pi := PG(2,q)$ over a finite field $GF(q)$; see [5]. These are:

(1) an irreducible conic, having precisely $q + 1$ points;

(2) two distinct lines, having $2q + 1$ points;

(3) two lines in the quadratic extension $PG(2, q^2)$, intersecting in a point of π, and so having only one point;

(4) a repeated line.

When q is even (and so $q = 2^h$) the first three types of conics define a certain point called the *nucleus*, which lies on the conic if and only if the conic is reducible. In the first case it is the point of intersection of all the tangents; in the second it is the intersection of the two lines; in the third case the nucleus is the only point of π that is on the conic.

A *nonic* is defined to be a conic plus its nucleus: there are three kinds of nonics

(1) an irreducible conic plus its nucleus, having precisely $q+2$ points, which form a $(q + 2)$-arc — an *arc* has no three points collinear;

(2) two distinct lines (a *line-pair*), minus their point of intersection, having $2q$ points;

(3) two lines in the quadratic extension $PG(2, q^2)$, minus their point of intersection, and so having no points in π.

---

*. This research was supported by a von Humboldt Fellowship at the University of Tübingen, and gratitude is due to a delayed rapido between Florence and Bologna.

Let us distinguish the three types of nonics, of types (1), (2), and (3), by calling them *irreducible*, *degenerate*, and *imaginary* respectively. Thus these nonics have $q + 2$, $2q$, and 0 points respectively.

The *dual-line code* of the plane is the vector space (over $GF(q)$) of all functions from the points of π to $GF(q)$ such that the sum of the values on any line is zero — any such function is called a 1-*good function*. The rank (over $GF(p)$ or $GF(q)$, where $q = p^h$, p prime) of the point/line incidence matrix of the Galois plane $PG(2, q)$ is $\left(\frac{p+1}{2}\right)^h + 1$; see [4] or [7]. (Also see [3] for an interpretation and proof of this by counting terms of algebraic 1-good functions.) Since the dimension of a code plus the dimension of its dual is the dimension of the space that contains both codes it follows that the dimension of the dual-line code (with $p = 2$) is

$$q^2 + q + 1 - \left[\left(\tfrac{2+1}{2}\right)^h + 1\right] = q^2 - 3^h + q.$$

Thus the number of 1-good functions of the plane is $q^{q^2 - 3^h + q}$, and there exist sets of $q^2 - 3^h + q$ independent 1-good functions such that every 1-good function is the sum of these in a unique way.

All the codes (over $GF(q)$ or $GF(p)$, where $q = p^h$, p prime) defined as sets of functions that have zero sums for all subspaces of fixed dimension μ in $PG(n, q)$, have been shown to consist of algebraic μ-good functions having terms with certain properties; see [3].

From now on we consider any set of points to be the same as its characteristic function, which is one on the set and zero off it. The interesting thing about nonics is that their characteristic functions are 1-good. To see this it is necessary to show that every line intersects a nonic in an even number of points. (This is obvious with the imaginary nonic, and there are only a few cases to check for the other types: we leave this as an exercise.) Hence we have the following main questions:

(1) Which collections of nonics generate the dual-line code, and can we find collections which are fairly small?

(2) Given any 1-good function $\mathcal{O}$ of π denote by $n(\mathcal{O})$ the minimum number n of nonics such that the function is a sum of n nonics. The problem is to calculate or give bounds for the value of $n(\mathcal{O})$, in particular for the case where $\mathcal{O}$ is an oval (or $(q + 2)$-arc).

2. Cubic curves and nets

As an aid for an answer to the previous questions let us define an *independent net of nonics* to be the set of $q^2 + q + 1$ nonics generated algebraically by three given independent nonics, which have nuclei lying in a triangle of the plane. The net $\mathcal{N}$ can be written $\{\mathcal{N}(n) \mid n \in \pi\}$, where $\mathcal{N}(n)$ is the unique nonic of

156

the net having nucleus n, and where also $\mathcal{N}(n) = \mathcal{C}(n) + n$, where $\mathcal{C}(n)$ is the conic of the net with nucleus n. Note that we are considering characteristic functions when we write $\mathcal{C}(n)$ and n.

The *simplest example* of an independent net is that generated by the three line-pairs of a triangle. Then the three nuclei are the vertices of the triangle. Naturally, any minimal sum of nonics can be assumed to not contain imaginary nonics because these have no points in π.

Here we include some of the relevant parts of the theory of cubic curves in a plane π over a field K of characteristic two; see, for example, [2].

A general *cubic curve* of π is defined to be a set of points

$$C(A, a) := \{(x, y, z) \in \pi \mid (x, y, z)A(x^2, y^2, z^2)^t + axyz = 0\},$$

where A is a 3×3 matrix over K and $a \in K$, such that A and a are not both zero.

The *first polar* with respect to $C = C(A, a)$ of a point (x, y, z) of π is the conic

$$(x, y, z)C := Q[(x, y, z)A, a(x, y, z)] :=$$

$$\{(r, s, t) \mid (x, y, z)A(r^2, s^2, t^2)^t + a(xst + yrt + zrs) = 0\},$$

if $(x, y, z)A$ and a are not both zero.

The tangents of C passing through a general point (x, y, z) of π are found by finding the intersection of the first polar of (x, y, z) with C, then joining these points to (x, y, z). If $a \neq 0$, the nucleus of the first polar conic at (x, y, z), $Q[(x, y, z)A, a(x, y, z)]$, is (x, y, z), and clearly the set of all first polars to such a cubic with $a \neq 0$ is an independent net. Conversely, three conics with independent nuclei (e.g. $(1, 0, 0)$, $(0, 1, 0)$, $(0, 0, 1)$) determine a cubic curve, which must have $a \neq 0$. Thus, a plane cubic curve with $a \neq 0$ determines and is determined by an independent net. The cubic curve of the net is given by the nuclei of the reducible conics (or of the degenerate and imaginary nonics) in the net. In the previous example the corresponding cubic curve is the degenerate cubic which is the product of the three lines of the triangle.

3. Some answers

Now let us return to the finite world where $\pi = PG(2, q)$, q even.

Lemma 3.1 *If f is a 1-good function of π and $\mathcal{N} = \{\mathcal{N}(n) \mid n \in \pi\}$ is an independent net of nonics then*

$$f = \Sigma_{n \in \pi} f(n).\mathcal{N}(n).$$

(Note that $\mathcal{N}(n)$ is the characteristic function of the nonic with nucleus n in $\mathcal{N}$.)

Proof. The conics of the net which pass through any fixed point P of π form a linear pencil, the nuclei of which lie on a line. Hence the sum of conics, which have values assigned from a function in the dual-line code, is zero. The corresponding sum of nonics gives the 1-good function. Algebraically we have the following:

$$\Sigma_{n \in \pi} f(n).\mathcal{N}(n) = \Sigma f(n).(\mathcal{C}(n) + n)$$
$$= \Sigma f(n).\mathcal{C}(n) + \Sigma f(n).n = 0 + \Sigma f(n).n = f.$$

$\square$

Theorem 3.2 *Any independent net $\mathcal{N}$ of nonics generates the dual-line code of the plane.*

Proof. Any nonic has a characteristic function that is 1-good. Hence the net generates a sub-code. Now let f be any 1-good function. The above Lemma shows that f is contained in the sub-code generated by the nonics of $\mathcal{N}$, and so the sub-code is the whole code of 1-good functions. $\square$

From this theorem we obtain a series of results which shed a lot of light on the main questions posed above.

Corollary 3.3 *There is a set of $q^2 + q + 1$ irreducible nonics generating every 1-good function.*

Proof. Consider a cubic curve made up of three lines of a triangle in the cubic extension of π, and having no points of π. All the nonics in the polar net are irreducible because reducible conics in the net of a cubic curve come from points of the cubic in π. (The nonics of such a net form an orbit under a *Singer cycle* of the plane.) $\square$

Theorem 3.4 $n(\mathcal{O}) \leq q - 1$, *for all ovals $\mathcal{O}$ of π.*

Proof. Consider the polar net corresponding to any triangle of points on $\mathcal{O}$. Since any nonic has a characteristic function that is 1-good, we see by the Lemma that the oval is the sum of the $q + 2$ nonics of the net having nuclei on the oval. Of these nonics $q - 1$ are irreducible because their nuclei do not lie on the triangle. The remaining three nonics are reducible. However, it is perhaps unexpected that the sum of these three nonics is zero, because each point on a line of the triangle is on two of the three nonics, and each point not in the triangle is obviously on none of the three degenerate nonics of the net. Hence the oval is the sum of $q - 1$ irreducible nonics. $\square$

Note that when $q \geq 4$ it is possible to find a triangle of lines that are all external to a given oval. Then taking the polar net of the triangle would give a representation of the oval as the sum of $q + 2$ irreducible nonics. Also, the set of nuclei of the imaginary nonics of any independent net form an arc ; see [2]. If a cubic curve (with xyz term) can be found so that it intersects the oval in a fairly large number of points of this imaginary type, then the number of nonics (of the other two types) needed to sum to the oval is reduced. This observation is the stimulus for the following.

Theorem 3.5 *For the irregular oval (18-arc) $\mathcal{I}$ of $PG(2,16)$ there holds $n(\mathcal{I}) \leq 9$. This oval is the sum of 9 irreducible nonics.*

Proof. There is a nice representation of $\mathcal{I}$ as a sum modulo 2 of two cubic curves, one of which has equation $x^3 + y^3 + z^3 + \delta xyz = 0$, and the other with equation $x^3 + y^3 + z^3 + \delta^4 xyz = 0$, for some $\delta \in GF(16) \backslash GF(4)$; see [2]. (In the notation of §2 the matrix A is the 3×3 identity matrix and $a = \delta$ or δ^4.) These cubics both have 18 points, and they intersect in 9 points (the inflections) of the Baer subplane $PG(2,4)$ consisting of the points with coordinates in $GF(4)$. Consider one of these curves. The polar conics at points of the curve in the subplane are all line-pairs, while the polar conics at points not in the Baer subplane are imaginary — these 9 points are in $\mathcal{I}$. Using Lemma 3.1 we see that $\mathcal{I}$ is the sum of 9 imaginary nonics and 9 irreducible nonics. However we can omit the 9 imaginary nonics as they have no points in $PG(2,16)$. $\square$

It is to be expected that even fewer nonics can be found to add to a given oval, because in the cases above we have only used 3 independent nonics. More than 3 (up to 6) independent nonics could be used. The question about the minimum number of nonics is applicable to any element of the dual-line code of the plane, but the author hasn't tried to answer that more general question. Other linear collections of algebraic curves (or varieties) could be used to generate codes in $PG(n, q)$, but nonics do appear to be a very nice example of this kind of phenomenon.

References

[1] **D. G. Glynn.** A condition for the existence of ovals in PG(2,q), q even. *Geom. Dedicata*, 32, pp. 247–252, 1989.

[2] **D. G. Glynn.** On cubic curves in projective planes of characteristic two. Preprint, 1992.

[3] **D. G. Glynn and J. W. P. Hirschfeld.** On the classification of projective geometry codes by algebraic functions, 1992. In preparation.

[4] **R. L. Graham and J. MacWilliams.** On the number of information symbols in difference-set cyclic codes. *Bell Sys. Tech. J.*, 45, pp. 1057–1070, 1966.

[5] **J. W. P. Hirschfeld**. *Projective Geometries over Finite Fields*. Clarendon Press, Oxford, 1979.

[6] **D. R. Hughes and F. C. Piper**. *Projective Planes*. Springer Verlag, Berlin, 1973.

[7] **K. J. C. Smith**. Majority decodable codes derived from finite geometries. Institute of Statistics Mimeo Series 561, University of North Carolina, 1967.

D. G. Glynn, Department of Mathematics, University of Canterbury, Christchurch, New Zealand. e-mail: dgg@math.canterbury.ac.nz

Orbits of arcs in projective spaces

C. E. Gordon

Abstract

An $M \times N$ matrix is associated with each ordered k-arc in a finite projective space of order N ($k = M + N + 2$.) The matrix is a projective invariant for ordered k-arcs in the space. The set of these matrices is denoted by Ω. Elements of the symmetric group S_k act on ordered arcs by permuting points. This induces a definition of S_k as a group of operators on Ω whose orbits correspond to projectively distinct unordered k-arcs. Application of theorems of Burnside and Cauchy leads to results concerning the number of orbits of k-arcs in $PG(N,q)$ under projectivity and under collineation. A subset of Ω is defined which contains representatives of each orbit under S_k. The reduced set of "normal" matrices is used by a counting algorithm. The results of this paper are applied to counting the projectively distinct unordered k-arcs (all k) in $PG(2,11)$ and $PG(2,13)$.

1. Introduction

This paper is concerned with the problem of counting equivalence classes of ordered k-arcs, unordered k-arcs, and k-gons with respect to either projectivity or collineation. Taking into account the three types of arc and two equivalence relations, there are six distinct but related problems to be solved. The six problems can each be stated in terms of the orbits of *ordered arcs* under some group of operators. The solutions, themselves, fall into two categories: formulas for the number of orbits and algorithms for counting the orbits. The main idea is to select a set Ω which is in one-to-one correspondence with the projectivity classes of ordered k-arcs and then to find groups of operators on Ω whose orbits are in one-to-one correspondence with the sets in which we are interested. Burnside's formula for counting orbits can then be applied to these particular groups. The set Ω will be chosen so that its elements can be easily represented in computer language and so that the groups of operations on these representations can be easily implemented. The ideas presented here appear in a restricted context and from a different perspective in [4] and [3].

The following notation and terminology will be used. Let p be a prime, $q = p^h$, and $\mathcal{F} = GF(q)$. A nonzero vector X in $V(N+1, \mathcal{F})$ which generates a point X of the N-dimensional projective space $PG(N, \mathcal{F}) = PG(N, q)$ is a *homogeneous coordinate representation* of the point. Throughout the paper, M and N represent arbitrary positive integers and $k = N + 2 + M$. The sets of ordered k-arcs, unordered k-arcs, and k-gons in $PG(N, \mathcal{F})$ will be denoted by $\mathcal{O}_k$, $\mathcal{U}_k$, and $\mathcal{C}_k$ respectively. The projectivity and the collineation groups of $PG(N, \mathcal{F})$, when treated as groups of operators on any of the sets $\mathcal{O}_k$, $\mathcal{U}_k$, $\mathcal{C}_k$ will be denoted ambiguously by *PROJ* and *COLL* respectively. Equivalence of arcs α and β with respect to projectivity or collineation will be denoted by $\alpha \underset{\text{proj}}{=} \beta$ and $\alpha \underset{\text{coll}}{=} \beta$. The existence of a bijection between sets X and Y will be denoted by $X \sim Y$. The set of orbits in a set X under a group of operators G will be denoted by X/G.

2. From operators on ordered arcs to operators on projective invariants

This section provides some development which is independent of the particular choice of the set Ω, discussed in the introduction.

Definition 2.1 A complete projective invariant on $\mathcal{O}_k$ is a pair (χ, Ω), where Ω is a set and χ is a function from $\mathcal{O}_k$ onto Ω such that, for $\alpha, \beta \in \mathcal{O}_k$, $\alpha \underset{\text{proj}}{=} \beta$ if and only if $\alpha^\chi = \beta^\chi$.

For the remainder of this section, let (χ, Ω) be an arbitrary complete projective invariant for $\mathcal{O}_k$.

Definition 2.2 A group G of operators on $\mathcal{O}_k$ is compatible with projective equivalence if for any $\alpha, \beta \in \mathcal{O}_k$ and any $g \in G$,

$$\alpha \underset{\text{proj}}{=} \beta \implies \alpha^g \underset{\text{proj}}{=} \beta^g.$$

The function χ can be viewed as a one-to-one correspondence between Ω and the set of projectivity classes in $\mathcal{O}_k$. Thus a group of operators on $\mathcal{O}_k$ which is compatible with projective equivalence can be viewed as a group of operators on Ω. This is restated by the following.

Theorem 2.3 *If G is a group of operators on $\mathcal{O}_k$, which is compatible with projective equivalence, and if (χ, Ω) is a complete projective invariant on $\mathcal{O}_k$, then the group G can be made to act as a group of operators on Ω such that, for any $\alpha \in \mathcal{O}_k$,*

$$\alpha^{\chi g} = \alpha^{g \chi}. \tag{1}$$

The next task is to define the six groups of operators on $\mathcal{O}_k$ mentioned in the introduction. It will be left to the reader to verify that each definition describes a well-defined homomorphism from the specified group into the group of bijections on Ω. The six groups are:

(1) The trivial group $\{e\}$ with e fixing each ordered k-arc.

(2) The symmetric group S_k with $g \in S_k$ acting on an ordered k-arc to permute its points. If $g^{-1} = h$, then

$$(X_1, \ldots, X_k)^g = (X_{1^h}, \ldots, X_{k^h})$$

(3) The cyclic group Z_k with generator, 1, acting on an ordered arc in the same way that the cycle $(1 \cdots k)$ of S_k does.

(4) The cyclic group Z_h with generator, 1, acting on an ordered arc α to produce α^φ, where φ is the field automorphism: $x \mapsto x^p$. An automorphism acts on an ordered arc by acting coordinatewise on homogeneous coordinate representations of its points.

(5) The direct product $Z_h \times S_k$, where $\alpha^{(g,h)} = (\alpha^g)^h$.

(6) The direct product $Z_h \times Z_k$, where $\alpha^{(g,h)} = (\alpha^g)^h$.

Theorem 2.4 *For each group G in the first row of the following table, Ω/G is in one-to-one correspondence with the corresponding set in the second row.*

$G =$	$\{e\}$	S_k	Z_k	Z_h	$Z_h \times S_k$	$Z_h \times Z_k$
$\Omega/G \sim$	$\mathcal{O}_k/PROJ$	$\mathcal{U}_k/PROJ$	$\mathcal{C}_k/PROJ$	$\mathcal{O}_k/COLL$	$\mathcal{U}_k/COLL$	$\mathcal{C}_k/COLL$

Proof. For any $T \in PROJ$ and $f \in Z_h$, let T^f be the projectivity such that, for $\alpha \in \mathcal{O}_k$,

$$\alpha^{Tf} = \alpha^{fT^f}.$$

Let $Z_h \otimes PROJ$ be the group with elements from $Z_h \times PROJ$ and operation defined by

$$(g, T) \cdot (f, S) = (gf, T^f S).$$

Now if G_1 is either S_k or Z_k and G_2 is either $PROJ$ or $Z_h \otimes PROJ$, then, as operators on $\mathcal{O}_k$, elements of G_1 commute with elements of G_2 so $G_1 \times G_2$ is well defined by $\alpha^{(g,h)} = (\alpha^g)^h$. In addition, this defines G_1 as a group of operators on $\mathcal{O}_k/G_2$ and G_2 as a group of operators on $\mathcal{O}_k/G_1$. We now have the following:

(1) $\Omega/\{e\} \sim \Omega \sim \mathcal{O}_k/PROJ$.

(2) Ω/S_k
$$\sim (\mathcal{O}_k/PROJ)/S_k \sim \mathcal{O}_k/(S_k \times PROJ) \sim (\mathcal{O}_k/S_k)/PROJ \sim \mathcal{U}_k/PROJ.$$

(3) Ω/Z_k
$$\sim (\mathcal{O}_k/PROJ)/Z_k \sim \mathcal{O}_k/(Z_k \times PROJ) \sim (\mathcal{O}_k/Z_k)/PROJ \sim \mathcal{C}_k/PROJ.$$

(4) $\Omega/Z_h \sim \mathcal{O}_k/(Z_h \otimes PROJ) \sim \mathcal{O}_k/COLL$.

(5) $\Omega/(Z_h \times S_k) \sim \mathcal{O}_k/(S_k \times (Z_h \otimes PROJ)) \sim (\mathcal{O}_k/S_k)/COLL \sim \mathcal{U}_k/COLL$.

(6) $\Omega/(Z_h \times Z_k) \sim \mathcal{O}_k/(Z_k \times (Z_h \otimes PROJ)) \sim (\mathcal{O}_k/Z_k)/COLL \sim C_k/COLL.$

$\square$

3. A complete projective invariant on ordered arcs

In this section a particular complete projective invariant (χ, Ω) is defined and the operations on Ω associated with the five nontrivial groups of Theorem 2.4 are described. Let $U_0, \ldots, U_{N+1}$ be the points with homogeneous coordinate representations $U_0 = (1, 0, \ldots, 0), \ldots, U_N = (0, \ldots, 0, 1), U_{N+1} = (1, \ldots, 1)$.

Definition 3.1 An ordered k-arc in $PG(N, \mathcal{F})$ is standard if its first $N + 2$ points are $U_0, \ldots, U_{N+1}$ respectively.

Given any ordered k-arc, there is a unique projectivity which maps it to a standard ordered arc. Any point of a standard arc other than $U_1, \ldots, U_{N+1}$ has a unique affine coordinate representation $(x_1, \ldots, x_N)$ that is, it has homogeneous coordinate representation $(x_1, \ldots, x_N, 1)$. These facts justify the following definition and theorem.

Definition 3.2 (a) χ is the function which assigns to each ordered k-arc α in $PG(N, \mathcal{F})$ the $M \times N$ matrix $[x_{i,j}]$ over $\mathcal{F}$, whose ith row $(i = 1, \ldots, M)$ is the affine coordinate representation of the $(N + 2 + i)$th point of the unique standard ordered arc which is projectively equivalent to α. (b) Ω is the range of χ.

Theorem 3.3 The pair (Ω, χ) is a complete projective invariant for ordered k-arcs.

The following provides a useful criteria for determining whether a given matrix is in Ω. It appears in [3] as Theorem 1.3 and also, in the context of coding theory, as Theorem 8 on page 321 of [6].

Theorem 3.4 An $M \times N$ matrix A over $\mathcal{F}$ is an element of Ω if and only if an $(M + 1) \times (N + 1)$ matrix A^+ formed from it by adjoining a row and column of 1's has the property that all its minors are nonzero.

The next theorem provides a formula for α^χ in terms of a homogeneous coordinate representation for α. For convenience, the following notation is introduced.

Definition 3.5 Given any sequence η of k vectors from $V(N+1, \mathcal{F})$, $|\eta|$ is the determinant of the first $N + 1$ vectors in η. If a permutation $\sigma \in S_k$ acts on η by permuting vectors, the determinant of the first $N + 1$ vectors of the result will be denoted by $|\eta\sigma|$.

Theorem 3.6 *If α' is any sequence of vectors representing the points of an ordered k-arc α, then α^χ is the $M \times N$ matrix $[x_{i,j}]$, where*

$$x_{i,j} = \frac{|\alpha'(j\ N + 2 + i)|/|\alpha'(j\ N + 2)|}{|\alpha'(N + 1\ N + 2 + i)|/|\alpha'(N + 1\ N + 2)|} \tag{2}$$

Proof. Let $\alpha' = (A_1, \ldots, A_k)$. Let β be the standard ordered k-arc which is projectively equivalent to α. Let T be the unique projectivity which maps β to α and let $\mathcal{T}$ be the matrix for the unique nonsingular linear transformation which is associated with T and which maps vector U_{N+1} to vector A_{N+2}. The matrix $\mathcal{T}$ is found as follows. Let $h_1, \ldots, h_{N+1}$ be such that T maps $U_0, \ldots, U_N$ to

$$h_1 A_1, \ldots, h_{N+1} A_{N+1}. \tag{3}$$

Now $\mathcal{T}$ has row vectors (3). Solving equation $U_{N+1}\mathcal{T} = A_{N+2}$ for $h_1, \ldots, h_{N+1}$ produces

$$h_j = \frac{|\alpha'(j\ N + 2)|}{|\alpha'|}, \qquad \text{for } j = 1, \ldots, N + 1. \tag{4}$$

Having determined $\mathcal{T}$, we can find vectors $Y_1, \ldots, Y_M$ such that, for $i = 1, \ldots, M$,

$$Y_i \mathcal{T} = A_{N+2+i}. \tag{5}$$

Letting $Y_i = (y_{i,1}, \ldots, y_{i,N+1})$ and solving (5) for Y_i we get

$$y_{i,j} = \frac{|\alpha'(j\ N + 2 + i)|}{h_j |\alpha'|}, \qquad \text{for } j = 1, \ldots, N + 1. \tag{6}$$

Converting to affine coordinates by setting $x_{i,j} = y_{i,j}/y_{i,N+1}$, we get (2). $\quad\square$

Remark 3.7 If $M = N = 1$, then α is an ordered tetrad on the projective line and α^χ is the familiar cross-ratio; so this projective invariant is, in some sense, a generalized cross-ratio.

The next order of business is to investigate the action of S_k on Ω. The following easily verified facts will be useful.

Lemma 3.8 *Let η be a sequence of k vectors from $V(N + 1, \mathcal{F})$.*

(a) If $1 \leq i \leq N+1, 1 \leq j \leq N+1, 1 \leq m \leq k, 1 \leq n \leq k$, and i, j, m, and n are distinct, then $|\eta(i \; j)(m \; n)| = -|\eta(m \; n)|$.

(b) If $1 \leq i \leq N+1, 1 \leq j \leq N+1, 1 \leq m \leq k$, and i, j, and m are distinct, then $|\eta(i \; j)(i \; m)| = -|\eta(j \; m)|$.

(c) If $N+2 \leq i \leq k, N+2 < j \leq k, 1 \leq m \leq k, 1 \leq n \leq k$, and i, j, m, and n are distinct, then $|\eta(i \; j)(m \; n)| = +|\eta(m \; n)|$.

(d) If $N+2 \leq i \leq k, N+2 \leq j \leq k, 1 \leq m \leq k$, and i, j, and m are distinct, then $|\eta(i \; j)(i \; m)| = +|\eta(j \; m)|$.

If α and α' are as above and $g \in S_k$, then, by Theorem 3.6, $\alpha^{gx} = [y_{i,j}]$, where

$$y_{i,j} = \frac{|\alpha'g(j \; N+2+i)|/|\alpha'g(j \; N+2)|}{|\alpha'g(N+1 \; N+2+i)|/|\alpha'g(N+1 \; N+2)|}. \tag{7}$$

In order to make the notation of this paper consistent with that of earlier papers [3] and [4], rename the symbols $1, \ldots, k$ on which S_k acts as $c_1, \ldots, c_N, c_0, r_0, r_1, \ldots, r_M$ respectively.

Definition 3.9 (a) $R_i = (r_0 \; r_i)$, (for $i = 1, \ldots, M$). (b) $C_j = (c_0 \; c_j)$, (for $j = 1, \ldots, N$). (c) $J = (r_0 \; c_0)$.

S_k is generated by transpositions $R_1, \ldots, R_M, C_1, \ldots, C_N, J$. The actions of these operators on Ω are described in the next theorem.

Theorem 3.10 Let $A = [x_{i,j}]$ be an element of Ω. Let $n = 1, \ldots, N$ and $m = 1, \ldots, M$.

(a) $A^{R_m} = [y_{i,j}]$, where $y_{m,j} = 1/x_{m,j}$ and, if $i \neq m$, then $y_{i,j} = x_{i,j}/x_{m,j}$.

(b) $A^{C_n} = [y_{i,j}]$, where $y_{m,j} = 1/x_{m,j}$ and, if $i \neq m$, $y_{i,j} = x_{i,j}/x_{m,j}$.

(c) $A^{J} = [y_{i,j}]$ where $y_{i,j} = 1 - x_{i,j}$.

Proof. Pick α', a homogeneous coordinate representation for some ordered arc α such that $\alpha^x = A$. By (1), $A^g = \alpha^{gx}$. For part (a), let $g = (N+2 \; N+2+m)$ in equation (7). Using Lemma 3.8 to simplify yields

$$i \neq m \implies y_{i,j} = \frac{|\alpha'(j \; N+2+i)|/|\alpha'(j \; N+2+m)|}{|\alpha'(N+1 \; N+2+i)|/|\alpha'(N+1 \; N+2+m)|} = \frac{x_{i,j}}{x_{m,j}};$$

$$y_{m,j} = \frac{|\alpha'(j \; N+2)|/|\alpha'(j \; N+2+m)|}{|\alpha'(N+1 \; N+2)|/|\alpha'(N+1 \; N+2+m)|} = \frac{1}{x_{m,j}}. \tag{8}$$

Part (b) is similar to part (a). Part (c) is more difficult. Let $g = (N+1 \; N+2)$. In this case, Lemma 3.8 and equation (7) give

$$y_{i,j} = \frac{|\alpha'(N+1 \; N+2)(j \; N+2+i)|/-|\alpha'(j \; N+2)|}{|\alpha'(N+1 \; N+2+i)|/|\alpha'|}. \tag{9}$$

It is required to show that $x_{i,j} + y_{i,j} = 1$. Using equations (2) and (9), this condition becomes

$$|\alpha'||\alpha'(N+1\ N+2)(j\ N+2+i)| - |\alpha'(j\ N+2+i)||\alpha'(N+1\ N+2)|$$
$$+|\alpha'(N+1\ N+2+i)||\alpha'(j\ N+2)| = 0$$
$$\tag{10}$$

Let the vectors in positions j, $N+1$, $N+2$, and $N+2+i$ in α' be A, B, C, and D respectively and the first $N+1$ vectors of α', other than A and B be $H_1, \ldots, H_{N-1}$. For notational convenience, perform identical row permutations on all the determinants involved in (10) to move vectors in positions j and $N+1$ to positions 1 and 2. The equation then becomes

$$\begin{vmatrix} A \\ B \\ H_1 \\ \vdots \\ H_{N-1} \end{vmatrix} \begin{vmatrix} D \\ C \\ H_1 \\ \vdots \\ H_{N-1} \end{vmatrix} - \begin{vmatrix} D \\ B \\ H_1 \\ \vdots \\ H_{N-1} \end{vmatrix} \begin{vmatrix} A \\ C \\ H_1 \\ \vdots \\ H_{N-1} \end{vmatrix} + \begin{vmatrix} A \\ D \\ H_1 \\ \vdots \\ H_{N-1} \end{vmatrix} \begin{vmatrix} C \\ B \\ H_1 \\ \vdots \\ H_{N-1} \end{vmatrix} = 0 \tag{11}$$

To verify this determinant identity, let Q be the $(2N+2) \times (2N+2)$ matrix shown in (12). Then Q' is obtained from Q by elementary row and column operations. Since every $(N+1) \times (N+1)$ minor in the last $N+1$ columns of Q' has value 0, a Laplace expansion along those rows produces a determinant of 0. Hence the determinant of Q is 0.

$$Q = \begin{bmatrix} A & A \\ B & B \\ C & C \\ D & D \\ H_1 & 0 \\ \vdots & \vdots \\ H_{N-1} & 0 \\ 0 & H_1 \\ \vdots & \vdots \\ 0 & H_{N-1} \end{bmatrix} \qquad Q' = \begin{bmatrix} A & 0 \\ B & 0 \\ C & 0 \\ D & 0 \\ H_1 & 0 \\ \vdots & \vdots \\ H_{N-1} & 0 \\ 0 & H_1 \\ \vdots & \vdots \\ 0 & H_{N-1} \end{bmatrix} \tag{12}$$

Now consider a Laplace expansion of Q along its first $N+1$ columns. If an $(N+1 \times N+1)$ minor in the first N rows is nonzero and its complementary minor is also nonzero, then the minor and its complement must each contain rows $H_1, \ldots, H_{N-1}$, as well as two of the first four rows. The Laplace expansion for $|Q|$ produces

$$2\left(\begin{vmatrix} A \\ B \\ H_1 \\ \vdots \\ H_{N-1} \end{vmatrix} \begin{vmatrix} C \\ D \\ H_1 \\ \vdots \\ H_{N-1} \end{vmatrix} - \begin{vmatrix} A \\ C \\ H_1 \\ \vdots \\ H_{N-1} \end{vmatrix} \begin{vmatrix} B \\ D \\ H_1 \\ \vdots \\ H_{N-1} \end{vmatrix} + \begin{vmatrix} A \\ D \\ H_1 \\ \vdots \\ H_{N-1} \end{vmatrix} \begin{vmatrix} B \\ C \\ H_1 \\ \vdots \\ H_{N-1} \end{vmatrix}\right). \qquad (13)$$

Since $|Q| = 0$, equation (11) follows. $\qquad\square$

Corollary 3.11 *A permutation of the symbols $c_1, \ldots, c_N$ acts as a column permutation on a matrix in Ω while a permutation of the symbols $r_1, \ldots, r_M$ is a row permutation.*

Proof. It is sufficient to consider transpositions. For distinct $i1$ and $i2$ in $1, \ldots, M$, $(r_{i1}\ r_{i2}) = R_{i1}R_{i2}R_{i1}$. It follows from Theorem 3.10(a) that this acts on a matrix $A \in \Omega$ to permute rows $i1$ and $i2$. Similarly $(c_{j1}\ c_{j2})$ permutes columns $j1$ and $j2$. $\qquad\square$

Theorem 3.10 associates with each permutation in S_k a certain bijection on Ω. This association is a homomorphism but not necessarily an isomorphism. In particular, it is well known that if $M = N = 1$, then the 4! permutations of an ordered tetrad correspond to only 3! distinct operations on cross-ratios. This is however an exceptional case, arising because when $k = 4$, S_k has an "extra" normal subgroup. The general situation is described in the next theorem.

Theorem 3.12 *Assume that $\Omega \neq \emptyset$. Let Γ be the group of bijections on Ω associated with elements of S_k.*
   (a) If $|\mathcal{F}| \geq 5$ and $k \geq 5$ then $\Gamma \cong S_k$.
   (b) If $|\mathcal{F}| \geq 5$ and $k = 4$, then $\Gamma \cong S_3$.
   (c) If $|\mathcal{F}| = 4$, then $\Gamma \cong S_2$.
   (d) If $|\mathcal{F}| = 3$, then $\Gamma \cong S_1$.

Proof. The group Γ is a homomorphic image of S_k. If $k \neq 4$, the only possible homomorphic images of S_k are (up to isomorphism) S_1, S_2, and S_k. If $k = 4$ then, in addition to these, S_3 is possible.

If $|\mathcal{F}| = 3$, then $|\Omega| = 1$ and $|\Gamma| = 1$ so $\Gamma \cong S_1$. If $|\mathcal{F}| = 4$, then $|\Omega| = 2$ and $|\Gamma| = 2$ so $\Gamma \cong S_2$. If $|\mathcal{F}| \geq 5$ and $k = 4$, then $M = N = 1$ and $R_1 = C_1$; but I (the identity element of S_k), J, and R_1 are distinct; so $\Gamma \cong S_3$. Assume the hypothesis of part (a). If $M \geq 3$, then $I, (c_1\ c_2)$, and $(c_2\ c_3)$ are distinct; so $\Gamma \cong S_k$. Similarly, if $N \geq 3$, then $\Gamma \cong S_k$. If $M = N = 2$, then $I \neq J$, $I \neq C_1$, $I \neq R_1$, and $C_1 \neq R_1$; so $|\Gamma| \geq 3$ and $\Gamma \cong S_k$. If $M = 1$ and $N = 2$, then I, J, and R_1 are distinct so $\Gamma \cong S_k$. Similarly, if $N = 1$ and $M = 2$, then $\Gamma \cong S_k$. □

So far this section has only dealt with S_k, the most interesting of the six groups described in Section 1. Z_h is essentially the automorphism group on $\mathcal{F}$. An automorphism g acts on an element $[x_{i,j}]$ of Ω to produce $[x_{i,j}^g]$. The group Z_k is isomorphic to the subgroup of S_k generated by the k-cycle

$$(1 \ \ldots \ k) = (r_0 \ r_M)(r_0 \ c_0)(c_0 \ c_N)(c_1 \ \ldots \ c_N)(r_1 \ \ldots \ r_M).$$

This acts by applying $R_M J C_N$ followed by cyclic permutations of the columns and rows. For example, if $M = N = 3$, then

$$\begin{bmatrix} a & b & c \\ d & e & f \\ g & h & i \end{bmatrix} \longrightarrow \begin{bmatrix} \frac{i}{i-1} & i \cdot \frac{g-1}{i-1} & i \cdot \frac{h-1}{i-1} \\ \frac{i}{i-c} & i \cdot \frac{g-a}{i-c} & i \cdot \frac{h-b}{i-c} \\ \frac{i}{i-f} & i \cdot \frac{g-d}{i-f} & i \cdot \frac{h-e}{i-f} \end{bmatrix}$$

The action of each of the two direct products groups is determined by the action of its component groups.

If one is willing to restrict attention to those (unordered) k-arcs extending a fixed $(N+2)$-arc, one can view an element of Ω as a representation for an arc and one can view a "projectivity" on arcs as a member of the group Γ generated by the operations of Theorem 3.10. This is valid in the sense that two of these arcs are projectively equivalent if and only if some group element maps one to the other. However, if one projectivity maps a pair of arcs to another pair, it may require two different elements of Γ to effect the same result. The "collineations" of this theory are much like the collineations of $PG(N, \mathcal{F})$, in that each is a product of operation from S_k with an operation from Z_h. A difference however is that each member of S_k commutes with each member of Z_h so that the group is a direct product. Projectivities do not commute with field automorphisms.

4. Duality

Given $k = M + 2 + N$, let $\Omega(M, N, \mathcal{F})$ denote the set Ω of Definition 3.2. From Theorem 3.4, an $M \times N$ matrix A is in $\Omega(M, N, \mathcal{F})$ if and only if its transpose A^{T} is in $\Omega(N, M, \mathcal{F})$. The symmetric group on symbols $r_0, \ldots, r_M, c_0, \ldots c_N$, as the group of operators on $\Omega(M, N, \mathcal{F})$ generated by $R_1, \ldots, R_M, C_1, \ldots, C_N, J$, is isomorphic to the symmetric group on symbols $c_0, \ldots, c_M, r_0, \ldots r_N$, as the group of operators on $\Omega(N, M, \mathcal{F})$ generated by $C_1, \ldots, C_M, R_1, \ldots, R_N, J$. This provides a one-to-one correspondence between each of the six sets of orbits ($\mathcal{U}_k/PROJ$, $\mathcal{O}_k/PROJ$, $\mathcal{C}_k/PROJ$, $\mathcal{U}_k/COLL$, $\mathcal{O}_k/COLL$, and $\mathcal{C}_k/COLL$) in $PG(N, \mathcal{F})$ and the corresponding set of orbits in $PG(M, \mathcal{F})$.

Leo Storme has pointed out that, in regard to $\mathcal{U}_k/PROJ$, this is the same one-to-one correspondence provided by the "duality" relationship of [8] and [7].

5. Formulas for numbers of orbits

The number of orbits of ordered k-arcs under projectivity is the cardinality of Ω. This is the same as the total number of k-arcs extending a given $(N+2)$-arc. For $k \leq 8$ this value can be derived from formulas presented in [2].

A well known result of Burnside gives a nice formula for the number of orbits in a set which is acted upon by a group of operators. This formula is in terms of the number of elements of the set which are fixed by each element of the group. The essence of this result can be found in Theorem VII, Chapter X of [1]. A variant of Burnside's result (Theorem 5.1) utilizes the fact that if two elements of a group are conjugates then there is a one-to-one correspondence between their sets of fixed points.

Theorem 5.1 (Burnside) *If G is a group of operators on a set X and S is a subset of G which contains exactly one element from each conjugacy class of G, then the number of orbits in X is*

$$\frac{\sum_{g \in S} \mathrm{Fix}_g \cdot \mathrm{Conj}_g}{|G|},$$

where Fix_g is the number of elements of X fixed by g and Conj_g is the number of conjugates of g in G.

Two elements of a symmetric group are conjugates if and only if they have the same cycle structure ([5], p.173). A formula for the number of conjugates of a given permutation is provided by Cauchy ([5], p.73). In this formula, each element g of the symmetric group S_k is to be thought of as a product of disjoint cycles, including cycles of length 1. For example, the identity permutation just consists of k-cycles of length 1.

Definition 5.2 For each element g of S_k,

$$A_g = \prod_{i=1}^{k} \left(A_g(i) \cdot i^{A_g(i)} \right),$$

where $A_g(i)$ is the number of cycles of length i in g.

Theorem 5.3 (Cauchy) *The number of conjugates of an element g of S_k is given by;*

$$\mathrm{Conj}_g = \frac{k!}{A_g}.$$

170

Definition 5.4 (a) For each $g \in S_k$ and each $t \in Z_h$, $\Phi(t, g)$ is the number of elements A of Ω such that $A^t = A^g$. (b) $F_g = \Phi(0, g)$ (c) $H_g = \frac{1}{h} \cdot \sum_{t=0}^{h-1} \Phi(t, g)$

Theorem 5.5 *Let S be any subset of S_k containing exactly one element from each conjugacy class. The numbers of orbits of k-arcs under projectivity and under collineation in $PG(N, \mathcal{F})$ are, respectively,*

$$\sum_{g \in S} \frac{F_g}{A_g} \quad and \quad \sum_{g \in S} \frac{H_g}{A_g}.$$

Proof. The orbits referred to in the theorem are in one-to-one correspondence with the orbits in Ω under the S_k and $Z_h \times S_k$. The result follows easily from Theorems 5.1 and 5.3 and the fact that, if S is any subset of S_k which contains exactly one representative from each conjugacy class, then, since Z_h is commutative, $Z_h \times S$ contains exactly one representative of each conjugacy class in $Z_h \times S_k$.

6. Normalization

In this section, a set of "normal" matrices from Ω is considered along with a set of representatives of the left cosets of a certain subgroup Π of S_k. This reduced set of matrices and reduced set of operations have application in algorithms for counting orbits.

Definition 6.1 (a) $\rho_{i1,i2} = (r_{i1} \; r_{i2})$ and $\kappa_{i1,i2} = (c_{i1} \; c_{i2})$, where $i1$ and $i2$ are distinct elements of $\{1, \ldots, M\}$ and $j1$ and $j2$ are distinct elements of $\{1, \ldots, N\}$. (b) Π is the subgroup of S_k generated by all $\rho_{i1,i2}$ and $\kappa_{j1,j2}$.

Let I be the identity element of S_k. The following lemmas are easily verified. Parts (a), (b), (c), and (d) of Lemma 6.2 say that in considering products of generators, one can assume that blocks involving just $\mathbf{R}$'s and $\mathbf{C}$'s are of the form $R_i C_j$. The remaining parts of 6.2 show how elements of Π migrate to the right in products. Lemma 6.3 says that in a product of the form $R_{i1} C_{j1} J R_{i2} C_{j2} J$, one can assume that $i1 < i2$ and $j1 < j2$.

Lemma 6.2 *Let $i, i1, i2$ be in the range $1, \ldots, M$ and let $j, j1, j2$ be in the range $1, \ldots, N$.*
(a) $R_i R_i = C_j C_j = JJ = I$,   (b) $C_j R_i = R_i C_j$,
(c) $R_{i1} R_{i2} = R_{i2} \rho_{i1,i2}$,   (d) $C_{j1} C_{j2} = C_{j2} \kappa_{j1,j2}$,
(e) $J R_i J = R_i J R_i$ (f) $J C_j J = C_j J C_j$
(g) $\rho_{i1,i2} J = J \rho_{i1,i2}$,   (h) $\kappa_{j1,j2} J = J \kappa_{j1,j2}$,

(i) $\rho_{i1,i2}C_j = C_j\rho_{i1,i2}$,

(j) $\kappa_{j1,j2}R_i = R_i\kappa_{j1,j2}$,

(k) $i \notin \{i1,i2\} \Longrightarrow$

(l) $j \notin \{j1,j2\} \Longrightarrow$

$$\rho_{i1,i2}R_i = R_i\rho_{i1,i2},$$

$$\kappa_{j1,j2}C_j = C_j\kappa_{j1,j2},$$

(m)$\rho_{i1,i2}R_{i1} = R_{i2}\rho_{i1,i2}$,

(n) $\kappa_{j1,j2}C_{j1} = C_{j2}\kappa_{j1,j2}$

Lemma 6.3 *Let $i, i1, i2$ be in the range $1, \ldots, M$ and let $j, j1, j2$ be in the range $1, \ldots, N$.*

(a) $R_iJR_iC_jJ = JR_iC_jJC_j$.

(b) $i1 \neq i2 \Longrightarrow R_{i2}JR_{i1}C_jJ = R_{i1}JR_{i2}C_jJC_j\rho_{i1,i2}$.

(c) $C_jJR_iC_jJ = JR_iC_jJR_i$.

(d) $j1 \neq j2 \Longrightarrow C_{j2}JR_iC_{j1}J = C_{j1}JR_iC_{j2}JR_i\kappa_{j1,j2}$.

Let $R_0 = C_0 = I$ and consider products of the following forms. *Form 0* consists of all products R_iC_j with $i \in \{0,\ldots,M\}$ and $j \in \{0,\ldots,N\}$. *Form 1* consists of products $R_{i1}C_{j1}JR_iC_j$ with $i1,i \in \{0,\ldots,M\}$ and $j1,j \in \{0,\ldots,N\}$. *Form 2* consists of products

$$R_{i1}C_{j1}JR_{i2}C_{j2}JR_iC_j$$

with

$$0 \leq i1 < i2 \leq M, \quad 0 \leq j1 < j2 \leq N, \quad i \in \{0,\ldots,M\} \text{ and } j \in \{0,\ldots,N\}.$$

For $t > 2$, *form t* consists of products

$$R_{i1}C_{j1}J \cdots JR_{it}C_{jt}JR_iC_j$$

with

$$0 \leq i1 < \cdots < it \leq M, \quad 0 \leq j1 < \cdots < jt \leq N,$$
$$i \in \{0,\ldots,M\} \quad \text{and} \quad j \in \{0,\ldots,N\}.$$

Lemma 6.4 *Let L be the minimum of M and N. Then the total number of products of forms $0,\ldots,l$ is $[S_k : \Pi]$.*

Proof. There are no products of form t, for $t > L$. For $t = 1, \ldots, L$, the number of products of that form is

$$(M+1)(N+1)\binom{M+1}{t}\binom{N+1}{t}.$$

Equation (14) is a well known identity (see, for example identity 10, page 207 of [11].)

$$\binom{k}{r} = \sum_{t=0}^{r}\binom{r}{t}\binom{k-r}{t}. \tag{14}$$

172

The result follows by letting $r = M + 1$ in (14) and multiplying by $(M + 1)(N + 1)$. □

Theorem 6.5 *Let L be the minimum of M and N. Then each left coset of Π contains one and only one representative element of one of forms $0, \ldots, L$.*

Proof. Let $g \in S_k$. Applying Lemmas 6.2 and 6.3, one can find an element of $g\Pi$ of one of the desired forms. By Lemma 6.4, there cannot be more than one. □

Now suppose that a linear ordering has been imposed on the elements of $\mathcal{F}$. Extend this ordering lexicographically to affine coordinate representations of points and then to elements of Ω. For each $A \in \Omega(M, N, \mathcal{F})$, let A^* be the member of $\Omega(M - 1, N, \mathcal{F})$ obtained from A by eliminating the last row. Notice that if A and B are members of $\Omega(M, N, \mathcal{F})$ with last rows a and b respectively, then

$$A < B \iff A^* < B^* \text{ or } (A^* = B^* \text{ and } a < b). \tag{15}$$

Lemma 6.6 *Assume that $M > 1$. If G is any of the four groups of operators on $\Omega(M, N, \mathcal{F})$ from Theorem 2.4 associated with $\mathcal{O}_k$ and $\mathcal{U}_k$ (but not $\mathcal{C}_k$) and G^* is the corresponding group of operators on $\Omega(M - 1, N, \mathcal{F})$, then G^* can be embedded in G in such a way that, for any $A \in \Omega(M, N, \mathcal{F})$ and $g \in G^*$,*

$$A^{g*} = A^{*g}. \tag{16}$$

Proof. If G is trivial then so is G^*. If $G = S_k$, then $G^* = S_{k-1}$. Also G^* is generated by $R_1, \ldots, R_{M-1}, C_1, \ldots, C_N, J$. It is easy to verify that if g is any of these generators, then g satisfies (16). If $G = Z_h$, then G and $G*$ are both essentially the group of field automorphisms, acting coordinatewise on matrices of $\Omega(M, N, \mathcal{F})$ and $\Omega(M - 1, N, \mathcal{F})$ respectively. Equation (16) is immediate. It is also immediate that since S_k and Z_h have the desired property, then so must $Z_h \times S_k$. □

Theorem 6.7 *Assume that $M > 1$. Let G and G^* be as in Lemma 6.6. If $A \in \Omega(M, N, \mathcal{F})$ is the least element of its orbit under G, then A^* is the least element of its orbit under G^**

Proof. Suppose that A is the least element of its orbit under G but A^* is not the least element of its orbit under G^*. Pick $g \in G^*$ such that $A^{*g} < A^*$. By (16), $A^{g*} < A^*$ so by (15), $A^g < A$. This is a contradiction. The situation when $N < M$ is symmetric. □

Definition 6.8 (a) Norm is the operator which maps each $A \in \Omega$ to the least element of its orbit under Π. (b) A matrix $A \in \Omega$ is Normal if $A^{\text{Norm}} = A$.

From each of the operators g of forms $0, \ldots, L$ (L the minimum of M, N) produce a new operator gNorm by following g by normalization. Given a normal matrix $A \in \Omega$, application of these $k!/(M!\,N!)$ operators produces all the normal matrices in the orbit of A under S_k.

Considering unordered arcs in $PG(2,11)$ and $PG(2,13)$ and using a computer program to separate the normal elements of Ω by orbit, one obtains the results below.

$$PG(2,11)$$

k	5	6	7	8	9	10	11	12
k-arcs	2	15	21	21	5	2	1	1
Complete k-arcs	0	0	1	9	3	1	0	1

$$PG(2,13)$$

k	5	6	7	8	9	10	11	12	13	14
k-arcs	3	26	80	181	110	27	2	2	1	1
Complete k-arcs	0	0	0	2	30	21	0	1	0	1

Numbers of Projectively Distinct Arcs

References

[1] **W. S. Burnside.** *The Theory of Groups.* Cambridge University Press, Cambridge, 2nd edition, 1911.

[2] **D. G. Glynn.** Rings of geometries II. *J. Combin. Theory Ser. A*, 49, pp. 26–66, 1988.

[3] **C. E. Gordon.** Orbits of arcs in $PG(N, K)$ under projectivities. *Geom. Dedicata*, 42, pp. 187–203, 1992.

[4] **C. E. Gordon and R. Killgrove.** Representative arcs in field planes of prime order. *Congr. Numer.*, 71, pp. 73–86, 1990.

[5] **W. Ledermann.** *Finite Groups.* Oliver and Boyd, Edinburgh, London, 1961.

[6] **F. J. MacWilliams and N. J. A. Sloane.** *The Theory of Error-Correcting Codes.* North Holland, Amsterdam, 1977.

[7] **L. Storme and J. A. Thas.** k-arcs and dual k-arcs. To appear.

[8] **J. A. Thas.** Connection between the Grassmannian $G_{k-1;n}$ and the set of the k-arcs of the Galois space $S_{n,q}$. *Rend. Mat.*, 6, pp. 121–134, 1969.

C. E. Gordon, Department of Mathematics and Computer Science, California State University, Los Angeles, CA 90032-8204-3, U.S.A.
e-mail: cgordon@atss.calstatela.edu

There exists no (76,21,2,7) strongly regular graph

W. H. Haemers

Abstract

It is shown that a strongly regular graph with parameters $(v, k, \lambda, \mu) = (76, 21, 2, 7)$ cannot exist.

Let Γ be a strongly regular graph with parameters $(v, k, \lambda, \mu) = (76, 21, 2, 7)$. Then Γ is pseudo geometric to a generalized quadrangle of order (3,6). It was proved by Dixmier and Zara [1] (see also Payne and Thas [2]) that such a generalized quadrangle does not exist. In this note we show in a short and elementary way that Γ has to be geometric, and therefore cannot exist.

Lemma 1 *The neighbourhood Γ_x of a vertex x of Γ consists of a disjoint union of cycles with sizes divisible by 3.*

Proof. $\lambda = 2$ implies that Γ_x is the disjoint union of cycles. Let C be such a cycle with c vertices. Let c_i denote the number of vertices at distance 2 from x, adjacent to precisely i vertices of C. Then $\sum_{i=0}^{c} c_i = v - k - 1 = 54$, $\sum_{i=0}^{c} i c_i = c(k-1-\lambda) = 18c$ and $\sum_{i=0}^{c} i(\mu-i)c_i = \sum_{i=0}^{c} i(7-i)c_i = 6c(k-c) = 6(21c-c^2)$. The last identity follows by counting the paths of length 2 between C and the other vertices of Γ_x. By use of these formulas we obtain $\sum_{i=0}^{c}(i - \frac{c}{3})^2 c_i = 0$. Therefore $3i = c$.

Lemma 2 *Let S denote the complete graph on 4 vertices from which one edge is deleted. Then S is not an induced subgraph of Γ.*

Proof. Suppose S is a subgraph of Γ. Let S_i denote the set of vertices outside S adjacent to precisely i points of S. Clearly $S_4 = \emptyset$ (since $\lambda = 2$). By Lemma 1, Γ has no subgraph isomorphic to W_5, the wheel on 5 vertices, hence also $S_3 = \emptyset$. Now we easily have $|S_2| = 4(\lambda - 1) + (\lambda - 2) + (\mu - 2) = 9$. Moreover, $|S_1| + 2|S_2| + 10 = 4k = 84$ yields $|S_1| = 56$, so $|S_0| = 7$. Let s_i denote the number of edges between S_i and S_0. Counting paths of length 2 between S and S_0 gives $s_1 + 2s_2 = 28\mu = 196$. Clearly $2s_0 + s_1 + s_2 = 7k = 147$. Hence $s_2 = 49 + 2s_0$ and $s_1 = 98 - 4s_0$. Let p_i denote the number of paths of length

2 between distinct vertices of S_0 via a vertex of S_i. Then

$$p_0 + p_1 + p_2 = \lambda s_0 + \mu(21 - s_0) = 147 - 5s_0. \tag{1}$$

From $s_1 = 98 - 4s_0$ and $|S_1| = 56$ it follows that $p_1 \geq 42 - 4s_0$ (indeed, p_1 is minimal if the numbers of neighbours in S_0 of points of S_1 are as equal as possible). Similarly, $s_2 \geq 49$ and $|S_2| = 9$ imply $p_2 \geq 5 \times \binom{5}{2} + 4 \times \binom{6}{2} = 110$. So $p_1 + p_2 \geq 152 - 4s_0$, contradicting equation (1).

Theorem *There exists no (76,21,2,7) strongly regular graph.*

Proof. By Lemma 2, the two common neighbours of two adjacent points of Γ are always adjacent. So every edge is in a unique 4-clique. This implies that the vertices and the 4-cliques of Γ form a generalized quadrangle of order (3,6). So, by the result of Dixmier and Zara the assertion follows.

References

[1] **S. Dixmier and F. Zara.** Etude d'un quadrangle généralisé autour de deux de ses points non liés. Unpublished manuscript, 1976.

[2] **S. E. Payne and J. A. Thas.** *Finite Generalized Quadrangles*, volume 104 of *Research Notes in Mathematics*. Pitman, Boston, 1984.

W. H. Haemers, Tilburg University, Dept. of Econometrics, P. O. Box 90153, 5000 LE Tilburg, The Netherlands. e-mail: t460haemers@kub.nl

Group-arcs of prime power order on cubic curves

J. W. P. Hirschfeld   J. F. Voloch

Abstract

This article continues the characterization of elliptic curves among sets in a finite plane which are met by lines in at most three points. The case treated here is that of sets of prime-power cardinality.

1. Notation

$GF(q)$	the finite field of q elements
$PG(2,q)$	the projective plane over $GF(q)$
$PG^{(1)}(2,q)$	the set of lines in $PG(2,q)$
$P(X)$	the point of $PG(2,q)$ with coordinate vector X
PQ	the line joining the points P and Q
$\ell(P,Q)$	PQ
$\langle P \rangle$	the group generated by P.

2. Introduction

This article continues the work of [5] in considering sufficient conditions for a set of points in a finite plane to be embedded in a cubic curve. Similar results to those in [5] were obtained independently by Ghinelli, Melone and Ott [1].

For completeness the main results in [5] need to be summarized.

Definition 2.1 A $(k;n)$-arc in $PG(2,q)$ is a set of k points with at most n points on any line of the plane.

The fundamental problem is to decide when a $(k;n)$-arc $\mathcal{K}_n$ lies on an absolutely irreducible algebraic curve $\mathcal{C}_n$ of degree n. Here we consider the problem for $n = 3$.

A crucial point is the number of points $\mathcal{K}_3$ contains and the number of rational points on $\mathcal{C}_3$. Let $m_3(2,q)$ be the maximum number of points on $\mathcal{K}_3$. Then

$$m_3(2,q) \leq 2q + 1 \text{ for } q > 3, \tag{1}$$

[10], [2, p.331], and the exact values known are given in Table 1.

For $q = 11$, $2q - 1 \leq m_3(2,q) \leq 2q + 1$.

q	2, 3	4, 5, 7	8, 9
$m_3(2,q)$	$2q+3$	$2q+1$	$2q-1$

Table 1. Values for $m_3(2,q)$

For an elliptic curve, $N_q(1)$ is the maximum number of points it can contain. Its value, for $q = p^h$ with p prime, is

$$N_q(1) = \begin{cases} q + [2\sqrt{q}] & \text{when } h \text{ is odd, } h \geq 3 \text{ and } p|[2\sqrt{q}] \\ q + 1 + [2\sqrt{q}] & \text{otherwise,} \end{cases}$$

where $[t]$ denotes the integer part of t, [12], [3, p. 273]. The precise values that are achieved by the number of points of an elliptic curve over $GF(q)$ are also known [12], as well as the number of isomorphism classes and the number of plane projective equivalence classes for a given value, [9]. For such a value the possible structures of the abelian group the points form is also known, [8], [11].

3.   Axioms

Now, we recall the axioms imposed on a $(k; 3)$-arc in [5], and then solve the main case unresolved there. For further motivation and details concerning the axioms, see [5, section 2].

Let $\mathcal{K}$ be a $(k; 3)$-arc in $PG(2, q)$. Four axioms (E1) - (E4) are required. For each axiom, the property that it gives to $\mathcal{K}$ is mentioned in parentheses.
(E1) There exists O in $\mathcal{K}$ such that $\ell \cap \mathcal{K} = \{O\}$ for some line ℓ. (INFLEXION)
(E2) There exists an injective map $\tau : \mathcal{K}\backslash\{O\} \to PG^{(1)}(2, q)$ such that $P \in P\tau$ and $|P\tau \cap \mathcal{K}| \leq 2$, for all $P \in \mathcal{K}\backslash\{O\}$. (TANGENT)
(E3) If $P, Q \in \mathcal{K}$ and $PQ \neq P\tau$ or $Q\tau$, then $|PQ \cap \mathcal{K}| = 3$. (FEW BISECANTS)
(E4) For $P \in \mathcal{K}$, define $\overline{P}$ to be the third point of $\mathcal{K}$ on OP. For $P, Q \in \mathcal{K}$, define $P + Q = \overline{R}$, where R is the third point of $\mathcal{K}$ on PQ. Now, let $\mathcal{K}$ be an abelian group under the operation $+$ with identity O and $-P = \overline{P}$. (ABELIAN GROUP)

Definition 3.1 A $(k; 3)$-arc $\mathcal{K}$ satisfying (E1) - (E4) is called a group-arc or k-group-arc.

It follows from the axioms that
(a) any subgroup of a group-arc is a group-arc,

178

(b) $P + Q + R = O$ if and only if P, Q, R are collinear.

Definition 3.2 *In $PG(2, q)$, the point set S is linearly determined by the set $\mathcal{T}$ of points and lines if every point of S is the intersection of two lines each of which is in $\mathcal{T}$ or is the join of two points of $\mathcal{T}$ or is the join of two points iteratively determined in this way.*

Lemma 3.3 *If P is a point of an arbitrary group-arc, then the cyclic group $\langle P \rangle$ is linearly determined by $\{O, \pm P, \pm 2P, 3P, (-2P)\tau\}$.*

Lemma 3.4 *Let P be a point of order at least six of a group-arc. Then $\langle P \rangle$ is a subgroup of a unique cubic curve with inflexion O.*

Lemma 3.5 *Let $\mathcal{E}_1$ and $\mathcal{E}_2$ be cubic curves and $\mathcal{K}$ a k-group-arc which is a subgroup of both $\mathcal{E}_1$ and $\mathcal{E}_2$. If $k > 5$, then $\mathcal{E}_1 = \mathcal{E}_2$.*

Lemma 3.6 *Let $\mathcal{K}$ be a group-arc contained in a cubic curve $\mathcal{E}$ such that any cyclic subgroup of $\mathcal{K}$ is a subgroup of $\mathcal{E}$. Then $\mathcal{K}$ is a subgroup of $\mathcal{E}$.*

Theorem 3.7 *Let $\mathcal{K}$ be a k-group-arc in $PG(2, q)$ such that one of the following hold:*
(a) $k = p_1 p_2 r$ where p_1 and p_2 are distinct primes ≥ 7;
(b) $k = 2^a 3^b 5^c p_1^d$, where p_1 is a prime ≥ 7, $d \geq 1$ and $2^a 3^b 5^c \geq 6$.
Then $\mathcal{K}$ is a subgroup of the group of non-singular points of a cubic curve.

The theorem leaves the following values of k to be considered:
(i) $k = 2^a 3^b 5^c$, with $a, b, c \geq 0$; (ii) $k = e p_1^d$, with p_1 prime $\geq 7, d \geq 1, 1 \leq e \leq 5$.

In the next section we consider case (ii).

4.   The main theorem

Lemma 4.1 *Suppose P, Q are elements of a group-arc $\mathcal{K}$ both of prime order $p_1 \neq 2, 3$ generating a subgroup G of order $(p_1)^2$. Then G is uniquely* *determined by*

$$O, \pm P, \pm Q, P \pm Q, 2P.$$

Proof. First,

$$\begin{aligned} -P - Q &= \ell(P, Q) \cap \ell(O, P + Q), \\ -P + Q &= \ell(P, -Q) \cap \ell(O, P - Q). \end{aligned}$$

Now assume, by induction on $m < p_1 - 1$, that we know

$$\pm(iP + Q), \pm iP$$

for $i = 0, \ldots, m$. This is true for $i = 1$. Now we determine these points for $i = m + 1$ as follows:

$$\begin{aligned} -(m+1)P - Q &= \ell(P, mP + Q) \cap \ell(2P, (m-1)P + Q), \\ (m+1)P + Q &= \ell(-P, -mP - Q) \cap \ell(O, -(m+1)P - Q), \\ (m+1)P &= \ell(-P, -mP) \cap \ell(Q, -(m+1)P - Q), \\ -(m+1)P &= \ell(P, mP) \cap \ell(O, (m+1)P). \end{aligned}$$

The last equality works providing the two lines are distinct; that is, providing $(m+1)P \neq O$ or $(2m+2)P \neq O$. However, the first is true since otherwise the induction would have been finished at the previous step.

In particular, $\langle P \rangle$ has been determined. Now $\langle P_1 \rangle$, where $P_1 = P + Q$, is found. From the previous step,

$$O, \pm P_1, \pm Q, P_1 \pm Q, 2P_1$$

are required. Of these, the only ones lacking are $P_1 + Q$ and $2P_1$. These are determined as follows:

$$\begin{aligned} P_1 + Q &= P + 2Q &= \ell(-P - Q, -Q) \cap \ell(-2P - Q, P - Q), \\ 2P_1 &= 2P + 2Q &= \ell(-2P - Q, -Q) \cap \ell(-3P - Q, P - Q). \end{aligned}$$

Now, with P_1 instead of P, we can determine $\langle P_2 \rangle$, where $P_2 = P_1 + Q = P + 2Q$. Continuing this process, $\langle P + mQ \rangle$ for $m = 0, 1, \ldots, p_1 - 1$ can be determined. To complete the proof, only $\langle Q \rangle$ needs to be found. By reversing the initial roles of P and Q, we require

$$O, \pm P, \pm Q, Q \pm P, 2Q.$$

Of these, only $2Q$ is missing; this is given by

$$2Q = \ell(P, -P - 2Q) \cap \ell(-P, P - 2Q).$$

□

Corollary 4.2 *A group-arc $\mathcal{K}$ isomorphic to $(Z_{p_1})^2, p_1 \geq 5$, is a subgroup of a unique cubic curve.*

Proof.   Given $O, \pm P, \pm Q, P \pm Q, 2P$, where $\mathcal{K} = \langle P \rangle \oplus \langle Q \rangle$, the conditions that a cubic passes through these points and has an inflexion at O are nine independent conditions and determine the cubic uniquely. $\square$

Theorem 4.3 *Let $\mathcal{K}$ be a k-group-arc in $PG(2,q)$ such that k is divisible by a prime $p_1 \geq 7$. Then $\mathcal{K}$ is a subgroup of a unique cubic curve.*

Proof.   By Theorem 3.7, it suffices to consider the case that $k = ep_1^d$ with $1 \leq e \leq 5$.

Consider first the case that the p_1-Sylow subgroup $\mathcal{P}_1$ of $\mathcal{K}$ is cyclic so that $\mathcal{P}_1 = \langle P_1 \rangle$. Now, $\mathcal{K} = \mathcal{P}_1 \oplus G$, where $|G| = e$ and $|\mathcal{P}_1| = p_1^d$. As $\mathcal{P}_1$ is cyclic it is contained in a cubic curve $\mathcal{E}_1$. For any point P in $\mathcal{P}_1$, the subgroup $\langle P \rangle$ is contained in a cubic curve $\mathcal{E}$, which coincides with $\mathcal{E}_1$ by Lemma 3.5. If Q is any point of $\mathcal{K}$, then $Q = P + R$ for some $P \in \mathcal{P}_1$ and $R \in G$. By Lemma 3.4, $\langle Q \rangle$ is contained in an cubic curve $\mathcal{E}'$; also, since the orders of P and R are coprime, $\langle Q \rangle$ contains both $\langle P \rangle$ and $\langle R \rangle$. Again, by Lemma 3.5, $\mathcal{E}' = \mathcal{E}_1$. Hence $\mathcal{K} \subset \mathcal{E}_1$.

Now consider the non-cyclic case and let $\mathcal{K}_1 \subset \mathcal{K}$ with $\mathcal{K}_1$ isomorphic to $(Z_{p_1})^2$. Then, by the previous corollary, $\mathcal{K}_1$ is contained in a cubic $\mathcal{E}$. As in the previous case, $\mathcal{K} = \mathcal{K}_0 \oplus G$ where $|G| = e$ and $|\mathcal{K}_0| = p_1^d$. If P in $\mathcal{K}_0 \backslash \mathcal{K}_1$ has order p_1^λ, then $\langle P \rangle$ is contained in a cubic $\mathcal{E}'$ and, for $Q \in \mathcal{K}_1 \backslash \{O\}$, the sum $\langle p_1^{\lambda-1} P \rangle \oplus \langle Q \rangle$ is contained in a cubic $\mathcal{E}''$. Now $\mathcal{E}'' \cap \mathcal{E} \supset \langle Q \rangle$, whence $\mathcal{E}'' = \mathcal{E}$ by Lemma 3.5. Also, $\mathcal{E}' \cap \mathcal{E}'' \supset \langle p_1^{\lambda-1} P \rangle$ and so $\mathcal{E}' = \mathcal{E}''$. Hence $\mathcal{E} = \mathcal{E}'$ and therefore $\mathcal{K}_0 \subset \mathcal{E}$.

Now, let $R \in G$. Then there is a cubic $\mathcal{E}'''$ containing $\langle R + Q \rangle$. As $e(R+Q) = eQ \in \mathcal{E}$ and $eQ \neq O$, so $\langle eQ \rangle = \langle Q \rangle$ and $\mathcal{E}''' \cap \mathcal{E} \supset \langle Q \rangle$. Therefore $\mathcal{E}''' = \mathcal{E}$ by Lemma 3.5 and $\langle R + Q \rangle \subset \mathcal{E}$, whence $p_1(R + Q) = p_1 R \in \mathcal{E}$. So $R \in \mathcal{E}$. It has now been shown that both G and $\mathcal{K}_0$ lie in $\mathcal{E}$, whence $\mathcal{K} \subset \mathcal{E}$. $\square$

5.   Small cases

I. k = 8

Lemma 5.1 *An 8-group-arc $\mathcal{K}$ isomorphic to $Z_2 \times Z_2 \times Z_2$ exists in $PG(2,q)$ if and only if $q = 2^h, h \geq 3$. Such a group-arc lies on a unique cuspidal cubic.*

Proof.   Let $O = P(1,0,0), P = P(0,1,0), Q = P(0,0,1), R = P(1,1,1)$ be points of $\mathcal{K}$. Then

$$
\begin{aligned}
R + Q &= P(t, t, 1), t \neq 0, 1; \\
P + R &= P(1, s, 1), s \neq 1.
\end{aligned}
$$

Also

$$
\begin{aligned}
P + Q &= \ell(P, Q) \cap \ell(Q + R, P + R) &= P(0, t - ts, 1 - t); \\
P + Q + R &= \ell(P + R, Q) \cap \ell(P, Q + R) &= P(1, s, t^{-1}).
\end{aligned}
$$

Now, $P + Q + R \in \ell(P + Q, R) \Rightarrow$

$$
\begin{vmatrix}
1 & 1 & 1 \\
0 & t - ts & 1 - t \\
1 & s & t
\end{vmatrix} = 0
$$

$\Rightarrow 1 - s + 1 - t - (t - ts) - s(1 - t) = 0$
$\Rightarrow 2(1 - s)(1 - t) = 0$
$\Rightarrow 2 = 0.$

Since O is on none of the lines

$$
\ell(Q, P + R), \ell(R + Q, P + R), \ell(P + Q, P + Q + R),
$$

it follows that $s \neq 0, s \neq t, s \neq t^{-1}$; hence $q > 4$. Also the 7 points of $\mathcal{K} \backslash \{O\}$ form a $PG(2, 2)$. The 8 points lie on the unique cubic $\mathcal{C}$ with equation

$$
(s + 1)x^2y + s(t + 1)x^2z + (t + 1)y^2z + t(s + 1)yz^2 = 0.
$$

This is irreducible when $(s + t)(st + 1) \neq 0$, which is satisfied in this case. It has a cusp at $P(\sqrt{t}, \sqrt{st}, 1)$ and all tangents to $\mathcal{C}$ are concurrent at O.

$\square$

For more on cuspidal cubics, see [2, section 11.3].

Lemma 5.2 *An 8-group-arc $\mathcal{K}$ isomorphic to $Z_2 \times Z_4$ exists in $PG(2, q)$ if and only if q is odd with $q \geq 5$. Such a group-arc lies on a unique cubic curve, which is elliptic.*

Proof. The eight points of $\mathcal{K}$ written as elements of $Z_2 \times Z_4$ are

$$
\begin{aligned}
O = (0, 0), \quad &P_1 = (0, 2), \quad P_2 = (1, 0), \quad P_3 = (1, 2), \\
Q_1 = (0, 1), \quad &Q_2 = (0, 3), \quad Q_3 = (1, 1), \quad Q_4 = (1, 3).
\end{aligned}
$$

Hence

$$
2P_1 = 2P_2 = 2P_3 = O, \; P_1 + P_2 + P_3 = O, \; 2Q_1 = 2Q_2 = 2Q_3 = 2Q_4 = P_1.
$$

So P_1, P_2, P_3 are the points of contact of the tangents through O, and Q_1, Q_2, Q_3, Q_4 the points of contact of the tangents through P_1.

Let $O = P(0,0,1)$ with tangent $y = 0$. Let $P_1 = P(0,1,0), P_2 = P(1,1,1), P_3 = P(\alpha,1,\alpha)$ with respective tangents $x = 0, x = y, x = \alpha y$; so $\alpha \neq 0,1$. Then, if $\mathcal{K}$ lies on the cubic curve $\mathcal{E}$, consider the intersection divisors in which $\mathcal{E}$ meets the two curves with equations

$$y(x - z)^2 = 0 \text{ and } x(x - y)(x - \alpha y) = 0.$$

In both cases the divisor is

$$O \oplus O \oplus O \oplus P_1 \oplus P_1 \oplus P_2 \oplus P_2 \oplus P_3 \oplus P_3,$$

where $\oplus$ has been used to denote the formal sum to distinguish it from the sum on a cubic curve elsewhere in this paper. So $\mathcal{E}$ has equation

$$y(x - z)^2 + \lambda x(x - y)(x - \alpha y) = 0. \tag{2}$$

The common points of a line $z = tx$ through P_1 and C are determined by

$$(1 - t)^2 x^2 y + \lambda x(x - y)(x - \alpha y) = 0; \tag{3}$$

that is, apart from P_1, the points defined by

$$\lambda x^2 + xy\{(1 - t)^2 - \lambda(1 + \alpha)\} + \lambda \alpha y^2 = 0. \tag{4}$$

Since there are four tangents through P_1, so q is odd. For a tangent, the discriminant $\Delta = 0$. Here

$$\Delta = \{(1 - t)^2 - \lambda(1 + \lambda)\}^2 - 4\lambda^2 \alpha = (1 - t)^4 - 2\lambda(1 + \alpha)(1 - t)^2 + \lambda^2(1 - \alpha^2).$$

Since $\Delta = 0$ has four solutions for t, so the discriminant Δ' of Δ considered as a quadratic in $(1 - t)^2$ is a square. Now,

$$\Delta' = \lambda^2(1 + \alpha)^2 - \lambda^2(1 - \alpha)^2 4\lambda^2 \alpha.$$

Hence $\alpha = \beta^2$; this incidentally means that $GF(q)$ contains a square other than 0 and 1, whence $q \neq 3$. Solving $\Delta = 0$ for $(1 - t)^2$ gives

$$(1 - t)^2 = \lambda(1 + \beta^2) \pm 2\lambda\beta = \lambda(1 \pm \beta)^2.$$

Hence $\lambda = \gamma^2$. Thus
$$1 - t = \pm\gamma(1 \pm \beta).$$

Therefore, (4) becomes $(x \pm \beta y)^2 = 0$. This gives for Q_1, Q_2, Q_3, Q_4 the points

$$P(e\beta, 1, e\beta + f\beta\gamma - ef\beta^2\gamma)$$

where $e, f = \pm 1$. Also C has equation

$$y(x - z)^2 + \gamma^2 x(x - y)(x - \beta^2 y) = 0,$$

which is elliptic. □

For the calculation of the equations of cubic curves with a precise number of points, see also [4], [6], [7].

II. k = 25

Each case not covered in this paper can be reduced to a finite calculation. An arbitrary group-arc $\mathcal{K}$ of a given order is given by a set of points, where some of the coordinates are elements of $GF(q)$ and some are indeterminates. The necessary collinearities are given by a set of polynomial equations in the indeterminates. An algebraic manipulation programme can then determine the consistency of these equations, and check whether or not $\mathcal{K}$ lies on a cubic curve. For example, A. Simis (personal communication) has verified that if $\mathcal{K}$ is isomorphic to $(Z_5)^2$, then this works, as one expects from Corollary 4.2.

References

[1] **D. Ghinelli, N. Melone, and U. Ott**. On abelian cubic arcs. *Geom. Dedicata*, 32, pp. 31–52, 1989.

[2] **J. W. P. Hirschfeld**. *Projective Geometries over Finite Fields*. Clarendon Press, Oxford, 1979.

[3] **J. W. P. Hirschfeld**. *Finite Projective Spaces of Three Dimensions*. Oxford University Press, Oxford, 1985.

[4] **J. W. P. Hirschfeld and J. A. Thas**. Sets with more than one representation as an algebraic curve of degree three. In *Finite Geometries and Combinatorial Designs*, pages 99–110, Providence, 1990. American Mathematical Society.

[5] **J. W. P. Hirschfeld and J. F. Voloch**. The characterization of elliptic curves over finite fields. *J. Austral. Math. Soc. Ser. A*, 45, pp. 275–286, 1988.

[6] **A. Keedwell**. Simple constructions for elliptic cubic curves with specified small numbers of points. *European J. Combin.*, 9, pp. 463–481, 1988.

[7] **A. Keedwell**. More simple constructions for elliptic cubic curves with small numbers of points. Preprint, 1991.

[8] **H.-G. Rück**. A note on elliptic curves over finite fields. *Math. Comput.*, 49, pp. 301–304, 1987.

[9] **R. Schoof**. Non-singular plane cubic curves over finite fields. *J. Combin. Theory Ser. A*, 46, pp. 183–211, 1987.

[10] **J. A. Thas**. Some results concerning $\{(q+1)(n-1); n\}$–arcs and $\{qn - q + n; n\}$–arcs in finite projective planes of order q. *J. Combin. Theory Ser. A*, 19, pp. 228–232, 1975.

[11] **J. F. Voloch**. A note on elliptic curves over finite fields. *Bull. Soc. Math. France*, 116, pp. 455–458, 1988.

[12] **W. G. Waterhouse**. Abelian varieties over finite fields. *Ann. Sci. Ecole Norm. Sup.*, 2, pp. 521–560, 1969.

J. W. P. Hirschfeld School of Mathematical and Physical Sciences, University of Sussex, Brighton BN1 9QH, United Kindom.
e-mail: mmfd4@central.sussex.ac.uk

J. F. Voloch, Department of Mathematics, University of Texas at Austin, Austin, TX 78712, USA. e-mail: voloch@math.utexas.edu

Planar Singer groups with even order multiplier groups

C. Y. Ho*

Abstract

We completely determine the subgroups, which also are subplanes, of a Singer group of planar order 81. We prove that each subgroup of a Singer group is invariant under the involution of the multiplier group, except possibly if the Singer group is non abelian of planar order 16. If the subgroup is a subplane of non square order, then this subplane is centralized by the involution of the multiplier group. We study $v(n) = v(x)v(y)v(z)$ from a geometrical point of view, where n is the order of a projective plane and $v(r) = r^2 + r + 1$ for any r.

1. Introduction

A *Singer group* of a projective plane is a collineation group acting regularly on the points of the plane. In 1938, Singer proved that a finite Desarguesian plane admits a cyclic Singer group. On the other hand, in 1964, Karzel proved that a plane admitting an infinite cyclic Singer group is not Desarguesian. Projective planes and Singer groups in this article are of finite cardinalities. An automorphism of a Singer group is a *multiplier* if it is also a collineation when we identify the points of the plane with the elements of the group. The set of all multipliers is called the multiplier group of the Singer group. The importance of the multiplier group can be seen from Ott's result [12] that a plane admitting a cyclic Singer group is Desarguesian or its full collineation group is a semi-direct product of a cyclic Singer group with its multiplier group.

The *planar order* of a Singer group is defined to be the order of the projective plane in which this Singer group acts on. Up to isomorphism, a cyclic Singer group and its multiplier group are uniquely determined by the planar order of the Singer group. In general, two Singer groups of the same planar order might not be isomorphic to each other and their multiplier groups might have different orders. For example, the multiplier group of a nonabelian Singer group of planar order 4 has order 3, but the multiplier group of an abelian Singer group of planar order 4 has order 6.

---

*. Partially supported by a grant from NSA.

For an abelian Singer group, Hall (2.3 below) proved that any divisor of its planar order yields a multiplier. The lack of such existence theorems for multipliers of nonabelian Singer groups explains the difficulty of studying the general case.

As in group theory, Sylow 2-subgroups (especially, the involutions) of the multiplier group play a special role. We know that a Sylow 2-subgroup T of the multiplier group of a Singer group is cyclic (2.1 below). In a forthcoming paper we will prove that T is a direct factor of the multiplier group [5]. In [7], we completely determine the structure of an abelian Singer group when T is maximally permitted in the sense that its index in a Sylow 2-subgroup of $Aut(S)$ is 2. In this article we are interested in the case $|T| \neq 1$.

A geometrical structure can be given to the elements of a group of order 73 (respectively, 13 and 7) so that it becomes a Singer group of planar order 8 (respectively, 3 and 2). When these three groups are embedded, as abstract subgroups, in a Singer group S of planar order 81, we prove that only the structure of a plane of order 3 is compatible with S. Observe that Bruck's conditions (2.2 below): $8^2 + 8 = 72 < 81$ and $2^2 + 2 = 6 < 81$ are satisfied by possible group subplanes of order 8 and 2, where a *group subplane* of a Singer group is a subplane which is also a subgroup. Note that a subplane need not be a subgroup (e.g., the subplane of order 2 in a cyclic plane of order 8), and a subgroup need not be a subplane (e.g., the subgroup of order 7 in a cyclic Singer group of planar order 9). However, group subplanes seem to be important in the study of Singer groups. The following completely determines the group subplanes of a Singer group of planar order 81.

Theorem 1.1 *The group subplanes of a Singer group of planar order 81 are subgroups of order* $13, 7 \cdot 13$ *and* $7 \cdot 13 \cdot 73$, *whose planar orders are 3, 9 and 81 respectively.*

Suppose the multiplier group $M = M(S)$ has even order. Let α be the involution of M. If S is cyclic, then each subgroup of S is α-invariant. The following generalizes this and provides information about group subplanes.

Theorem 1.2 *Suppose the multiplier group $M(S)$ of a Singer group S has even order. Then the following conclusions hold.*

1. *Each subgroup of S is α-invariant except possibly when $n = 16$ and S is non-abelian.*

2. *A group subplane not in the Baer subplane $C_S(\alpha)$ must have square planar order.*

For any positive integer r, let $v(r) = r^2 + r + 1$. As $v(6)$ dividing $v(3^{28})$, it is possible that $v(pq)$ divides $v(p^a)$ for two distinct primes p and q. The fact that $v(2)$ does not divide $v(8)$ shows that the condition $x|y$ is not a sufficient condition for $v(x)|v(y)$. From $v(2)v(11) = v(30)$, we see that $v(x) = v(y)v(z)$ does not imply $x = y^2$ or $x = z^2$. (The fact that $v(2)|v(11)$ and $v(1)|v(10)$ shows that the conditions $v(x)|v(y)$ and $v(x-1)|v(y-1)$ are not sufficient conditions for inferring $x^2|y^2$ or $x|y$.) Since $v(n)$ is the number of points of a projective plane of order n, one expects that the factorization $v(n) = v(a)v(b)$ has some geometrical meaning. In Theorem 2.4 of [8], we prove that if $max\{a, b\} = m$ is the order of a proper subplane, then $n = m^2$. For a Singer group of planar order n such that $v(n) = v(a)v(b)$, the number $i(a, b)$ of elements of $\{a, b\}$ which are orders of group subplanes can be 0 or 1 as the example $v(16) = v(1)v(9) = v(3)v(4)$ shows.

Using Hall's multiplier theorem we see that $M(S)$ has even order for an abelian Singer group S if and only if n is a square. For an abelian Singer group of planar order n, $i(a, b) < 2$. If in addition we assume that $max\{a, b\} = m$ is the order of a proper subplane, then $n = m^2$, $i(a, b) = 1$, and m is the order of a group subplane but $m - 1$ cannot be the order of a group subplane. The existence of the group subplane of order m is provided by the centralizer of an involutory multiplier. In general, the centralizer of a multiplier provides a group subplane. This shows once again how the existence of non trivial multipliers influences the study of Singer groups. The non-existence of a group subplane of order $m - 1$ follows from Theorem 2.1 below. When $n = m^2$, the question whether a subgroup of order $v(m)$ (or $v(m - 1)$) is a subplane or not is still not settled. The question whether a group subplane can always be realized as a centralizer of a multiplier is still open.

For an abelian Singer group, Theorem 1.2 yields several results on the nonexistence of certain group subplanes. Sections 4 and 5 contain some general results in this direction. One of the results in 4.5 is the following.

Theorem 1.3 *Let S be an abelian Singer group of square planar order n. Suppose $v(n) = v(x)v(y)v(z)$ where $x < y < z$. Then x, y, z cannot all be orders of group subplanes.*

The author would like to thank Professor Feit for his interest in the subject and encouraging conversations. The author also would like to express his gratitude to the Department of Mathematics, Yale University, where the major part of this article has been written.

2. Definitions, notations and some known results

Let p be a prime number. For any integer m, m_p denotes the highest power of p dividing m. For a Singer group S of a finite projective plane, the multiplier group of S is denoted by $M(S)$, and the *planar order* of S is defined to be the order of the projective plane. A *group subplane* of a Singer group is a subplane which is also a subgroup. For an integer t and $s \in S$, if $\mu(t) : s \longrightarrow s^t$ is a multiplier, then we call t a numerical multiplier. Sometimes we say that t is a multiplier to mean that t is a numerical multiplier.

For a set of collineations H, let $\mathcal{P}(H)$ (respectively, $\mathcal{L}(H)$) be the set of common fixed points (respectively, lines) of the elements of H and $Fix(H)$ be the set of fixed point-line substructure $(\mathcal{P}(H), \mathcal{L}(H))$ of H. A collineation α is planar if $Fix(\alpha)$ is a subplane.

Our terminology in group theory is taken from [4], that of projective planes is taken from [10], and that of difference sets is taken from [1] or [11]. For the convenience of the reader we recall some known results.

Theorem 2.1 (Ho)[9] *Let S be a Singer group of a projective plane of order n and let $M = M(S)$ be the multiplier group of S. Then a Sylow 2- subgroup of M is cyclic. Suppose M has an involution α. Then n is a square, $S = AB$, where $A = [S, \alpha] = \{s \in S | s^\alpha = s^{-1}\}$ is an abelian normal Hall subgroup of order $v(\sqrt{n}-1)$, which is an arc (i.e., no three points of A are collinear); and $B = C_S(\alpha)$ is a Hall subgroup, which is a Baer subplane. Further, $S = A \times B$ except possibly for $n = 16$.*

Theorem 2.2 (Bruck)[10] *Let m be the order of a proper subplane of a projective plane of order n. Then $n = m^2$ or $m^2 + m \leq n$.*

Theorem 2.3 (Hall)[10] *If S is an abelian Singer group of planar order n, then any divisor of n is a multiplier (in the sense that $s \longrightarrow s^n$ is a multiplier). In particular, if $n = m^2$, then m^3 is an involutory multiplier.*

Lemma 2.4 *A nontrivial multiplier cannot be a perspectivity. Thus an involutory multiplier is a Baer involution (i.e., its fixed point-line substructure is a Baer subplane). In particular, the map $s \to s^{-1}$ is not a multiplier.*

Proof. The first conclusion is known. (See, for example, 2.1 of [9]) The second conclusion follows from the first and the fact that an involutory collineation is a quasiperspectivity. (See, for example, [10].) The last conclusion can be deduced from the second conclusion. $\square$

3. Singer groups with an even order multiplier group

In this section, we assume that the multiplier group $M = M(S)$ of a Singer group S of planar order n has even order. Let α be the involution of M. We write $A = [S, \alpha]$ and $B = C_S(\alpha)$.

Proof of Theorem 1.2 (1) We now prove that each subgroup of S is α-invariant except possibly when $n = 16$ and S is non abelian. Let ω be the set of prime divisors of $|A|$ and ω' be the complement of ω. Suppose $n \neq 16$ or S is abelian. Then $S = A \times B$, where A is a Hall ω-subgroup and B is a Hall ω'- subgroup of S by 2.1. Let G be a subgroup of S. Since $|S|$ is odd, S is solvable by [3]. Thus G is solvable. So $G = HK$, where H is a Hall ω-subgroup and K is Hall ω'-subgroup of G. The normality of A implies that A is the unique Hall ω-subgroup of S. Hence $H \leq A$. Since α inverts every element of A, $H^\alpha = H$. Next $K \leq B$ as B is the unique Hall ω'-subgroup of S. Since α centralizes B, $K^\alpha = K$. Therefore $G^\alpha = H^\alpha K^\alpha = HK = G$ and G is α-invariant as desired.

(2) We now prove that a group subplane not in B must have square planar order. Let R be a group subplane of S of planar order r not contained in B. If $n \neq 16$, then R is α-invariant by (1). So α induces a nontrivial involutory multiplier of R. This implies that r is a square by 2.1.

Next assume $n = 16$. If R is a Baer subplane, then $r = 4$ is a square. Suppose R is not a Baer subplane, then $r^2 + r \leq 16$ by 2.2. So $2 \leq r \leq 3$. If $r = 2$ (respectively, 3), then $|R| = 7$ (respectively, 13). A moment of thought shows that R is a normal Sylow subgroup of S in both cases. Hence R is characteristic. Since α does not centralize R as R is not contained in B, this implies that α inverts R, as R has prime order. This contradicts 2.4, as α is a multiplier of R. This completes the proof of (2). $\qquad\qquad$ □

Remarks

(1) For a Singer group S of planar order 16, if $|M(S)|_2 = 2$, then $M(B)$ has odd order. (See for example [5].) Let H be a subgroup of order 3. If 13 divides $|C_S(H)|$, then there are exactly 7 subgroups of order 3 and all these subgroups are in B. A moment of thought shows that each subgroup of S is α-invariant in this case.

Assume now that 13 does not divide $|C_S(H)|$. Since there are at most 7 subgroups of order 3 in B, there are at least two subgroups H_1 and H_2 of order 3 in HA not in B. Note that $A = [H_1, H_2]$, and H_1 is conjugate to H_2 in HA. So we may assume $A = < a >$, such that $a = h_1 h_2$, where $H_1 = < h_1 >$ and $H_2 = < h_2 >$. By the definition of A, $a^\alpha = a^{-1}$. Suppose α leaves invariant H_1 and H_2. Then $h_1^\alpha = h_1^{-1}$ and $h_2^\alpha = h_2^{-1}$. This implies that $h_2^{-1} h_1^{-1} = a^{-1} = a^\alpha = h_1^\alpha h_2^\alpha = h_1^{-1} h_2^{-1}$. So $[h_1, h_2] = 1$, which is impossible. This contradiction shows that at least one subgroup of order 3 is not α-invariant.

(2) In a cyclic Singer group of a Desarguesian plane of order $n = 5^{10} = (5^2)^5 = (5^5)^2$, the involutory multiplier $\alpha : x \to x^{(5^5)^3}$ fixes the Baer subplane B of order 5^5. Note that B contains a group subplane of order 5, which is not

of square order. There is also a group subplane V of order 5^2, which is $C_S(\beta)$, where $\beta : x \rightarrow x^{(5^2)^3}$ is a multiplier of order 5. Note that V is of square order. Also $V \not\subset B$. (One way to see this is by considering the restriction of α.)

Corollary 3.1 *Let $n = m^2$. Suppose $m = kl$, where $1 < k < l$ and l is not a square. Then l cannot be the order of a group subplane.*

Proof. By way of contradiction, let L be a subplane of order l. Since l is not a square, $L \subset B$ by the second conclusion of 1.2. The order of the Baer subplane is m and $m > l$ as $k > 1$. By 2.2, this implies that $l^2 \leq m$. However $m = kl < l^2$ as $k < l$. This contradiction proves the corollary. $\qquad\square$

Remark

Let $n = 2^2 \cdot 8^2$, where $m = 2 \cdot 8 = kl$. Then a group subplane of order 8 cannot exist when S admits an involutory multiplier α (for example when S is abelian). On the other hand, there certainly exists a group subplane of order 2 in a Desarguesian plane of order $2^2 \cdot 8^2$ even though 2 is not a square. Thus the condition $k < l$ is essential in 3.1. We now improve the above corollary. In the following theorem the commutativity of the Singer group is used to provide enough multipliers in order to apply induction.

Theorem 3.2 *Let S be an abelian Singer group of planar order $n = m^{2^a}$ with $a \geq 1$. Suppose $m = kl$, where $1 < k < l$ and l is not a square. Then l^j cannot be the order of a group subplane for any integer $j \geq 1$.*

Proof. By way of contradiction, let L be a group subplane of order l^j. We use induction on n and j. Since S is abelian and n is a square , S admits an involutory multiplier α. Let $B = C_S(\alpha)$. We divide the proof into the following steps.

Step 1. S does not contain any group subplane of order l, i.e., the case $j = 1$ cannot occur.
Proof. The case $a = 1$ is treated by 3.1. Assume $a > 1$. Since l is not a square, a group subplane of order l is in B by 1.2. Since $a - 1 \geq 1$ and the planar order of B is $m^{2^{a-1}}$, induction on n shows that B does not contain any group subplane of order l. This establishes the claim in Step 1.

Step 2. j is odd.
Proof. This follows from an induction on j and the fact that a group subplane of order l^j for some even j contains a group subplane of order $l^{\frac{j}{2}}$.

Step 3. Final contradiction.
Proof. Since l is not a square and j is odd by Step 2, l^j is not a square. Thus $l^j \neq n$. Then L is a proper subplane of order l^j, so $(l^j)^2 \leq n = (kl)^{2^a} < l^{2^{a+1}}$

as $k < l$. Hence $2j < 2^{a+1}$ and $j < 2^a$. As j is odd and $j > 1$ by Step 1, this last inequality forces $a > 1$. Since l^j is not a square, 1.2 implies that $L \subset B$. Now applying induction on the planar order $m^{2^{a-1}}$ of B yields the final contradiction. $\quad\square$

Corollary 3.3 *Let $n = (pq)^{2^a}$, where $p \neq q$, and both p and q are not squares. If n is the planar order of an abelian Singer group, then p^{2^i} and q^{2^j} cannot both be orders of group subplanes for $1 \leq i,j \leq a$.*

Proof. This is because either $p > q$ or $q > p$, and the fact that a group subplane of order r^2 contains a group subplane of order r. $\quad\square$

Proof of Theorem 1.1 We want to prove that the group subplanes of a Singer group of planar order 81 are the subgroups of order 13, $7 \cdot 13$ and $7 \cdot 13 \cdot 73$, whose planar orders are 3, 9 and 81 respectively.

Observe that a group of order $v(81) = v(8)v(3)v(2) = 73 \cdot 13 \cdot 7$ is abelian. One way to see this is to show consecutively that subgroups of order 73, 13 and 7, respectively, are central. (Hence a Singer group S of planar order 81 is cyclic. Therefore the corresponding plane is Desarguesian by a result of V.K. Keiser on p.209 of [2].) The following proof does not use the fact that the plane is Desarguesian but uses the commutativity of S to provide the multipliers needed. Since 81 is a square and S is abelian, S admits an involutory multiplier α. Since 8 and 2 are not squares, 1.2 implies that a group subplane V of order 8 or 2 is in the Baer subplane $B = C_S(\alpha)$. Applying 1.2 again, this time to B, we see that V is in the Baer subplane of B, which has order 3. Since both 8^2 and 2^2 are bigger than 3, Bruck's result (2.2) yields a contradiction if V exists. This proves that S does not contain a group subplane of order 8 or 2.

Next, we determine all group subplanes of S. The orders of subgroups are 1, 7, 13, 73, $7 \cdot 13$, $7 \cdot 73$, $13 \cdot 73$, and $7 \cdot 13 \cdot 73$. Let H be a group subplane of order r of S. We just proved that $r \neq 8$ and 2. Hence $|H| \neq 7$ and 73. If 73 divides $|H| = v(r)$, then $r \equiv 8$ or 64 $(mod\ 73)$. This rules out the possibilities that $v(r) = v(2)v(8)$ and $v(3)v(8)$. (Clearly $r \neq 8$. Since $v(64) = v(8^2) = v(7)v(8) > v(2)v(8)$ and $v(3)v(8)$, so $r < 64$. Now $8+73 > 64$ proves that r cannot exist.) Hence $H = S$ if 73 divides $|H|$. Therefore a group subplane is a subgroup of order 13, $7 \cdot 13$ or $7 \cdot 13 \cdot 73$. Since a subgroup of one of these orders is the set of fixed points of a multiplier, it is a subplane of planar order 3, 9 or 81. This completes the proof of Theorem 1.1. $\quad\square$

Remark

The reason why 2, 3, and 8 cannot all be orders of group subplanes is easily obtained by Theorem 1.3, which we will prove in the following section.

4. The function $v(n)$ and group subplanes

For any real number x, let $v(x) = x^2 + x + 1 = N(x - \omega)$, where ω is a complex primitive third root of unity. The example $v(2)v(4)v(9) = v(1)v(3)v(2)^3 = v(1)v(3)v(18)$ convinces us that some conditions are needed in order to study the product $v(x)v(y)v(z)$ in a reasonable way. (Note that $v(2)v(4)v(9) \neq v(n)$ for any n.) Our aim is to relate this product to the study of projective planes. We remark that in a cyclic Singer group of a Desarguesian plane of order 16^2, the equation $v(16^2) = v(16)v(15) = v(3)v(4)v(15) = v(1)v(9)v(15)$ holds. Therefore if we let $x = 1$, $y = 9$, $z = 15$ and $n = 16^2$, then $v(n) = v(x)v(y)v(z)$ and none of x, y, z is the order of a group subplane. (Here we distinguish triangles and subplanes.) Thus Theorem 1.3 says that for a square n, if $v(n) = v(x)v(y)v(z)$ such that n is the planar order of an abelian Singer group and $x < y < z$, then the number of elements in the set $\{x, y, z\}$ which are orders of group subplanes can be 0, or 1, or 2 but not 3.

In this section we are interested in the situation where $v(n) = v(x)v(y)v(z)$ with $x < y < z$. Reasonable conditions imply $y^2 \neq z$. (See, for example, 4.2 and 4.3 below.)

Whenever $n = m^4$, $v(n) = v(m-1)v(m)v(m^2-1)$. In this case if we let $x = m - 1$, $y = m$, $z = m^2 - 1$, then $y^2 > z$. Note that in an abelian Singer group of planar order $n = m^4$, m is the order of a Baer subplane of a Baer subplane, and both $m - 1$, $m^2 - 1$ cannot be the order of a group subplane. (Looking at the Baer subplane of order m^2, the equation $v(m^2) = v(m-1)v(m)$ rules out the possibility for $m - 1$ to be the order of a group subplane by Theorem 2.1. The possibility for $m^2 - 1$ is ruled out by the same reason and the equation $v((m^2)^2) = v(m^2 - 1)v(m^2)$.)

Next the example $v(17^2) = v(16)v(17) = v(3)v(4)v(17)$ is of the form $v((y^2+1)^2) = v(y-1)v(y)v(y^2+1)$, where $y = 4$. If we let $x = 3$, $y = 4$, $z = 17$ and $n = 17^2$, then $v(n) = v(x)v(y)v(z)$ and $y^2 = 16 < z = 17$. In a cyclic Singer group of a Desarguesian plane of order 17^2, 17 is the order of a Baer subplane, none of 3 and 4 is the order of a group subplane.

Again the example $v(17^2) = v(3)v(4)v(17) = v(1)v(9)v(17)$ shows that in $v(n) = v(x)v(y)v(z)$ with $x < y < z$, it is possible that $y^2 > z$ and z is the order of a group subplane but both x and y are not. Note also that in this example $y = 9$ is a square.

We begin by studying some Diophantine equations.

Theorem 4.1 *Let* e, f, c, d, j, *and* m *be positive integers.*

(1) If $v(m) = ev(c)v(d)$ and $v(m-1) = fv(c-1)v(d-1)$, then $f > e$.

(2) The simultaneous equations $v(m) = v(c)v(d)$ and $v(m-1) = v(c-1)v(d-1)$ have no solutions.

(3) The simultaneous equations $v(m) = v(j)v(c)v(d)$ and $v(m-1) = v(j-1)v(c-1)v(d-1)$ have no solutions.

Proof.

(1) By way of contradiction, assume $e \geq f$. Without loss of generality, we may assume that $c \leq d$. Suppose $fv(d-1) < m$. Then $v(d-1) < m$ as $f \geq 1$. From $c \leq d$, we obtain $v(c-1) \leq v(d-1) < m$. So $v(c-1) \leq m-1$. Hence $m^2 - m + 1 = v(m-1) = fv(c-1)v(d-1) = (fv(d-1))v(c-1) < m(m-1) = m^2 - m$, a contradiction. Hence $fv(d-1) \geq m$. However, this implies that $v(m-1) + 2m = v(m) = ev(c)v(d) = e(v(c-1)+2c)(v(d-1)+2d) = e(v(c-1)v(d-1)+2(v(c-1)d+cv(d-1))+4cd) \geq f(v(c-1)v(d-1))+2f(v(c-1)d+cv(d-1))+4fcd \geq v(m-1)+2fv(c-1)d+2cfv(d-1)+4cfd > v(m-1)+2m$. This contradiction establishes (1).

(2) This follows from (1) by taking $e = 1$ and $f = 1$.

(3) This follows from (1) as $v(x) > v(x-1)$ for $x \in \{j, c, d\}$. □

In the rest of this section, n, x, y, z will always denote positive integers.

Lemma 4.2 (1) If $v(y)$ does not divide $v(z)$, then $z \neq y^2$.

(2) Suppose $v(n) = v(x)v(y)v(z)$ where $x < y < z$. Then $n \neq y^2$. If in addition we assume that $y^2 > z$, then $n \leq y^4$.

Proof.

(1) This is because $v(y^2) = v(y)v(y-1)$.

(2) Assume $v(n) = v(x)v(y)v(z)$ where $x < y < z$. If $n = y^2$, then $v(y-1)v(y) = v(y^2) = v(n) < v(x)v(y)v(z)$, a contradiction. Hence $n \neq y^2$. Suppose $y^2 > z$. Then $v(n) = v(x)v(y)v(z) \leq v(y-1)v(y)v(z) = v(y^2)v(z) \leq v(y^2)v(y^2-1) = v(y^4)$. Therefore $n \leq y^4$ as desired. □

Theorem 4.3 Let $v(n) = v(x)v(y)v(z)$ where $x < y < z$. If n is the order of a projective plane and z is the order of a subplane, then $y^2 \neq z$. Furthermore, $n = z^2, z = y^2 + 1, x = y - 1$ if and only if $y^2 < z$.

Remark

$\{3, 4, 17\}$ and $\{9, 10, 101\}$ are two sets of examples for $\{y-1, y, y^2+1\}$.

Proof. Suppose $y^2 = z$. Then $v(z) = v(y-1)v(y)$. If $x = y-1$, then $v(n) =$

$v(y-1)^2 v(y)^2$, which is a square. This is impossible. Therefore $x \neq y-1$. Since $x < y$, this implies that $x < y-1$. So $x+1 \leq y-1$. This implies $v(n) = v(x)v(y)v(z) \leq v((x+1)y)v(z) \leq v((y-1)y)v(z) = v(y^2-y)v(z)$. However, $y > 1$. So $v(y^2-y)v(z) < v(y^2-1)v(z) = v(z-1)v(z) = v(z^2)$. Therefore $n < z^2$. However z ,being the order of a subplane, satisfies $z^2 \leq n$ by 2.2. This contradiction proves that $y^2 \neq z$.

If $n = z^2$, $y^2 = z-1$, then of course $y^2 < z$.

We now assume $y^2 < z$. Then $v(n) = v(x)v(y)v(z) \leq v(y-1)v(y)v(z) = v(y^2)v(z) \leq v(z-1)v(z) \leq v(z^2)$. Hence $n \leq z^2$. On the other hand $z^2 \leq n$ by 2.2. Therefore $n = z^2$. This implies, by the expression of $v(n)$, that $v(z-1) = v(x)v(y) \leq v(y^2)$ as $x < y$. So $z-1 \leq y^2$. From $y^2 < z$, we obtain $y^2 = z-1$. This together with the equation $v(z-1) = v(x)v(y)$ implies $x = y-1$ as desired. $\qquad\square$

The following lemma studies the case $y^2 > z$.

Lemma 4.4 *Assume $v(n) = v(x)v(y)v(z)$ where $x < y < z$ and $y^2 > z$. If n is the order of a projective plane, then the following conclusions hold.*
- *(1) If z is the order of a subplane of a Baer subplane, then $n = z^2$.*
- *(2) If y is the order of a subplane of a Baer subplane, then $n = y^4$, $x = y-1$, and $z = y^2 - 1$.*

Proof.
- (1) Consider a subplane of order z inside the Baer group subplane of order $\sqrt{n}$. If these two subplanes coincide, then $z^2 = n$. If they are distinct, then $z^2 \leq \sqrt{n}$ by 2.2. Since $n \leq y^4 < z^4$ by the second conclusion of 4.2, $\sqrt{n} < z^2$, a contradiction. Hence $n = z^2$.
- (2) As in (1), either $y^2 = n$ or $y^2 \leq \sqrt{n}$ by 2.2. However, if $y^2 = n$, then $v(y-1)v(y) = v(n) = v(x)v(y)v(z)$ will imply $v(y-1) = v(x)v(z)$, which contradicts the fact that $z > y-1$. Hence $y^2 \leq \sqrt{n}$. On the other hand, $\sqrt{n} \leq y^2$ by 4.2. Therefore $n = y^4$ as desired. So $v(y-1)v(y)v(y^2-1) = v(y^4) = v(n) = v(x)v(y)v(z)$ implies $v(y-1)v(y^2-1) = v(x)v(z)$. Since $x < y$ and $y^2 > z$, we have $x \leq y-1$, $z \leq y^2-1$. This together with the last equation implies that $x = y-1$ and $z = y^2-1$ as desired. $\qquad\square$

Remarks
- (1) In the first statement of 4.4, we cannot prove that $x = y-1$ and $y^2 - 1 = z$ as the example $v(1)v(9)v(17) = v(17^2)$, with $x = 1, y = 9, z = 17, n = 17^2$, and $9^2 = y^2 > 17 = z$ shows.
- (2) The example $v(2)v(3)v(8) = v(81)$ shows that case 2 of 4.4 really occurs.

196

We now apply the above results to abelian Singer groups. Justification for $y^2 \neq z$ can be found in (1) of 4.2 or 4.3.

Theorem 4.5 *Let S be an abelian Singer group of planar order $n = m^2$. Suppose $v(n) = v(x)v(y)v(z)$ where $x < y < z$.*

(1) *If $y^2 < z$ and z is the order of a subplane, then $n = z^2, y^2 = z - 1$, $x = y - 1$, and x and y cannot both be orders of group subplanes.*

(2) *If $y^2 > z$ and z is the order of a subplane inside a Baer subplane, then $n = z^2$ and x and y cannot both be orders of group subplanes.*

(3) *If $y^2 > z$ and y is the order of a subplane inside a Baer subplane, then $n = y^4, x = y - 1, z = y^2 - 1$, and x and y cannot both be orders of group subplanes.*

(4) *Suppose $y^2 > z$. Let $s \in \{y, z\}$. If s is not a square and s is the order of a group subplane, then this subplane is a subplane of a Baer subplane and the corresponding conclusions in (2) or (3) hold.*

(5) *The numbers x, y, z cannot all be orders of group subplanes.*

Proof. Since S is abelian, S admits an involutory multiplier α as n is a square. Let $B = C_S(\alpha)$ and $A = [S, \alpha]$.

(1) By 4.3, it suffices to prove that x and y cannot both be orders of group subplanes. A group subplane G of order x or y is in A by an elementary argument using group orders. But A is an arc by 2.1. This proves that G cannot exist and establishes (1).

(2) By (1) of 4.4, $n = z^2$. This implies that $v(x)v(y) = v(z - 1)$. An argument using group orders shows that a subgroup of order $v(x)$ or $v(y)$ is inside the Hall subgroup A. We now reach the conclusion as in the proof of (1).

(3) By (2) of 4.4, $n = y^4, x = y - 1$, and $z = y^2 - 1$. A subgroup of order $v(z)$ must be A, which is an arc. So z cannot be the order of a group subplane. A subgroup of order $v(y - 1)$ coincides with $B_1 = [B, \beta]$, where β is the involutory multiplier of B (β exists as the planar order of B is y^2). Again B_1 is an arc. So $x = y - 1$ cannot be the order of a group subplane.

(4) We reach the hypothesis imposed on s in (2) or (3) by Theorem 1.2.

(5) By way of contradiction, suppose x, y, z are orders of group subplanes X, Y, Z respectively. By 4.3, $y^2 \neq z$. By (1) $y^2 < z$ cannot occur. So $y^2 > z$. Both y and z must be squares by (4), say $y = c^2$ and $z = d^2$. Also we may assume that $Y \not\subset B$ and $Z \not\subset B$ by (2) and (3). By 1.2, X is α- invariant. Hence $X = [X, \alpha] \times C_X(\alpha)$. If $X \subset B$, then $X = C_X(\alpha)$. If $X \not\subset B$, then x is a square by 1.2, say $x = h^2$, and $|[X, \alpha]| = v(h - 1)$ and $|C_X(\alpha)| = v(h)$. Let $e = |C_X(\alpha)|$ and $f = |[X, \alpha]|$. Then whether

197

or not $X \subset B$ or $X \not\subset B$, we have $f < e$, e divides $v(m)$ and f divides $v(m-1)$.

Since $Y \not\subset B$, $Y = [Y, \alpha] \times C_Y(\alpha)$ is a nontrivial product. So $v(y) = v(c-1)v(c)$ and $v(c-1)|v(m-1)$ as $|[Y, \alpha]| = v(c-1)$ and $[Y, \alpha] \subset A$, and $v(c)|v(m)$ as $C_Y(\alpha) \leq B$. Similarly from $Z = [Z, \alpha] \times C_Z(\alpha)$, we obtain $v(z) = v(d-1)v(d)$ and $v(d-1)|v(m-1)$, and $v(d)|v(m)$. Hence $v(m-1)v(m) = v(n) = v(x)v(y)v(z) = (fv(c-1)v(d-1))(ev(c)v(d))$. Now $(v(m-1), v(m)) = 1$ and $f|v(m-1)$; $v(c-1)|v(m-1)$; $v(d-1)|v(m-1)$; $e|v(m)$; $v(c)|v(m)$; $v(d)|v(m)$. So $(fv(c-1)v(d-1), v(m)) = 1$. Thus $fv(c-1)v(d-1)|v(m-1)$. Similarly $(ev(c)v(d), v(m-1)) = 1$. So $ev(c)v(d)|v(m)$. This together with the equation $v(m-1)v(m) = (fv(c-1)v(d-1))(ev(c)v(d))$ implies that $v(m) = ev(c)v(d)$ and $v(m-1) = fv(c-1)v(d-1)$. However, this contradicts 4.1 as $e \geq f$. This contradiction establishes (5).

$\square$

References

[1] L. D. Baumert. *Cyclic Difference Sets*, volume 182 of *Lecture Notes in Math*. Springer Verlag, Berlin, 1971.

[2] P. Dembowski. *Finite Geometries*. Springer Verlag, Berlin, 1968.

[3] W. Feit and J. G. Thompson. Solvability of groups of odd order. *Pacific J. Math.*, 13, pp. 755–1029, 1963.

[4] D. Gorenstein. *Finite Groups*. Harper and Row, New York, 1968.

[5] C. Y. Ho. Planar Singer groups. In preparation.

[6] C. Y. Ho. Projective planes with a regular collineation group and a question about powers of a prime. To appear in *J. Algebra*.

[7] C. Y. Ho. Singer groups, an approach from a group of multiplier of even order. To appear in *AMS Proceedings*.

[8] C. Y. Ho. Some remarks on orders of projective planes, planar difference sets and multipliers. *Designs, Codes and Cryptography*, 1, pp. 69–75, 1991.

[9] C. Y. Ho. On bounds for groups of multipliers of planar difference sets. *J. Algebra*, 148, pp. 325–336, 1992.

[10] D. R. Hughes and F. C. Piper. *Projective Planes*. Springer Verlag, Berlin, 1973.

[11] E. S. Lander. *Symmetric Designs: an Algebraic Approach*, volume 74 of *London Math. Soc. Lecture Note Ser.* Cambridge University Press, Cambridge, 1983.

[12] U. Ott. Endliche zyklische Ebenen. *Math. Z.*, 144, pp. 195–215, 1975.

C. Y. Ho, Department of Mathematics, University of Florida, Gainesville, Florida 32611, USA. e-mail: cyh@math.ufl.edu

On a footnote of Tits concerning D_n-geometries

C. Huybrechts*

Abstract

If Γ is a thick and residually connected D_n-geometry, $n \geq 4$, it is well known that Γ is defined over a unique ground division ring which is commutative. Here we give an elementary proof of the commutativity based on the construction of null polarities in the projective subspaces of Γ, for $n = 4$.

1. Introduction

Geometries over a diagram of type D_n, $n \geq 4$ are completely classified and well known. They are often discussed together with polar spaces since they stem essentially from quadrics of maximal Witt index. However, the theory of polar spaces is rather intricate and it requires well over a hundred printed pages as we can see from [11] or [4]. A straightforward theory for D_n geometries is much simpler and shorter. Actually, it may come next to the theory of projective geometries by its simplicity. This was observed quite early ([10]).

We are developing a selfcontained theory of D_n-geometries from the definition to the classification. To the best of our knowledge such a treatment is not yet available. It may be useful in various directions, in particular the preparation of extensions to geometries over slightly more general diagrams.

The main result of the theory is as follows.

If Γ is a thick D_n-geometry, $n \geq 4$, then there is a field F such that Γ is isomorphic to the geometry of all singular subspaces of the quadric Q in the projective space $P_{2n-1}(F)$ of dimension $2n - 1$ over the field F, defined by the equation $x_1 x_2 + \ldots + x_{2n-1} x_{2n} = 0$. This result is due to Tits ([9], [11], [12]) up to some improvements by Meixner (see [8] and [2])

One of the first goals of the theory is to show that the division ring underlying the residual projective geometries of a thick D_n-geometry is commutative. This is the purpose of the present paper. In order to do so, it suffices to deal with the case $n = 4$. We make use of a footnote in [10], hinting at the presence of symplectic polarities in the singular 3- subspaces of a D_4-

---

*. Aspirant du Fonds National Belge de la Recherche Scientifique

geometry, Γ. These polarities are inherited from Γ. They force the required commutativity of the ground division ring.

Here is the observation made by Tits.

> La commutativité du corps de base résulte du fait que les géométries projectives à 3 dimensions résiduelles d'une géométrie Γ de type D_4, sont porteuses de polarités nulles, qu'on obtient au moyen de constructions explicites au sein de Γ. ([10]).

It is not necessarily obvious to check this property. We are giving a way to do so. The idea is to choose three maximal pairwise disjoint 3-dimensional singular projective subspaces of the geometry Γ and to study the structure induced by the collinearity relation on their union.

This material is part of a dissertation [7] written under the supervision of F.Buekenhout who provided helpful conversations on it.

2. Main results

We assume some knowledge of basic facts from the theory of diagram geometries (see for instance [3] or [4]).

We recall the following statements.

Direct sum theorem. *Let Γ be a residually connected geometry of finite rank over I. Let i, j be elements of I which are contained in distinct connected components of the basic diagram of Γ. Then every i-element of Γ is incident with every j-element of Γ.*

Theorem 2.1 *Let Γ be a residually connected geometry over I. If i, j are two distinct elements of I and if p, q are elements of Γ, then there is a path from p to q in the incidence graph of Γ, all of whose elements different from p and q are of type i or j.*

We also recall that a geometry is said to be firm (resp. *thick*) if any of its comaximal flag is contained in at most two (resp. three) chambers.

Let Γ be a thick and residually connected D_n-geometry ($n \geq 4$). Let the set of types be $I = \{0, 1, 2, \ldots, n-3, r, b\}$, these elements being used as on the diagram

$$
\begin{array}{ccccccc}
\underset{0}{\circ} & \text{——} & \underset{1}{\circ} & \cdots & \underset{n-4}{\circ} & \text{——} & \underset{n-3}{\circ} < \begin{array}{l} \circ\, r \\ \circ\, b \end{array}
\end{array}
$$

We call the elements of type 0, 1, r and b respectively *points*, *lines*, *reds* and *blues*.

If two elements x, y of Γ have no common incident point, then we say that x, and y are *disjoint*.

The geometry Γ has residues of type A_{n-1}, namely

$$
\overset{0}{\circ} \underline{\hspace{1.5cm}} \overset{1}{\circ} \underline{\hspace{1.5cm}} \overset{2}{\circ} \cdots \overset{n-3}{\circ} \underline{\hspace{1.5cm}} \circ
$$

and these are projective geometries. In such a geometry, a *null polarity* is a polarity π such that any point p is incident with its image $\pi(p)$.

Since $n \geq 4$, these thick projective geometries are Desarguesian and they are defined over a *ground division ring* $D(\Gamma)$ which is uniquely defined for Γ, in view of the fact that Γ is residually connected.

In this paper, we shall first show that if $n = 4$, the following properties hold in Γ.

(3DR) (Three Disjoint Reds). *For any red R of Γ, there are two reds R', R'' such that R, R'' and R'' are pairwise disjoint.*

(D) (Duality). *If R and R' are disjoint reds, then Γ induces a duality of $\Gamma_{R'}$ onto Γ_R.*

(NP) (Null Polarity). *If R is a red, then Γ_R carries a null polarity.*

Next, we get the following theorem on D_n.

Theorem. *The ground division ring of a thick and residually connected D_n-geometry is a field.*

3. Preliminaries

We recall some definitions. The *shadow* $\sigma_0(x)$ of an element x of Γ is the set of points incident with x. Two points incident with a line are said to be *collinear*. We put $p^\perp$ for the set of points collinear with p. Let $F = (F_j)_{j \in J}$ be a nonempty flag. We define $F^\perp$ as the set of points which are collinear with any point of $\sigma_0(F_j)$ for all j in J. We list elementary properties of firm and residually connected D_n-geometries ($n \geq 4$) that we need in order to prove our main results. These properties are well known. Their proofs are gathered in [7] and in [6].

(LL). *Two different points are incident with at most one line [8].*

(OS). *If x, y are distinct elements of the same type or if x is a red and y is a blue, then there is a point incident to x and not incident with y.*

(O). *Distinct elements of the same type have different shadows [8].*

(IBR). *If B is a blue and R is a red, then one of the following holds :*
   a) the elements B and R are incident,
   b) there exists an element z_k incident with B and R such that
      $\sigma_0(z_k) = \sigma_0(B) \cap \sigma_0(R)$ with $0 \leq k \leq n - 4$ and $n - k$ even,
   c) n is odd and B is disjoint from R.
If R, R' are distinct reds, then one of the following holds :
   d) there exists an element z_k incident with R and with R' such that
      $\sigma_0(z_k) = \sigma_0(R) \cap \sigma_0(R')$ with $0 \leq k \leq n - 3$ and $n - k$ odd,
   e) n is even and R is disjoint from R'.

(SPCP). *Let p be a point and let R be a red not incident with p. Then, there is a unique blue B incident with R and such that $p^{\perp} \cap \sigma_0(R) = \sigma_0(B) \cap \sigma_0(R)$. Moreover, this blue is the unique blue incident with p and with R.*

(SPCB). *Assume that $n = 4$. Let R be a red and let B be a blue incident to R. If R' is a red disjoint from R, then there is a unique point p incident with B and such that $\{B, R\}^{\perp} \cap \sigma_0(R') = \{p\}$. This point is the unique point incident with B and with R'.*

(SPCL). *Assume that $n = 4$. Let R be a red and let N be a line disjoint from R. Then there is a unique line L such that $N^{\perp} \cap \sigma_0(R) = \sigma_0(L)$. Moreover, this line is incident with R.*

Here are some properties concerning polarities in projective geometries needed afterwards.

Property (B). *[1] Let Γ be a projective geometry of rank at least 3 over a division ring D. If Γ carries a null polarity, then the division ring D is commutative.*

Assume now that Γ is a thick geometry over $I = \{0, 1, \ldots, n - 1\}$ belonging to A_n, where $n \geq 3$ and that δ is a duality of Γ. It is well known that δ is a polarity if and only if for any two points p, q the point q is incident with $\delta(p)$ whenever p is incident with $\delta(q)$. [1].
The following statement is perhaps well known.

Property (P). *Let δ be a duality of Γ such that any point p is incident with $\delta(p)$. Then δ is a polarity.*

Proof. Assume that p, q are two different points such that q is incident with $\delta(p)$ and p is not incident with $\delta(q)$. As $\delta(p)$ and $\delta(q)$ are distinct $(n - 1)$-elements in the residue Γ_q, there is a $(n - 2)$-element x_{n-2} incident with $\delta(p)$, with $\delta(q)$ and with q. Using the direct sum theorem in the residue of x_{n-2} and the fact that p is not incident with $\delta(q)$, we deduce that p is not incident with x_{n-2}. As Γ is a projective geometry, there is a line L incident with p and with q. By the thickness of the geometry, there is a point r different from the points p and q and incident with the line L . As p, q, r are incident with L and as δ is a morphism, the elements $\delta(p)$, $\delta(q)$, $\delta(r)$ are incident with $\delta(L)$. However, the elements x_{n-2}, $\delta(L)$ are of type $n - 2$ and are both incident with $\delta(p)$, $\delta(q)$. Hence the elements $\delta(L)$ and x_{n-2} are equal, which implies that x_{n-2} is incident with $\delta(r)$. By the direct sum theorem in the residue of x_{n-2}, the point q is incident with $\delta(r)$. As the points q, r are in the residue of $\delta(r)$ which is a projective geometry, the line L is also in this residue and by the direct sum theorem, the point p is incident with $\delta(r)$. So the elements p, x_{n-2} are incident with the elements $\delta(p), \delta(r)$, which means that $\delta(p)$ and $\delta(r)$ are equal because p is not incident with x_{n-2}. This contradicts the fact that δ is bijective. We thus have shown that if q is incident with $\delta(p)$, then p is incident with $\delta(q)$, for any points p, q. We deduce that δ is a polarity.   □

4.   Existence of three disjoint reds in D_4

In this section, Γ denotes a thick and residually connected D_4-geometry.
Definition Let x, y be elements of Γ. If there is a line L incident with x and with y such that $\sigma_0(x) \cap \sigma_0(y) = \sigma_0(L)$, then we say that x intersects y in the line L.

Property (3DR) (Three Disjoints Reds). *For any red R of Γ, there are two reds R', R'' such that R, R' and R'' are pairwise disjoint.*

Proof.   We start constructing R' By definition of a geometry, there is a line L incident with R. Using the firmness of the geometry, we deduce that there is a red R^0 different from R and incident with L. As Γ_{R^0} is a projective geometry, there is a line N disjoint from L in this residue. By the firmness of the geometry, there is a red R' different from R^0 and incident with N. The direct sum theorem and properties (IBR), (OS) of section 3 imply that the red R^0 intersects R (resp. R') in the line L (resp. N). Hence, as the lines L and N are disjoint, the line L is disjoint from R' and the line N is disjoint from R. We now show that the reds R and R' are disjoint. Suppose by way of contradiction that there is a point p incident with R and with R'. Hence, as p and N are in the projective geometry $\Gamma_{R'}$, the point p is in $N^\perp$. However, using the fact that the lines L and N are in the projective geometry Γ_{R^0}, the direct sum theorem and properties (SPCL), (OS) of section 3, we deduce that $\sigma_0(R) \cap N^\perp = \sigma_0(L)$. Hence the point p is incident with L. This is impossible because the point p is also incident with R' and we showed earlier that L and R' are disjoint. We now construct R''. As the lines L and N are in the 3-dimensional and thick projective geometry Γ_{R^0}, there is a point not incident with L and N in this residue. There is thus a point p not incident with R and with R' in the geometry. By property (SPCP) of section

3, there is a blue B (resp. B') incident with p and R (resp. R') such that $\sigma_0(B) \cap \sigma_0(R) = p^\perp \cap \sigma_0(R)$ (resp. $\sigma_0(B') \cap \sigma_0(R') = p^\perp \cap \sigma_0(R')$). As B, B' are in the 3-dimensional and thick projective geometry Γ_p, there is a red R'' in this residue, not incident with B and with B'. Property (IBR) implies that p is the unique point incident with R'' and with B. The reds R and R'' are disjoint. Indeed, assume that there is a point q incident with R and with R''. Hence, as the points p, q are in the projective geometry $\Gamma_{R''}$, the point q is collinear with p. The point q is thus incident with B. However, the point q is also incident with R''. The points p, q are then equal, which is impossible because p is not incident with R. By symmetry, the reds R' and R'' are also disjoint. $\square$

5. Construction of null polarities in D_4

In this section, Γ denotes a thick and residually connected D_4-geometry.

Assume that R and R' are pairwise disjoint reds. Thanks to properties (SPCP), (SPCL) and (SPCB) of section 3, we may introduce a mapping $\delta_{R'R}$ of $\Gamma_{R'}$ onto Γ_R, defined in the following way. Let p be a point of $\Gamma_{R'}$, let N be a line of $\Gamma_{R'}$ and let B' be a blue of $\Gamma_{R'}$. Then

- $\delta_{R'R}(p)$ is the blue B incident with R and with p,
- $\delta_{R'R}(N)$ is the line L such that $\sigma_0(L) = N^\perp \cap \sigma_0(R)$,
- $\delta_{R'R}(B')$ is the point p incident with B' and with R.

Property (D) (Duality). *If R and R' are disjoint reds, then Γ induces a duality $\delta_{R'R}$ of $\Gamma_{R'}$ onto Γ_R. Its inverse is $\delta_{RR'}$.*

Proof. By properties (SPCP), (SPCL) and (SPCB) of section 3, the mappings $\delta_{R'R}.\delta_{RR'}$ and $\delta_{RR'}.\delta_{R'R}$ are equal to the identity. The mapping $\delta_{R'R}$ is thus a bijective mapping and δ_I (namely the permutation induced on I) has order 2. Properties (SPCP), (SPCL), (SPCB) and (LL) of section 3 imply that $\delta_{R'R}$ maps two incident elements of $\Gamma_{R'}$ on two incident elements of Γ_R and two elements of the same type on two elements of the same type. Hence, using the symmetry between R' and R, we deduce that $\delta_{R'R}$ is a duality. $\square$

Notation. If R, R' and R'' are pairwise disjoint reds, then we put π_R for $\delta_{R''R}.\delta_{R'R''}.\delta_{RR'}$.

Property (NP) (Null Polarity). *If R is a red, then Γ_R carries a null polarity π_R.*

Proof. By property (3DR) of section 4, there are two reds R' and R'' such that R, R' and R'' are pairwise disjoint. We first show that any point p is incident with $\pi_R(p)$. By definition, the blue $\delta_{RR'}(p)$ is incident with the points p and $\delta_{R'R''}.\delta_{RR'}(p)$. The points p and $\delta_{R'R''}.\delta_{RR'}(p)$ are then collinear. Hence the point p is incident with the blue $\delta_{R''R}.\delta_{R'R''}.\delta_{RR'}(p)$, which is $\pi_R(p)$. We *now show that π_R is a polarity.* By property (D), this mapping is the product of three dualities. It is thus a duality and property (P) of section 3 implies that the mapping π_R is a polarity. $\qquad\Box$

For more information concerning dualities and polarities of D_n-geometries, $n \geq 4$, see [5].

6. Proof of the theorem on D_n.

Let Γ be a thick and residually connected D_n-geometry.

Theorem. *The ground division ring of Γ is a field.*

Proof. We proceed by induction on n. *We first assume that $n = 4$.* By definition of a geometry, there is a red R in Γ. Properties (B) of section 3 and (NP) of section 5 imply that the ground division ring of Γ_R is a field. Hence, the ground division ring $D(\Gamma)$ is a field. *We now assume that $n \geq 5$.* If F is a flag of type $\{0, \ldots, n-5\}$, then the residue Γ_F belongs to the diagram D_4. Hence, by the induction hypothesis, the ground division ring of Γ_F is a field, which shows that $D(\Gamma)$ is a field. $\qquad\Box$

References

[1] **R. Baer.** *Linear Algebra and Projective Geometry.* Academic Press Inc., New York, 1952.

[2] **A. E. Brouwer and A. M. Cohen.** Some remarks on Tits geometries. *Indag. Math.*, 45, pp. 393–402, 1983.

[3] **F. Buekenhout.** The basic diagram of a geometry. In *Geometries and Groups, Proceedings of the Conference in Honour of H. Lenz*, volume 893 of *Lecture Notes in Math.*, pages 1–29, Berlin, 1981. Springer Verlag.

[4] **F. Buekenhout and A. M. Cohen.** Diagram geometry. Book manuscript.

[5] **C. Huybrechts.** Construction of dualities and null polarities in a D_n-geometry. In preparation.

[6] **C. Huybrechts.** Elementary properties of D_n-geometries. To appear in *Bull. Soc. Math. Belg.*

[7] **C. Huybrechts.** Approche élémentaire des immeubles de type D_n et généralisations. Master's thesis, Université Libre de Bruxelles, 1992.

[8] **F. G. Timmesfeld.** Tits geometries and parabolic systems in finite groups. *Math. Z.*, 184, pp. 377–396, 1983.

[9] **J. Tits.** Géométries polyédriques et groupes simples. In *Atti della 2a Riunione del Groupement Math. d'Expression Latine (Firenze 1961)*, pages 66–88, Rome, 1962. Cremonese.

[10] **J. Tits.** Géométries polyédriques finies. In *Simp. Geom. Finite (Roma 1963)*, Rend. Mat. Appl., pages 156–165, 1964.

[11] **J. Tits**. *Buildings of Spherical Type and Finite BN-pairs*, volume 386 of *Lect. Notes in Math.* Springer Verlag, Berlin, 1974.

[12] **J. Tits**. A local approach to buildings. In *The Geometric Vein (Coxeter Festschrift)*, pages 519–547. Springer Verlag, Berlin, 1981.

C. Huybrechts, Dép. de Mathématique, C.P.216, Université Libre de Bruxelles, Boulevard du Triomphe, B-1050 Bruxelles, Belgique.

The structure of the central units of a commutative semifield plane

V. Jha   G. P. Wene

Abstract

Let π^{ℓ_∞} be a semifield plane of order q^N, with middle nucleus $GF(q)$. Relative to any fixed natural autotopism triangle, every fixed affine point I (not on the triangle) determines up to isomorphism a unique coordinatising semifield D_I. I is called a *central unit* if D_I is commutative. We determine the geometric distribution of the central units of π^{ℓ_∞} and hence show that the plane has precisely $(q^N - 1)(q - 1)$ central units.

1. Introduction

Let π^{ℓ_∞} be a semifield plane with an autotopism triangle OXY: where $XY = \ell_\infty$ is the translation axis, and OY is a shears axis, with $O \in \pi^{\ell_\infty}$. Now each choice of a "unit point" I, off the chosen autotopism triangle (assumed fixed from now on) determines uniquely up to isomorphism a semifield D_I that coordinatises π^{ℓ_∞}. We shall call I a *central unit* relative to the chosen frame if D_I is a commutative semifield. By a criterion of Ganley [2] [theorem 3] the finite semifield planes admitting central units are precisely the finite translation planes that admit orthogonal polarities. However, no geometric characterisation of the set of central units in a given semifield plane has ever been recorded. The purpose of this note is to provide a geometric description of the distribution of the central units of a given commutative semifield plane. Our result implies:

Corollary 1.1 *A commutative semifield plane of order q^N with middle nucleus $GF(q)$ has precisely $(q^N - 1)(q - 1)$ central units (relative to a fixed autotopism triangle whose sides include the translation axis and a shears axis).*

This combinatorial result is a consequence of our main theorem which gives a geometric description of the distribution of the central units. This is given in terms of the $K := (q^N - 1)/(q - 1)$ rational Desarguesian nets of degree $q + 1$ that are left fixed by the full group of middle-nucleus homologies

H (with axis OY and coaxis OX). We show that there is a set of middle-nucleus planes $\{\pi_1 \ldots \pi_K\}$, exactly one in each of the K middle-nucleus nets fixed by H, such that the set $\mathcal{C}$ of central units is given by

$$\mathcal{C} = \bigcup_{i=1}^{N} \pi_i - (OX \cup OY).$$

Thus the set $\mathcal{C}$, of central units, is a disjoint union of the "interiors" (excluding their points on the autotopism triangle OXY) of K middle-nucleus planes, any two of which share only two points on ℓ_∞, viz., $\{X, Y\}$.

We would like acknowledge that the ideas used in this article are inspired by an old result of Albert [1, p. 703 16 eqn (3)], later rediscovered by Ganley [2, theorem 4] in answering a question of P. Dembowski.

2. Central units for semifield spreads

In this section we state a spread-theoretic version of the above indicated result, and mention some simple corollaries. For general background on semifields see [3]; for the basics on spreads, and particularly Desarguesian nets, see [5]. We shall always use the following notation.

Notations

1. $V = \mathcal{F}_p^n \oplus \mathcal{F}_p^n$, where $\mathcal{F}_p = GF(p)$, for a prime p; we often write $D = \mathcal{F}_p^n$, especially when a multiplicative structure on $\left(\mathcal{F}_p^n, +\right)$ is being considered.

2. When $(D, +, \circ)$ is any finite semifield then its middle nucleus is defined to be the field

$$N_m(D, +, \circ) = \{f \in D : x \circ (f \circ y) = (x \circ f) \circ y \ \forall x, y \in D\}$$

3. $\pi = (V, \Gamma)$ is a *shears* or *semifield* spread on V whose component set Γ consists of $p^n + 1$ mutually disjoint additive subgroups of V, each of order p^n, that include $X = \mathcal{F}_p^n \oplus O$ and $Y = O \oplus \mathcal{F}_p^n$ such that Y is the shears axis (fixed by all the affine elations in $\mathrm{Aut}(V, \Gamma)$).

The standard way of obtaining the shears spread $\pi = (V, \Gamma)$ is to start with any semifield $(D, +, \circ)$ of order p^n, where $(D, +) = (\mathcal{F}_p^n, +)$, and then to define Γ on $V = D \oplus D$ by

$$\Gamma := \{\text{"}y = m \circ x\text{"} : m \in D\} \cup \{Y\}$$

where "$y = m \circ x$" always means $\{(x, m \circ x) : x \in D\}$. Throughout the article, we denote this spread by $\pi(D, +, \circ)$. We can now define central units in spread-theoretic terms.

Definition 2.1 $\pi = (V, \Gamma)$ *is coordinatised by a semifield* $(D, +, \circ)$, *relative to the shears frame,* (X, Y), *and unit point* $u \in V - (X \cup Y)$ *if there is an additive bijection* $\psi : V \to V$ *such that* ψ *leaves* X *and* Y *invariant, and induces a spread isomorphism from* π *to* $\pi(D, +, \circ)$, *such that* $\psi(u) = (e, e)$, *where* e *is the multiplicative identity of* $(D, +, \circ)$. *If* $(D, +, \circ)$ *turns out to be commutative, we call* u *a central unit.*

Now suppose $GF(q)$ is the middle nucleus of π, and so we may write the order of π as q^N, for some integer $N \geq 1$. Let $H < \mathrm{Aut}\,\pi$ denote the group of middle-nucleus homologies of π, i.e., $\mathrm{Fix}(H) = Y$, and H leaves X invariant. So π contains a set of $K = (q^N - 1)/(q - 1)$ distinct rational Desarguesian partial spreads of degree $q + 1$, say $(\Delta_1, \Delta_2, \ldots, \Delta_K)$, uniquely determined by the following (inter-related) conditions:

1. $\{X, Y\}$ are components of every Δ_i;
2. $\Delta_i \cap \Delta_j = \{X, Y\}$ if $i \neq j$;
3. For each choice of $i = 1 \ldots K$, H fixes each of the $(q^N - 1)/(q - 1)$ Desarguesian planes of order q that pass through the origin O and have the same slope set as Δ_i; (in particular the orbits of H on ℓ_∞ union $\{X, Y\}$ define the components of each Δ_i).

The Desarguesian planes of order q mentioned in 3 are precisely the *middle nucleus* subplanes of the spread π. If δ is the affine part of any such subplane, we define its *interior* by $\breve{\delta} = \delta - (X \cup Y)$, and its X-edge by $\overline{\delta}$. Thus H, the group of middle-nucleus homologies, is transitive on the slopes of $\breve{\delta}$ and on the points of $\overline{\delta} = \delta \cap X - \{o\}$; thus the X-edges are precisely the non-trivial orbits on X, of the middle-nucleus group H. We can now state our main result.

Theorem 2.2 *Let* $\pi = (V, \Gamma)$ *be a commutative semifield spread (i.e the corresponding translation plane admits an orthogonal polarity), relative to a shears frame* (X, Y), *with middle-nucleus* $GF(q)$ *and order* q^N.

Let $(\Delta_1, \Delta_2, \ldots, \Delta_K)$ *be the middle-nucleus partition of* $\Gamma - \{X, Y\}$ *into* $K = (q^N - 1)/(q - 1)$ *rational Desarguesian partial spreads of degree* $q + 1$. *Let* $\mathcal{C}$ *be the set of central units of* π, *relative to the given shears frame* (X, Y). *Then there exists a unique system of middle-nucleus planes* $(\delta_1, \delta_2, \ldots, \delta_K)$ *with* $\delta_i \in \Delta_i$ *such that*

1.
$$\mathcal{C} = \bigcup_{i=1}^{K} \breve{\delta}_i$$

2. *Every* X-*edge* X_i *(i.e. non-trivial middle-nucleus orbit on* X*) is of form* $X_i = \breve{\delta}_i \cap X$, *for exactly one* $i \in 1 \ldots K$.

Remark 2.3 *We do not know whether the analogue of item (2.2.2) for the Y-axis holds; i.e., whether the middle-nucleus planes whose interiors lie in C induce a partition of the non-zero elements of the Y-axis.*

The following result is an immediate consequence of the theorem (and actually arises in its proof).

Corollary 2.4 *Let $\pi = (V, \Gamma)$ denote a commutative semifield spread of order q^N, with middle nucleus $GF(q)$. Then every component $W \in \Gamma$, distinct from the frame components X and Y, contains exactly $q - 1$ central units and these together with O form an additive subgroup of W, of order q.*

A special case is the following result, which could have been deduced long before the completion of the proof of the theorem.

Corollary 2.5 *Let G be the autotopism group of an affine commutative semifield plane π^{ℓ_∞}, fixing the affine point O, and points X and Y on the translation axis ℓ_∞. Let ℓ be any other line through O. Then G_ℓ is transitive $\ell - O$ only if the plane π^{ℓ_∞} is Desarguesian.*

More general versions of the above corollary may be easily deduced from the theorem. For instance the transitivity hypothesis of G_ℓ can be replaced by $|G_\ell|$ being divisible by a p-primitive divisor of $p^n - 1$, where p^n is the order of the plane. In particular,

Corollary 2.6 *Suppose an affine translation plane π^{ℓ_∞} of order p^n admits a collineation of order u, where u is a p-primitive divisor of $p^n - 1$, that fixes at least three slopes. Then the plane is Desarguesian, whenever π, the projective closure of π^{ℓ_∞}, admits an orthogonal polarity.*

3. Proof of the main result

We begin by listing our conventions regarding spread-sets. We only use additive spread-sets since we are only concerned with semifields; we always assume that our spread-sets contain the identity matrix.

Definition 3.1 *An additive spread-set $\mathcal{M}$, of order p^n, is an additive group of p^n $n \times n$ matrices, over $GF(p)$, such that $I_n \in \mathcal{M}$ and all the non-zero matrices in $\mathcal{M}$ are non-singular. The spread coordinatised by $\mathcal{M}$ is $\pi_{\mathcal{M}} = (V, \Gamma_{\mathcal{M}})$, where the component-set $\Gamma_{\mathcal{M}}$ contains, besides Y, precisely the elements of form "$y = Mx$",viz., , $\left\{(x, Mx) : x \in \mathcal{F}_p^n\right\}$, whenever $M \in \mathcal{M}$; so X and $I =$*

$\left\{(x,x) : x \in \mathcal{F}_p^n\right\}$ are always in $\mathcal{M}$.

The following definition recalls the essentially standard way of obtaining a semifield from an additive spread-set $\mathcal{M}$, and conversely of recovering this additive spread-set from the semifield; $\mathcal{M}$ is often called the set of *slope maps* of the semifield.

Definition 3.2 *Given an additive spread-set $\mathcal{M}$, of order p^n, and $x, e \in \mathcal{F}_p^n - \{o\}$, the unique member M of $\mathcal{M}$ such that $Me = x$ is denoted by $\mathcal{M}_x^{(e)}$, and $Div_e(\mathcal{M}) = (D, +, \circ_e)$ where $a \circ_e b = \mathcal{M}_a^{(e)}(b) \; \forall a, b \in D$. Conversely, the slope set of a semifield $(D, +, \circ)$ is defined to be the additive spread-set*

$$\{M \in GL(n, p) \cup \{O\} : \exists m \in D \text{ such that } Mx = m \circ x \; \forall x \in D\}$$

We now collect together some well-known and easily verifiable facts related to the above definition.

Result 3.3 *Let $\mathcal{M}$ be an additive spread-set of order p^n, and $e \in D - \{o\}$. Then*

1. *$Div_e(\mathcal{M})$ is a semifield, with multiplicative identity e, coordinatising the spread $\pi_{\mathcal{M}} = (V, \Gamma_{\mathcal{M}})$, when the unit point corresponding to the coordinatisation is chosen to be $(e, e) \in I_n$.*
2. *The slope-set of $Div_e(\mathcal{M}) = \mathcal{M}$.*
3. *If $(D, +, \circ)$ is a semifield, with multiplicative identity e, and $\mathcal{M}$ is its slope set, then $Div_e(\mathcal{M}) = (D, +, \circ)$.*

Lemma 3.4 *Let $\mathcal{M}$ be an additive spread-set. Then $Div_e(\mathcal{M})$ is a commutative semifield if and only if:*

$$(AB - BA)e = o \; \forall \, A, B \in \mathcal{M}.$$

Proof. By result 3.3(1) $Div_e(\mathcal{M})$ is a semifield; it remains to show that this semifield is commutative precisely when the given condition holds. Since any matrix $X \in \mathcal{M}$ is of form $\mathcal{M}_x^{(e)}$ (see definition 3.2), for some $x \in D$, the given condition is clearly equivalent to

$$(\mathcal{M}_a^{(e)} \mathcal{M}_b^{(e)} - \mathcal{M}_b^{(e)} \mathcal{M}_a^{(e)})e = o \; \forall \, a, b \in D$$

and hence to

$$\mathcal{M}_a^{(e)} b = \mathcal{M}_b^{(e)} a \; \forall \, a, b, \in D$$

211

and by definition 3.2 this is equivalent to the desired result

$$a \circ_e b = b \circ_e a \; \forall \, a, b \in D.$$

$\square$

Lemma 3.5 *Let $\mathcal{M}$ be an additive spread-set. If $Div_e(\mathcal{M})$ is commutative then*

$$(AB - BA)f = o \; \forall \, A, B \in \mathcal{M} \Leftrightarrow f \in N_m(Div_e(\mathcal{M})).$$

Proof. Letting the commutative semifield $Div_e(\mathcal{M}) = (D, +, \circ)$, result 3.3 implies $\mathcal{M}$ is its slope-set. Hence we obtain the following chain of equivalences:

$$(AB - BA)f = o \; \forall \, A, B \in \mathcal{M}$$
$$\Leftrightarrow \; (M_x^{(e)} M_y^{(e)} - M_y^{(e)} M_x^{(e)})f = o \; \forall \, x, y \in D$$
$$\Leftrightarrow \; x \circ (y \circ f) - y \circ (x \circ f) = o \; \forall \, x, y \in D$$
$$\Leftrightarrow \; x \circ (f \circ y) - (x \circ f) \circ y = o \; \forall \, x, y \in D, \text{ by commutivity of } \circ$$
$$\Leftrightarrow \; f \in N_m(D_e)$$

and the lemma follows.

$\square$

Lemma 3.6 *Let $(D, +, \circ)$ be a commutative semifield with multiplicative identity e, and let π_e be the unique middle-nucleus subplane of $\pi(D, +, \circ)$ that contains the point (e, e), (with both X and Y among its components). Then $\breve{\pi}_e$, the interior of π_e, consists entirely of central units, and $\ell \cap \breve{\pi}_e$ is the full set of central units on the component ℓ of $\pi(D, +, \circ)$, whenever $\ell \cap \pi_e \neq o$, i.e., whenever ℓ is a component of the subplane π_e.*

Proof. Let $\mathcal{M}$ denote the slope set of $(D, +, \circ)$, and so by result 3.3(3) $Div_e(\mathcal{M}) = (D, +, \circ)$. Now because $(D, +, \circ)$ is commutative, lemmas 3.4 and 3.5 together imply that $Div_f(\mathcal{M})$ is commutative precisely when $f \in N_m(D, +, \circ) - \{o\}$. But according to result 3.3 $Div_f(\mathcal{M})$ coordinatises the spread associated with $\mathcal{M}$ when the unit point is (f, f), on the component "$x = y$". Thus the central units on the component "$x = y$", of the spread $\pi(D, +, \circ)$, are precisely the elements of the set $\{(f, f) : f \in N_m(D, +, \circ), \; f \neq o\} = (\pi_e - \{O\}) \cap (\text{"}y = x\text{"})$. Now the full group of middle-nucleus homologies H, of $\pi(D, +, \circ)$, leaves invariant the middle nucleus subplane π_e, and the H-orbit of "$y = x$" is precisely the set of all components ℓ (distinct from X and Y) that meet π_e non-trivially. Thus $\ell \cap \pi_e - \{O\}$ consists of the full set of central units on ℓ, and the lemma follows.

$\square$

212

Lemma 3.7 Let $(D, +, *)$ be a finite semifield. If $\pi(D, +, *)$ can be recoordinatised by a commutative semifield, then $\pi(D, +, *)$ has a central unit of type (c, c), for some $c \in D - \{o\}$.

Proof. By hypothesis, there is a commutative semifield $(D, +, \circ)$ such that we have an additive bijection $\alpha : D \oplus D \to D \oplus D$, fixing the subspaces X and Y, that induces a spread isomorphism from $\pi(D, +, \circ)$ to $\pi(D, +, *)$ and hence satisfying the requirement:

$$\alpha : (x, m \circ x) \mapsto (G(x), H(m \circ x)) \ \forall \, x, m \in D \tag{1}$$

where G, H are resp. the action of α on X and Y [1]. But letting M be defined so that "$y = m \circ x$" maps to "$y = M(m) * x$" under α we also have

$$(x, m \circ x) \mapsto (G(x), M(m) * G(x)) \ \forall \, x, m \in D \tag{2}$$

Hence by equations (1) and (2):

$$H(m \circ x) = M(m) * G(x)$$

and by the commutativity of $m \circ x$ we therefore have

$$M(m) * G(x) = M(x) * G(m) \tag{3}$$

Now choosing μ so that $M(\mu) = e$, the identity for $*$, we have

$$G(x) = M(x) * G(\mu) \ \forall \, x \in D$$

and hence equation (3) can be rewritten

$$M(m) * (M(x) * G(\mu)) = M(x) * (M(m) * G(\mu)) \ \forall \, x, m \in D \tag{4}$$

and now letting T_d represent the slope of the generic element $d \in D$, relative to $*$, we have:

$$(T_m T_x - T_x T_m) G(\mu) = o \ \forall \, x, m \in D$$

and now by lemma 3.4 $(G(\mu), G(\mu))$ is a central unit of $\pi(D, +, *)$, as required.
□

Now consider any semifield spread $\pi = (V, \Gamma)$, coordinatisable by a commutative semifield, and let $u \in V - (X \cup Y)$. Now regarding u as the unit point, we have a coordinatising isomorphism $\psi : V \to V$, from π to some $\pi(D, +, *)$, where $(D, +, *)$ is a semifield (but not necessarily commutative). By the lemma above $\pi(D, +, *)$ has a central unit on the component "$y = x$", and so π has a central unit on the component containing u, for every u not in $X \cup Y$. Thus we have established the following consequence of the lemma above:

---

1. It may be worth recalling that X and Y are components of both $\pi(D, +, *)$ and $\pi(D, +, \circ)$.

Corollary 3.8 *If* $\pi = (V, \Gamma)$ *is a spread coordinatisable by a commutative semifield, then every component* $\ell \in \Gamma - \{X, Y\}$ *contains a central unit.*

Now, under the hypothesis of this corollary, each component $\ell \in \Gamma - \{X, Y\}$, contains a central unit e, and by lemma 3.6 the interior of the unique middle nucleus plane π_e, through e and having X and Y as components, consists of central units. Also if π_f is another middle-nucleus plane, whose interior consists entirely of central units, then π_e and π_f cannot share any components, distinct from X and Y, without contradicting the last part of lemma 3.6. Thus the set of central units C, of the commutative semifield plane π, is given by

$$C = \bigcup_{i=1}^{K} \check{\delta}_i$$

where each $\check{\delta}_i$ is the interior of a middle-nucleus plane δ_i, and every component of π, distinct from X and Y, lies in exactly one δ_i. Thus we have established part 1 of theorem 2.2; in particular, we have $K = (q^N - 1)/(q - 1)$,c.f., corollary 1.1.

We now turn our attention to determining the intersection of two such δ_i 's on X, and establish that $(\delta_i \cap X)$ and $(\delta_j \cap X)$ cannot share any non-zero affine point, whenever $i \neq j$. It is not clear to us whether a similar condition applies to the "Y-edges" defined by the central planes.

Lemma 3.9 *Let* $(D, +, \circ)$ *be a commutative semifield with multiplicative identity* e. *If* (e, f) *is a central unit of* $\pi(D, +, \circ)$, *then* $f \in N_m(D, +, \circ)$.

Proof. Let $\mathcal{M}$ be the slope set of $(D, +, \circ)$, and F the slope of f, in $\mathcal{M}$. Now the linear bijection

$$\psi : D \oplus D \;\rightarrow\; D \oplus D$$
$$(x, y) \;\mapsto\; (Fx, y)$$

maps the component "$y = Mx$", where $M \in \mathcal{M}$, onto the subspace "$y = MF^{-1}x$" and hence defines an isomorphism from the spread $\pi_{\mathcal{M}}$ onto the spread $\pi_{\mathcal{N}}$, where $\mathcal{N} = \mathcal{M}F^{-1}$, such that $(e, f) \mapsto (Fe, Fe)$. Thus (Fe, Fe) is a central unit of $\pi_{\mathcal{N}}$, on the component "y=x". So by lemma 3.4 applied to $\mathcal{N}$ we have

$$\left[MF^{-1}NF^{-1} - NF^{-1}MF^{-1} \right] Fe = o \; \forall M, N \in \mathcal{M}$$

Thus

$$MF^{-1}Ne = NF^{-1}Me \; \forall M, N \in \mathcal{M}$$

214

and now writing $W = F^{-1}$ we obtain

$$m \circ W(n \circ e) = n \circ W(m \circ e) \ \forall m, n \in D$$

and hence

$$m \circ W(n) = n \circ W(m) \ \forall m, n \in D \tag{5}$$

Now writing $w_1 = W(e)$ and choosing $m = e$ we obtain

$$W(n) = n \circ w_1 \ \forall n \in D \tag{6}$$

and so equation (5) becomes:

$$m \circ (n \circ w_1) = n \circ (m \circ w_1) \ \forall m, n \in D$$

and by the commutativity of $\circ$ we have

$$m \circ (w_1 \circ n) = (m \circ w_1) \circ n \ \forall m, n \in D$$

Now $w_1 \in N_m(D)$, and so equation (6) yields

$$W(n) = n \circ w_1 \in N_m(D, +, \circ) \ \forall n \in N_m(D, +, \circ)$$

Thus W maps $N_m(D, +, \circ)$ onto itself, and hence so does $F = W^{-1}$. Now

$$(e, f) = (e, Fe) = (e, n) \ \exists n \in N_m(D, +, \circ)$$

as required. $\qquad \square$

Corollary 3.10 *Let $\pi = (V, \Gamma)$ be a spread coordinatisable by a commutative semifield. Suppose π_0 and π_1 are distinct middle nucleus subplanes, through O, such that their interiors $\mathring{\pi}_0$ and $\mathring{\pi}_1$ are contained in the central units of π. Then*

$$\pi_0 \cap \pi_1 \cap X = O$$

Proof. Without loss of generality we may assume that the component I, i.e.,"$y = x$", meets π_0 non-trivially; if not then replace the spread set $\mathcal{M}$ defining π by a spread set $\mathcal{M}L^{-1}$, where L is the slope of some component of π_0, distinct from X and Y. Now choose $e \neq o$ so that $(e, e) \in \pi_0 \cap I$. Thus, since the interior of π_0 consists of central units, π may be identified with $\pi(D, +, \circ)$, a commutative semifield with identity e such that $\pi_0 = N_m \oplus N_m$, where $N_m = N_m(D, +, \circ)$, since any point (e, e) lies in a unique middle-nucleus plane (relative to the axis X and Y). Now, if the lemma were false, we would have some $(\alpha, o) \in \pi_1$, where $\alpha \in N_m$. Hence π_1 also contains a point (α, β) where $\beta \in D - N_m$, since the interiors of the distinct central planes π_0 and π_1

215

are disjoint (c.f., theorem 2.2(1)). Now applying the middle-nucleus homology
$\hat{\alpha} : (x,y) \mapsto (\alpha^{-1} \circ x, y)$, to the plane coordinatised by the commutative
semifield $(D,+,\circ)$ we find that $(e, \alpha^{-1} \circ \beta) \in \pi_1$, since π_1 is invariant under
the middle nucleus homologies. But now, since $(e, \alpha^{-1}\beta)$, is a central unit,
the lemma above yields $\alpha^{-1} \circ \beta \in N_m(D,+,\circ)$, contradicting the fact that
$\beta \notin N_m(D,+,\circ)$. Hence the corollary follows. $\qquad\square$

Thus in any commutative semifield spread π, of order q^N and with
middle-nucleus $GF(q)$, distinct middle-nucleus subplanes, whose interiors are
central, have disjoint X-edges. Since we have already seen (c.f., theorem
2.2(1)) that the number of "central planes" is precisely $(q^N-1)/(q-1)$, and
this is also the number of non-trival middle-nucleus orbits on X, we conclude
that every X-edge is contained in a unique middle-nucleus subplane whose
interior consists of central units. This completes the proof of the second part
of theorem 2.2. $\qquad\square$

We end by remarking that in a sequel [4] we shall demonstrate that,
in the even order case, the central units of commutative semifield planes are
partitioned by a set of translation ovals which, together with X and Y, define
a rational Desarguesian net.

References

[1] A. A. Albert. Non-associative algebras. *Pacific J. Math.*, 43, pp. 547–551, 1943.

[2] M. J. Ganley. Polarities in translation planes. *Geom. Dedicata*, 1, pp. 103–106, 1972.

[3] D. R. Hughes and F. C. Piper. *Projective Planes*. Springer Verlag, Berlin, 1973.

[4] V. Jha and G. P. Wene. On a partial oval cover for the central units of a commutative semifield spread. Preprint, 1992.

[5] T. G. Ostrom. *Finite Translation Planes*, volume 158 of *Lecture Notes in Math.* Springer Verlag, Berlin, New York, 1970.

V. Jha,Mathematics Department, Glasgow Polytechnic, Cowcaddens Road, Glasgow G4 OBA, Scotland. e-mail: mat.jha@glasgow-poly.ac.uk

G. P. Wene, Mathematics Department, University of Texas at San Antonio, San Antonio, Texas TX 78249-0600, U.S.A. e-mail: wene@ringer.cs.utsa.edu

Partially sharp subsets of $P\Gamma L(n, q)$

N. L. Johnson

Abstract

In this article, certain translation nets which are unions of subplane covered nets are shown to be equivalent to partially sharp subsets of $P\Gamma L(n, q)$. In particular, translation planes of order q^2 and arbitrary kernel that admit two distinct Baer groups of order $q-1$ with identical component orbits are shown to be equivalent to partially sharp subsets of cardinality q of $P\Gamma L(2, K)$ for some field K isomorphic to $GF(q)$.

1. Introduction

Recently, a number of various connections have been established between translation planes and other geometric or combinatorial incidence structures. In particular, there are connections between flocks of hyperbolic quadrics in $PG(3, q)$ and translation planes with spreads in $PG(3, q)$ such that the spread is a union of reguli that share two lines. Similarly, partial flocks correspond to translation nets whose partial spreads are unions of reguli that share two lines.

It is well known that a Miquelian Minkowski plane can be defined using a hyperbolic quadric in $PG(3, q)$ and when this connection is made, the points of the Minkowski plane are identified with the elements of $PG(1, q) \times PG(1, q)$, the circles (conics) correspond to the elements of $PGL(2, q)$ and a flock corresponds to a sharply transitive subset of $PGL(2, q)$.

Recently, Knarr [10] generalized this idea and showed that a sharply transitive subset of $P\Gamma L(2, q)$ produces a translation plane of order q^2 whose spread is a union of derivable partial spreads that contain two lines but all of these do not necessarily lie in the same projective space.

In [8], it was shown that a partial hyperbolic flock of deficiency one (missing exactly one conic) corresponds to a translation plane with spread in $PG(3, q)$ that admits a Baer group of order $q - 1$. Note that a partial hyperbolic flock of deficiency one corresponds to a partially sharp subset of $PGL(2, q)$ of cardinality q (see definition 2.1). One of the main objectives of this article is to show that partially sharp subsets of $P\Gamma L(2, q)$ of cardinality q produce translation planes that admit a Baer group of order $q - 1$ thus extending the work of Knarr mentioned above. Note that the constructed translation planes will not necessarily have spreads in $PG(3, q)$.

More generally, we want to study translation nets that can be constructed from partially sharp subsets in $P\Gamma L(n, q)$ and consider their possible extensions. One main problem would be to determine conditions to impose on a translation plane of order q^2 in order that there is a corresponding partially sharp subset of $P\Gamma L(2, K)$ for some field K isomorphic to $GF(q)$. We shall state our main results in the next sections. Concerning characterizations of translation planes by groups, our nicest result involves planes admitting Baer groups.

Theorem 1.1 (see theorem 3.6) *(i) Let π denote a translation plane of order q^2 and arbitrary kernel. If π admits two distinct Baer groups of order $q-1$ that have the same orbits on the line at infinity then there is a field K isomorphic to $GF(q)$ such that there is a corresponding partially sharp subset of $P\Gamma L(2, K)$ of cardinality q.*

(ii) If S is a partially sharp subset of $P\Gamma L(2, q)$ of cardinality q then there is a translation plane of order q^2 that admits two distinct Baer groups of order $q-1$ such that the groups have the same orbits on the line at infinity.

In section 2, we give the general connections between translation nets of a particular type and partially sharp sets. In section 3, we show how to connect translation nets that admit certain collineation groups with partially sharp sets of a particular type. In section 4, we note how results on net extension produce results on extensions of partially sharp sets. In particular, we show the following theorem.

Theorem 1.2 (see theorem 4.2). *If P is a partially sharp subset of $P\Gamma L(n, q)$ for $n \geq 4$ of deficiency one, then P may be extended to a sharply 1-transitive subset of $P\Gamma L(n, q)$.*

Acknowledgement

The author is indebted to Professor Kathleen O'Hara for helpful conversations with regards to this article. In particular, the idea for the proof of theorem 2.6 is due to the interaction.

2. Partially sharp subsets of $P\Gamma L(n, q)$ and their corresponding translation nets

In this section, we lay down the general foundation for the connections between partially sharp subsets of $P\Gamma L(n, q)$ and translation nets admitting certain collineation groups. In particular, we are interested in showing that partially sharp subsets of $P\Gamma L(2, q)$ of cardinality q produce translation planes of order q^2 that admit two Baer groups whose orbits on the line at infinity are equal.

Definition 2.1 *Let S be a set of permutations on a set X. S shall be said to be partially sharp if and only if for all x, y in X and for all g, h in S then $xg = xh$ if and only if $g = h$ and $xg = yg$ if and only if $x = y$. If $|X|$ is finite, a partially sharp subset of cardinality $|X|$ is said to be sharply transitive.*

Theorem 2.2 *(1) Given a partially sharp subset S of $P\Gamma L(n, q)$ acting on the points of $PG(n-1, q)$ of cardinality $t \leq (q^n - 1)/(q - 1)$, there is a partially sharp subset of $\Gamma L(n, q)$ of cardinality $t(q - 1)$ which defines a translation net N_S of order q^n and degree $t(q - 1) + 2$. The net N_S is the union of t translation nets N_i, $i = 1, 2, \ldots, t$ of degree $q + 1$ and order q^n which mutually share two components L, M.*

(2) Furthermore, each net N_i is a subplane covered translation net and corresponds to a regulus in some projective space $PG(n - 1, K_i)$ where $K_i \cong GF(q)$.

(3) The net N_S admits two collineation groups H_L, H_M of order $q - 1$ which fix the two common components where H_L fixes L pointwise, and H_M fixes M pointwise. $\langle H_L, H_M \rangle$ fixes each net N_i so the groups H_L and H_M have the same orbits on the components of N_S.

Proof. The proof is similar to the one of (2.1) in [7], so we shall give only a sketch. For each element $g \in S$ choose a preimage g^+ within $\Gamma L(n, q)$. Represent g^+ by the mapping

$$(x_1, x_2, \ldots, x_n) \to (x_1^{\sigma_M}, x_2^{\sigma_M}, \ldots, x_n^{\sigma_M})M$$

where $\sigma_M \in \operatorname{Aut} GF(q)$ and $M \in GL(n, q)$.

Note that any other preimage would be represented by

$$(x_1, x_2, \ldots, x_n) \to (x_1^{\sigma_M}, \ldots, x_n^{\sigma_M})MuI_n$$

for some $u \in GF(q)^*$.

Form the partial spread $N_g = \{y = x^{\sigma_M}MvI_n, \ x = 0, \ y = 0 \| v \in GF(q)^*\}$ where $x = (x_1, x_2, \ldots, x_n)$, $x^{\sigma_M} = (x_1^{\sigma_M}, \ldots, x_n^{\sigma_M})$, $y = (y_1, y_2, \ldots, y_n)$.

It follows from Johnson [7] that this is a subplane covered net of degree $q + 1$ and order q^n. It follows from [4] that this net is a regulus net and corresponds to some regulus in an associated projective space $PG(n - 1, K_g)$ for some field $K_g \cong GF(q)$.

Furthermore, $N_S = \cup\{N_g \| g \in S\}$ is a translation net of degree $|S|(q - 1) + 2$ and degree q^n. Note that the indicated semilinear maps acting on the nonzero vectors of a n-space over $GF(q)$ form a partially sharp set in $\Gamma L(n, q)$.

Finally, note that the components L, M are $x = 0$, $y = 0$ and $H_L = H_x$ is given by $\langle (x,y) \rightarrow (x,y)(uI_n, I_n)\|u \in GF(q)^*\rangle$ and that $H_M = H_y = \langle (x,y) \rightarrow (x,y)(I_n, uI_n)\|u \in GF(q)^*\rangle$. Note that each group fixes each net N_g for $g \in S$. Also, note that another set of preimages of elements of S produces the same translation net N_S. □

Before we give our main results, we prove two results on sets of mutually disjoint subspaces which may be of independent interest.

Theorem 2.3 *Let V_{2n} denote a $2n$-dimensional vector space over $GF(q)$ for n a positive integer. Let L and M be two disjoint n dimensional vector subspaces and enumerate the 1-dimensional subspaces on L and M by $\{X_i\|i = 1, 2, \ldots, (q^n - 1)/(q - 1)\}$ and $\{Y_i\|i = 1, 2, \ldots, (q^n - 1)/(q - 1)\}$ respectively. Let σ be any permutation of $\{1, 2, \ldots, (q^n - 1)/(q - 1)\}$. Then $R_\sigma = \{\langle X_i, Y_{\sigma(i)} \rangle\|i = 1, 2, \ldots, (q^n - 1)/(q - 1)\}$ is a set of mutually disjoint subspaces which cover L and M.*

Proof. Choose a basis for L and M so that L is represented by $\{(x_1, x_2, \ldots, x_n, 0, 0, \ldots, 0)\}$ and M by $\{(0, 0, \ldots, 0, y_1, y_2, \ldots, y_n)\}$ for all $x_i, y_i \in GF(q)$ for $i = 1, 2, \ldots, n$. Let H_L and H_M be represented exactly as in the proof to theorem 2.2. Note that H_L and H_M fix each 2-dimensional subspace $\langle X_i, Y_j \rangle$ generated by the 1-dimensional subspaces X_i and Y_j so that $\langle X_i, Y_j \rangle - (L \cup M)$ is an orbit of length $(q-1)^2$ under the group of linear transformations $\langle H_L, H_M \rangle$. This implies that any two distinct such 2-dimensional subspaces can intersect only on L or M. Hence, if σ is a permutation of $\{1, 2, \ldots, (q^n - 1)/(q - 1)\}$ then $\langle X_i, Y_{\sigma(i)} \rangle$ and $\langle X_j, Y_{\sigma(j)} \rangle$ for $i \neq j$ are disjoint 2-dimensional subspaces. So, $R_\sigma = \{\langle X_i, Y_{\sigma(i)} \rangle\|i = 1, 2, \ldots, (q^n - 1)/(q - 1)\}$ is a set of mutually disjoint 2-dimensional subspaces which covers L and M. □

Theorem 2.4 *(1) Let V_{2n} be a vector space of dimension $2n$ over $K \cong GF(q)$. Let N_1 be a translation net whose points are the vectors of V_{2n}, is of degree $q + 1$ and order q^n such that the components correspond to a $(n - 1)$-regulus in $PG(2n - 1, K_1)$ for some field $K_1 \cong GF(q)$) possibly distinct from K. Assume that L, M are two components of N_1 which are K-subspaces. If the subplanes of N_1 incident with the zero vector are 2-dimensional K-subspaces, enumerate the 1-dimensional subspaces on L, M as in (2.2). Then the set of subplanes of the net N_1 which contain the zero vector is $\{\langle X_i, Y_{\tau(i)} \rangle\|i = 1, 2, \ldots, (q^m - 1)/(q - 1)\}$ where τ is some permutation of $\{1, 2, \ldots, (q^n - 1)/(q - 1)\}$.*

(2) Let $\{N_k\|k = 1, 2, \ldots, t\}$ be a set of translation nets of degree $q + 1$ and order q^n whose points are the vectors of a $2n$-dimensional subspace over

a given field $K \cong GF(q)$ and where N_k corresponds to a $(n-1)$-regulus in some projective space $PG(2n-1, K_k)$ for $K_k \cong GF(q)$ possibly distinct from K. Assume that the nets mutually share exactly the components L, M, that these components are n-dimensional K-subspaces, and that the subplanes of N_k incident with the zero vector are 2-dimensional K-subspaces for each $k = 1, 2, \ldots, t$. Let $\{\tau_k \| k = 1, 2, \ldots, t\}$ denote the set of permutations of $\{1, 2, \ldots, (q^n - 1)/(q - 1)\}$ corresponding to $\{N_k\}$ as in (1). Then $\{\tau_k\}$ is a partially sharp subset of permutations of $\{1, 2, \ldots, (q^n - 1)/(q - 1)\}$.

Proof. The distinct 2-dimensional K-subspaces which are subplanes of N_k are disjoint. Hence, given a 1-dimensional subspace X_i on L, there is a unique 1-dimensional subspace Y_j on M such that $\langle X_i, Y_j \rangle$ is a subplane of N_k. Since the set of subplanes of N_k incident with the zero vector partitions L and M, the mapping $i \to j$ defined as above is onto and thus produces a permutation τ_k of $\{1, 2, \ldots, (q^n - 1)/(q - 1)\}$.

Now let N_k and N_z denote any two translation nets and assume that the associated permutations τ_k and τ_z share an image; that there exists an integer i such that $\tau_k(i) = \tau_z(i)$. Then the nets also share a subplane $\langle X_i, Y_{\tau_k(i)} \rangle = \langle X_i, Y_{\tau_z(i)} \rangle$ contrary to the assumption. □

Theorem 2.5 Let S be a partially sharp subset of $P\Gamma L(n, K)$, for $K \cong GF(q)$, of cardinality t and N_S the corresponding translation net of degree $t(q-1) + 2$, where $N_S = \cup N_i$ for $i = 1, 2, \ldots, t$ and the nets N_i are the associated translation nets of degree $q+1$ which share two components L, M. Then there are $(q^n - 1)/(q - 1) - t = s$ distinct sets R_i, $i = 1, 2, \ldots, s$ such that, for each i, R_i is a set of $(q^n - 1)/(q - 1)$ mutually disjoint 2-dimensional K-subspaces which cover L and M and such that the subspaces are disjoint from the components of $N_S - \{L, M\}$.

Proof. By theorem 2.4(2), corresponding to the nets N_k is a set of t permutations τ_k for $k = 1, 2, \ldots, t$ of $\{1, 2, \ldots, (q^n - 1)/(q - 1)\}$ such that $\{\tau_k\}$ is a partially sharp set of permutations. Note that this is equivalent to the fact that the matrix $[\tau_k(i) = a_{ki}]$ is a $t \times (q^n - 1)/(q - 1)$ partial Latin square. This partial Latin square may be extended to a Latin square (see e.g. [1]) so that there is a set of s permutations τ_k for $k = t+1, \ldots, (q^n - 1)/(q - 1)$ such that each set $T_j = \{\tau_i \| i = 1, 2, \ldots, t, \text{ and } i = j\}$ for $t + 1 \le j \le (q^n - 1)/(q - 1)$ forms a partially sharp set of permutations of cardinality t+1.

Choose any permutation τ_j and form the corresponding set R_j of 2-dimensional subspaces $= \{\langle X_i, Y_{\tau_j(i)} \rangle \| i = 1, 2 \ldots, (q^n - 1)/(q - 1)\}$. The subplanes are mutually disjoint by theorem 2.4. Since the set T_j is partially sharp, the only possibly intersections of subplanes are on L or M. □

221

Theorem 2.6 *Assume the conditions of theorem 2.5. If a set R_i is the set of subplanes incident with the zero vector of a translation net N_{t+1}^i of degree $q + 1$ and order q^n which contain L and M as components then there is a partially sharp set S^+ of cardinality $t + 1$ of $P\Gamma L(n, q)$ which contains S.*

Proof. The groups H_L, H_M fix each subplane of R_i. Let $g \in \langle H_L, H_M \rangle$ and consider the translation net $N_{t+1}^i g$. This is a translation net whose components are covered by the same subplanes as N_{t+1}^i. This means the set of components of N_{t+1}^i and of $N_{t+1}^i g$ cover the same affine points. Hence, this implies that the two subnets of degree $q - 1$ defined by the respective components not equal to L or M must cover each other. This implies that the two subnets of degree $q - 1$ and order q^n are replacements of each other. However, by [3], this can occur only if the nets are equal. Hence $\langle H_L, H_M \rangle$ is a collineation group of N_{t+1}^i.

Represent the components L and M by $(x = 0)$ and $(y = 0)$ respectively. Furthermore, represent the components of N_{t+1}^i by $(x = 0)$, $(y = 0)$ and $y = xW$ where $\{W\}$ is a set of cardinality $q - 1$ of $nr \times nr$ matrices over the prime field $GF(p)$ where $K \cong GF(q)$ and $q = p^r$. Clearly, the matrices and their distinct differences are nonsingular.

Since H_L and H_M are both regular on $\{y = xW\}$, we assert that, as a linear transformation, W must normalize $\{uI_n \| u \in K\}$. Indeed, we still may represent $H_L = \langle \text{Diag}\{uI_n, I_n\} \| u \in K^* \rangle$ and $H_M = \langle \text{Diag}\{I_n, uI_n\} \| u \in K^* \rangle$ even though at this point, we do not know the connection of $y = xW$ with $\Gamma L(n, K)$. There is a subgroup of $\langle H_L, H_M \rangle$ of order $q - 1$ which leaves $y = xW$ invariant.

Considering uI_n as a matrix over $GF(p)$, we have that there exists a subgroup whose elements have the form $\text{Diag}\{uI_n, vI_n\}$ as matrices over $GF(p)$ which leave $y = xW$ invariant. This implies that $u^{-1}I_n W v I_n = W$ so that $W v I_n W^{-1} = uI_n$. Suppose, for some u, the corresponding $v = 1$ in the stabilizer of $y = xW$. Then this element fixes M pointwise so that no other component can be left invariant since otherwise there would be points fixed by H_M which lie on a component not equal to M. Similarly, no two u's or no two v's in distinct elements of the stabilizer of $y = xW$ can be equal. This implies that $W\{uI_n \| u \in K\}W^{-1} = \{vI_n \| v \in K\}$.

Hence, W normalizes the field $\{uI_n \| u \in K\}$. Since the automorphisms of K extend to the automorphisms of $\{uI_n\}$, it follows that $W u I_n = u^\sigma I_n W$ for all $u \in K$ and for some $\sigma \in \text{Aut } K$. Moreover, since $y = xW \to y = x \, W u I_n$ for all $u \in K^*$ by H_M, it follows that there is exactly one automorphism of K associated with the net N_{t+1}^i.

We have the n-vectors $(x_1, x_2, \ldots, x_n) = x$ over K and considering x as a nr-vector over $GF(p)$, we have that (x, xW) is a vector of some subplane and hence there is a unique n-vector $(y_1, \ldots, y_n) = xW$ over K. That is, W permutes the set of n-vectors over K. Hence, it follows that W is a semilinear mapping over K; W may be considered within $\Gamma L(n, K)$.

So, N_{t+1}^i may be represented by $y = 0$, $x = 0$, $y = x^{\sigma_J} J u I_n$ for all $u \in K$ and $\sigma_J \in \operatorname{Aut} K$ where $J \in GL(n, K)$. (Define $g(x) = xW$. Then $g(xu) = g(xuI_n) = xuI_n W = xWu^\sigma I_n = g(x)u^\sigma I_n = g(x)u^\sigma$. Since g is clearly additive, it follows that g is semilinear.)

It now follows that $S \cup \{x \to x^{\sigma_J} J\} = S^+$ represents a partially sharp subset of $P\Gamma L(n, q)$ since the corresponding set in $\Gamma L(n, K)$ determines a net of degree $(t+1)(q-1) + 2$. $\qquad\square$

Corollary 2.7 *(1) Given a sharply transitive set S in $P\Gamma L(2, q)$ of cardinality t, there corresponds a translation net N_S of degree $t(q-1) + 2$ and order q^2 which consists of t subplane covered translation nets of degree $q + 1$ and order q^2 that mutually share two components L and M. There exist $(q+1) - t$ distinct translation nets of degree $q+1$ and order q^2, R_i, $i = 1, \ldots, (q+1) - t$, such that $R_i \cup \{N_S - \{L, M\}\}$ is a translation net of degree $(t+1)(q-1) + 2$. This translation net contains L and M as Baer subplanes and admits two Baer groups of order $q - 1$, H_L, H_M which have the same component orbits.*

(2) If, for some i, R_i is a derivable net then there is a sharply transitive set S^+ in $P\Gamma L(2, q)$ containing S of cardinality $t + 1$ and corresponds to (produces) the translation net $E_i \cup N_S - L, M$, where E_i denotes the net derived from R_i.

(3) If there exists a partially sharp subset of $P\Gamma L(2, q)$ of cardinality q (deficiency one) then there corresponds a translation plane of order q^2 which admits two Baer groups of order $q - 1$ which have the same component orbits.

(4) If the translation plane of order q^2 of (3) is derivable (by the extended net) then the partially sharp subset may be extended to a sharply transitive subset of $P\Gamma L(2, q)$. Conversely, if the partially sharp subset may be extended then the corresponding translation plane is derivable.

Proof. The only remaining part is to prove (4). If a partially sharp subset of cardinality q can be extended then there is a translation plane corresponding to this extension and contains the translation net of degree $q(q - 1) + 2$. And, there is a translation plane admitting two Baer groups from (3). These two translation planes share the net minus the two common components L and M (these appear as subplanes in one translation plane) so that the two remaining nets of degree $q+1$ in the two planes must be replacements of each other. By [11], the two translation planes are either equal or one is derived from the other. The planes are not equal since L, M are components of one and Baer subplanes of the other. $\qquad\square$

3. Two Baer groups

In this section, we consider translation nets of degree $(t+1)(q-1)+2$ and order q^2 that admit two distinct Baer groups of order $q-1$ that have the same component orbits. We shall show that there exists a field $K \cong GF(q)$ and a corresponding partially sharp subset of cardinality t in $P\Gamma L(2, K)$. Our main result shows that translation planes of order q^2 that admit two Baer groups with the same orbits on the line at infinity correspond to partially sharp subsets of $P\Gamma L(2, K)$ of cardinality q for some field K isomorphic to $GF(q)$. We first note the following theorem.

Theorem 3.1 ([5], [6]) *(1) Let N be a translation net of degree $q+1$ and order q^2 which admits a Baer group B of order $q-1$. Then the subplane $\text{Fix}\, B$ which is pointwise fixed by B is Desarguesian and there is another Baer subplane incident with the zero vector which is fixed by B. We shall denote the second fixed subplane of B indicated above by $\text{coFix}\, B$.*

(2) If there exists a third subplane in N then N is derivable.

Theorem 3.2 *Let N be a translation net of degree $(t+1)(q-1)+2$ for t a positive integer which admits two distinct Baer groups B_1, B_2 of order $q-1$ in the translation complement. If the Baer groups have the same component orbits then $\text{Fix}\, B_i = \text{coFix}\, B_j$ for $i \neq j$, $i, j = 1, 2$.*

Proof. By the assumptions, the set of fixed components of B_1 and the set of fixed components of B_2 are equal. Let D denote the net of common fixed components. Assume that $\text{Fix}\, B_1 \neq \text{coFix}\, B_2$. Then $\text{Fix}\, B_1$ or $\text{coFix}\, B_1$ must be moved within D to a third Baer subplane. By theorem 3.1, it follows that the net D is derivable. Moreover, the $q+1$ Baer subplanes of D incident with the zero vector have orbit structure $\{1, q\}$ or $q+1$ under $\langle B_1, B_2 \rangle$.

If $\langle B_1, B_2 \rangle$ is transitive on the subplanes, we may use the argument of [2](see (1.1)), to show that that there is a Sylow p-subgroup of order q for $p^r = q$. Thus, in either case, there is a Baer group of order q. However, since the component orbits of B_1 and B_2 are the same, the stabilizer of a component C not in the net D has order $|\langle B_1, B_2 \rangle|/(q-1)$. Hence, there is a Baer group of order q which fixes a component C not in the net D. Since C is a vector space, there must be fixed points other than the zero vector in C. It follows that the Baer group of order q fixes all affine points, a contradiction. $\square$

Lemma 3.3 ([5] pp. 33-37) *Let N be a translation net of degree $q+1$ and order q^2 which admits two Baer groups B_i, for $i = 1, 2$ of order $q-1$ such that $\text{Fix}\, B_i = \text{coFix}\, B_j$ for $i \neq j$; $i, j = 1, 2$. Then there is a representation* for $N = \{(x_1, x_2, y_1, y_2) \| x_i, y_i$ are r-vectors over $GF(p)$ for $p^r = q$, $i = 1, 2\}$ such that

    (i) $\text{Fix}\, B_2 = \{(0, x_2, 0, y_2) \| x_2, y_2$ are r-vectors over $GF(p)\}$ and $\text{coFix}\, B_2 = \{(x_1, 0, y_1, 0) \| x_1, y_1$ are r-vectors over $GF(p)\}$;

    (ii) the components of N have the form

$$x = 0, \quad y = x \begin{bmatrix} A_1 & 0 \\ 0 & A_4 \end{bmatrix}$$

where A_1 and A_4 are $2r \times 2r$ matrices over $GF(p)$. The sets $\{A_1\}$ and $\{A_4\}$ corresponding to the components of N are both irreducible and the respective centralizers K_1, K_4 are fields of matrices isomorphic to $GF(q)$.

    (iii)

$$B_1 = \langle \begin{bmatrix} I & 0 & 0 & 0 \\ 0 & C & 0 & 0 \\ 0 & 0 & I & 0 \\ 0 & 0 & 0 & C \end{bmatrix} \| C \in K_4 \rangle \quad \text{and} \quad B_2 = \langle \begin{bmatrix} D & 0 & 0 & 0 \\ 0 & I & 0 & 0 \\ 0 & 0 & D & 0 \\ 0 & 0 & 0 & I \end{bmatrix} \| D \in K_1 \rangle.$$

K_1 is the kernel of $\text{Fix}\, B_1 = \text{coFix}\, B_2$ and K_4 is the kernel of $\text{coFix}\, B_1 = \text{Fix}\, B_2$.

Assumptions

In what follows, we assume that N is a translation net of order q^2 and degree $(t+1)(q-1)+2$ for t a positive integer which admits two Baer groups B_1, B_2 of order $q - 1$ such that the component orbits of the two groups are the same.

    We started with a $4r$-dimensional vector space over $GF(p)$ for $p^r = q$ and determined a basis $E = E_1 \cup E_2 \cup E_3 \cup E_4$ with $E_1 = \{e_1, \ldots, e_r\}$, $E_2 = \{e_{r+1}, \ldots, e_{2r}\}$, $E_3 = \{e_{2r+1}, \ldots, e_{3r}\}$, $E_4 = \{e_{3r+1}, \ldots, e_{4r}\}$. In this basis, $\text{Fix}\, B_1 = \langle E_1 \cup E_3 \rangle$ and $\text{Fix}\, B_2 = \langle E_2 \cup E_4 \rangle$. Moreover, $\text{Fix}\, B_1$ may be considered as a 2-dimensional vector space over an $r \times r$ matrix field $K_4 \cong GF(q)$ so that $\text{Fix}\, B_1 = \{(x_1, 0, y_1, 0)\}$ with respect to E such that $x_1 = \sum_1^r x_{1i} e_i$ and $y_1 = \sum_1^r y_{1i} e_{2r+i}$ for x_{1i}, $y_{1i} \in GF(p)$. We are taking a basis $\{f_1, f_3\}$ for $\text{Fix}\, B_1$ over K_4 so that $f_i \in \langle E_i \rangle$ for $i = 1, 3$ so we may simultaniously consider a vector $(x_1, 0, y_1, 0)$ as a K_4-vector or a $GF(p)$-vector.

    Similarly, we may consider $\text{Fix}\, B_2 = \{(0, x_2, 0, y_2)\}$ with basis $E_2 \cup E_4$ and consider a vector $(0, x_2, 0, y_2)$ as a K_1-vector or as a $GF(p)$-vector.

    Now consider the ordered basis $B = E_1 \cup E_3 \cup E_2 \cup E_4$. In this basis, $\text{Fix}\, B_1 = \{(x_1, x_2, 0, 0) \| x_i \in K_4$, for $i = 1, 2\}$ and $\text{Fix}\, B_2 = \{(0, 0, y_1, y_2) \| y_i \in K_1$ for $i = 1, 2\}$.

    Let σ be any isomorphism from K_1 onto K_4. We then can consider the $4r$-dimensional vector space over $GF(p)$ as a 4-dimensional vector space

225

over K_4 by taking as vectors the $GF(p)$-vectors (x_1, x_2, y_1, y_2) such that $x_i \in K_4$, $y_i \in K_1$, for $i = 1, 2$ and define scalar multiplication by $D \in K_4$ by $(x_1, x_2, y_1, y_2)D = (x_1D^\sigma, x_2D^\sigma, y_1D, y_2D)$.

In particular, since K_1 and K_4 are $r \times r$ matrix fields over $GF(p)$, there is a matrix T such that $T^{-1}K_1T = K_4$.

Let W be a fixed $2r$-dimensional vector space over $GF(p)$ and identify Fix B_1 and Fix B_2 as $2r$-subspaces with W. Then the underlying $4r$-dimensional space is $W \oplus W$.

Let Γ denote the subnet whose partial spread is the set of components which are not fixed by B_1 or B_2. Then $\Gamma \cup \{\text{Fix } B_1, \text{Fix } B_2\} = \Gamma^+$ is a net upon which the linear group $\langle B_1, B_2 \rangle$ acts as a collineation group.

Now change bases by $\begin{bmatrix} I & 0 \\ 0 & T \end{bmatrix} = g$. The translation net Γ^+ is mapped onto an isomorphic net Γ^+g which admits the collineation groups B_1 and B_2^T.

Hence, we have shown the following lemma.

Lemma 3.4 *Let Γ denote the subnet which does not contain the fixed components of B_1 or B_2 and let $\Gamma^+ = \Gamma \cup \{Fix\,B_1, Fix\,B_2\}$. Change basis by B as above. Then, by identifying the $GF(p)$-subspaces $Fix\,B_1$ and $Fix\,B_2$, there is a basis change g so that we may take $K_1 = K_4$ and the net Γ^+g is isomorphic to Γ^+.*

Theorem 3.5 *Let N be a translation net of order q^2 and degree $t(q-1)+2$ for t a positive integer that admits two distinct Baer groups B_1, B_2 of order $q - 1$ and which have the same component orbits. Let Γ^+ denote the subnet with partial spread the union of $\{Fix\,B_i, i = 1, 2\}$ and the set of components not fixed by B_i. Then there is a net R isomorphic to Γ^+ and a matrix field $K \cong GF(q)$ such that the net R admits the collineation groups*

$$B_1 = \langle \begin{bmatrix} C & 0 & 0 & 0 \\ 0 & C & 0 & 0 \\ 0 & 0 & I & 0 \\ 0 & 0 & 0 & I \end{bmatrix} \|C \in K \rangle$$

$$B_2 = \langle \begin{bmatrix} I & 0 & 0 & 0 \\ 0 & I & 0 & 0 \\ 0 & 0 & C & 0 \\ 0 & 0 & 0 & C \end{bmatrix} \|C \in K \rangle$$

and the net R has components of the form $x = 0$, $y = 0$ (Fix B_1 and Fix B_2) and $y = xM$ where $\{M\}$ is a set of nonsingular $2r \times 2r$ matrices over $GF(p)$ such that the differences are also nonsingular. Corresponding to each com-

ponent $y = xM$ is a semilinear mapping in $\Gamma L(2, K)$ and the set of all such semilinear mappings forms a partially sharp subset of cardinality $t(q - 1)$ in $\Gamma L(2, K)$. Modulo the K-scalar mappings, there is a corresponding partially sharp subset of cardinality t of $P\Gamma L(2, K)$, $K \cong GF(q)$.

Proof. It is easy to verify that the components have the basic form $y = xM$. Let $y = xJ$ be any such component. Since we have assumed that the component orbits of the Baer groups acting on the original translation net are equal, it follows that the groups have the same orbits on Γ^+. Since we define the components of Γ^+g to be the images of the components of Γ^+, it follows that the groups have the same orbits as acting on Γ^+g but with respect to a different basis.

Hence, for a component $y = xJ$, there is a subgroup of $\langle B_1, A_2 = B_2^T \rangle$ of order $q - 1$ which leaves $y = xJ$ invariant. If an element

$$\begin{bmatrix} D & 0 & 0 & 0 \\ 0 & D & 0 & 0 \\ 0 & 0 & C & 0 \\ 0 & 0 & 0 & C \end{bmatrix}$$

leaves $y = xJ$ invariant then

$$\begin{bmatrix} D^{-1} & 0 \\ 0 & D^{-1} \end{bmatrix} J \begin{bmatrix} C & 0 \\ 0 & C \end{bmatrix} = J.$$

Now $D = I$ if and only if $C = I$ in such an element as such elements fix exactly $x = 0$ or $y = 0$ pointwise or are trivial. Similarly, it follows that no two distinct elements in the subgroup which fixes $y = xJ$ can have the same $(1, 1) = (2, 2)$ entries. Thus, there is a 1-1 and onto mapping σ_J from K^* onto K^* defined by

$$J \begin{bmatrix} C & 0 \\ 0 & C \end{bmatrix} J^{-1} = \begin{bmatrix} D & 0 \\ 0 & D \end{bmatrix}$$

where

$$\begin{bmatrix} D & 0 & 0 & 0 \\ 0 & D & 0 & 0 \\ 0 & 0 & C & 0 \\ 0 & 0 & 0 & C \end{bmatrix}$$

leaves $y = xJ$ invariant. It is easy to verify that if we define $\sigma_J(0) = 0$ then σ_J becomes an automorphism of K. For example, that $\sigma_J(CD) = \sigma_J(C)\sigma_J(D)$ follows from the observation that

$$J \begin{bmatrix} CD & 0 \\ 0 & CD \end{bmatrix} J^{-1} = J \begin{bmatrix} C & 0 \\ 0 & C \end{bmatrix} J^{-1} J \begin{bmatrix} D & 0 \\ 0 & D \end{bmatrix} J^{-1}.$$

Since J acts as a linear transformation on the $2r$–space over $GF(p)$, W, and we are identifying $x = 0$ and $y = 0$ with W, J permutes the K-vectors of $x = 0$ and acts additively. Hence, if $\{f_1, f_2\}$ is a K-basis of W, we consider f_i as a $2r$-vector over $GF(p)$ in some suitable basis. The images $f_i J$ are $2r$-vectors which, in turn, define 2-vectors over K and in the same basis $\{f_1, f_2\}$ provide a 2×2 matrix over K.

Now, define $h(x) = xJ$. Then

$$h(xD) = h((x_1, x_2) \begin{bmatrix} D & 0 \\ 0 & D \end{bmatrix}) = x \begin{bmatrix} D & 0 \\ 0 & D \end{bmatrix} J$$

which equals

$$xJ \begin{bmatrix} D^\tau & 0 \\ 0 & D^\tau \end{bmatrix}$$

for some automorphism τ of K so that $h(xD) = h(x)D^\tau$, $h(x+z) = h(x)+h(z)$ so that h is a semilinear mapping. Furthermore, h is in $\Gamma L(2, K)$ since J is nonsingular. Note that the images of $y = xJ$ under the group are

$$y = xJ \begin{bmatrix} D & 0 \\ 0 & D \end{bmatrix}$$

for all $D \in K^*$ so that there are exactly $t(q-1)$ semilinear mappings which form a partially sharp subset of $\Gamma L(2, K)$ since there is an associated net and we may represent the net R in the form

$$x = 0, \quad y = 0, \quad y = x^{\sigma_J} J \begin{bmatrix} D & 0 \\ 0 & D \end{bmatrix}$$

for all $D \in K^*$ and for exactly t nonsingular matrices J. It follows immediately that the set $\{X \to X^{\sigma_J} J\}$ defines a partially sharp subset of cardinality t of $P\Gamma L(2, K)$ for $K \cong GF(q)$. $\qquad \square$

Theorem 3.6 *There is a 1-1 correspondence between partially sharp subsets of $P\Gamma L(2, q)$ of deficiency one and translation planes of order q^2 that admit two distinct Baer groups of order $q - 1$ which have the same component orbits. If B is one of the Baer groups of order $q - 1$ then any component orbit union the subspaces $\mathrm{Fix}\,B$ and $\mathrm{coFix}\,B$ is a derivable partial spread. The net containing $\mathrm{Fix}\,B$ is derivable if and only if the partially sharp subset of $P\Gamma L(2, q)$ can be extended to a sharply transitive set.*

Proof. We have seen that given a partially sharp set of cardinality q in $P\Gamma L(2, q)$, there is a corresponding translation net of degree $q(q - 1) + 2$ and order q^2 consisting of t subplane covered translation nets that share two components L and M. Furthermore, there is a set Σ of mutually disjoint 2-dimensional subspaces such that the subspaces are disjoint from the components different from the common components L and M. Hence, there is actually a spread of $q(q-1) + q + 1 = q^2 + 1$ components. Note that this translation plane admits two Baer groups which have the same component orbits. If the added net is derivable, we have seen above that the partially sharp set of cardinality q can be extended to a set of cardinality $q+1$ which is then sharply transitive.

If π is a translation plane of order q^2 which admits two Baer groups of order $q-1$ with the same component orbits, let Γ^+ denote the net which consists of the components which are not fixed by a Baer group B union $\{\text{Fix}\, B, \text{coFix}\, B\}$. By theorem 3.5, there is an isomorphic net R and a field $K \cong GF(q)$ which produces a partially sharp subset in $P\Gamma L(2, K)$ of cardinality q. Also, recall the isomorphism g leaves Fix B and coFix B invariant. Now assume that the plane π is derivable by the net D containing Fix B. Derive π by D to obtain the translation plane π_D containing Fix B and coFix B as components and note that π_D contains the net Γ^+ as a subnet. Let E denote the derived net of D. It follows that $\pi_D g$ is also derivable by the net Eg and contains the net $R = \Gamma^+ g$ as a subnet. The group B acts as a homology group of $\pi_D g$ and the $(q-1)$ components of $\pi_D g$ which are not in R are in an orbit under B. The argument in theorem 3.5 shows that corresponding to this orbit of components is a set of $q-1$ semilinear mappings over K of the form

$$X \to X^r J \begin{bmatrix} C & 0 \\ 0 & C \end{bmatrix}$$

for all $C \in K^*$.

Hence, modulo the scalar mappings, there is exactly one additional semilinear mapping in $P\Gamma L(2, K)$ and this mapping extends the original set corresponding to R. $\qquad\square$

By the results of sections 2 and 3, we have completed the proof to theorem 1.1.

4. Extensions of partially sharp subsets

There are non-Hall translation planes of order $q^2 = 16$, 25, and 81 which admit two Baer groups of order $q-1$ which have the same component orbits. These planes produce partially sharp subsets of $PGL(2, q)$ of deficiency one and are studied more generally in [2]. It is not always the case that the subsets can be extended to sharply transitive subsets of $P\Gamma L(2, q)$ and none of these can be extended within $PGL(2, q)$. We shall see that the situation for partially sharp subsets of deficiency one in $P\Gamma L(n, q)$ is quite different for $n > 3$.

Let S be a partially sharp subset of $P\Gamma L(n, q)$ of cardinality t. We have seen that to extend S requires that some set of 2-dimensional subspaces of the associated translation net should be the set of subplanes incident with the zero vector of a subplane covered net of degree $q + 1$. But, we can say a bit more on the extension of the associated translation net.

Theorem 4.1 *Let S be a partially sharp subset of $P\Gamma L(n, q)$ of cardinality t and let Γ^+ denote the associated translation net of degree $t(q-1)+2$ which is the union of t subplane covered translation nets of degree $q + 1$ that mutually share two components L and M. Let the two groups (homology groups) of order $q - 1$ be denoted by H_L and H_M. If Γ^+ can be extended to a translation net of degree at least $t(q-1)+2$ then there is an extension $\Gamma^+(L)(\Gamma^+(M))$ of degree $(t+1)(q-1)+2$ which admits $H_L(H_M)$ as a collineation group and such that the net is a union of $(t+1)$ subplane covered nets. If $\Gamma^+(L) = \Gamma^+(H)$ then this translation net corresponds to an extension S^+ of S which is a partially sharp subset of $P\Gamma L(n, q)$ of cardinality $t + 1$.*

Note, for example, if $y = xM$ is an added component and M normalizes the field of matrices defined by $GF(q)$ then the net would admit both groups so that $\Gamma^+(L) = \Gamma^+(M)$ in this case.

Proof. Suppose that N is any net extension of Γ^+. Choose L and M as $x = 0$ and $y = 0$ respectively and decompose as in section 3 so that the components of Γ^+ have the form $y = x^{\sigma_W} W u I_n$ where W is an $n \times n$ matrix over $GF(q)$, $\sigma_W \in \text{Aut } GF(q)$ and $u \in GF(q)^*$. If there is a net which extends Γ^+ then there is a component which may be represented by $y = xT$ for some matrix T over the prime field $GF(p)$ where $p^r = q$. It is easy to verify that the net $N_M = \{x = 0,\ y = 0,\ y = xTuI_n, \forall u \in GF(q)^*\}$ is a subplane covered translation net of degree $q + 1$ so that $\Gamma^+(M) = \Gamma^+ \cup N_M$ is an extension of degree $(t+1)(q-1)+2$ which admits H_M. Similarly, $N_L = \{x = 0,\ y = 0,\ y = xu I_n T \forall u \in GF(q)^*\}$ is a subplane covered translation net and $\Gamma^+(L) = \Gamma^+ \cup N_L$ admits H_L as a collineation group. If these two nets are equal then the net admits the groups H_L and H_M so that the groups have the same component orbits. We may then apply the main results of sections 2 and 3 to extend the partially sharp subset of $P\Gamma L(n, q)$. $\square$

Suppose we consider the problem of whether a partially sharp subset of cardinality t of $P\Gamma L(n, q)$ can be extended to a sharply transitive set. Since a sharply transitive subset corresponds to a translation plane, we see that this problem is essentially a question of extending a partial spread defining a translation net to a spread defining a translation plane. The corresponding translation net has deficiency $q - 1$ in this case.

Let $f(x) = (x^3 + 3 + 2x(x + 1))x/2$. By [3], if $q^n > f(q - 2)$ then the translation net may be extended to a unique affine plane. Furthermore, by

[9](3.2), (3.3) the affine plane must be a translation plane.

By theorem 4.1, if Γ^+ is the translation net of order q^n and degree $(q^n-1)/(q-1)-1)(q-1)+2$ (deficiency $q-1$) corresponding to the partially sharp subset of $P\Gamma L(n,q)$, assume that $q^n > f(q-2)$ and let π be the unique translation plane which extends Γ^+. If $y = xJ$ is any component of $\pi - \Gamma^+$ then there are two extensions $\Gamma^+(M)$ and $\Gamma^+(L)$ containing $y = xJ$ which then must be equal so that π admits the collineation groups H_M and H_L. By theorem 4.1, there is a corresponding extension of the partially sharp subset of $P\Gamma L(n,q)$ to a sharply transitive set. Thus, we have the following theorem.

Theorem 4.2 *Let S be a partially sharp subset of $P\Gamma L(n,q)$ of deficiency one. If $q = 2,3,4$ or 5 and $n = 3$ or if $n > 3$ then S may be extended to a sharply transitive subset of $P\Gamma L(n,q)$.*

Proof. We need only to show that $q^n > f(q-2) = (q-2)((q-2)^3 + 3 + 2(q-2)(q-1))/2$ which is equal to $(q^4 - 6q^3 + 14q^2 - 15q + 6)/2$. Clearly, q^n is larger than this number if $n \geq 4$, and a short calculation shows that q^3 is larger when $q = 2,3,4$ or 5. $\qquad\square$

This proves also theorem 1.2.

References

[1] I. **Anderson**. *A First Course in Combinatorial Mathematics.* Oxford Appl. Math. Comput. Sci. Ser. Clarendon Press, Oxford, 1974.

[2] M. **Biliotti and N. L. Johnson**. Maximal Baer groups in translation planes and compatibility with homology groups. Submitted to *Geom. Dedicata.*

[3] R. H. **Bruck**. Finite nets II: uniqueness and embedding. *Pacific J. Math.*, 13, pp. 421–457, 1963.

[4] F. **De Clerck and N. L. Johnson**. Subplane covered nets and semi-partial geometries. *Discrete Math.*, 106/107, pp. 127–134, 1992.

[5] D. A. **Foulser**. Subplanes of partial spreads in translation planes. *Bull. London Math. Soc.*, 4, pp. 32–38, 1972.

[6] V. **Jha**. *On the automorphisms of quasifields.* PhD thesis, Westfield College, University of London, 1972.

[7] N. L. **Johnson**. Translation planes covered by subplane covered nets. To appear in *Simon Stevin.*

[8] N. L. **Johnson**. Translation planes admitting Baer groups and partial flocks of quadric sets. *Simon Stevin*, 63, pp. 167–188, 1989.

[9] D. **Jungnickel**. Maximal partial spreads and translation nets of small deficiency. *J. Algebra*, 90, pp. 119–132, 1984.

[10] **N. Knarr.** Sharply transitive subsets of $P\Gamma L(2, F)$ and spreads covered by derivable partial spreads. *J. Geom.*, 40, pp. 121–124, 1991.

[11] **T. G. Ostrom.** Nets with critical deficiency. *Pacific J. Math.*, 14, pp. 1381–1387, 1964.

N.L. Johnson, Mathematics Dept. University of Iowa, Iowa City, Iowa 52242, USA. e-mail: njohnson@umaxc.weeg.uiowa.edu

Partial ovoids and generalized hexagons

G. Lunardon [*]

Abstract

An alternative construction for the $^3D_4(q)$-hexagon (q odd) is given using the algebraic variety which represents a regular spread onto the grassmannian of the planes of $PG(5,q)$.

1. Introduction

Let V be a vector space over the field F. We will denote by $PG(V,F)$ the projective space associated with V. For each vector subspace A of V, we will denote by $P(A)$ the projective subspace of $PG(V,F)$ associated with A. If $A = <v>$ has dimension 1, we will indicate with $P(v)$ the point of $PG(V,F)$ associated with A. If V has finite dimension n over the field $F = GF(q)$, we will write $PG(n,q)$ instead of $PG(V,F)$.

Let $W(2n+1,q)$ be the polar space arising from the symplectic polarity of $PG(2n+1,q)$ associated with the non-singular alternating bilinear form $(\ ,\ )$ of V. A partial ovoid $\mathcal{O}$ of $W(2n+1,q)$ is a set of pairwise non-orthogonal points. A partial spread $\mathcal{P}$ of $W(2n+1,q)$ is a set of pairwise disjoint maximal totally isotropic subspaces. Assume $\mathcal{P}$ is of size $q^n + 1$ such that each point of $\mathcal{O}$ belongs to an element of $\mathcal{P}$. We will denote by T_p the unique element of $\mathcal{P}$ incident with the point p of $\mathcal{O}$. If $p_i = P(v_i)$ ($i = 1,2,3,4$) are mutually distinct points of $\mathcal{O}$, then let

$$Q(p_1,p_2,p_3,p_4) = \{P(\sum_{i=1}^{4} x_i v_i) : \sum_{i,j=1,i\neq j}^{4} x_i x_j(v_i,v_j) = 0\},$$

$$C(p_1,p_2,p_3) = Q(p_1,p_2,p_3,p_4)\cap <p_1,p_2,p_3> .$$

Let us suppose that:

(a) $<T_{p_1},p_2> \cap\mathcal{O} = \{p_1,p_2\}$;

(b) no plane contains four points of $\mathcal{O}$;

(c) a line of $W(2n+1,q)$ joining a point of T_{p_1} to a point of T_{p_2} is not incident with p_3;

---

[*]. The author has a partial financial support from the Italian M.U.R.S.T.

(d) if $r \in\ < p_1, p_2, p_3 > \cap T_{p_4}$, then r does not belong to $C(p_1, p_2, p_3)$;

(e) no five points of $\mathcal{O}$ belong to $Q(p_1, p_2, p_3, p_4)$.

Let $W(2n + 3, q)$ be the polar space arising from a symplectic polarity $\perp$ of $PG(2n + 3, q)$. If x and y are two points of $PG(2n + 3, q)$ not collinear in $W(2n + 3, q)$, then $x^{\perp} \cap y^{\perp} = T$ is a $(2n + 1)$-dimensional subspace of $PG(2n+3, q)$ and $T \cap W(2n+3, q) = W(2n+1, q)$. Let us define a point-line geometry $H(\mathcal{O}, \mathcal{P})$ in the following way:

Points:

(1) the point x;

(2) the points different from x but contained in one of the lines $< x, p >$ where p belongs to $\mathcal{O}$;

(3) the maximal totally isotropic subspaces of $W(2n + 3, q)$ not contained in $x^{\perp}$ and meeting one of the $(n+1)$-dimensional subspaces $< x, T_p >$ $(p \in \mathcal{O})$ in a n-dimensional subspace;

(4) the points of $PG(2n+3, q) \setminus x^{\perp}$.

Lines:

(A) the lines $< x, p >$ where $p \in \mathcal{O}$;

(B) the n-dimensional subspaces not incident with x, and contained in one of the $(n+1)$-dimensional subspaces $< x, T_p > (p \in \mathcal{O})$;

(C) the totally singular lines not contained in $x^{\perp}$ and meeting $x^{\perp}$ in a point belonging to one of the lines $< x, p > (p \in \mathcal{O})$.

Incidences:

Points of type (2) and lines of type (C) are never incident. All other incidences are inherited from $PG(2n + 3, q)$.

In Section 2 we will prove that

Theorem 1.1 *If q is odd, $H(\mathcal{O}, \mathcal{P})$ is a generalized hexagon with parameters* (q^n, q).

In [1] we have proved that if q is a power of an odd prime number different from 3 and $n = 1$, then $\mathcal{O}$ is a twisted cubic, $\mathcal{P}$ is the set of tangents of $\mathcal{O}$, and $W(3, q)$ is the polar space associated with the symplectic polarity of $PG(3, q)$ interchanging a point of $\mathcal{O}$ with the osculating plane at $\mathcal{O}$ in that point. Moreover, the generalized hexagon $H(\mathcal{O}, \mathcal{P})$ is isomorphic to the dual $G_2(q)$-hexagon. Let us denote by $\mathcal{O}_3$ the algebraic variety which represents a regular spread of $PG(5, q)$ onto the grassmannian of the planes of $PG(5, q)$ (see [7]). If T_p is the tangent space of $\mathcal{O}_3$ at the point p of $\mathcal{O}_3$, then $\mathcal{S} = \{T_p : p \in \mathcal{O}_3\}$ is a partial spread of $PG(7, q)$ (see [8]).

In Section 4, we will prove the following theorem.

Theorem 1.2 *If q is odd, then $\mathcal{O}_3$ is a partial ovoid and S is a partial spread of the polar space $W(7, q)$; also $H(\mathcal{O}_3, S)$ is isomorphic to the $^3D_4(q)$-hexagon.*

Our construction only holds for q odd. For q even, we have found the same type of difficulties we had in [1].

2. Proof of Theorem 1.1

Let $s, t > 1$ be natural numbers. Let G be a group of order $s^2 t^3$. For any $i = 1, 2, ..., s + 1$, let us fix the subgroups $A_1(i), A_2(i), A_3(i), A_4(i)$ of G such that $A_1(i) \leq A_2(i) \leq A_3(i) \leq A_4(i)$, where $|A_1(i)| = t$, $|A_2(i)| = st$, $|A_3(i)| = st^2$ and $|A_4(i)| = s^2 t^2$. Let us define a point-line geometry $H = (P, L, I)$ as follows:

$$P = \{\mathcal{I}, A_4(i)g, A_2(i)g, g : g \in G, i \in \{1, 2..., s + 1\}\},$$
$$L = \{[i], A_3(i)g, A_1(i)g : g \in G, i \in \{1, 2..., s + 1\}\}.$$

where $\mathcal{I}$ and $[i]$ are symbols and the incidences are $\mathcal{I}I[i]$, $A_4(i)I[i]$, $gIA_1(i)g$ for all $g \in G$ and $i = 1, 2, ..., s + 1$, while $A_r(i)gIA_{r+1}(j)h$ if and only if $i = j$ and $g \in A_{r+1}(j)h$ with $r = 1, 2, 3$. In [1] we have proved the following theorem

Theorem 2.1 $H = (P, L, I)$ *is a generalized hexagon with parameters (s, t) if and only if, for all distinct i, j, h, m, n in $\{1, 2..., s + 1\}$, the following conditions hold:*

1) $A_4(i) \cap A_1(j) = 1$,
2) $A_3(i) \cap A_1(j)A_1(h) = 1$,
3) $A_3(i) \cap A_2(j) = 1$,
4) $A_2(i)A_2(j) \cap A_1(h) = 1$,
5) $A_2(i) \cap A_1(j)A_1(h)A_1(m) = 1$,
6) $A_2(i) \cap A_1(j)A_1(i)A_1(h) = A_1(i)$,
7) $A_1(i) \cap A_1(j)A_1(h)A_1(m)A_1(n) = 1$,
8) $A_1(i) \cap A_1(j)A_1(h)A_1(m)A_1(h) = 1$,
9) $A_1(i)A_1(j) \cap A_1(j)A_1(i) = A_1(i) \cup A_1(j)$,
10) $A_1(i) \cap A_1(j)A_1(h)A_1(j)A_1(h) = 1$.

In $PG(V, F) = PG(2n + 3, q)$, we can choose a basis $\{e_0, e_1, ..., e_{2n+3}\}$ of V in such a way that the bilinear form

$$\left(\sum_{r=0}^{2n+3} x_r e_r, \sum_{r=0}^{2n+3} y_r e_r \right) = \sum_{r=0}^{n+1} x_r y_{2n+3-r} - x_{2n+3-r} y_r$$

is the alternating bilinear form, which defines the symplectic polarity $\perp$ associated with $W(2n + 3, q)$. For each element $v = \sum_{r=1}^{n+1} (a_r e_r + b_r e_{n+1+r})$ of the subspace $< e_1, e_2, ..., e_{2n+2} >$ of V, let $M(v, \gamma)(\gamma \in F)$ be the non-singular linear map of V in itself defined by

$$e_0 \mapsto e_0 + \sum_{r=1}^{n+1}(a_r e_r + b_r e_{n+1+r}) + \gamma e_{2n+3},$$

$$e_r \mapsto e_r + b_r e_{2n+3}, \quad r = 1, ..., n+1,$$

$$e_{n+1+r} \mapsto e_{n+1+r} - a_r e_{2n+3}, \quad r = 1, ..., n+1,$$

$$e_{2n+3} \mapsto e_{2n+3}.$$

In the following, we will denote by $M(v, \gamma)$ also the collineation of the projective space $PG(2n+3, q)$ defined by the linear map $M(v, \gamma)$. We can prove with a direct calculation that $(w_1 M(v, \gamma), w_2 M(v, \gamma)) = (w_1, w_2)$ for all vectors w_1 and w_2 of V. This implies that the collineation $M(v, \gamma)$ stabilizes the polarity $\perp$. Moreover, $M(v, \gamma)$ fixes the point $x = P(e_{2n+3})$ and all subspaces of $x^\perp$ incident with x. Let us denote by G the group of all collineations $M(v, \gamma)$.

If $y = P(e_0)$, then $T = x^\perp \cap y^\perp = P(< e_1, e_2, ..., e_{2n+2} >)$. If $\mathcal{O} = \{p_i = P(v_i) : i = 1, 2, ..., q^n + 1\}$, let us define

$$A_4(i) = \{g \in G : p_i g = p_i\},$$
$$A_3(i) = \{g \in G : T_i g = T_i\},$$
$$A_2(i) = \{g \in G :< y, T_i > g =< y, T_i >\},$$
$$A_1(i) = \{g \in G :< y, p_i > g =< y, p_i >\}.$$

for $i = 1, 2, ..., q^n + 1$. We will denote by i, j, h, m, n five mutually distinct indexes between 1 and $q^n + 1$.

1) $A_4(i) \cap A_1(j) = 1$.

The point y belongs to $p_i^\perp$. If a non identity element g of G belongs to $A_4(i) \cap A_1(j)$, then the point $z = yg \neq y$ belongs to $p_i^\perp$ and to $< y, p_j >$. Hence $< y, p_j >$ is contained in $p_i^\perp$ and $p_j \in p_i^\perp \cap T$. As $\mathcal{O}$ is a partial ovoid of $W(2n + 1, q)$, this is impossible.

2) $A_3(i) \cap A_1(j)A_1(h) = 1$.

We notice that $A_1(j) = \{M(av_j, 0) : a \in F\}$ and $A_3(i) = \{M(w, \gamma) : P(w) \in T_{p_i}\}$. If $g_j = M(av_j, 0)$ and $g_h = M(bv_h, 0)$, then $g_j g_h = M(av_j + bv_h, ab(v_j, v_h))$. If $g_j g_h \neq 1$ belongs to $A_3(i)$, then $P(av_j + bv_h)$ is a point of T_i. This implies that $< p_j, p_h > \cap T_i \neq \emptyset$. Therefore $< T_i, p_j >$ contains three points of $\mathcal{O}$.

3) $A_3(i) \cap A_2(j) = 1$.

If an element g different from 1 belongs to $A_3(i) \cap A_2(j)$, then the point $z = yg$ belongs to $T_i^\perp \cap < y, T_j >$. As $z \neq y$, the line $< y, z >$ is contained in $T_i^\perp \cap < y, T_j >$. Then the point $r =< y, z > \cap T$ belongs to $T_i \cap T_j$. As $\mathcal{P}$ is a partial spread this is impossible.

4) $A_2(i)A_2(j) \cap A_1(h)$.

By way of contradiction, let us suppose that a non-identity element $g_i g_j$ of $A_2(i)A_2(j)$ belongs to $A_1(h)$. Let $r = yg_j^{-1}$ and $s = yg_i$. As $g_i \in A_2(i)$, the point s belongs to the subspace $< T_i, y >$. Similarly, $r \in < T_j, y >$. Therefore the line $< r, s >$ is contained in $y^{\perp}$ because $< T_i, y >$ and $T_j, y >$ are subspaces of $y^{\perp}$. As $g_i g_j \in A_1(h)$, we have $< r, s > = < y, p_h > g_j^{-1}$. Then $p_h g_j^{-1}$ belongs to $y^{\perp}$ and to $< x, p_h >$ because g_j^{-1} fixes all lines of $x^{\perp}$ incident with x. This implies $p_h g_j^{-1} = p_h$ and $< r, s > = < r, p_h >$. The line $< r, s >$ is a totally isotropic line because $< y, p_h >$ is totally isotropic and g_j is an automorphism of $W(2n + 3, q)$. Moreover, $y \notin < r, s >$ because 1 does not belong to $A_1(h)g_j^{-1}$. Hence the plane $< y, r, s >$ is totally isotropic. The line $l = < y, r, s > \cap x^{\perp}$ is totally isotropic and incident with p_h. As $< y, r >$ is a line of $< T_j, y >$ and $< y, s >$ is a line of $< T_i, y >$, the line l is incident with a point of T_j and with a point of T_i. As this is impossible because of property (c), we have the required contradiction.

5) $A_2(i) \cap A_1(j)A_1(h)A_1(m) = 1$.

If $g_j = M(av_j, 0) \in A_1(j)$, $g_h = M(bv_h, 0) \in A_1(h)$, $g_m = M(cv_m, 0) \in A_1(m)$, then $g_j g_h g_m = M(av_j + bv_h + cv_m, ab(v_j, v_h) + ac(v_j, v_m) + bc(v_h, v_m))$. As $A_2(i) = \{M(w, 0) : P(w) \in T_i\}$, the element $g_j g_h g_m$ belongs to $A_2(i)$ if and only if $w = av_j + bv_h + cv_m$ and $ab(v_j, v_h) + ac(v_j, v_m) + bc(v_h, v_m) = 0$. This is impossible because of property (d).

6) $A_2(i) \cap A_1(j)A_1(i)A_1(h) = A_1(i)$.

If $g_j g_i g_h = M(av_j + bv_i + cv_h, ab(v_j, v_i) + ab(v_j, v_h) + bc(v_i, v_h))$ belongs to $A_2(i)$, then $P(av_j + bv_i + cv_h)$ belongs to T_i. If $P(av_j + bv_i + cv_h) \neq P(v_i)$, then the plane $P(< v_j, v_i, v_h >)$ intersects T_i in a line. Therefore $< T_i, p_j >$ contains three points of $\mathcal{O}$. This is impossible because of property (a).

7) $A_1(i) \cap A_1(j)A_1(h)A_1(m)A_1(n) = 1$.

If $g_j = M(av_j, 0)$, $g_h = M(bv_h, 0)$, $g_m = M(cv_m, 0)$ and $g_n = M(dv_n, 0)$, then $g_j g_h g_m g_n = M(av_j + bv_h + cv_m + dv_n, ab(v_j, v_i) + ac(v_j, v_h) + bc(v_i, v_h) + ad(v_j, v_n) + bd(v_h, v_n) + cd(v_m, v_n))$. Therefore, $g_j g_h g_m g_n \neq 1$ belongs to $A_1(i)$ if and only if $p_i \in Q(p_j, p_h, p_m, p_n)$. This is impossible because of property (e).

8) $A_1(i) \cap A_1(j)A_1(h)A_1(m)A_1(h) = 1$.

If $g_j = M(av_j, 0)$, $g_h = M(bv_h, 0)$, $g_m = M(cv_m, 0)$ and $g_h' = M(dv_h, 0)$, then $g_j g_h g_m g_h' = M(av_j + bv_h + cv_m + dv_h, ab(v_j, v_i) + ac(v_j, v_h) + bc(v_i, v_h) + ad(v_j, v_h) + cd(v_m, v_h))$. If $g_j g_h g_m g_h' \in A_1(i)$ is different from 1, then p_i belongs to the plane $< p_j, p_h, p_m >$. Because of property (b), this implies $g_j g_h g_m g_h' = 1$.

9) $A_1(i)A_1(j) \cap A_1(j)A_1(i) = A_1(i) \cup A_1(j)$.

If $g_i = M(av_i, 0)$ and $M(bv_j, 0)$, then $g_i g_j = M(av_i + bv_j, ab(v_i, v_j))$ and $g_j g_i = M(av_i + bv_j, ab(v_j, v_i))$. As $\mathcal{O}$ is a partial ovoid, $(v_i, v_j) = -(v_j, v_i) \neq 0$.

Therefore $g_i g_j = g_j g_i$ if and only if either $g_i = 1$ or $g_j = 1$.

    10) $A_1(i) \cap A_1(j) A_1(h) A_1(j) A_1(h) = 1$.

    If $M(av_j + bv_h + cv_j + dv_h, (ab - bc + ad - cd)(v_j, v_h)) \neq 1$ belongs to $A_1(i)$, then $p_i = P(av_j + bv_h + cv_j + dv_h) \in< p_j, p_h >$. This implies that $< T_i, p_j >$ contains three points of $\mathcal{O}$. This is impossible because of property (a).   □

3.   Segre varieties and regular spreads

In this section we will recall some of the results contained in [7] [1] and [8]. Let Σ^* be a projective space. A subset Σ of points of Σ^* is a *subgeometry* of Σ^* if there is a set $\mathcal{L}$ of subsets of Σ with the following properties:

(1) each element of $\mathcal{L}$ is contained in a line of Σ^*;

(2) $(\Sigma, \mathcal{L})$ is a projective space;

(3) if a line l of Σ^* contains two points of Σ, then $l \cap \Sigma \in \mathcal{L}$;

(4) no line of Σ^* belongs to $\mathcal{L}$.

Let Σ be a subgeometry of $\Sigma^* = PG(n, q^r)$ isomorphic to $PG(m, q)$. We will say that Σ is a *canonical* subgeometry of Σ^* if $\Sigma^* =< \Sigma >$ and $n = m$. In this case, if $\Sigma = PG(V, F)$ and $\Sigma^* = PG(V^*, K)$, a basis of V over F is also a basis of V^* over K (i.e. $V^* = K \otimes V$).

    Let Σ be a canonical subgeometry of Σ^*. For each subspace S of Σ^*, the set $S \cap \Sigma$ is a subspace of Σ. We will say that a subspace S of Σ^* is a subspace of Σ whenever S and $S \cap \Sigma$ have the same dimension.

    Let $K = GF(q^3)$, and let V_1, V_2, V_3 be three vector spaces over K. If V_i $(i = 1, 2, 3)$ has dimension 2, the vector space $W = V_1 \otimes V_2 \otimes V_3$ has dimension 8 over K. Let $PG(7, q^3) = PG(W, K)$, and let

$$S_{2,2,2}(q^3) = \{P(v_1 \otimes v_2 \otimes v_3) \in PG(7, q^3) : v_i \in V_i (i = 1, 2, 3)\}.$$

Thus, $S_{2,2,2}(q^3)$ is a Segre-variety in $PG(7, q^3)$ of type $(2, 2, 2)$. Let us define

$$S_1 = \{P(V_1 \otimes Kv_2 \otimes Kv_3) : v_2 \in V_2, v_3 \in V_3\},$$

$$S_2 = \{P(Kv_1 \otimes V_2 \otimes Kv_3) : v_1 \in V_1, v_3 \in V_3\},$$

$$S_3 = \{P(Kv_1 \otimes Kv_2 \otimes V_3) : v_1 \in V_1, v_2 \in V_2\}.$$

The elements of S_i $(i = 1, 2, 3)$ are lines of $PG(7, q^3)$ contained in $S_{2,2,2}(q^3)$. Moreover, two lines of S_i are disjoint and each point of $S_{2,2,2}(q^3)$ belongs to a line of S_i. If $p = P(v_1 \otimes v_2 \otimes v_3)$ is a point of $S_{2,2,2}(q^3)$, let r_i be the line of S_i incident with p $(i = 1, 2, 3)$. The 3-dimensional subspace $T_p^* =< r_1, r_2, r_3 >$ is the tangent space of $S_{2,2,2}(q^3)$ at the point p. We can prove by a direct calculation that $T_p^* \cap S_{2,2,2}(q^3) = r_1 \cup r_2 \cup r_3$.

---

1. In [7] there are many misprints due to technical problems at the printers.

Let $\Sigma^* = PG(V^*, K) = PG(5, q^3)$. Let Σ be a canonical subgeometry of Σ^* isomorphic to $PG(5, q) = PG(V, F)$. Then $V^* = K \otimes V$, and a basis of V is also a basis of V^*. Let us denote by $\{t_1, t_2, t_3, t_4, t_5, t_6\}$ a fixed basis of V. Let σ be the semilinear regular map of V^* in itself, which maps the vector $\sum_{i=1}^{6} x_i t_i$ to the vector $\sum_{i=1}^{6} x_i^q t_i$. The subgeometry Σ is the set of the fixed points of the collineation of Σ^* associated with σ, and $\sigma^3 = 1$. A subspace $S = P(A)$ of Σ^* is a subspace of Σ if and only if $A\sigma = A$ (see [7] Section 3). Let $p = P(v)$ be a point of Σ^*. We will say that p is an *imaginary* point if the subspace $\alpha(p) = P(<v, v\sigma, v\sigma^2>)$ has dimension two. It is easy to verify that $\alpha(p)$ is a subspace of Σ. If all the points on a line l are imaginary, we will say that l is an *imaginary* line.

If $l_0 = P(L_0)$ is an imaginary line of Σ^*, let us define $L_i = L_0\sigma^i$ and $l_i = P(L_i)$ for $i = 1, 2$. Let $\mathcal{U}$ be the set of all the planes α of Σ^* such that $l_i \cap \alpha = P(v_i)$ is a point for all $i \in \{0, 1, 2\}$. As l_0 is an imaginary line, $\alpha = P(<v_0, v_1, v_2>)$. By construction, if $\alpha = P(A)$ is a plane of $\mathcal{U}$, then $P(A\sigma) \in \mathcal{U}$. Let us define $\mathcal{F} = \{\alpha \in \mathcal{U} : \alpha \text{ is a subspace of } \Sigma\}$. We can prove that a plane α of $\mathcal{U}$ belongs to $\mathcal{F}$ if and only if $\alpha = P(<v, v\sigma, v\sigma^2>)$ where $P(v) = l_0 \cap \alpha$, and that $\mathcal{F}$ is a regular spread of Σ (see [7] (3.11)).

If $\{e_1, e_2\}$ is a fixed basis of L_0, let $e_3 = e_1\sigma$, $e_4 = e_2\sigma$, $e_5 = e_1\sigma^2$ and $e_6 = e_2\sigma^2$. Then, $\{e_3, e_4\}$ is a basis of L_1, $\{e_5, e_6\}$ is a basis of L_2, and $\{e_1, e_2, e_3, e_4, e_5, e_6\}$ is a basis of V^* because the line l_0 is imaginary. For all non singular matrices $A = \begin{pmatrix} a & b \\ c & d \end{pmatrix}$ with coefficients in K, we will denote by $\bar{A}$ the linear map of V^* into itself defined by the matrix

$$\begin{pmatrix} a & b & 0 & 0 & 0 & 0 \\ c & d & 0 & 0 & 0 & 0 \\ 0 & 0 & a^q & b^q & 0 & 0 \\ 0 & 0 & c^q & d^q & 0 & 0 \\ 0 & 0 & 0 & 0 & a^{q^2} & b^{q^2} \\ 0 & 0 & 0 & 0 & c^{q^2} & d^{q^2} \end{pmatrix}$$

with respect to the basis $\{e_1, e_2, e_3, e_4, e_5, e_6\}$. We will denote by τ_A the linear collineation of Σ^* defined by $\bar{A}$. As $\tau_A\tau_B = \tau_{AB}$, the set $G = \{\tau_A : A \in GL(2, q^3)\}$ is a group isomorphic to $PGL(2, q^3)$. It is easy to verify that $\bar{A}\sigma = \sigma\bar{A}$. This implies that τ_A defines a collineation of Σ. Moreover, $\mathcal{U}\tau_A = \mathcal{U}$ because τ_A fixes the lines l_i ($i = 0, 1, 2$). Consequently G fixes the spread $\mathcal{F}$ and acts 3-transitively on $\mathcal{F}$.

Let $U^* = \bigwedge^3 V^*$ be the third exterior power of V^*. Then U^* has dimension 20. We will call grassmannian of the planes of $\Sigma^* = PG(5, q^3) = PG(V^*, K)$ the set of points $G(q^3) = \{P(v_1 \wedge v_2 \wedge v_3) : v_1, v_2, v_3 \in V^*, v_1 \wedge v_2 \wedge v_3 \neq 0\}$ of $PG(19, q^3) = PG(U^*, K)$. We can define a Grassmann map

239

g between the set of all planes of $\Sigma^* = PG(5, q^3)$ and $G(q^3)$ in the following way. For each plane $\alpha = P(< v_1, v_2, v_3 >)$ of Σ^*, let $g(\alpha) = P(v_1 \wedge v_2 \wedge v_3)$. The map g is a bijection. If α and β are two planes of Σ^*, let $a = g(\alpha)$ and $b = g(\beta)$. Then $\alpha \cap \beta$ is a line if and only if the line of $PG(19, q^3)$ joining a and b is contained in $G(q^3)$. For more details on the grassmannians see [2] or [3].

If $U = \wedge^3 V$, we have $U^* = K \otimes U$ because $V^* = K \otimes V$. Then, $PG(U, F) = PG(19, q)$ is a canonical subgeometry of $PG(U^*, K)$. Moreover, $G(q^3) \cap PG(19, q) = G(q)$ is the grassmannian of the planes of $\Sigma = PG(5, q)$ (see [7] Section 4). Let I be the identity map of U, and s the automorphism of K defined by $x \mapsto x^q$. If $\hat{\sigma} = s \otimes I$, then $\hat{\sigma}$ is a regular semilinear map of U^* in itself, which fixes all the vectors of U. So the subgeometry $PG(19, q)$ is the set of fixed points of the collineation of $PG(19, q^3)$ defined by $\hat{\sigma}$. We can prove that $(v_1 \wedge v_2 \wedge v_3)\hat{\sigma} = v_1\sigma \wedge v_2\sigma \wedge v_3\sigma$. Therefore, a plane $\alpha = P(< v_1, v_2, v_3 >)$ of Σ^* is a plane of Σ if and only if $(v_1 \wedge v_2 \wedge v_3)\hat{\sigma} = v_1 \wedge v_2 \wedge v_3$ (see [7] (4.4)). By [7] Section 4, $g(\mathcal{U}) = S_{2,2,2}(q^3)$ and there is a 7-dimensional subspace T^* of $PG(19, q^3)$ such that $g(\mathcal{U}) = T^* \cap G(q^3)$. Moreover, $T = T^* \cap PG(19, q)$ is a canonical subgeometry of T^*, and $g(\mathcal{F}) = T \cap G(q) = T \cap g(\mathcal{U}) = \mathcal{O}_3$ (see [7] (4.6)). A regulus $\mathcal{R}$ of $PG(5, q)$ is a set of $q + 1$ mutually disjoint planes such that, if a line ℓ meets three planes of $\mathcal{R}$, then $\ell \cap \alpha$ is a point for all α in $\mathcal{R}$. If $q \geq 3$ and $\mathcal{R}$ is a regulus of $\mathcal{F}$, then there is a 3-dimensional subspace S of T such that $g(\mathcal{R}) = S \cap \mathcal{O}_3$ and $g(\mathcal{R})$ is a twisted cubic of S (see [7] (2.6)). In [8] Teorema 1, we have proved that for each point p of $S_{2,2,2}(q^3)$, the tangent space T_p^* of $S_{2,2,2}(q^3)$ at p is a subspace of T (i.e. $T_p = T_p^* \cap T$ has dimension three), and, if p and r are two points of $\mathcal{O}_3$, the subspaces T_p and T_r are disjoint. Notice that $S = \{T_p : p \in \mathcal{O}_3\}$ is a partial spread of $T = PG(7, q)$.

We put $w_1 = e_1 \wedge e_3 \wedge e_5$, $w_2 = e_1 \wedge e_3 \wedge e_6$, $w_3 = e_1 \wedge e_4 \wedge e_5$, $w_4 = e_1 \wedge e_4 \wedge e_6$, $w_5 = e_2 \wedge e_3 \wedge e_5$, $w_6 = e_2 \wedge e_3 \wedge e_6$, $w_7 = e_2 \wedge e_4 \wedge e_5$, $w_8 = e_2 \wedge e_4 \wedge e_6$. We will denote by $(x_1, x_2, x_3, x_4, x_5, x_6, x_7, x_8)$ the homogeneous coordinates of a point of T^* with respect to the basis $\{w_1, w_2, w_3, w_4, w_5, w_6, w_7, w_8\}$. A point p of $T = PG(7, q)$ belongs to $\mathcal{O}_3$ if and only if either $p = P(w_1)$ or there is an element $a \in K$ such that $p = (a^{1+q+q^2}, a^{1+q}, a^{1+q^2}, a, a^{q+q^2}, a^q, a^{q^2}, 1)$ (see [8]).

For each regular linear map $\bar{A}$ of V^* in itself, let $\hat{A} = \wedge^3 \bar{A}$ and let $\hat{\tau}_A$ be the linear collineation of $PG(19, q^3)$ defined by $\hat{A}$. As $\hat{\tau}_{AB} = \hat{\tau}_A\hat{\tau}_B$, the set $\hat{G} = \{\hat{\tau}_A : A \in GL(2, q^3)\}$ is a group isomorphic to $PGL(2, q^3)$. By construction, $\hat{\tau}_A$ fixes $G(q^3)$. As $\bar{A}\sigma = \sigma\bar{A}$, it is easy to prove that $\hat{A}\hat{\sigma} = \hat{\sigma}\hat{A}$. Hence $\hat{\tau}_A$ fixes $PG(19, q)$ and $G(q)$. The collineation $\hat{\tau}_A$ fixes $g(\mathcal{U}) = S_{2,2,2}(q^3)$, because τ_A fixes $\mathcal{U}$. Therefore $\mathcal{O}_3$ is fixed by $\hat{\tau}_A$. This implies that each element of $\hat{G}$ is a collineation of T and $\hat{G}$ acts 3-transitively on $\mathcal{O}_3$ (see [7] (4.8)).

4.   Proof of Theorem 1.2

We will denote by $\perp$ the polarity of $T^* = PG(7, q^3) = PG(W, K)$ associated with the non-singular alternating bilinear form $(\ ,\ )$ of W defined by

$$(\sum_{i=1}^{8} x_i w_i, \sum_{i=1}^{8} y_i w_i) = x_1 y_8 - x_2 y_7 - x_3 y_6 + x_4 y_5 - x_5 y_4 + x_6 y_3 + x_7 y_2 - x_8 y_1$$

Lemma 4.1 *Let S be a subspace of T^*. Then, $S^\perp$ is a subspace of T if and only if S is a subspace of T.*

Proof.   It is easy to verify that $(v\widehat{\sigma}, u\widehat{\sigma}) = (v, u)^q$. If p is a point of $PG(7, q)$, then there is a vector $v \in W$ such that $p = P(v)$ and $v\widehat{\sigma} = v$. If $P(u)$ belongs to $p^\perp = \{P(w) : (w, v) = 0\}$, then $(u\widehat{\sigma}, v) = (u\widehat{\sigma}, v\widehat{\sigma}) = (u, v)^q = 0$. Hence, $P(u\widehat{\sigma}) \in p^\perp$. We have proved that if p is a point of T, then $p^\perp$ is an hyperplane of T. $\square$

Remark

By Lemma 4.1, $\perp$ is also a polarity of $T = PG(7, q)$.

Lemma 4.2 *If α and β are two planes of $\mathcal{U}$, then let $P(a) = g(\alpha)$, $P(b) = g(\beta)$. The planes α and β are disjoint if and only if $(a, b) \neq 0$.*

Proof.   If $\alpha = P(< v_1, v_2, v_3 >)$ and $\beta = P(< u_1, u_2, u_3 >)$ then $a = v_1 \wedge v_2 \wedge v_3$ and $b = u_1 \wedge u_2 \wedge u_3$. The planes α and β are disjoint if and only if $a \wedge b = v_1 \wedge v_2 \wedge v_3 \wedge u_1 \wedge u_2 \wedge u_3 \neq 0$. If $a = \sum_{i=1}^{8} x_i w_i$ and $b = \sum_{i=1}^{8} y_i w_i$, then $v_1 \wedge v_2 \wedge v_3 \wedge u_1 \wedge u_2 \wedge u_3 = (a, b) e_1 \wedge e_3 \wedge e_5 \wedge e_2 \wedge e_4 \wedge e_6$. Hence, α and β are disjoint if and only if $(a, b) \neq 0$. $\square$

Lemma 4.3 *For each collineation $\widehat{\tau}_A$ of $\widehat{G}$, and for each subspace S of T, we have $(S^\perp)\widehat{\tau}_A = (S\widehat{\tau}_A)^\perp$.*

Proof.   Let $X = \begin{pmatrix} 0 & 1 \\ 1 & 0 \end{pmatrix}$, $B_a = \begin{pmatrix} a & 0 \\ 0 & 1 \end{pmatrix}$, and $C_b = \begin{pmatrix} 1 & 0 \\ b & 1 \end{pmatrix}$, where $a, b \in K$ and $a \neq 0$. For each matrix $A = \begin{pmatrix} a & b \\ c & d \end{pmatrix}$, if $b \neq 0$, then $A = B_b C_d X B_e C_f$ where $fb = a$ and $fd + e = c$. If $b = 0$, we can suppose $d = 1$ and $A = B_a C_c$. Then, it is enough to prove that τ_X, τ_{B_a}, and τ_{C_b} preserve the

241

polarity $\perp$ of $T^* = PG(7, q^3)$ associated with $(\ , \ )$. We have

$$\hat{X} = \begin{pmatrix} 0 & 0 & 0 & 0 & 0 & 0 & 0 & 1 \\ 0 & 0 & 0 & 0 & 0 & 0 & 1 & 0 \\ 0 & 0 & 0 & 0 & 0 & 1 & 0 & 0 \\ 0 & 0 & 0 & 0 & 1 & 0 & 0 & 0 \\ 0 & 0 & 0 & 1 & 0 & 0 & 0 & 0 \\ 0 & 0 & 1 & 0 & 0 & 0 & 0 & 0 \\ 0 & 1 & 0 & 0 & 0 & 0 & 0 & 0 \\ 1 & 0 & 0 & 0 & 0 & 0 & 0 & 0 \end{pmatrix},$$

$$\hat{B}_a = diag(a^{1+q+q^2}, a^{1+q}, a^{1+q^2}, a, a^{q+q^2}, a^q, a^{q^2}, 1),$$

$$\hat{C}_b = \begin{pmatrix} 1 & 0 & 0 & 0 & 0 & 0 & 0 & 0 \\ b^{q^2} & 1 & 0 & 0 & 0 & 0 & 0 & 0 \\ b^q & 0 & 1 & 0 & 0 & 0 & 0 & 0 \\ b^{q+q^2} & b^q & b^{q^2} & 1 & 0 & 0 & 0 & 0 \\ b & 0 & 0 & 0 & 1 & 0 & 0 & 0 \\ b^{1+q^2} & b & 0 & 0 & b^{q^2} & 1 & 0 & 0 \\ b^{1+q} & 0 & b & 0 & b^q & 0 & 1 & 0 \\ b^{1+q+q^2} & b^{1+q} & b^{1+q^2} & b & b^{q+q^2} & b^q & b^{q^2} & 1 \end{pmatrix}.$$

With a direct calculation, we can prove that $(v\hat{X}, u\hat{X}) = (v, u)$, $(v\hat{C}_b, u\hat{C}_b) = (v, u)$ and $(v\hat{B}_a, u\hat{B}_a) = a^{1+q+q^2}(v, u)$. This completes the proof. $\square$

Remark

When q is even, $\mathcal{O}_3$ is a very particular ovoid of $Q^+(7, q)$, which is called *desarguesian* in [5], and the spread S is the spread of $Q^+(7, q)$ constructed by W.M. Kantor in [5] Section 8 (see [8] Teorema 4).

Lemma 4.4 *If q is odd, then let $W(7, q)$ be the symplectic polar space arising from the polarity $\perp$ of $T = PG(7, q)$. Then,*

(a) $\hat{H} = \{\hat{\tau}_A : A \in GL(2, q^3), \det A = 1\} = \hat{G} \cap PSp(8, q)$ is 2-transitive on $\mathcal{O}_3$ and on S.

(b) $\mathcal{O}_3$ is a partial ovoid of $W(7, q)$.

(c) S is a partial spread of $W(7, q)$.

Proof. As $H = \{\tau_A : A \in GL(2, q^3), \det A = 1\} \simeq PSL(2, q^3)$ is 2-transitive on the regular spread $\mathcal{F}$ of $PG(5, q)$, property (a) follows from Lemma 4.3. By Lemma 4.2, $\mathcal{O}_3$ is a partial ovoid of $W(7, q)$. If $p = P(w_1)$, then $T_p^* = P(< w_1, w_2, w_3, w_5 >)$ is a maximal totally isotropic subspace of the polarity $\perp$ of $T^* = PG(7, q^3)$. Because of Lemma 4.1, $T_p = T_p^* \cap T$ is a maximal totally isotropic subspace of $W(7, q)$. As $\hat{H}$ is a subgroup of $PSp(8, q)$, which acts 2-transitively on S, all elements of S are maximal totally isotropic subspaces of $W(7, q)$. Therefore, S is a partial spread of $W(7, q)$. $\qquad\square$

Remark

By Lemma 4.3, the group $\hat{G}$ is contained in the stabilizer of the polarity $\perp$ in $PGL(8, q)$.

Lemma 4.5 *Let $q > 3$ be odd. Let p_1, p_2, p_3 be three points of $\mathcal{O}_3$ and let T_{p_i} be the tangent space of $\mathcal{O}_3$ at p_i.*

1) There is a unique 3-dimensional subspace S of T, which intersects $\mathcal{O}_3$ in a twisted cubic and is incident with p_1, p_2 and p_3. If $q \neq 3^n$, then S in non singular with respect to the polarity $\perp$. If p is a point of the twisted cubic $S \cap \mathcal{O}_3$, then $T_p \cap S$ is the tangent line of $S \cap \mathcal{O}_3$ at p.

2) If a 3-dimensional subspace $U \neq S$ of T contains p_1, p_2 and p_3, then U is incident with at least four points of $\mathcal{O}_3$.

3) $< T_{p_1}, p_2 > \cap \mathcal{O}_3 = \{p_1, p_2\}$.

4) A line of $W(7, q)$ joining a point of T_{p_1} to a point of T_{p_2} cannot be incident with p_3.

5) Let p_4 be a point of $\mathcal{O}_3$ different from p_1, p_2, p_3. If $p_4 \notin S$, then $T_{p_1} \cap < p_2, p_3, p_4 > = \emptyset$. If $p \in T_{p_1} \cap < p_2, p_3, p_4 >$, then p does not belong to $C(p_2, p_3, p_4)$.

6) If p_5 is a point different from p_1, p_2, p_3, p_4 and $p_1 \in < p_2, p_3, p_4, p_5 >$, then p_1 does not belong to $Q(p_2, p_3, p_4, p_5)$.

Proof. As a regulus is represented by a twisted cubic, no four points of $\mathcal{O}_3$ are coplanar. Suppose that a three dimensional subspace S of T contains five points p_1, p_2, p_3, p_4, p_5 of $\mathcal{O}_3$. Let $\alpha_i = g^{-1}(p_i)$ for $i = 1, 2, 3, 4, 5$ and let $\mathcal{R}$ be the regulus of $PG(5, q)$ containing the planes α_1, α_2 and α_3. By Lemma (2.1) of [7], each transversal line of $\mathcal{R}$ is incident with α_4 and α_5. As the α_i ($i = 1, 2, 3, 4, 5$) are elements of a regular spread, α_4 and α_5 belong to $\mathcal{R}$. Then $S \cap \mathcal{O}_3 = g(\mathcal{R})$. As $\hat{G}$ stabilizes $\perp$ and $\hat{G}$ acts 3-transitively on $\mathcal{O}_3$, we can suppose that the points $p_1 = P(w_1)$, $p_2 = P(w_8)$ and $p_3 = P(\sum_{i=1}^{8} w_i)$. Therefore $S \cap \mathcal{O}_3 = g(\mathcal{R})$ where $\mathcal{R}$ is the regulus of $PG(5, q)$ containing the planes $P(< e_1, e_3, e_5 >)$, $P(< e_2, e_4, e_6 >)$, $P(< e_1 + e_2, e_3 + e_4, e_5 + e_6 >)$. As the plane $P(< -e_1 + e_2, -e_3 + e_4, -e_5 + e_6 >)$ belongs to $\mathcal{R}$, the point $P(-w_1 + w_2 + w_3 - w_4 + w_5 - w_6 - w_7 + w_8)$ of $\mathcal{O}_3$ is incident with S. Hence S has equations $x_2 = x_3 = x_5$ and $x_4 = x_6 = x_7$. Thus, $S \cap \mathcal{O}_3 = \{(t^3, t^2u, t^2u, tu^2, t^2u, tu^2, tu^2, u^3) : t, u \in F\}$, the line $T_{p_1} \cap S$ is the tangent line of $S \cap \mathcal{O}_3$ at p_1, and $T_{p_1} \cap S = \{(\alpha, \beta, \beta, 0, \beta, 0, 0, 0) : \alpha, \beta \in F\}$. The group $G_S = \{\tau_A : A \in GL(2, q)\}$ is the stabilizer of the regulus $\mathcal{R}$ in the group G and acts 3-transitively on $\mathcal{R}$. If $\hat{G}_S = \{\tau \in \hat{G} : S^\tau = S\} = \{\hat{\tau}_A : A \in GL(2, q)\}$,

243

then $\hat{G}_S$ is 3-transitive on $S \cap \mathcal{O}_3$. Therefore, $T_p \cap \mathcal{O}_3$ is the tangent line of $S \cap \mathcal{O}_3$ at p for each point p of $S \cap \mathcal{O}_3$. The subspace $S^\perp$ has equations (in T^*)

$$\begin{cases} x_1 = x_8 = 0, \\ x_1 - x_2 - x_3 + x_4 - x_5 + x_6 + x_7 - x_8 = 0, \\ x_1 + x_2 + x_3 + x_4 + x_5 + x_6 + x_7 + x_8 = 0. \end{cases}$$

When $q \neq 3^n$, we can prove with a direct calculation that S and $S^\perp$ are disjoint.

As $T_{p_1} = \{(\alpha, b^{q^2}, b^q, 0, b, 0, 0, 0) : \alpha \in F, b \in K\}$, it is an easy calculation to prove that $< T_{p_1}, p_2 > \cap \mathcal{O}_3 = \{p_1, p_2\}$.

Let l be a line intersecting T_{p_i} $(i = 1, 2)$ and let $x_i = T_{p_i} \cap l$. Then $x_1 = (\alpha, b^{q^2}, b^q, 0, b, 0, 0, 0)$ and $x_2 = (0, 0, 0, c, 0, c^q, c^{q^2}, \beta)$. If the point $p_3 = (1, 1, 1, 1, 1, 1, 1, 1)$ belongs to the line l, we can suppose that x_1 and x_2 are respectively the points $(1, 1, 1, 0, 1, 0, 0, 0)$ and $(0, 0, 0, 1, 0, 1, 1, 1)$. As $(x_1, x_2) = -2 \neq 0$, the line l does not belong to $W(7, q)$.

Let $p_4 = (a^{1+q+q^2}, a^{1+q}, a^{1+q^2}, a, a^{q+q^2}, a^q, a^{q^2}, 1)$. A point of the plane $< p_2, p_3, p_4 >$ belongs to T_{p_1} if and only if $a \in F$. This is equivalent to saying that p_4 belongs to S. Let $p_4 = (a^3, a^2, a^2, a, a^2, a, a, 1)$ and let $p_5 = (b^3, b^2, b^2, b, b^2, b, b, 1)$ with a, b, 1, 0 mutually distinct elements of F. We have $(p_2, p_3) = -1$, $(p_2, p_4) = -a^3$, $(p_3, p_4) = (1 - a)^3$, $(p_2, p_5) = -b^3$, $(p_3, p_5) = (1 - b)^3$, $(p_4, p_5) = (a - b)^3$. Let $p_i = P(v_i)$ for $i = 1, 2, 3, 4,$.

We have $C(p_2, p_3, p_4) = \{P(\alpha v_2 + \beta v_3 + \gamma v_4) : -\alpha\beta - a^3\alpha\gamma + (1-a)^3\beta\gamma = 0\}$. Then $p = T_{p_1} \cap < p_2, p_3, p_4 > = P(\alpha v_2 + \beta v_3 + \gamma v_4)$, where $\beta + a\gamma = 0$ and $\alpha + \beta + \gamma = 0$. Notice that $\gamma \neq 0$. As $-\alpha\beta - a^3\alpha\gamma + (1 - a)^3\beta\gamma = 2a^2\gamma^2(a - 1) \neq 0$, p does not belong to $C(p_2, p_3, p_4)$.

By definition, $Q(p_2, p_3, p_4, p_5)$ is the set of all points $P(\alpha v_2 + \beta v_3 + \gamma v_4 + \delta v_5)$ such that $-\alpha\beta - a^3\alpha\gamma + (1-a)^3\beta\gamma - b^3\alpha\delta + (1-b)^3\beta\delta + (a-b)^3\gamma\delta = 0$. As $p_1 = P(\alpha v_2 + \beta v_3 + \gamma v_4 + \delta v_5)$, where $\alpha = (a-1)(b-1)(a-b)$, $\beta = ab(b-a)$, $\gamma = -b(b-1)$ and $\delta = a(a-1)$, p_1 does not belong to $Q(p_2, p_3, p_4, p_5)$ because $-\alpha\beta - a^3\alpha\gamma + (1-a)^3\beta\gamma - b^3\alpha\delta + (1-b)^3\beta\delta + (a-b)^3\gamma\delta = -2a^2b(a-1)(b-1)(a-b)^2 \neq 0$. □

Remark

By lemma 4, we have just proved that the properties (a), (b), (c), (d), (e) are verified when q is odd and $q > 3$.

Lemma 4.6 $H(\mathcal{O}_3, S)$ is isomorphic to the $^3D_4(q)$-hexagon.

Proof. Using Tits' description of Moufang polygons given in [10], W.M. Kantor has given the following construction of the $^3D_4(q)$-hexagon ([6]). Let $Q = F \times K \times F \times K \times F$. If we put $(\alpha, b, \gamma, d, \epsilon) \cdot (\alpha', b', \gamma', d', \epsilon') = (\alpha + \alpha', b +$

$b', \gamma + \gamma' + \alpha'\epsilon - tr(b'd), d + d', \epsilon + \epsilon')$, then $Q = Q(\cdot)$ is a group of order q^9 whose centre $Z = \{(0,0,\gamma,0,0) : \gamma \in F\}$ has order q. Moreover, Q/Z is a vector space over F. Let $\bar{K} = K \cup \{\infty\}$. Define

$$A_4(\infty) = \{(0,b,\gamma,d,\epsilon) : b,d \in K; \gamma, \epsilon \in F\},$$
$$A_3(\infty) = \{(0,0,\gamma,d,\epsilon) : d \in K; \gamma, \epsilon \in F\},$$
$$A_2(\infty) = \{(0,0,0,d,\epsilon) : d \in K; \epsilon \in F\},$$
$$A_1(\infty) = \{(0,0,0,0,\epsilon) : \epsilon \in F\}.$$

and, for all $t \in K$, let[2]

$$A_4(t) = \{(\alpha,b,\gamma,d,\alpha t^{1+q+q^2} - tr(bt^{q+q^2} - dt)) : \alpha, \gamma \in F; b,d \in K\},$$
$$A_3(t) = \{(\alpha,b,\gamma,-\alpha t^{q+q^2} + b^q t^{q^2} + b^{q^2} t^q, -2\alpha t^{1+q+q^2} + tr(bt^{q+q^2})) :$$
$$\alpha, \gamma \in F; b \in K\},$$
$$A_2(t) = \{(\alpha,b,-\alpha^2 t^{1+q+q^2} - tr(b^{q+q^2}t - \alpha bt^{q+q^2}), -\alpha t^{q+q^2} + b^q t^{q^2} + b^{q^2} t^q,$$
$$-2\alpha t^{1+q+q^2} + tr(bt^{q+q^2})) : \alpha \in F; b \in K\},$$
$$A_1(t) = \{(\alpha,\alpha t,-\alpha^2 t^{1+q+q^2}, \alpha t^{q+q^2}, \alpha t^{1+q+q^2}) : \alpha \in F\}.$$

Let us define a point-line geometry $H = (P, L, I)$ as in Section 2. Then H is isomorphic to the $^3D_4(q)$ hexagon (see [10] and [6]).

Let $\bar{W} = Fw_0 \oplus W \oplus Fw_9$ be a vector space of dimension 10 over $F = GF(q)$. If $\bar{W}^* = K \otimes \bar{W}$, then $\bar{W}^* = Kw_0 \oplus W^* \oplus Kw_9$. Thus, w_0, w_1, w_2, w_3, w_4, w_5, w_6, w_7, w_8, w_9 form a basis of $\bar{W}^*$, and $PG(\bar{W}, F) = PG(9,q)$ is a canonical subgeometry of $PG(\bar{W}^*, K) = PG(9, q^3)$. We will denote by $(x_0, x_1, x_2, x_3, x_4, x_5, x_6, x_7, x_8, x_9)$ the homogeneous coordinates of the point $P(\sum_{i=0}^{9} x_i w_i)$. The semilinear map ρ of $\bar{W}^*$ into itself, described by $\sum_{i=0}^{9} x_i w_i \mapsto x_0^q w_0 + x_1^q w_1 + x_3^q w_2 + x_5^q w_3 + x_7^q w_4 + x_2^q w_5 + x_4^q w_6 + x_6^q w_7 + x_8^q w_8 + w_9^q w_9$, defines a collineation of $PG(9, q^3)$, which fixes all the points of $PG(9,q)$. Notice that ρ induces $\hat{\sigma}$ on the subspace $T = P(W)$ of $PG(9,q)$.

Let $v = \sum_{i=0}^{9} x_i w_i$ and $u = \sum_{i=0}^{9} y_i w_i$. Let us denote by $\times$ the alternating bilinear form of $\bar{W}^*$ defined by $v \times u = x_0 y_9 + x_1 y_8 - x_2 y_7 - x_3 y_6 + x_4 y_5 - x_5 y_4 + x_6 y_3 + x_7 y_2 - x_8 y_1 - x_9 y_0$. Notice that $\times$ induces on W^* the bilinear form $(,)$ defined in Section 3. As $v\rho \times u\rho = (v \times u)^q$, the polarity of $PG(9, q^3)$ associated with $\times$ is also a polarity of $PG(9,q)$, which induces on $T = P(W)$ the polarity $\perp$ associated with the bilinear form $(,)$ defined in Section 3. For this reason, we will denote by $\perp$ also the polarity of $PG(9,q)$ associated with $\times$. Let $W(9,q)$ be the polar space arising from $\perp$. If $x = P(w_9)$ and $y = P(w_0)$, then $x^\perp \cap y^\perp = T$ and $T \cap W(9,q) = W(7,q)$ is the polar space

---

2. There are some misprints in the definition of the automorphism t_6 given in [6] p. 100. The right expression is $t_6 : (\alpha, b, \gamma, d, \epsilon) \mapsto (\alpha, b + \alpha t, c - \alpha^2 t^{1+q+q^2} - tr(b^{q+q^2}t) - tr(\alpha bt^{q+q^2}), d + \alpha t^{q+q2} + b^q t^{q^2} + b^{q^2} t^q, \epsilon + \alpha t^{1+q+q^2} + tr(bt^{q+q^2}) + tr(dt))$.

arising from the symplectic polarity of T associated with the bilinear form $(\,,\,)$.

If $P(w_1) = p_\infty$, then $T_{p_\infty} = \{P(\alpha w_1 + b^{q^2} w_2 + b^q w_3 + b w_5) : \alpha \in F; b \in K\}$. If $v_a = a^{1+q+q^2} w_1 + a^{1+q} w_2 + a^{1+q^2} w_3 + a w_4 + a^{q+q^2} w_5 + a^q w_6 + a^{q^2} w_7 + w_8$ with $a \in K$, then $T_{p_a}^* = P(< v_a, v_1, v_2, v_3 >)$ where

$$
\begin{aligned}
v_1 &= e_1 \wedge (a^q e_3 + e_4) \wedge (a^{q^2} e_5 + e_6) = a^{q+q^2} w_1 + a^q w_2 + a^{q^2} w_3 + w_4, \\
v_2 &= (ae_1 + e_2) \wedge e_3 \wedge (a^{q^2} e_5 + e_6) = a^{1+q^2} w_1 + a w_2 + a^{q^2} w_5 + w_6, \\
v_3 &= (ae_1 + e_2) \wedge (a^q e_3 + e_4) \wedge e_5 = a^{1+q} w_1 + a w_3 + a^q w_5 + w_7.
\end{aligned}
$$

Therefore $T_{p_a} = \{P(\alpha v_a + b v_1 + b^q v_2 + b^{q^2} v_3) : \alpha \in F, b \in K\}$.

For all $\alpha, \gamma, \epsilon \in F$ and $b, d \in K$, let $M(\alpha, b, d, \epsilon, \gamma) = M(\alpha w_1 + b^{q^2} w_2 + b^q w_3 + d w_4 + b w_5 + d^q w_6 + d^{q^2} w_7 + \epsilon w_8, \gamma)$ be the linear map of $\bar{W}^*$ defined in Section 2. We can prove with a direct calculation that

$$
v M(\alpha, b, d, \epsilon, \gamma) \times u M(\alpha, b, d, \epsilon, \gamma) = v \times u,
$$

$$
M(\alpha, b, d, \epsilon, \gamma)\rho = \rho M(\alpha, b, d, \epsilon, \gamma).
$$

This implies that the collineation $M(\alpha, b, d, \epsilon, \gamma)$ fixes $PG(9, q)$ and stabilizes the polarity $\perp$. Then $M(\alpha, b, d, \epsilon, \gamma)$ is a collineation of $PG(9, q)$, which fixes the point x and all subspaces of $x^\perp$ incident with x. If $\tilde{Q} = \{M(\alpha, b, d, \epsilon, \gamma) : \alpha, \gamma, \epsilon \in F, b, d \in K\}$, then $\tilde{Q}$ is a group of order q^9 transitive on the points of $W(9, q) \setminus x^\perp$. For each point p_t $(t \in \tilde{K})$, we put

$$
\begin{aligned}
\tilde{A}_4(t) &= \{h \in \tilde{Q} : p_t h = p_t\}, \\
\tilde{A}_3(t) &= \{h \in \tilde{Q} : T_{p_t} h = T_{p_t}\}, \\
\tilde{A}_2(t) &= \{h \in \tilde{Q} :< y, T_{p_t} > h =< y, T_{p_t} >\}, \\
\tilde{A}_1(t) &= \{h \in \tilde{Q} :< y, p_t > h =< y, p_t >\}.
\end{aligned}
$$

With a direct calculation, we can prove that

$$
\begin{aligned}
\tilde{A}_4(\infty) &= \{M(\alpha, b, d, 0, \gamma) : \alpha, \gamma \in F; b, d \in K\}, \\
\tilde{A}_3(\infty) &= \{M(\alpha, b, 0, 0, \gamma) : \alpha, \gamma \in F; b \in K\}, \\
\tilde{A}_2(\infty) &= \{M(\alpha, b, 0, 0, 0) : \alpha \in F; b \in K\}, \\
\tilde{A}_1(\infty) &= \{M(\alpha, 0, 0, 0, 0) : \alpha \in F\}, \\
\tilde{A}_4(0) &= \{M(0, b, d, \epsilon, \gamma) : \epsilon, \gamma \in F; b, d \in K\}, \\
\tilde{A}_3(0) &= \{M(0, 0, d, \epsilon, \gamma) : \epsilon, \gamma \in F; d \in K\}, \\
\tilde{A}_2(0) &= \{M(0, 0, d, \epsilon, 0) : \gamma \in F; d \in K\}, \\
\tilde{A}_1(0) &= \{M(0, 0, 0, \epsilon, 0) : \epsilon \in F\}.
\end{aligned}
$$

Moreover, for $t \in K \setminus \{0\}$ we have

$$\tilde{A}_4(t) = \{M(\alpha t^{1+q+q^2} - tr(bt^{q+q^2} - dt), d, b, \alpha, \gamma) : \alpha, \gamma \in F; b, d \in K\},$$
$$\tilde{A}_3(t) = \{M(-2\alpha t^{1+q+q^2} + tr(bt^{q+q^2}), -\alpha t^{q+q^2} + b^q t^{q^2} + b^{q^2} t^q, b, \alpha, \gamma) :$$
$$\alpha, \gamma \in F; b \in K\},$$
$$\tilde{A}_2(t) = \{M(-2\alpha t^{1+q+q^2} + tr(bt^{q+q^2}), -\alpha t^{q+q^2} + b^q t^{q^2} + b^{q^2} t^q, b, \alpha, 0) :$$
$$\alpha \in F; b \in K\},$$
$$\tilde{A}_1(t) = \{M(\alpha t^{1+q+q^2}, \alpha t^{q+q^2}, \alpha t, \alpha, 0) : \alpha \in F\}.$$

The map of Q into $\tilde{Q}$ defined by

$$g = (\alpha, b, \gamma, d, \epsilon) \mapsto \tilde{g} = M(\epsilon, d, b, \alpha, -2\gamma + \epsilon\alpha - tr(bd))$$

is an automorphism of Q into $\tilde{Q}$. We can prove directly that $A_i(t) = \{g \in Q : \tilde{g} \in \tilde{A}_i(t)\}$, for all $t \in \hat{K}$ and $i = 1, 2, 3, 4$. As the map $\theta : H \to H(\mathcal{O}_3, S)$ defined by

$$\theta : \mathcal{I} \mapsto x,$$
$$\theta : [t] \mapsto \langle x, p_t \rangle,$$
$$\theta : A_4(t)g \mapsto p_t \tilde{g},$$
$$\theta : A_3(t)g \mapsto T_{p_t} \tilde{g},$$
$$\theta : A_2(t)g \mapsto \langle T_{p_t}, y \rangle \tilde{g},$$
$$\theta : A_1(t)g \mapsto \langle p_t, y \rangle \tilde{g},$$
$$\theta : g \mapsto y\tilde{g},$$

is an isomorphism, $H(\mathcal{O}_3, S)$ is isomorphic to the ${}^3D_4(q)$-hexagon. $\square$

5. Subhexagons

Let S be a 3-dimensional subspace of T such that $S \cap \mathcal{O}_3$ is a twisted cubic. If p_a is a point of $\mathcal{O}_3$, then $T_{p_a} \cap S$ is a line if and only if $p_a \in S$. Let $U = \langle x, S, y \rangle = PG(5, q)$. Let us define a point-line geometry $H(S)$ in the following way. A point z of $H(\mathcal{O}_3, S)$ of type (1) or (2) or (4) is a point of $H(S)$ if and only if z is a point of U. A point X of $H(\mathcal{O}_3, S)$ of type (3) is a point of $H(S)$ if and only if $U \cap X$ is a plane. A line l of $H(\mathcal{O}_3, S)$ of type either (a) or (c) is a line of $H(S)$ if and only l is contained in U. A line X of $H(\mathcal{O}_3, S)$ of type (b) is a line of $H(S)$ if and only if $X \cap U$ is a line. Therefore, $H(S)$ can be regarded as a substructure of $H(\mathcal{O}_3, S)$. In particular, each chain in $H(S)$ defines a chain of $H(\mathcal{O}_3, S)$ of the same length.

Theorem 5.1 $H(S)$ is a generalized hexagon isomorphic to the dual $G_2(q)$-hexagon.

Proof. $H(S)$ contains exactly $(1+q)(1+q+q^2)$ points and $(1+q)(1+q+q^2)$ lines. By construction, there are no circuits of length less than 12. Moreover, if p and r are two distinct points of $\mathcal{O}_3 \cap S$, then $\{x, < x, p >, p, T_p \cap S, < T_p \cap S, y >, < p, y >, y, < y, r >, < T_r \cap S, y >, T_r \cap S, r, < r, x >, x\}$ is a circuit of length 12 of $H(S)$. By [9] p. 5, $H(S)$ is a generalized hexagon with parameters (q, q). In [4] W.M. Kantor has proved that there is a canonical way to construct the dual $G_2(q)$-hexagon starting from a twisted cubic (see [4] Remark 2). Thus, $H(S)$ is isomorphic to the dual $G_2(q)$-hexagon. $\square$

By Theorem 5.1, we can extend the construction of the dual $G_2(q)$-hexagon given in [1] to the case $q = 3^r$.

References

[1] **L. Bader and G. Lunardon**. Generalized hexagons and BLT-sets. In A. Beutelspacher, F. Buekenhout, F. De Clerck, J. Doyen, J. W. P. Hirschfeld, and J. A. Thas, editors, *Finite Geometry and Combinatorics*, pages 5–15, Cambridge, 1993. Cambridge University Press.

[2] **W. Burau**. *Mehrdimensionale Projektive und Höhere Geometrie*. VEB Deutscher Verlag der Wissenschaften, Berlin, 1961.

[3] **J. W. P. Hirschfeld and J. A. Thas**. *General Galois Geometries*. Oxford University Press, Oxford. Oxford Science Publications, 1991.

[4] **W. M. Kantor**. Generalized quadrangles associated with $G_2(q)$. *J. Combin. Theory Ser. A*, 29, pp. 212–219, 1980.

[5] **W. M. Kantor**. Spreads, translation planes and Kerdock sets. I. *SIAM J. Algebraic Discrete Methods*, 3, pp. 151–165, 1982.

[6] **W. M. Kantor**. Generalized polygons, SCABs and GABs. In L. A. Rosati, editor, *Buildings and the Geometry of Diagrams, Proceedings Como 1984*, volume 1181 of *Lecture Notes in Math.*, pages 79–158, Berlin, 1986. Springer Verlag.

[7] **G. Lunardon**. Fibrazioni planari e sottovarietà algebriche della varietà di Grassmann. *Geom. Dedicata*, 16, pp. 291–313, 1984.

[8] **G. Lunardon**. Varietà di Segre e ovoidi dello spazio polare $Q^+(7,q)$. *Geom. Dedicata*, 20, pp. 121–131, 1986.

[9] **S. E. Payne and M. G. Tinsley**. On $v_1 \times v_2(n, s, t)$ configurations. *J. Combin. Theory*, 7, pp. 1–14, 1969.

[10] **J. Tits**. Classification of buildings of spherical type and Moufang polygons: a survey. In *Teorie Combinatorie, Proc. Intern. Colloq. (Roma 1973)*, volume I, pages 229–246. Accad. Naz. Lincei, 1976.

G. **Lunardon** Dipartimento di Matematica e Applicazioni, Complesso di Monte S. Angelo, Edificio T, Via Cintia, I-80134 Napoli (Italy). e-mail: Lunardon@napoli.infn.it

A census of known flag-transitive extended grids

T. Meixner   A. Pasini

Abstract

We list all examples of flag-transitive (possibly repeated) extensions of grids that we presently know. Many of the families we will describe have never appeared before in the literature; we also give new constructions for some non-new examples. Some characterization theorems will be proved in a forthcoming paper [13], carrying on the classification work of [8], [11], [12], [16].

1. Introduction

1.1. Notation

The symbol $c^n.C_2(s,1)$ will always denote the following diagram of rank $n+2$, with $n \geq 1$:

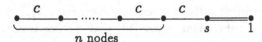

equivalently:

where $s, 1$ are orders and $s < \infty$. We also write $c.C_2(s,1)$ instead of $c^1.C_2(s,1)$ when $n = 1$. We write $c^n.C_2$ or $c.C_2$ instead of $c^n.C_2(s,1)$ or $c.C_2(s,1)$ respectively, when the order s is clear from the context or when we are not interested in it.

Residually connected geometries belonging to the diagram $c^n.C_2(s,1)$ will be called n-times extended grid of order s. A 1-time extended grid of order s will be called an extended grid of order s, for short.

We use the nonnegative integers $0, 1,..., n, n + 1$ to denote types, as follows:

$$\underset{0}{\bullet}\!\!\!-\!\!\!-\!\!\!-\!\!\!\underset{1}{\bullet}\!\!\!-\!\!\cdots\!\!-\!\!\!\underset{n-2}{\bullet}\!\!\!-\!\!\!-\!\!\!\underset{n-1}{\bullet}\!\!\!\overset{c}{-\!\!\!-\!\!\!-}\!\!\!\underset{n}{\bullet}\!\!\!=\!\!\!=\!\!\!\underset{n+1}{\bullet}$$

The elements of type 0 will be called *points*, those of type $n+1$ *blocks* and those of type 1 *line* or *edges* or 1-*faces*. More generally, the elements of type $i = 1, 2, ..., n$ will be called i-*faces* (also *faces*, for short).

We are interested in flag-transitive geometries. Assuming Γ flag-transitive, we denote by Γ_k ($k = 0, 1, ..., n-1$) the isomorphism type of the residues of Γ of type $\{k+1, k+2, ..., n+1\}$; in particular Γ_0 is the isomorphism type of residues of points. We say that Γ is a $(k+1)$-*times extension* of Γ_k; trivially, Γ_k is an $(n-1-k)$-times extended grid, when $k < n-1$.

1.2. Geometric properties

Let Γ be an n-times extended grid of order s. It follows from [15] (Proposition 2) that the *Intersection Property* holds in Γ if and only if distinct faces are incident with distinct sets of points. If this is the case, then distinct blocks are also incident with distinct sets of points, whence the blocks can be viewed as sets of points (of size $s+1+n$) and the i-faces are the $(i+1)$-subsets of the blocks ($i = 1, 2, ..., n$); furthermore, the incidence relation can be interpreted as (symmetrized) inclusion. Therefore, if the Intersection Property holds in Γ, the geometry Γ is uniquely determined by its rank and its point-block system.

At the opposite side we may place the class of flat geometries. We say that Γ is *flat* if every point of Γ is incident with all blocks of Γ. It is an easy exercise to prove that, if $n \geq 2$ and all $c.C_2$ residues of Γ are flat, then Γ is flat and all i-faces of Γ are incident with all blocks, for every $i = 1, 2, ..., n-1$ (however, we will prove in a subsequent paper [13] that no flag-transitive example of this kind exist).

The *point-graph* $\mathcal{G}(\Gamma)$ of Γ is defined as follows: the vertices of $\mathcal{G}(\Gamma)$ are the points of Γ; two points of Γ are adjacent in $\mathcal{G}(\Gamma)$ precisely when they are incident with a common block of Γ.

Trivially, the blocks of Γ form $(s+1+n)$-cliques in $\mathcal{G}(\Gamma)$. If the Intersection Property holds in Γ and the blocks of Γ are precisely the maximal cliques of $\mathcal{G}(\Gamma)$, then we say that Γ is *determined* by its point-graph $\mathcal{G}(\Gamma)$ (when $n = 1$, this corresponds to the triangular extensions of [6]).

We also define a *block-graph* $\mathcal{G}^*(\Gamma)$, taking the blocks as vertices and stating that two blocks are adjacent in this graph when they are incident with a common n-face. We say that the set of blocks of Γ *splits* when $\mathcal{G}^*(\Gamma)$ is bipartite. It follows from [17] that the set of blocks of Γ splits if Γ is 2-simply connected.

We say that Γ *admits unfolding* if Γ can be obtained as shadow geometry (see [19]) from a geometry belonging to the following diagram of rank $n+2$ (shadows are taken with respect to the initial node of this diagram, circled in the picture):

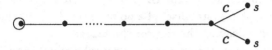

Trivially, Γ admits unfolding if and only if the set of blocks of Γ splits and, for every point or face x and every n-face y, the elements x and y are incident iff the two blocks incident with y are also incident with x.

We will mention which of the above properties hold in the examples we will consider, but we will not prove them in every case, leaving that job for the reader. A number of different parameters can be chosen to extimate the size of Γ; we will use the diameter of the point-graph $\mathcal{G}(\Gamma)$ to this purpose; we will denote it by δ.

1.3. Covers and quotients

In some cases we will only describe the 2-simply connected representatives of the family of examples we consider. We will not list all possible quotients of those 2-simply connected geometries (in some cases, we are unable to do that). We will only mention some quotients that either have been described in a different way in the literature or look particularly interesting for some respect.

In other cases, we know that the examples we describe are not 2-simply connected, but we do not know their universal 2-covers. In those cases, we give the information we have on proper 2-covers.

When the Intersection Property holds in Γ, the points, the lines and the 3-subsets of the blocks of Γ (namely the 2-faces, when $n \geq 2$) form a 2-dimensional simplicial complex. We call it the *shadow complex* of Γ and we denote it by $\mathcal{K}(\Gamma)$. It follows from [18] that the $n+1$-homotopy group of Γ and the homotopy group of $\mathcal{K}(\Gamma)$ are isomorphic. Therefore, if the Intersection Property holds in Γ, we can safely substitute $\mathcal{K}(\Gamma)$ for Γ when we want to check if Γ is $n+1$-simply connected.

1.4. Notation for groups

Given a flag-transitive subgroup $G \leq \operatorname{Aut}(\Gamma)$, we denote by B the Borel subgroup of G, namely the stabilizer in G of a chamber of Γ. The stabilizer in G of an element of type $i = 0, 1, ..., n+1$ is denoted by G_i. The symbol K_i denotes the elementwise stabilizer of the residue Γ_x of an element x of type i and $\overline{G}_i = G_i/K_i$ is the action of G_i on Γ_x. The subgroup K_i is called the *kernel* of G_i; we will often write kernels in square brackets; for instance, writing $G_{n+1} = [S_k] \times S_m$ we mean that $G_{n+1} = K_{n+1} \times \overline{G}_{n+1}$ with $K_{n+1} = S_k$ and $\overline{G}_{n+1} = S_m$.

We will never describe all subgroups G_i ($i = 0, 1, ..., n + 1$). We often only give some information on G_0 (stabilizer of a point) or on G_{n+1} (stabilizer of a block), or on G_{n+1} only, leaving the rest for the reader.

Furthermore, in many cases the examples we describe admit more than one flag-transitive automorphism groups. We will not list all of them in every case. We often only describe the full automorphism group Aut(Γ) or a (often unique) minimal flag-transitive automorphism group, leaving the rest for the reader.

We use the notation of [7] when dealing with finite groups. We also use the symbol $\wr$ to denote wreath products. Furthermore, given a Coxeter group W, by W^+ we mean the subgroup of W consisting of all elements of even length of W. In particular, $(S_h \times S_k)^+$ is the subgroup of the (reducible) Coxeter group $W = S_h \times S_k$, consisting of all pairs of permutations (f, g) with $f \in S_h$, $g \in S_k$ and where f, g are either both even or both odd.

2. The examples

2.1. The Coxeter family $\mathrm{Cox}(n, s)$

The examples we are going to describe in this subsection may be considered as the 'standard' ones.

Let $D_{n,s}$ be the following Coxeter diagram of rank $n + 2s$ ($n \geq 1$, $s \geq 1$):

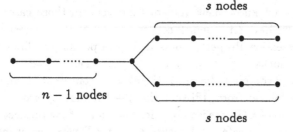

In particular, we have $D_{n,1} = D_{n+2}$ if $n \geq 2$, $D_{1,s} = A_{2s+1}$, $D_{2,2} = E_6$, $D_{3,2} = \tilde{E}_6$ and $D_{2,3} = \tilde{E}_7$.

Let C be the Coxeter complex belonging to $D_{n,s}$. Truncating the last $s-1$ nodes in each of the two right horns of the diagram, we obtain a geometry, call it $tr(C)$, belonging to the following diagram of rank $n + 2$:

(of course, $tr(C) = C$ when $s = 1$, namely no truncation is done in this case). Let Γ be the shadow geometry of $tr(C)$, with respect to the initial node of the diagram (circled in the next picture):

Then Γ is an n-times extended grid of order s. We denote Γ by the symbol $\mathrm{Cox}(n, s)$.

Geometric properties. The geometry $\mathrm{Cox}(n, s)$ satisfies the Intersection Property and it is determined by its point-graph. The diameter δ of the point-graph is infinite when either $n > 2$ and $s \geq 2$ or $s > 2$ and $n \geq 2$. When $n = 1$, we have $\delta = s + 1$. When $s = 1$, then $\delta = 2$. When $n = s = 2$, then $\delta = 3$.

Trivially, $\Gamma_0 = \mathrm{Cox}(n - 1, s)$ if $n \geq 2$, where Γ_0 is the isomorphism type of the residues of points of $\Gamma = \mathrm{Cox}(n, s)$.

2-simple connectedness. It follows from [14] (Theorem 1) that $tr(\mathcal{C})$ is 2-simply connected. By [17] and since $tr(\mathcal{C})$ is the unfolding of $\mathrm{Cox}(n, s)$, the universal 2-cover of $\mathrm{Cox}(n, s)$ admits unfolding and its unfolding is a 2-cover of $tr(\mathcal{C})$. Therefore, $\mathrm{Cox}(n, s)$ is 2-simply connected, since $tr(\mathcal{C})$ is 2-simply connected.

Studying all possible quotients of $\mathrm{Cox}(n, s)$ seems to be hopeless, in general (this would be the same as studying all possible quotients of the Coxeter complex $\mathcal{C}$). However, something can be said in some special cases, as we will see later.

Flag-transitive automorphism groups. We have $\mathrm{Aut}(\mathrm{Cox}(n, s)) = W_{n,s} : 2$, where $W_{n,s}$ denotes the Coxeter group of type $D_{n,s}$ and the factor 2 on top is contributed by a diagram automorphism of $W_{n,s}$. The Borel subgroup of $G = W_{n,s} : 2$ has the following structure: $B = S_s \times S_s$. The group G_0 acts faithfully on Γ_0 (notation as in subsections 1.1 and 1.4) and it is isomorphic to $W_{n-1,s} : 2$ if $n \geq 2$. When $n = 1$, then $G_0 = S_{s+1} \wr 2$.

We have $G_{n+1} = [S_s] \times S_{n+s+1}$ (notation as in subection 1.4).

When $s \geq 2$, then $W_{n,s}^+ : 2$ also acts flag-transitively on $\mathrm{Cox}(n, s)$ (notation as in subsection 1.4); in fact, $W_{n,s}^+ : 2$ is the minimal flag-transitive automorphism group of $\mathrm{Cox}(n, s)$. If we now assume $G = W_{n,s}^+ : 2$ instead of $G = W_{n,s} : 2$, then we obtain a description similar to the above for G_0, but substituting $W_{n-1,s}^+ : 2$ for $W_{n-1,s} : 2$ when $n \geq 2$ (we have $G_0 = A_{s+1} \wr 2$ when $n = 1$). The Borel subgroup B now has the following structure: $B = (S_s \times S_s)^+$.

We now have $G_{n+1} = [A_s] : S_{n+s+1} = (S_s \times S_{n+s+1})^+$.

Special cases. $\mathrm{Cox}(1, s)$ is the so-called *Johnson geometry* [6]. In particular, $\mathrm{Cox}(1, 2)$ is the affine polar space [6] obtained deleting a secant hyperplane from $Q_5^+(2)$.

We have $W_{1,s} = S_{2s+2} \times 2$ and $\mathrm{Cox}(1, s)$ admits just one flag-transitive proper quotient, obtained factorizing over the direct factor 2 of the splitting $S_{2s+2} \times 2$ (that factor is contributed by a suitable involutory diagram automorphism of the Coxeter complex C of type A_{2s+1}). We denote this quotient by the symbol $\text{Cox}(1, s)/2$ (it is called the 'halved Johnson geometry' in [6]). Trivially, the alternating group A_{2s+2} and the symmetric group S_{2s+2} are the two flag-transitive automorphism groups of $\text{Cox}(1, s)/2$.

Cox$(n, 1)$ is the Coxeter complex of type C_{n+2} (trivially, we have $W : 2 = (2^{n+1} : S_{n+2}) \times 2$ is its unique flag-transitive automorphism group). It admits just one flag-transitive proper quotient, obtained factorizing over the center Z_2 of the Coxeter group $2^{n+1} : S_{n+2} \times 2$. This quotient is flat.

Remark

A 2-times extended grid of order 3, call it Γ, is mentioned in [2] (example (30')) admitting the symplectic group $S_6(2)$ as flag-transitive automorphism group. The Borel subgroup and the stabilizers of elements of Γ are just the same as in Cox$(2, 3)$. Since $S_6(2)$ is simple, Γ does not admit unfolding, hence it is not 2-simply connected. It will turn out from a theorem of [13] that Γ is in fact a quotient of Cox$(2, 3)$.

2.2. The permutation family $\text{Sym}(n + 3)$

Let $\mathcal{G}^+$ be the square-grid graph naturally defined on the $(n + 3)^2$ pairs (i, j) $(i, j = 1, 2, ..., n + 3; n \geq 1$; two pairs (i, j) and (h, k) are adjacent in $\mathcal{G}^+$ when either $i = h$ or $j = k$). Let $\mathcal{G}^-$ be the complement of $\mathcal{G}^+$.

An n-times extended grid Γ of order 2 can be defined taking the vertices as points, the maximal cliques of $\mathcal{G}^-$ as blocks, the $i + 1$-cliques of $\mathcal{G}^-$ as i-faces $(i = 1, 2, ..., n)$ and defining the incidence relation in the natural way, as (symmetrized) containment. Trivially, $\mathcal{G}^-$ is the point-graph of Γ; the set of blocks of Γ can also be viewed as the group of permutational matrices of rank $n + 3$. Because of this, we denote Γ by the symbol $\text{Sym}(n + 3)$.

Geometric properties. The Intersection Property holds in $\text{Sym}(n+3)$ and it is quite evident that $\text{Sym}(n + 3)$ is determined by its point-graph. The diameter of the point graph is 2.

The residues of $\text{Sym}(n+3)$ of type $\{k+1, k+2, ..., n+1\}$ $(k = 0, 1, ..., n-2)$ are isomorphic with $\text{Sym}(n + 2 - k)$.

Furthermore, the set of blocks of $\text{Sym}(n + 3)$ splits; this follows from the 2-simple connectedness of $\text{Sym}(n + 3)$ (see below), using [17]. However, that splitting can also be recognized directly: we have remarked above that the blocks of $\text{Sym}(n + 3)$ can be viewed as permutational matrices of rank $n + 3$, whence as permutations of degree $n + 3$; the set of even permutations and the set of odd permutations give us that splitting of the set of blocks.

The geometry $\text{Sym}(n+3)$ admits unfolding, since its set of blocks splits and the Intersection Property holds in it.

2-simple connectedness. The Intersection Property holds in the geometry $\text{Sym}(n + 3)$. Hence the $n + 1$-homotopy group of $\text{Sym}(n + 3)$ is the homotopy group of the shadow complex of $\text{Sym}(n + 3)$, which is easily seen to be trivial. Therefore $\text{Sym}(n + 3)$ is $n + 1$-simply connected; in particular, $\text{Sym}(4)$ is 2-simply connected. By an easy inductive argument, we obtain that $\text{Sym}(n + 3)$ is 2-simply connected for every $n \geq 1$.

Flag-transitive automorphism groups. Let S be the set of blocks of $\text{Sym}(n + 3)$. As we have remarked before, S can naturally be identified with the symmetric group S_{n+3}, in its natural action on the set of symbols $X = \{1, 2, ..., n+3\}$. Let S_L and S_R be the two actions of the symmetric group S on itself, by left and right multiplication respectively. An action of $S_L \times S_R$ on the set of blocks of $\text{Sym}(n+3)$ is defined in this way. The action of $S_L \times S_R$ on points and faces of $\text{Sym}m(n + 3)$ is uniquely determined by the action of $S_L \times S_R$ on the set of blocks, since points and faces are uniquely determined by the sets of blocks which they belong to. In particular, the set of points of $\text{Sym}(n + 3)$ can be identified with $X \times X$, $X = \{1, 2, ..., n + 3\}$; given $f \in S$, let f_L and f_R be the elements of S_L and S_R, respectively, corresponding to f. Then, given $f, g \in S$ and $i, j \in X$, the element $(f_L, g_R) \in S_L \times S_R$ acts on the point $(i, j) \in X \times X$ as follows: $(f_L, g_R)(i, j) = (g^{-1}(i), f(j))$.

The full automorphism group $G = \text{Aut}(\text{Sym}(n + 3))$ of $\text{Sym}(n + 3)$ has the following structure: $G = (S_L \times S_R) : \langle \phi \rangle$, where ϕ acts as follows on $S_L \times S_R$: $\phi(f_L, g_R) = (g_L, f_R)$ for $f, g \in S$, the symbols f_L, f_R, g_L, g_R having the meaning stated above.

Stabilizers of blocks have the following isomorphism type:

$G_{n+1} = [Z_2] \times S_{n+3}$. The stabilizer G_0 of a point acts faithfully on the residue Γ_0 of the point (we have $G_0 = \text{Aut}(\text{Sym}(n+2))$ if $n \geq 2$ and $G_0 = S_3 \wr 2$ if $n = 1$). As for the Borel subgroup, we have $B = 2^2$.

Smaller flag-transitive automorphism groups also exist: for instance, there are two minimal flag-transitive automorphism groups, each of them isomorphic to $A_{n+3} \times S_{n+3}$ and obtained as $G = A_L \times S_R$ or $G = S_L \times A_R$, where A_L (respectively, A_R) is the alternating subgroup of the symmetric group S_L (of S_R, respectively).

In these minimal automorphism groups we have $G_0 = A_{n+2} \times S_{n+2}$ ($= Z_3 \wr 2$ when $n = 1$), $G_{n+1} = A_{n+3}$ and $B = 1$.

A special case. $\text{Sym}(4)$ ($n = 1$) is the affine polar space [6] obtained deleting a tangent hyperplane of $Q_5^+(2)$. Furthermore, $\text{Sym}(4)$ can also be viewed as a Minkowski plane (of order 3, that is with 4 points on each circle). Indeed the set of blocks of $\text{Sym}(4)$ is a sharply 3-transitive group of permutations (of degree 4).

Sym(4) also admits a proper flag-transitive quotient, which we denote by $\mathrm{Sym}(4)/2^2$ and is obtained factorizing over the Sylow 2-subgroup 2^2 of A_4, viewed as a subgroup of S_L or of S_R.

The geometry $\mathrm{Sym}(4)/2^2$ is flat and its full automorphism group is $S_4 \times S_3$ ($S_4 \times Z_3$ and $A_4 \times S_3$ are the two minimal flag-transitive automorphism groups of $\mathrm{Sym}(4)/2^2$).

Other proper quotients of $\mathrm{Sym}(n+3)$ can be constructed (in particular, flat quotients), for every $n \geq 1$. For instance, if $X = \{1, 2, ..., n+3\}$, we can take a non-trivial subgroup E of S acting fixed-point-freely on X, realizing E as a subgroup of S_L or of S_R. The full automorphism group of $\mathrm{Sym}(n+3)/E$ will be $S_{n+3} \times N(E)/E$, where $N(E)$ is the normalizer of E in S (we obtain flat quotients when E acts regularly on X). However, $\mathrm{Sym}(4)/2^2$ is the only flag-transitive quotient that we can obtain in this way. In fact, it is the only flag-transitive proper quotient that we know for a geometry of type $\mathrm{Sym}(n+3)$.

Links with other topics. The construction of $\mathrm{Sym}(n+3)$ can be generalized for any sharply k-transitive finite permutation group G, with $k \geq 2$ (see [4]). When $k = 2$ we obtain flag-transitive nets. When $k \geq 3$, we obtain flag-transitive geometries belonging to the following diagram of rank k:

where N denotes the class of nets and $s-1+k$ is the degree of the permutation group G. When $k = 3$, the geometries defined above are Minkowski planes (enriched with the pairs of distinct points contained in a common circle) and it is well known that the possibilities for G are the following ones: S_4, A_5 ($= L_2(4) = L_2(5)$), $PGL_2(q)$ and $L_2(q^2).2$ (q odd).

When $G = S_{1+k}$ ($k \geq 3$), then we obtain $\mathrm{Sym}(n+3)$ ($k = n+2$). When $G = A_{2+k}$ ($k \geq 4$), then we obtain a $k - 3$-times extension of the Minkowski plane associated to $A_5 = L_2(4)$. It follows from the well known classification of sharply k-transitive finite permutation groups with $k \geq 2$ that, apart from n-times extended grids $\mathrm{Sym}(n+3)$ and the above examples from alternating groups, only two more examples exist for $k \geq 4$, related to the Minkowski plane obtained from $L_2(9).2 = M_{10}$. Indeed that Minkowski plane can be extended twice, extending M_{10} to the Mathieu group M_{11} first ($k = 4$), next extending M_{11} to M_{12} ($k = 5$).

2.3. A tower for $\mathrm{Aut}(M_{22})$

A 3-times extended grid Γ of order 3 is constructed in [6] (example 9.12(i)), starting from two points a, b of the 5-$(24, 8, 1)$ design S for the Mathieu group M_{24}. The points of Γ are the 22 points of S other than a or b and the blocks of Γ are the blocks of S containing exactly one of a or b. The i-faces ($i = 1, 2$)

are the $i+1$-subsets of the set of points of Γ. The 3-faces of Γ are the 4-subsets of the blocks of Γ (viewed as sets of points of Γ). The incidence relation is the natural one, defined by (symmetrized) containment.

Geometric properties. It is clear from the above that the Intersection Property holds in Γ and that the set of blocks of Γ splits (in fact, the blocks are partitioned in two class according to which of the points a or b they contained in S). Therefore Γ admits unfolding.

The point-graph $\mathcal{G}(\Gamma)$ is trivial (diameter $\delta = 1$), whence Γ is not determined by its point-graph. In fact, there are 4-cliques of the point graph that do not belong to any block of Γ: these are the 4-cliques contained in blocks of S containing both a and b.

The residue Γ_0 of a point of Γ can be obtained mimicking the above construction in the 4-$(23, 7, 1)$ design for M_{23} and it has properties quite similar to those of Γ. In particular, the point-graph of Γ_0 is trivial. The residue Γ_1 of a point-line flag can also be obtained as above, now starting from the 3-$(22, 6, 1)$ design for M_{22}. Needless to say that Γ_1 satisfies the Intersection Property and admits unfolding (indeed these properties are inherited by residues) but the point-graph of Γ_1 is not trivial: it has diameter $\delta = 2$.

Notation. The geometry Γ_1 (see above) is a Cameron-Fisher extension of odd type (see [6]); we will examine that family of extended grids in subsection 2.5. According with the notation we will state there for that family, we denote Γ_1 by the symbol $CF^-(4)$. Consistently with this, we denote Γ_0 and Γ by $c.CF^-(4)$ and $c^2.CF^-(4)$, respectively.

Covers. Since the point-graph of $c^2.CF^-(4)$ is trivial and every 3-clique of that graph is a 2-face, the shadow complex of $c^2.CF^-(4)$ is simply connected. Therefore, $c^2.CF^-(4)$ is 4-simply connected. It will turn out from a result of [13] that $c.CF^-(4)$ is 3-simply connected, whence $c^2.CF^-(4)$ is also 3-simply connected.

However, $CF^-(4)$ is not 2-simply connected; a 2-fold 2-cover of $CF^-(4)$ is constructed in [3]. Let us denote it by $2 \cdot CF^-(4)$. The reader may find another construction for the extended grid $2 \cdot CF^-(4)$ in [1]. The 2-simple connectedness of $2 \cdot CF^-(4)$ follows from [9], where it is implicitly proved that the Johnson geometry Cox(3, 1), its quotient Cox(3, 1)/2, the Cameron-Fisher extension $CF^-(4)$ and its double cover $2 \cdot CF^-(4)$ are the only flag-transitive extended grids of order 3.

We do not know if $2 \cdot CF^-(4)$ can be extended further. In particular, we do not know if $c^2.CF^-(4)$ and $c.CF^-(4)$ are 2-simply connected.

Flag-transitive automorphism groups. The geometry $c^2.CF^-(4)$ admits $G = \mathrm{Aut}(M_{22})$ as unique flag-transitive automorphism group. The Borel subgroup has order 3.

The stabilizer G_4 of a block acts faithfully as A_7 on the 7 points of that

257

block. The stabilizer G_0 of a point acts faithfully as $L_3(4) : 2_2$ on the residue $\Gamma_0 = c.CF^-(4)$ of that point. In fact, $L_3(4) : 2_2$ is the unique flag-transitive automorphism group of $c.CF^-(4)$.

Let $G_{0,1}$ be the stabilizer in $G_0 = L_3(4) : 2_2$ of a point of $\Gamma_0 = c.CF^-(4)$. Then $G_{0,1} = (2^4 : L_2(4)) : 2$ acts faithfully on $\Gamma_1 = CF^-(4)$, inducing on it its minimal flag-transitive automorphism group.

Trivially, the stabilizer in G_0 (respectively, in $G_{0,1}$) of a block of Γ_0 (of Γ_1) acts faithfully as A_6 (as A_5) on the 6 (respectively, 5) points of that block.

2.4. Four families from quadrics in $PG(3,q)$ (q odd)

Let Q be a non-degenerate quadric, embedded in $PG(3,q)$, q odd. As usual, we say that a line L of $PG(3,q)$ is tangent to Q if $L \cap Q$ is a single point and that a plane α of $PG(3,q)$ is secant for Q if $\alpha \cap Q$ is a non-degenerate conic (equivalently, if $\alpha^\perp \notin Q$).

Let P^+ (respectively, P^-) be the set of non-singular points of $PG(3,q)$ of square norm (respectively, non-square norm). If L is a line of $PG(3,q)$ tangent to Q, then $L - (L \cap Q)$ is contained either in P^+ or in P^-. Therefore the lines of $PG(3,q)$ tangent to Q are partitioned in two families, call them $\mathcal{L}^+$ and $\mathcal{L}^-$, where a tangent line L belongs to $\mathcal{L}^+$ (respectively, to $\mathcal{L}^-$) if $L - (L \cap Q) \subseteq P^+$ (respectively, if $L - (L \cap Q) \subseteq P^-$). The tangent lines contained in a secant plane belong to the same family. Therefore, the secant planes are also partitioned in two families, call them Π^+ and Π^-, where a plane α is in Π^+ (respectively in Π^-) if the tangent lines on α belong to $\mathcal{L}^+$ (respectively, to $\mathcal{L}^-$). In fact, if $Q = Q_3^-(q)$ and $q \equiv 1 \pmod 4$ or $Q = Q_3^+(q)$ and $q \equiv 3 \pmod 4$, then Π^+ consists of the planes α such that $\alpha^\perp \in P^-$; on the other hand, when $Q = Q_3^-(q)$ with $q \equiv 3 \pmod 4$ or $Q = Q_3^+(q)$ with $q \equiv 1 \pmod 4$, then Π^+ consists of the planes α with $\alpha^\perp \in P^+$.

Let us take $\mathcal{L}^+$ as set of points and $P^+ \cup \Pi^+$ as set of blocks, defining the incidence relation between a point and a block in the natural way, by (symmetrized) containment. As 1-faces we take the (unordered) pairs of lines $L, M \in \mathcal{L}^+$ meeting in a point of P^+ (namely, contained in a plane of Π^+), stating that such a pair $\{L, M\}$ is incident with L, M, with $L \cap M \in P^+$ and with the plane of Π^+ spanned by L and M. An extended grid Γ of order $q-1$ is obtained in this way. We denote it by $Tg(Q)$.

Geometric properties. The Intersection Property holds in $Tg(Q)$ and $Tg(Q)$ is determined by its point-graph. The point-graph of $Tg(Q)$ has diameter $\delta = 3$, except when $Q = Q_3^+(3)$; in this case we have $\delta = 2$. The set of blocks of $Tg(Q)$ splits and $Tg(Q)$ admits unfolding.

Covers. We do not know if $Tg(Q)$ is simply connected when $q \geq 5$. Note that, since the Intersection Property holds in $Tg(Q)$ and $Tg(Q)$ is determined by its point graph, we should only check if all closed paths of the point-graph split in triangles.

When $q = 3$, we have $Tg(Q_3^-(3)) = \text{Cox}(1,2)$ and $Tg(Q_3^+(3)) = \text{Sym}(4)$ and these geometries are simply connected.

Flag-transitive automorphism groups. We will only examine the (unique) minimal flag-transitive automorphism group G of $Tg(Q)$ and describe the Borel subgroup B, the block stabilizer G_2 in G and the stabilizer $G_{0,1}$ of a point-line flag. The reader may easily obtain by himself further information on automorphism groups and parabolic subgroups, exploiting the description of $Tg(Q)$.

Henceforth, π will always denote the polarity associated to Q. When $Q = Q_3^-(q)$ with $q \equiv 1 \pmod 4$ and when $Q = Q_3^+(q)$ with $q \equiv 3 \pmod 4$, then there is an element $f \in PGL(3,q)$ leaving Q invariant and interchanging the two families P^+ and P^- of external points. We set $\phi = f\pi$.

We have 4 cases to examine.

$Q = Q_3^-(q)$ with $q \equiv 1 \pmod 4$.

We have $G = O_4^-(q)\langle\phi\rangle = L_2(q^2).2$ (for instance, when $q = 5$ we have $G = L_2(25).2_3$).

The stabilizer G_2 of a block acts faithfully as $PGL(2,q)$ on the $q+1$ points of that block. Whence the Borel subgroup is isomorphic to Z_{q-1}. The stabilizer of a point-line flag has the following structure: $G_{0,1} = [B]\langle\phi^a\rangle = Z_{2(q-1)}$, where ϕ^a is a suitable conjugate of ϕ, of period 4 and with $(\phi^a)^2 \in B$.

$Q = Q_3^-(q)$ with $q \equiv 3 \pmod 4$.

We now have $G = O_4^-(q) \times \langle\pi\rangle = L_2(q^2) \times 2$. As above, G_2 acts faithfully as $PGL(2,q)$ on the $q+1$ points of a block and $B = Z_{q-1}$. However, we now have $G_{0,1} = [B] \times Z_2$.

$Q = Q_3^+(q)$ with $q \equiv 3 \pmod 4$.

We have $G = (L_2(q) \times L_2(q))\langle\phi\rangle$, $G_2 = L_2(q)$ acting faithfully on the $q+1$ points of a block, $B = Z_{(q-1)/2}$ and $G_{0,1} = [B]\langle\phi^a\rangle = Z_{q-1}$, where $(\phi^a)^2 = -1$ and ϕ^a is a suitable conjugate of ϕ.

$Q = Q_3^+(q)$ with $q \equiv 1 \pmod 4$.

We have $G = (L_2(q) \times L_2(q)) \times \langle\pi\rangle$, $G_2 = L_2(q)$ acting faithfully on the $q+1$ points of a block, $B = Z_{(q-1)/2}$ and $G_{0,1} = [B] \times Z_2 = Z_{q-1}$.

Quotients. When $Q = Q_3^-(q)$ with $q \equiv 3 \pmod 4$ and when $Q = Q_3^+(q)$ with $q \equiv 1 \pmod 4$ we can factorize $Tg(Q)$ over π obtaining a flag-transitive quotient, which we denote by $Tg(Q)/2$. Trivially, the minimal flag-transitive automorphism group of $Tg(Q)/2$ is $O_4^-(q)$ or $L_2(q) \wr 2$, according to whether $Q = Q_3^-(q)$ or $Q = Q_3^+(q)$.

On the other hand, when $Q = Q_3^-(q)$ with $q \equiv 1 \pmod 4$ no flag-transitive proper quotients can be obtained. We do not know of any flag-transitive proper quotients of $Tg(Q)$ when $Q = Q_3^+(q)$ with $q \equiv 3 \pmod 4$, but for the exceptional case of $q = 3$, which will be examined later.

Special cases. Let $Q = Q_3^-(3)$. Then we have $Tg(Q) = \text{Cox}(1,2)$ and $Tg(Q)/2 = \text{Cox}(1,2)/2$. The geometry $Tg(Q)/2$ can also be constructed as follows: we take Q as set of points and Π^+ as set of blocks, the incidence relation being the natural one; the 1-faces are the pairs of distinct points in a common block.

Of course, we can repeat the above construction for $Tg(Q_3^-(3)/2)$ starting from any elliptic quadric $Q_3^-(q)$ (q odd) but we do not obtain an extended grid when $q \geq 5$; indeed we obtain an extended net of order $(q-1,(q-1)/2)$, which is nothing but 'a half' of the Möbius plane defined by $Q_3^-(q)$:

$$\overset{\displaystyle c}{\bullet} \underset{q-1}{\rule{0pt}{0pt}} \overset{\displaystyle N}{\underset{(q-1)/2}{\bullet}} \bullet$$

When $Q = Q_3^+(3)$ we have $Tg(Q) = \text{Sym}(4)$ (the smallest Minkowski plane). We know from subsection 2.2 that this geometry admits a flag-transitive (flat) proper quotient, namely $\text{Sym}(4)/2^2$.

When $Q = Q_3^+(5)$, the geometry $Tg(Q)$ is a 2-fold cover of the example obtained from $L_2(5)$ in [5], call it $\Gamma(L_2(5))$ (see also [6] (9.12(iii))); the reader may check that indeed we have $\Gamma(L_2(5)) = Tg(Q)/2$. It is worth recalling that the construction given in [5] for $\Gamma(L_2(5))$ exploits the following property of $L_2(5)$: the group $L_2(5)$ is a finite 2-transitive permutation group where the 2-point stabilizer has order 2 and fixes no additional points. It is proved in [6] (Lemma (9.13)) that S_4 and $L_2(5)$ are the only permutation groups with this property.

We will now give another construction for $Tg(Q_3^+(5))$, starting from $Q_3^+(4)$ instead of $Q_3^+(5)$ (needless to say that the exceptional isomorphism $L_2(4) \cong L_2(5)$ is involved in this possibility). Let $H = Q_3^+(4)$ embedded in $PG(3,4)$ and take the lines of $PG(3,4)$ not meeting H as points of the geometry Γ we are going to define. The points of $PG(3,4)$ not in H and the planes of $PG(3,4)$ secant for H will be taken as blocks of Γ. The incidence relation between points and blocks of Γ is defined by (symmetrized) containment. As 1-faces of Γ we take the (unordered) pairs of points of Γ in a common block (namely, intersecting pairs of lines of $PG(3,4)$ exterior to H). The reader may check by himself (comparing automorphism groups and parabolic subgroups, for instance) that the geometry Γ we have now constructed is nothing but another model for $Tg(Q_3^+(5))$.

Of course, the above construction could be repeated in general, starting from $H = Q_3^+(s)$ with $s = 2^m$, $m \geq 2$. We would obtain flag-transitive geometries belonging to the following diagram:

$$\overset{\displaystyle L}{\bullet} \underset{(s-2)/2\ s}{\rule{0pt}{0pt}} \overset{\rule{0pt}{0pt}}{\underset{1}{\bullet}} \bullet$$

2.5. Cameron-Fisher extensions of odd type

Let C be a quadratic cone in $PG(3, q)$, $q = 2^m$. Let v be the vertex of C and let R be the radical line of C, namely the line through v containing the nucleus of the conic $\pi \cap C$, for every plane π not on v. Let a and b be two distinct points in $R - \{v\}$.

We can define an extended grid Γ of order $q - 1$, as follows. $C - \{v\}$ is the set of points of Γ and the blocks of Γ are the planes of $PG(3, q)$ containing exactly one of a or b. The 1-faces of Γ are the (unordered) pairs of points in a common block.

We denote Γ by the symbol $CF^-(q)$. A motivation for this notation will be clear from the sequel.

Geometric properties. The Intersection Property holds in $CF^-(q)$ and it is evident that the set of blocks splits. Therefore $CF^-(q)$ admits unfolding. The point-graph has diameter $\delta = 2$, but $CF^-(q)$ is not determined by its point graph, except in the trivial case of $q = 2$ (note that $CF^-(2)$ is nothing but the Coxeter complex of type C_3). In fact, $CF^-(q)$ is an instance of a Cameron-Fisher extension of odd type, as described in Example (9.12) of [6].

Let us show how the construction of [6] (9.12) (see also [5]) can be recovered in the above. Let us identify the star of a in $PG(3, q)$ with the projective plane $\Pi = PG(2, q)$. Since the lines through a bijectively correspond to the points of C (v included), the points of $CF^-(q)$ can be viewed as points of Π. Let ∞ denote the point of Π corresponding to v (namely, ∞ is the radical line R of C). The planes of $PG(3, q)$ through a appear as lines in Π and those not containing v are lines of Π not through ∞. The q^2 planes of $PG(3, q)$ on b but not on a form a family of q^2 non-singular conics in Π, with nucleus ∞. Therefore $CF^-(q)$ can be realized in the dual affine plane $\Pi - \{\infty\}$, as in [6] and [5].

As we have remarked above, $CF^-(2)$ is the Coxeter complex of type C_3. Therefore, in the sequel we always assume $q \geq 4$, in order to avoid that trivial case.

Covers and quotients. It is proved in [3] that, for every proper divisor $d \neq 1$ of q, $CF^-(q)$ admits a d-fold cover. We denote it by $d \cdot CF^-(q)$. We have already remarked in Section 2.3 that $2 \cdot CF^-(4)$ is in fact the universal cover of $CF^-(4)$. We do not know if the same is true for $q/2 \cdot CF^-(q)$ in general, when $q \geq 8$. Proper quotients of $CF^-(q)$ can also be formed, but we do not if flag-transitive geometries can be obtained in this way (when $q \geq 4$).

Flag-transitive automorphism groups. Γ has just one minimal flag-transitive automorphism group, let us call it G. We have $G = 2^{2m+1} : L_2(q)$ $= (V : L_2(q)) : W$ where $V : L_2(q) = q^2 : L_2(q)$ is the natural 2-dimensional $GF(q)$-module for $L_2(q)$ and $W = Z_2$ is contributed by an automorphism ϕ

of C interchanging the two points a and b. Needless to say that, if $CF^-(q)$ is realized in a dual affine plane as explained above, then ϕ becomes a non-linear transformation. For instance, if the conics forming one of the two families of blocks are represented by equations of the following form:

$$x_3^2 + x_1 x_2 = u x_1^2 + v x_2^2,$$

with $u, v \in GF(q)$, then ϕ may act as follows:

$$\phi(x_1, x_2, x_3) = (x_1, x_2, x_3 + (x_1 x_2)^{q/2}).$$

The stabilizer G_2 of a block acts faithfully as $L_2(q)$ on the $q+1$ points of that block. Hence we have $B = Z_{q-1}$.

The reader may reconstruct by himself the structure of the full automorphism group of $CF^-(q)$.

Remark

Of course, the previous construction can be generalized starting from any set of lines of $PG(3, q)$ on a given point v, forming a hyperoval $\mathcal{I}$ in the star of v. Any of the lines forming $\mathcal{I}$ can be given the role that was of the radical line R of C. Call that line R, again. Then $\mathcal{I} - R$ will play the role of C. However, in order to have the flag-transitivity, we need an automorphism group of $\mathcal{I}$ transitive on $\mathcal{I} - R$. The only examples that we know of this kind are those arising from a cone C, giving rise to the geometries $CF^-(q)$ considered above.

Links with other topics. The construction we have given for $CF^-(q)$ can also be generalized taking, instead of two points in $R - \{v\}$, an orbit X on R of some subgroup of the automorphism group of C. In this way if $k+1$ is the size of X (and $k \geq 2$), we obtain flag-transitive extended nets of order $(q-1, k)$:

$$\overset{c}{\underset{q-1}{\bullet}} \quad \overset{N}{\underset{k}{\bullet}} \quad \bullet$$

In particular, when $X = R - \{v\}$ ($k = q - 1$) we obtain the Laguerre plane defined on C.

2.6. Cameron-Fisher extensions of even type

Let $\mathcal{I}$ be a set of lines on a point v of $PG(3, q)$ ($q = 2^m$, $m \geq 2$) forming a hyperoval in the star of v. For instance, we might have $\mathcal{I} = C \cup R$ where C is a cone of vertex v and R is its radical line.

Let L, M be two distinct lines of $\mathcal{I}$ and a, b points of L and M respectively, other than v. An extended grid Γ of order $q - 1$ can be defined as follows. The points of $PG(3, q)$ on $\mathcal{I} - (L \cup M)$ are the points of Γ. The blocks of Γ are the planes of $PG(3,q)$ not on v and containing just one of the points a or b. The 1-faces are the pairs of points of Γ in a common block.

We denote Γ by $\Gamma_{L,M}(\mathcal{I})$. When $\mathcal{I} - M$ (or $\mathcal{I} - L$) is a cone and M (respectively, L) is the radical line of that cone, then we write $CF^+(q)$ for $\Gamma_{L,M}(\mathcal{I})$.

When $q = 4$, we have $\Gamma_{L,M}(\mathcal{I}) \cong \mathrm{Sym}(4)$ (the reader may check this isomorphism directly, or he may obtain it from (8.3) of [6]). The properties of $\mathrm{Sym}(4)$ are known from subsection 2.2. They are fairly different from those of $\Gamma_{L,M}(\mathcal{I})$ when $q \geq 8$. Therefore, from now on we assume $q \geq 8$.

Geometric properties. The Intersection Property holds in $\Gamma_{L,M}(\mathcal{I})$ and the set of blocks splits. Hence $\Gamma_{L,M}(\mathcal{I})$ admits unfolding. The point graph has diameter $\delta = 2$, but $\Gamma_{L,M}(\mathcal{I})$ is not determined by its points graph (we recall that we have assumed $q \geq 8$).

$\Gamma_{L,M}(\mathcal{I})$ is a Cameron-Fisher extension of even type, in the sense of [6] (9.12(ii)). We can recognize the construction given in [6] (9.12(ii)) (see also [5]) by the following trick. Take a plane π on L not on M and project everything on π from b. Then the points of $\Gamma_{L,M}(\mathcal{I})$ are represented by the points of the affine plane $\pi - L$; the $q^2 - q$ blocks coming from planes on b are represented by the lines of π not containing any of the points v or a; the $q^2 - q$ blocks coming from planes on a are represented in the projective plane π by $q^2 - q$ hyperovals passing through both a and v. In particular, if $\mathcal{I} - M$ is a cone with radical line M, then these blocks form a family of $q^2 - q$ parabolas in the affine plane $\pi - L$.

Flag-transitivity. We do not know which conditions must be satisfied by $\mathcal{I}$, R and L, in general, in order to obtain the flag-transitivity for $\Gamma_{L,M}(\mathcal{I})$. We will only prove the flag-transitivity of $CF^+(q)$.

This property of $CF^+(q)$ is not clear at all if we consider the construction of $CF^+(q)$ in $PG(3,q)$ by means of a cone $\mathcal{C}$, of the radical line M of $\mathcal{C}$, a line L of $\mathcal{C}$ and points $a \in L$ and $b \in M$. However, we may consider the 'affine' model of $CF^+(q)$, obtained by projection from b onto a plane π on L not on M, as explained above. In this model the points of $CF^+(q)$ are the points of the affine plane $AG(2,q)$ and we can always assume that the two families of blocks of $CF^+(q)$, call them $\mathcal{B}^+$ and $\mathcal{B}^-$, consist of lines and parabolas respectively, represented by the following equations:

$y = rx + s$, $r, s \in GF(q)$, with $r \neq 0$ (family $\mathcal{B}^+$);
$y = rx^2 + s$, $r, s \in GF(q)$, with $r \neq 0$ (family $\mathcal{B}^-$).

It is now clear that what we need for the flag-transitivity of $CF^+(q)$ is a bijection of the set of points of $AG(2,q)$ interchanging the family of lines $\mathcal{B}^+$ with the family of parabolas $\mathcal{B}^+$. The following bijection does the job we need:

$$\sigma : (x,y) \longrightarrow (y, x^2).$$

Therefore, a flag-transitive automorphism group of $CF^+(q)$ exists, with the following structure: $G = ((AGL(1,q) \times AGL(1,q)) : Z_m).2$, where m is defined by the relation $2^m = q$.

The stabilizer G_2 of a block acts faithfully as $AGL(1,q) : Z_m$ on the q points of that block. The Borel subgroup is isomorphic to Z_m.

Quotients and covers. It is clear from the above that $CF^+(q)$ admits proper quotients (in particular, flat quotients); indeed, we can factorize over groups of translations of given direction of the affine plane in which $CF^+(q)$ is realized. However, we do not know any flag-transitive proper quotient of $CF^+(q)$ (except when $q = 4$, of course; but we have assumed $q \geq 8$).

As for covers, we do not know anything on them.

Links with other topics. The extended grids constructed in this section and in the previous one can also be embedded in the following larger 'Laguerre structure', call it $\mathcal{L}(\mathcal{I})$: take $\mathcal{I} - \{v\}$ as set of points and the planes not on v as blocks; pairs of points in the same block are 1-faces. $\mathcal{L}(\mathcal{I})$ belongs to the following diagram:

$$\overset{c}{\underset{q}{\bullet}} \rule{2cm}{0.4pt} \overset{Af^*}{\underset{q-1}{\bullet}} \bullet$$

where Af^* denotes the class of dual affine planes. When $q = 2$ or 4, then $\mathcal{L}(\mathcal{I})$ is flag-transitive.

2.7. An example in $PG(2m-1, 2)$

Let X be an m-dimensional subspace of $PG(2m-1, 2)$ ($m \geq 2$). We define a geometry $\overline{\Gamma}_m(2)$ over the set of types $\{+, 0, -\}$, as follows. The element of type 0 are the m-dimensional subspaces of $PG(2m-1, 2)$ that do not meet X. The elements of type $+$ (of type $-$) are the $m+1$-dimensional subspaces of $PG(2m-1, 2)$ (the $m-1$-dimensional subspaces of $PG(2m-1, 2)$) that meet X in just one point (that do not meet X). The incidence relation is the natural one, inherited from $PG(2m-1, 2)$. The geometry $\overline{\Gamma}_m(2)$ has diagram as follows:

$$\overset{c^*}{\underset{q-2}{\bullet}} \rule{1.5cm}{0.4pt} \overset{}{\underset{1}{\bullet}} \overset{c}{\underset{q-2}{\bullet}}$$

with $q = 2^m - 2$. Let $\Gamma_m(2)$ be the shadow geometry of $\overline{\Gamma}_m(2)$ with respect to the central node of the above diagram. The geometry $\Gamma_m(2)$ is an extended grid of order $q - 2$.

Note that $\Gamma_2(2) = \mathrm{Sym}(4)$.

Geometric properties. The Intersection Property holds in $\Gamma_m(2)$ and $\Gamma_m(2)$ is determined by its point-graph. The point-graph of $\Gamma_m(2)$ has diameter $\delta = 2m - 2$.

Flag-transitivity. The type-preserving automorphism group of $\bar{\Gamma}_2(m)$ is the stabilizer $2^{m^2} : (L_m(2) \times L_m(2))$ of X in $L_{2m}(2)$. The projective geometry $PG(2m - 1, 2)$ admits a non-degenerate polarity π fixing X. That polarity defines an automorphism of $\bar{\Gamma}_m(2)$ permuting the two types $+$ and $-$. Hence π is an automorphism of $\Gamma_m(2)$ and we have

$$\text{Aut}(\Gamma_m(2)) = 2^{m^2} : (L_m(2) \times L_m(2)).\langle \pi \rangle = 2^{m^2} : (L_m(2) \times L_m(2)).2$$

acting flag-transitively in $\Gamma_m(2)$. The full automorphism group of $\Gamma_m(2)$ is its unique flag-transitive automorphism group.

The stabilizer of a block of Γ is $2^{2m-1} : (L_{m-1}(2) \times L_m(2))$, acting as $ASL(m, 2) = 2^m : L_m(2)$ on the 2^m points of that block.

Links with other topics. The above construction can be repeated in $PG(2m, q)$, for every prime power q, producing a flag-transitive geometry $\Gamma_m(q)$ belonging to the following diagram:

$$\underset{q-1}{\bullet} \overset{L}{\underset{s}{\rule{3em}{0.4pt}}} \overset{}{\underset{}{\rule{3em}{0.4pt}}} \underset{1}{\bullet}$$

with $s + 1 = (q^m - 1)/(q - 1)$. The unfolding $\bar{\Gamma}_m(q)$ of $\Gamma_m(q)$ is a truncation of a geometry of rank $2m - 1$ belonging to the following diagram (see [10], Example 6):

Remark

The above example has been suggested to us by E. Shult (private communication).

Problem

We have $\Gamma_2(2) = \text{Sym}(4)$. Hence $\Gamma_2(2)$ admits a flat flag-transitive quotient (subsection 2.2). Is the same true in $\Gamma_m(2)$ if $m > 2$?

2.8. A flat family

Let $G_2 = AGL(1, q)$, with $q = 2^m$ and $m \geq 2$. Let $S = Z_{q-1} \leq G_2$ and let Z be another copy of Z_{q-1} with $Z \cap G_2 = 1$ and let us form the direct product $A = G_2 \times Z$.

Let t be an involutory automorphism of A such that $\langle S, S^t \rangle = SS^t = S \times Z$. Trivially, t normalizes the (elementary abelian) Sylow 2-subgroup of G_2, say V. Therefore it centralizes some non-trivial element $v \in V$.

Let us set $G_0 = (S \times Z)\langle t \rangle$, $G_1 = \langle v, t \rangle$ and $G = (G_2 \times Z)\langle t \rangle$. Define $P_i = G_j \cap G_k$, for $\{i, j, k\} = \{0, 1, 2\}$. It is easily seen that P_0, P_1, P_2 form a parabolic system in G, with trivial Borel subgroup $B = 1$. The associated chamber system is a geometry, in fact an extended grid of order $q - 2$, let us call it Γ. This extended grid is flat.

Furthermore, we have $G_i = \langle P_j, P_k \rangle$, for $\{i, j, k\} = \{0, 1, 2\}$. Namely, G_0, G_1, G_2 are stabilizers of a point, a line and a block of Γ, respectively, in the automorphism group G of Γ. In particular, the stabilizer of a block acts faithfully as $AGL(1, q)$ on the q points of that block.

Special cases. The flat quotient $\mathrm{Sym}(4)/2^2$ (see subsection 2.2) is evidently an instance of the above. In this case we have $G_2 = AGL(1, q) = A_4$, $Z = Z_3$ and t can be chosen to act as follows: if $u \in V$, $x \in S$, $y \in Z$ and α is an isomorphism from S to Z, then t maps uxy onto $u^2 x^2 \alpha(x^2) y^2$.

A similar choice can be made for t whenever m is even, say $m = 2k$. Indeed in that case we can define t as follows:

$$t : uxy \longrightarrow u^{2^k} x^{2^k} \alpha(x^{2^k}) y^{-1}$$

where $u \in V$, $x \in S$, $y \in Z$ and α is the isomorphism from S to Z, as above.

Therefore, a flat extended grid of order $q - 2$ exists for every $q = 2^m$, m even.

Problems

(1) How to construct an element t with the required properties when $q = 2^m$ with m odd ?

(2) We know only one flag-transitive flat extended grid that cannot be obtained by a construction as above, namely the proper quotient of the C_3 Coxeter complex. Are there other examples ? We will prove in [13] that the answer is negative if the automorphism group G is assumed to be solvable.

(3) What about covers of flat extended grids obtained as above ?

3.   Summary

We finish this paper summarizing some information on the examples of the previous section. For every example or family of examples we recall its name, the rank, the order, the size of the blocks and the action on blocks, but only considering stabilizers taken in minimal flag-transitive automorphism groups. We omit to mention flag-transitive quotients and covers, even when we know they exist.

$\mathrm{Cox}(n, s)$. Rank $n + 2$, order s. Blocks: size $n + s + 1$, action S_{n+s+1}.

$\mathrm{Sym}(n + 3)$. Rank $n + 2$, order 2. Blocks: size $n + 3$, action A_{n+3}.

$c^2 . CF^-(4)$. Rank 5, order 3. Blocks: size 7, action A_7.

$c . CF^-(4)$. Rank 4, order 3. Blocks: size 6, action A_6.

$Tg(Q)$. Rank 3, order $q - 1$ (q odd prime power). Blocks: size $q + 1$; action: $PGL(2, q)$ if $Q = Q_3^-(q)$ and $L_2(q)$ if $Q = Q_3^+(q)$.

$CF^-(q)$. Rank 3, order $q - 1$ ($q = 2^m$). Blocks: size $q + 1$, action $L_2(q)$.

$CF^+(q)$. Rank 3, order $q - 2$ ($q = 2^m$). Blocks: size q, action $AGL(1, q)$.

$\Gamma_m(2)$. Rank 3, order $q - 2$ ($q = 2^m$). Blocks: size q, action $ASL(m, 2)$.

Flat examples (subsection 2.7). Rank 3, order $q - 2$ ($q = 2^m$). Blocks: size q, action $AGL(1, q)$.

References

[1] A. Blokhuis and A. E. Brouwer. Locally 4-by-4 grid graphs. *J. Graph Theory*, 13, pp. 229–244, 1989.

[2] F. Buekenhout. Diagram geometries for sporadic groups. In J. McKay, editor, *Finite Groups Coming of Age*, volume 45 of *A.M.S. Series Contemp. Math.*, pages 1–32, 1985.

[3] P. J. Cameron. Covers of graphs and *EGQ*s, 1992. To appear in *Discrete Math.*

[4] P. J. Cameron and M. Deza. On permutation geometries. *J. London Math. Soc.*, 20, pp. 373–386, 79.

[5] P. J. Cameron and P. Fisher. Small extended generalized quadrangles. *European J. Combin.*, 11, pp. 403–413, 1990.

[6] P. J. Cameron, D. R. Hughes, and A. Pasini. Extended generalized quadrangles. *Geom. Dedicata*, 35, pp. 193–228, 1990.

[7] J. H. Conway, R. T. Curtis, S. P. Norton, R. A. Parker, and R. A. Wilson. *Atlas of Finite Groups*. Clarendon Press, Oxford, 1985.

[8] A. Del Fra, D. Ghinelli, T. Meixner, and A. Pasini. Flag-transitive extensions of C_n-geometries. *Geom. Dedicata*, 37, pp. 253–273, 1991.

[9] P. Fisher. Extended 4×4 grids. *European J. Combin.*, 12, pp. 383–388, 1991.

[10] D. R. Hughes. On some rank 3 partial geometries. In J. W. P. Hirschfeld, D. R. Hughes, and J. A. Thas, editors, *Advances in Finite Geometries and Designs*, pages 195–225, Oxford, 1991. Oxford University Press.

[11] T. Meixner. Two geometries related to the sporadic groups Co_1 and Co_3. To appear in *J. Algebra*.

[12] T. Meixner. Some polar towers. *European J. Combin.*, 12, pp. 397–416, 1991.

[13] T. Meixner and A. Pasini. Some classification for extended grids. In preparation.

[14] A. Pasini. Cover of finite geometries with non-spherical minimal circuit diagram. In *Buildings and the Geometry of Diagrams*, volume 1181 of *Lecture Notes in Math.*, pages 218–241. Springer Verlag, 1986.

[15] **A. Pasini.** Geometric and algebraic methods in the classification of geometries belonging to Lie diagrams. *Ann. Discrete Math.*, 37, pp. 315–356, 1988.

[16] **A. Pasini.** On extended grids with large automorphism groups. *Ars Combin.*, 29b, pp. 65–83, 1990.

[17] **S. Rinauro.** On some extensions of generalized quadrangles of grid type. *J. Geom.*, 38, pp. 158–164, 1990.

[18] **M. A. Ronan.** Covers of certain finite geometries. In P. J. Cameron, J. W. P. Hirschfeld, and D. R. Hughes, editors, *Finite Geometries and Designs*, volume 49 of *London Math. Soc. Lecture Note Ser.*, pages 316–331, Cambridge University Press, 1981.

[19] **R. Scharlau.** Geometric realizations of shadow geometries. *Proc. London Math. Soc.(3)*, 61, pp. 615–656, 1990.

T. Meixner, Math. Inst., Justus Liebig Univ., Arndtstrasse 2, Giessen, Germany. e-mail: thmxmbg@dgihrz01.bitnet

A. Pasini, Dept. Math., Univ. Siena, Via del Capitano 15, Siena, Italy. e-mail: pasini@sivax.cineca.it

Root lattice constructions of ovoids

G. E. Moorhouse

Abstract

Recently the author [5] has constructed new ovoids in $O_8^+(p)$ for p prime, using the E_8 root lattice, generalising a construction of Conway et al. [1]. Here we present a nine-dimensional lattice which greatly simplifies the description of these ovoids.

1. Introduction

An *orthogonal space* is a vector space V equipped with a quadratic form Q. We consider only finite-dimensional vector spaces over a finite field $F = GF(q)$. A *singular point* in such a space is a 1-dimensional subspace $\langle v \rangle$ such that $Q(v) = 0$. Usually we take Q to be nondegenerate, in which case (V, Q) is called an $O_{2m-1}(q)$-*space* if $\dim V = 2m-1$, or an $O_{2m}^\pm(q)$-*space* if $\dim V = 2m$, using superscript $+$ or $-$ according as Q has Witt defect 0 or 1. An *ovoid* in an orthogonal space (V, Q) is a set $\mathcal{O}$ consisting of singular points, such that every maximal totally singular subspace of V contains a unique point of $\mathcal{O}$. In a space of type $O_{2m}^+(q)$, $O_{2m-1}(q)$ or $O_{2m-2}^-(q)$, an ovoid is equivalently defined (see [3], [7]) as a set of $q^{m-1} + 1$ singular points of which no two are orthogonal. Ovoids are not known to exist in orthogonal spaces of 9 or more dimensions. Ovoids in $O_3(q)$ and in $O_4^-(q)$ necessarily consist of all singular points; viewed projectively, these are nondegenerate plane conics and elliptic quadrics in projective 3-space. We emphasise that the latter are discrete analogues of classical round objects in Euclidean space, and so the name 'ovoid' seems well-deserved. Ovoids in $O_6^+(q)$ (including ovoids in $O_5(q)$ as a special case under the natural embedding) are equivalent (see [4]) to translation planes of order q^2 with kernel containing $GF(q)$. These are known to exist in great abundance, and in general do not appear to originate from any Euclidean 'round' objects.

The known ovoids in $O_8^+(q)$ are listed in [4], [1] and [5]. The majority of these occur in $O_8^+(p)$ for p prime, and are constructed by taking lattice points on a certain Euclidean sphere, then reducing modulo p, as we shall describe in Sections 2 and 3. It is intriguing that such discrete geometric objects would appear to owe their existence to properties of the Euclidean metric (seemingly requiring the Cauchy-Schwarz inequality in $\mathbb{R}^8$ or $\mathbb{R}^9$), and again justice is done to the term 'ovoid'.

2. An Eight-Dimensional Description

We first indicate, without proof, the ovoid construction from 8-dimensional lattices. This description remains the most useful for computer implementation.

Let E be the root lattice of type E_8; that is, E consists of all vectors $\frac{1}{2}(a_1, a_2, \ldots, a_8)$ with $a_i \in \mathbb{Z}$ such that $a_1 \equiv a_2 \equiv \cdots \equiv a_8 \bmod 2$ and $\sum a_i \equiv 0 \bmod 4$. A detailed description of E, including the following properties, may be found in [2]. Let p be any prime. Then $\overline{E} = E/pE$ is an 8-dimensional vector space over $F = GF(p)$, and for $v \in E$ we write $\overline{v} = v + pE \in \overline{E}$. We call $\|v\|^2$ the *norm* of $v \in E$, and since E is an even lattice, $\|v\|^2 \in 2\mathbb{Z}$. For any positive integer m, the number of vectors in E of norm $2m$ is $240\sigma_3(m) = 240 \sum d^3$, summing over all positive integers d dividing m. In particular E has 240 vectors of norm 2, the *root vectors* of E. Define $Q : \overline{E} \to F$ by $Q(\overline{v}) = \frac{1}{2}\|v\|^2 \bmod p$. Then Q is a nondegenerate quadratic form on $\overline{E}$ with Witt defect 0, and Q is preserved by the Weyl group $W = W(E_8)$.

The *binary ovoids* of Conway et al. [1] are defined in $\overline{E}$ for p odd by

$$\mathcal{O}_{2,p}(x) = \mathcal{O}_{2,p}(\mathbb{Z}x + 2E) = \left\{ \langle \overline{v} \rangle \ : \ \|v\|^2 = 2p, \ v \in \mathbb{Z}x + 2E \right\}$$

where $x \in E$ such that $\frac{1}{2}\|x\|^2$ is odd. The sphere of norm $2p$ (radius $\sqrt{2p}$, centre 0) has exactly $2(p^3 + 1)$ points of the lattice $\mathbb{Z}x + 2E$, and these occur in $p^3 + 1$ antipodal pairs. Reducing modulo p, we obtain $p^3 + 1$ points (one-dimensional subspaces) $\langle \overline{v} \rangle$, which are singular since $Q(\overline{v}) \equiv \frac{1}{2}(2p) \equiv 0 \bmod p$. Moreover [1] no two points of $\mathcal{O}_{2,p}(x)$ are orthogonal, so $\mathcal{O}_{2,p}(x)$ is an ovoid. Since there are just 120 choices of sublattice $\mathbb{Z}x + 2E \subset E$ with $\frac{1}{2}\|x\|^2$ odd, all equivalent under W, we obtain 120 binary ovoids in $O_8^+(p)$, all of which are equivalent. We may take $x \in E$ to be our favourite root vector, and then the stabiliser $W_x \cong W(E_7) \cong 2 \times Sp_6(2)$ acts on the ovoid $\mathcal{O}_{2,p}(x)$. (Remark: if $x \in E$ is a root vector, then $\mathbb{Z}x + 2E = \mathbb{Z}x \oplus 2E_7^*$ where E_7^* is the dual of $E_7 = E \cap x^\perp$ in $x^\perp$. Thus the binary ovoids are computable from a knowledge [2] of the 'shells' of E_7^*.)

More generally, for primes $r \neq p$ we define

$$\mathcal{O}_{r,p}(x) = \mathcal{O}_{r,p}(\mathbb{Z}x + rE) = \bigcup_{1 \le i \le \lfloor \frac{r}{2} \rfloor} \left\{ \langle \overline{v} \rangle \ : \ \|v\|^2 = 2i(r-i)p, \ v \in \mathbb{Z}x + rE \right\}$$

where $x \in E$ such that $\frac{p}{2}\|x\|^2$ is a nonzero square modulo r. If $r > p$, it sometimes happens that $\mathcal{O}_{r,p}(x) = \{\langle 0 \rangle\}$, but in all other cases $\mathcal{O}_{r,p}(x)$ is an ovoid in $\overline{E}$, called an *r-ary ovoid* in $O_8^+(p)$. In Section 3 we will see an explanation for the 'failed ovoids' of the form $\{\langle 0 \rangle\}$. The cases $r \in \{2, 3\}$ give the binary

270

and ternary ovoids of Conway et al. [1]; for general r the above definition is a slight simplification of that given in [5]. By varying the choices of r and x, we expect from the computational evidence available that the number of isomorphism classes of r-ary ovoids in $O_8^+(p)$ is unbounded as $p \to \infty$, but this has not been proven.

The above definition of $\mathcal{O}_{r,p}(x)$ requires that we take lattice points on a union of $\lfloor \frac{r}{2} \rfloor$ spheres in $\mathbb{R}^8$. In Section 3 we shall interpret these spheres as hyperplane sections of a single sphere in $\mathbb{R}^9$, achieving a more concise definition of $\mathcal{O}_{r,p}(x)$ and a simplified proof that in fact we obtain ovoids.

3. A Nine-Dimensional Description

Throughout this section, r and p are distinct *odd* primes, which allows for a simpler presentation. The industrious reader will find that our presentation may be adapted to the general case; however the case $r = 2$ has already been treated by the description of the binary ovoids in Section 2, and the case $p = 2$ is trivial since $O_8^+(2)$ has a unique ovoid.

For each odd prime p, define a nine-dimensional Euclidean lattice by

$$\Lambda = \Lambda(p) = \sqrt{2}E \oplus \sqrt{p}\mathbb{Z}.$$

That is, Λ consists of vectors $\sqrt{2}e + \lambda z$ with $e \in E$ and $\lambda \in \mathbb{Z}$, where $z = (0, 0, \ldots, 0, \sqrt{p})$, and $\|\sqrt{2}e + \lambda z\|^2 = 2\|e\|^2 + p\lambda^2$. Note that Λ admits a group of isometries $G \cong 2 \times W$ generated by $W = W(E_8)$ acting naturally on the first eight coördinates and fixing z, together with the reflection in the hyperplane $z^\perp = \langle E \rangle$.

Now let r be an odd prime distinct from p. The quotients $\Lambda/p\Lambda$ and $\Lambda/r\Lambda$ are 9-dimensional vector spaces over $GF(p)$ and $GF(r)$, respectively. Each inherits from Λ a G-invariant quadratic form obtained by reducing $2\|e\|^2 + p\lambda^2 \in \mathbb{Z}$ modulo the corresponding prime. The quotient $\Lambda/r\Lambda$ is a (non-degenerate) $O_9(r)$-space.

However, the orthogonal space $\overline{\Lambda} = \Lambda/p\Lambda$ is degenerate, consisting of an $O_8^+(p)$-space over a 1-dimensional radical $\langle \overline{z} \rangle = \langle z + p\Lambda \rangle$; projectively, $\Lambda/p\Lambda$ is a 'hyperbolic cone over a point'. From the definition given in Section 1, we see that two types of ovoids are possible in $\overline{\Lambda} = \Lambda/p\Lambda$:

(i) The singleton $\{\langle \overline{z} \rangle\}$ is an ovoid in $\overline{\Lambda}$ since every maximal totally singular subspace of $\overline{\Lambda}$ is 5-dimensional and contains $\langle \overline{z} \rangle$. We call this the *degenerate ovoid* of $\overline{\Lambda}$.

(ii) Any set $\mathcal{O}$ consisting of $p^3 + 1$ mutually nonperpendicular singular points of $\overline{\Lambda}$ is an ovoid in $\overline{\Lambda}$. Such an ovoid does not contain $\langle \overline{z} \rangle$ and is called *nondegenerate*. For such an ovoid, $\{\langle \overline{v} \rangle + \langle \overline{z} \rangle \;:\; \langle \overline{v} \rangle \in \mathcal{O}\}$ is an ovoid in the $O_8^+(p)$-space $\overline{\Lambda}/\langle \overline{z} \rangle$, and conversely, ovoids in $O_8^+(p)$ lift to ovoids in $\overline{\Lambda}$.

Our construction in fact gives ovoids in $\overline{\Lambda} = \Lambda/p\Lambda$ of both types (although degenerate ovoids never occur for $r < p$), and thereby ovoids in $O_8^+(p)$ as described in (ii) above. Let π_r and π_p denote the natural maps from Λ to points of $\Lambda/r\Lambda$ and $\Lambda/p\Lambda$ respectively. That is, for $v \in \Lambda \smallsetminus r\Lambda$, we have $\pi_r(v) = \langle v + r\Lambda \rangle \leq \Lambda/r\Lambda$, and similarly for p in place of r. Consider the points of the lattice Λ which lie on the sphere of radius $r\sqrt{p}$, other than the 'poles' $\pm rz$, denoted thus:

$$\Lambda_{r^2p} = \left\{ v \in \Lambda \ : \ \|v\|^2 = r^2p \right\} \smallsetminus \left\{ \pm rz \right\}.$$

Our main result, as follows, will be proven later in this section.

Theorem 3.1   (i) $\pi_r(\Lambda_{r^2p})$ *is the set of singular points of $\Lambda/r\Lambda$ outside the hyperplane $H = \pi_r(E)$.*

   (ii) *Let $X = \langle x + r\Lambda \rangle$ be a singular point of $\Lambda/r\Lambda$ outside H, and let $\mathcal{X} = \left\{ v \in \Lambda_{r^2p} \ : \ \pi_r(v) = X \right\}$. Then $\pi_p(\mathcal{X})$ is an ovoid of $\Lambda/p\Lambda$.*

   (iii) *An ovoid of the form $\pi_p(\mathcal{X})$ as in (ii) is nondegenerate whenever $r < p$. If $r > p$ then $\pi_p(\mathcal{X})$ is nondegenerate for some $X, \mathcal{X}$.*

The situation of Theorem 3.1 may be appreciated from Figure 1, where typical points of the quadric in $\Lambda/r\Lambda$ outside the hyperplane H, are denoted by $\bullet$, $*$ and $\diamond$. These points are lifted back to the sphere Λ_{r^2p} and then projected down to the degenerate quadric in $\Lambda/p\Lambda$, obtaining in each case an ovoid, although the ovoid obtained from $\diamond$ is degenerate. Observe that, as pictured, the lattice points in Λ_{r^2p} lie on certain hyperplanes of $\mathbb{R}^9$ parallel to $z^\perp$.

We further illustrate the construction with an example in which $p = 3$ and $r = 5$. Now $\Lambda_{r^2p} = \left\{ \sqrt{2}e \pm z \ : \ e \in E, \ \|e\|^2 = 36 \right\} \cup \left\{ \sqrt{2}e \pm 3z \ : \ e \in E, \ \|e\|^2 = 24 \right\}$. For $X = \pi_5\left(\sqrt{2}(2^6, 0^2) + 3z \right)$ we obtain

$$\mathcal{X} = \left\{ \pm\left(\sqrt{2}(2^6; 0^2) + 3z \right), \ \pm\left(\sqrt{2} \cdot \tfrac{1}{2}(-7, 3^5; -5, 5) + z \right), \right.$$
$$\left. \pm\left(\sqrt{2}(4^2, -1^4; 0^2) + z \right), \ \dots \right\}$$

where '...' denotes similar vectors obtained by arbitrarily permuting the first six coördinates of E, and permuting the last two coördinates of E. Then $|\mathcal{X}| = 56$ and $\mathcal{X}$ projects to a nondegenerate ovoid of size 28 in $\Lambda/3\Lambda$, antipodal points of $\mathcal{X}$ giving the same ovoid point. Choosing $X' = \pi_5\left(\sqrt{2}(6, 0^7) + z \right)$, however, we obtain $\mathcal{X}' = \left\{ \pm\left(\sqrt{2}(6, 0^7) + z \right) \right\}$, which projects to the degenerate ovoid of $\Lambda/3\Lambda$.

Observe that by definition if $u = \sqrt{2}e + \lambda z \in \Lambda_{r^2p}$ then $\|u\|^2 = 2\|e\|^2 + p\lambda^2 = r^2p$, which implies that $|\lambda| < r$ and λ is odd, so that $\pi_r(u)$ is a singular

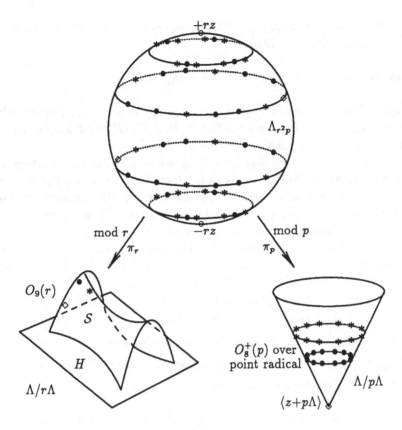

Figure 1. Two Projections of $\Lambda_{r^2 p}$

point of $\Lambda/r\Lambda$ which does not lie in the hyperplane $H = \pi_r(E)$; this proves half of conclusion (i) of Theorem 3.1.

Lemma 3.2 *If $u \cdot v \equiv 0 \bmod p$ for some $u, v \in \Lambda_{r^2 p}$ such that $\pi_r(u) = \pi_r(v)$, then $u = \pm v$.*

Proof. The hypotheses imply that $u - \alpha v \in r\Lambda$ for some $\alpha \in \mathbb{Z}$ not divisible by r. Thus $2\alpha u \cdot v = \|u\|^2 + \alpha^2 \|v\|^2 - \|u - \alpha v\|^2 \equiv 0 \bmod r^2$, so $u \cdot v \equiv 0 \bmod r^2$. Also $u \cdot v \equiv 0 \bmod p$ by hypothesis, so $u \cdot v \equiv 0 \bmod r^2 p$. But $|u \cdot v| \leq \|u\| \|v\| = r^2 p$ by Cauchy-Schwarz, so $|u \cdot v| = 0$ or $r^2 p$. If $|u \cdot v| = r^2 p$, then again by Cauchy-Schwarz, $u = \pm v$ and we are done. Otherwise $u \cdot v = 0$. But it is easy to see that $u \cdot v$ must be odd. For we have $u = \sqrt{2}e + \lambda z$, $v = \sqrt{2}e' + \mu z$ for some $e, e' \in E$ and odd integers λ, μ satisfying $2\|e\|^2 + p\lambda^2 = 2\|e'\|^2 + p\mu^2 = r^2 p$; thus $u \cdot v = 2e \cdot e' + p\lambda\mu \equiv 1 \bmod 2$, a contradiction. $\square$

Define $\Lambda'_{r^2p} = \Lambda_{r^2p} \cap \left(p\Lambda + \mathbb{Z}z\right)$, the set of all vectors in Λ_{r^2p} which project to the radical of $\Lambda/p\Lambda$.

Lemma 3.3 $|\Lambda_{r^2p}| + p^3|\Lambda'_{r^2p}| = 2r^3(r^4 - 1)(p^3 + 1)$.

Proof. This is proven in exactly the same way as Lemma 2.4 of [5], using the multiplicativity of σ_3, and the fact [6] that $E \oplus E$ has $480\sigma_7(m)$ vectors of norm $2m$ for every positive integer m. □

A *cap* in an orthogonal space is a set of singular points which are mutually nonperpendicular. Any cap in $O_8^+(p)$ has size at most $p^3 + 1$, and caps attaining this maximum size are ovoids (see [3], [7]). Consequently, caps in $\Lambda/p\Lambda$ have size at most $p^3 + 1$, and caps attaining this maximum size are nondegenerate ovoids; the radical point is a maximal cap of size 1.

Let S be the set of singular points of $\Lambda/r\Lambda$. Well-known counting arguments give $|S| = (r^8 - 1)/(r - 1)$ and $|S \cap H| = (r^3 + 1)(r^4 - 1)/(r - 1)$ since the hyperplane H is of type $O_8^+(r)$; thus $|S \smallsetminus H| = |S| - |S \cap H| = r^3(r^4 - 1)$. By Lemma 3.2, for each point $X \in S \smallsetminus H$, its preimage $\mathcal{X} = \{v \in \Lambda_{r^2p} : \pi_r(v) = X\}$ (which could conceivably be empty) projects to a cap $\pi_p(\mathcal{X})$ of $\Lambda/p\Lambda$. Also $|\pi_p(\mathcal{X})| = |\langle \bar{z} \rangle| = 1$ if $\mathcal{X} \subseteq \Lambda'_{r^2p}$; otherwise $\mathcal{X} \subseteq \Lambda_{r^2p} \smallsetminus \Lambda'_{r^2p}$ and $0 \le |\pi_p(\mathcal{X})| \le p^3 + 1$. Furthermore, Lemma 3.2 shows that the projection $\mathcal{X} \to \pi_p(\mathcal{X})$ is two-to-one. Therefore

$$\left|\Lambda_{r^2p}\right| - \left|\Lambda'_{r^2p}\right| = \left|\Lambda_{r^2p} \smallsetminus \Lambda'_{r^2p}\right| = \sum_{\substack{X \in S - H: \\ \mathcal{X} \subseteq \Lambda_{r^2p} - \Lambda'_{r^2p}}} |\mathcal{X}| \le \sum_{\substack{X \in S - H: \\ \mathcal{X} \subseteq \Lambda_{r^2p} - \Lambda'_{r^2p}}} 2(p^3 + 1)$$

$$= 2(p^3 + 1)\left|S \smallsetminus H\right| - 2(p^3 + 1)\left|\left\{X \in S \smallsetminus H : \mathcal{X} \subseteq \Lambda'_{r^2p}\right\}\right|$$

$$= 2(p^3 + 1)r^3(r^4 - 1) - (p^3 + 1)\left|\Lambda'_{r^2p}\right|,$$

in which equality holds by Lemma 3.3. Therefore $\left|\pi_p(\mathcal{X})\right| = p^3 + 1$ whenever $\mathcal{X} \subset \Lambda_{r^2p} \smallsetminus \Lambda'_{r^2p}$, thereby proving (i) and (ii) of Theorem 3.1. It is clear that $\Lambda'_{r^2p} = \emptyset$ whenever $r < p$, and that in any case $\Lambda_{r^2p} \supsetneq \Lambda'_{r^2p}$, whence not all ovoids $\pi_p(\mathcal{X})$ are degenerate, so (iii) follows as well, completing the proof of Theorem 3.1.

One checks without difficulty that for $x = \sqrt{2}e + \lambda z \in \Lambda_{r^2p}$, the ovoid of $\bar{\Lambda} = \Lambda/p\Lambda$ constructed from x as in Theorem 3.1, projects to the ovoid $\mathcal{O}_{r,p}(e)$ of $\bar{\Lambda}/\bar{z}$ described in Section 2, in the nondegenerate case ($e \notin pE$).

4. Further Remarks

Let X, $\mathcal{X}$, etc. be as in Theorem 3.1, and as before, let $G \cong 2 \times W$ be the isometry group of Λ, having natural orthogonal representations on both $\Lambda/r\Lambda$

and on $\Lambda/p\Lambda$. The stabiliser G_X acts on the ovoid $\pi_r(\mathcal{X})$, with kernel of order 2 or 4 in the nondegenerate case. In general, however, the stabilisers of these ovoids in the full orthogonal group, remain undetermined; cf. [5].

It is disappointing that the r-ary ovoid construction does not seem to work in $O_8^+(p^e)$ for $e > 1$. This contrasts sharply with the situation in $O_6^+(p^e)$, where ovoid constructions generally proliferate as e increases. The problem with $O_8^+(p^e)$ is more than a lack of inspiration: although $O_8^+(p)$ has at least one $Sp_6(2)$-invariant ovoid for every odd prime p (say, $\mathcal{O}_{2,p}\left(\frac{1}{2}(1^8)\right)$, and evidently many more as p increases), we have checked that no $Sp_6(2)$-invariant ovoids exist in $O_8^+(p^e)$ for $p^e \in \left\{2^2, 2^3, 2^4, 3^2, 3^3, 5^2\right\}$. (For $p^e = 9$, this is proven in [1].)

Can variations of the above constructions give new ovoids from other lattices, or perhaps even nonexistence results for higher-dimensional ovoids? Certainly any ovoid may be lifted back from L/pL to a lattice L, with great freedom in the choice of lifting and of L itself. We cannot expect all such preimages to be as elegant as the spheres arising in our construction; nevertheless can it be shown that every ovoid lifts to some subset of a lattice with high density? And could the apparent lack of ovoids in $O_{10}^+(q)$ be due to a lack of a suitably dense lattice packing in $\mathbb{R}^{10}$? These are mere speculations.

References

[1] **J. H. Conway, P. B. Kleidman, and R. A. Wilson**. New families of ovoids in O_8^+. *Geom. Dedicata*, 26, pp. 157–170, 1988.

[2] **J. H. Conway and N. J. A. Sloane**. *Sphere Packings, Lattices and Groups*. Springer Verlag, Berlin, New York, 1988.

[3] **J. W. P. Hirschfeld and J. A. Thas**. *General Galois Geometries*. Oxford University Press, Oxford. Oxford Science Publications, 1991.

[4] **W. M. Kantor**. Ovoids and translation planes. *Canad. J. Math.*, 34, pp. 1195–1207, 1982.

[5] **G. E. Moorhouse**. Ovoids from the E_8 root lattice. To appear in *Geom. Dedicata*.

[6] **J. P. Serre**. *A Course in Arithmetic*. Springer Verlag, Berlin, New York, 1973.

[7] **J. A. Thas**. Ovoids and spreads of finite classical polar spaces. *Geom. Dedicata*, 10, pp. 135–144, 1981.

G. E. Moorhouse, Department of Mathematics, University of Wyoming, Laramie, WY 82071–3036, U.S.A. e-mail: eric@silver.uwyo.edu

Coxeter groups in Coxeter groups

B. Mühlherr

Abstract

An automorphism of a Coxeter diagram M leads in a natural way to a Coxeter subgroup of the Coxeter group of type M. We introduce admissible partitions of Coxeter diagrams in order to generalize this situation. An admissible partition of a Coxeter diagram provides a Coxeter subgroup in a similar way. Our main result is a local criterion for the admissibility of a partition.

1. Introduction

We may ask in general which Coxeter groups arise as subgroups of a given Coxeter group. This question is of course far too general. However, there are Coxeter groups which arise canonically as subgroups of a given Coxeter group. Let for instance (W, S) be a Coxeter system and let S_1 be a subset of S, then $(\langle S_1 \rangle, S_1)$ is again a Coxeter system.

Our purpose here is to introduce another way to obtain Coxeter subgroups in a given Coxeter group. In the example above we considered *residues*; the procedure, which will be treated here, has also a geometric background. We will deal with subcomplexes of the Coxeter complex which behave like subcomplexes fixed by a polarity. We do not go into the details concerning these geometric aspects. However, our procedure is motivated by the following consideration:

Let I be a set, let M be a Coxeter diagram over I and let (W, S) be the associated Coxeter system. Let $l : W \longrightarrow N_0$ denote the length function. Let Γ be a group of automorphism of M and let $\tilde{I}$ be the set of orbits of Γ in I. Assume that for each $\alpha \in \tilde{I}$ we have that $M_\alpha := M\mid_{\alpha \times \alpha}$ is spherical and let r_α denote the element of maximal length in $W_\alpha = \langle s_i | i \in \alpha \rangle$. The group Γ acts on W as a group of automorphisms which preserves the length. It is easy to see that r_α is fixed under the action of Γ for each $\alpha \in \tilde{I}$. If we put $R = \{r_\alpha | \alpha \in \tilde{I}\}$ then one can show that $\tilde{W} = \langle R \rangle$ is the subgroup of W fixed by Γ. Actually we can prove the following important property:

(A) If $\tilde{w} \in \tilde{W}$ and $\alpha \in \tilde{I}$ then $l(\tilde{w}s_i) = l(\tilde{w}) - 1$ for each i in α or $l(\tilde{w}s_i) = l(\tilde{w}) + 1$ for each i in α.

It can be shown that $(\tilde{W}, R)$ is a Coxeter system. In the proof of this fact we do not need the group Γ, but only the property (A). So the property (A) will be taken as an axiom in our definition of *admissible partitions*.

Admissible partitions: Let I, M, W, S and $l : W \longrightarrow N_0$ be as above. Let $\tilde{I}$ be a spherical partition of I with respect to M; i.e. M_α is spherical for each $\alpha \in \tilde{I}$. For $\alpha \in \tilde{I}$ let r_α be the element of maximal length in W_α and put $R = \{r_\alpha | \alpha \in \tilde{I}\}$ and $\tilde{W} = \langle R \rangle$. The partition $\tilde{I}$ of I is said to be admissible with respect to M if the property (A) is satisfied.

Here are our main results:

Theorem 1.1 *Let I be a set, let M be a Coxeter diagram over I. Let $\tilde{I}$ be an admissible partition of I with respect to M. Then the pair $(\tilde{W}, R)$ is a Coxeter system, where R and $\tilde{W}$ are as above. The type of $(\tilde{W}, R)$ is the Coxeter diagram $\tilde{M}$ over $\tilde{I}$, where $\tilde{m}_{\alpha\beta} = |\langle r_\alpha r_\beta \rangle|$ for each pair $(\alpha, \beta) \in \tilde{I} \times \tilde{I}$.*

Remark: Observe that one only has to consider the $\alpha \cup \beta$-residue if one wants to compute $\tilde{m}_{\alpha\beta}$.

Theorem 1.2 *Let I be a set and let M be a Coxeter diagram over I. Let $\tilde{I}$ be a spherical partition of I with respect to M. The following are equivalent:*

 (1) *The partition $\tilde{I}$ is admissible with respect to M.*

 (2) *For each pair $(\alpha, \beta) \in \tilde{I} \times \tilde{I}$ the partition $\{\alpha, \beta\}$ of $\alpha \cup \beta$ is admissible with respect to $M_{(\alpha\cup\beta)}$.*

Theorem 1.3 *Let I be a set and let M be a Coxeter diagram over I. Let (W, S) be the Coxeter system of type M. Let Γ be a group of automorphisms of M and let $\tilde{I}$ be the set of orbits of Γ in I. Let $\tilde{J}$ be the set of all $\alpha \in \tilde{I}$ with the property that M_α is spherical. For $\beta \in \tilde{J}$ let r_β be the element of maximal length in W_β. The group Γ acts on W as a group of automorphisms which preserves the length. The subgroup $\tilde{W} := \{w \in W | w^g = w \text{ for all } g \in \Gamma\}$ is the group generated by the r_β where β runs through $\tilde{J}$. If we put $J = \bigcup_{\beta \in J} \beta$ then $\tilde{J}$ is an admissible partition with respect to M_J.*

This paper is organized as follows: In section 2 we recall some basic results about Coxeter systems. In section 3 we first prove some preliminary lemmas and fix some further notation concerning admissible partitions; the main goal of section 3 is to prove Proposition 3.5 which provides Theorem 1.1 and Theorem 1.2 as corollaries. In section 4 we will prove Theorem 1.3 . Examples and applications of admissible partitions will be given in section 5.

2. Preliminaries

Systems of involutions: Let I be a set. A system of involutions over I is a pair (W, S) consisting of a group W and a set S of involutions in W satisfying:

1. $\langle S \rangle = W$
2. There is a bijective mapping $i \longrightarrow s_i$ from I onto S.

Let (W, S) be a system of involutions over I and let F denote the free monoid on I. For $f = i_1 i_2 \ldots i_d \in F$ we put $L(f) = d$ and $s_f = s_{i_1} s_{i_2} \ldots s_{i_d} \in W$. Let $w \in W$. A representation of w in (W, S) is a word $f \in F$ with the property that $s_f = w$. The length of w in the system (W, S) is the minimum of all $L(f)$ where f runs through all representations of w. We denote the length of w by $l(w)$. A representation $f \in F$ of $w \in W$ is called reduced if we have $L(w) = l(w)$.

Coxeter diagrams: Let I be a set. A Coxeter diagram over I is a mapping $M : I \times I \longrightarrow N \cup \{\infty\}$ such that

1. $M(i,j) = M(j,i) \geq 2$ for $i \neq j$
2. $M(i,i) = 1$ for all $i \in I$

For $M(i,j)$ we will write m_{ij}. Let (W, S) be a system of involutions over I. Its type is defined to be the Coxeter diagram M over I where the m_{ij} are defined by $m_{ij} = |\langle s_i s_j \rangle|$. Now we will fix some notation concerning a system of involutions of a given type: Let I be a set. Let (W, S) be a system of involutions over I and let M be its type. Let F denote the free monoid on I. If $m_{ij} \neq \infty$, then $p(i,j) \in F$ denotes the word $\ldots ijij$ of length m_{ij}.

Let $J \subseteq I$. We put $S_J = \{s_j | j \in J\}$, $W_J = \langle S_J \rangle$ and $M_J = M \mid_{J \times J}$. It is clear that (W_J, S_J) is a system of involutions of type M_J.

Coxeter systems: Let I be a set and let M be a Coxeter diagram over I. A Coxeter system of type M is a system of involutions of type M which satisfies for each $w \in W$ the following axioms:

A For $i \in I$ we have $l(ws_i) = l(w) + 1$ or $l(ws_i) = l(w) - 1$.

B If $i, j \in I$ and $l(ws_i) = l(w) - 1 = l(ws_j)$, then $m_{ij} \neq \infty$ and there is a reduced representation of W ending with $p(i,j)$.

Though this characterization of Coxeter systems is probably well known, we will give a sketch of a proof:

Proposition 4 in [1] chap. 4 no 1.5 provides property A. That property B holds in Coxeter systems may be seen from [5] Theorem 2.16 and the observation that the elements of maximal length in finite rank 2 Coxeter groups have the reduced representations $p(i,j)$ and $p(j,i)$.

For the other direction one uses induction on $l(w)$ to see that two reduced representations of $w \in W$ can be transformed from into each other by elementary M-operations of type (II). Moreover, one shows by induction on $L(f)$ that a non-reduced word $f \in F$ can be shortened by elementary M-operations. This shows that (W, S) is a Coxeter system. (See [2] for the definitions and further details.)

In the rest of this section, we fix some notation and state some preliminary lemmas concerning Coxeter systems. For the proofs and further information we refer to [8] and to [5].

From now on let I be a set and let M be a Coxeter diagram over I. Up to isomorphism there exists exactly one Coxeter system of type M (see for instance [2]), so let (W, S) be 'the' Coxeter system of type M.

We start with a definition:

Definition 2.1 1. Let $w \in W$. We put $I^+(w) = \{i \in I | l(ws_i) = l(w)+1\}$ and $I^-(w) = \{i \in I | l(ws_i) = l(w) - 1\}$.

  2. Let $J \subseteq I$ and let $w_0 \in W$. We put $R_J(w_0) = w_0 W_J$. The set $R_J(w_0)$ is called the J-residue of w_0.

Lemma 2.2 *Let $w \in W_J$ and let $i_1 i_2 \ldots i_d \in F$ be a reduced representation of w. Then $i_k \in J$ for all $1 \leq k \leq d$. The system of involutions (W_J, S_J) is the Coxeter system of type M_J.*

The first assertion follows from [5] Lemma 2.10, the second follows from Corollary 2.14 in [5].

Lemma 2.3 *Let $J \subseteq I$ and $w_0 \in W$, then there exists a unique element $w^\star$ in $R_J(w_0)$ with the property that $l(w^\star) \leq l(w)$ for all $w \in R_J(w_0)$. For each $w_1 \in R_J(w_0)$ we have $l(w_1) = l(w^\star) + l(w^{\star-1}w_1)$.*

For a proof see [5] Theorem 2.9.
Remark: For given $w_0 \in W$ and $J \subseteq I$ we denote this element by $P_J(w_0)$. It is the 'projection' of the unit element onto the J-residue of w_0.

Lemma 2.4 *Let $J \subseteq I$ and let $w_0 \in W$. Then the following are equivalent:*
  P1 $P_J(w_0) = w_0$
  P2 $J \subseteq I^+(w_0)$
  P3 $l(w_0 w_1) = l(w_0) + l(w_1)$ for all $w_1 \in W_J$.
  P4 $l(w_2) = l(w_0) + l(w_0^{-1}w_2)$ for all $w_2 \in R_J(w_0)$.

Proof. P1 $\Rightarrow$ P2: $P_J(w_0) = w_0$ implies $l(w_0 s_j) \geq l(w_0)$ for each $j \in J$. From axiom A it is seen that $l(w_0 s_j) = l(w_0) + 1$ for each $j \in J$, hence $J \subseteq I^+(w_0)$.
P2 $\Rightarrow$ P1: Let $J \subseteq I^+(w_0)$ and put $w_1 = P_J(w_0)$. From lemma 2.3 it follows

that $l(w_0) = l(w_1) + l(w_1^{-1}w_0)$ and $l(w_0) + 1 = l(w_0 s_j) = l(w_1) + l(w_1^{-1}w_0 s_j)$ for each $j \in J$. We obtain $l(w_1^{-1}w_0 s_j) = l(w_1^{-1}w_0) + 1$ for each $j \in J$, hence $J \subseteq I^+(w_1^{-1}w_0)$. Since $w_1^{-1}w_0$ lies in W_J it follows that $w_1^{-1}w_0$ is the identity.

$P1 \Rightarrow P4$: This is the second assertion of lemma 2.3.

$P4 \Rightarrow P1$: P4 implies $l(w_2) \geq l(w_0)$ for each $w_2 \in R_J(w_0)$, hence $P_J(w_0) = w_0$.

$P4 \Leftrightarrow P3$: This is obvious. $\qquad\square$

Definition 2.5 Let $J \subseteq I$ and let $w \in W$. We say that J is admissible at w if $J \subseteq I^+(w)$ or $J \subseteq I^-(w)$.

Let $\tilde{I}$ be a partition of I and let $w \in W$. The partition $\tilde{I}$ is said to be admissible at w if α is admissible at w for each $\alpha \in \tilde{I}$.

Lemma 2.6 Let $J_1 \subseteq J \subseteq I$ and $w \in W$. Put $w' = P_J(w)$ and $w'' = w'^{-1}w$. Then we have:
$$J_1 \subseteq I^+(w) \Leftrightarrow J_1 \subseteq I^+(w'')$$
and
$$J_1 \subseteq I^-(w) \Leftrightarrow J_1 \subseteq I^-(w'')$$
In particular: J_1 is admissible at w if and only if it is admissible at w''.

Proof. This is an easy consequence of P4 in lemma 2.4. $\qquad\square$

Lemma 2.7 Let $J \subseteq I$. Then the following are equivalent:
S1 W_J is finite.
S2 There exists a unique element w^* in W_J such that $l(w^*) \geq l(w)$ for all $w \in W_J$.

Remark: If $J \subseteq I$ satisfies the equivalent conditions of lemma 2.7 then the element w^* of condition S2 is an involution. It will be denoted by r_J. If I satisfies the conditions of lemma 2.7, then the diagram M is called spherical.

Lemma 2.8 Let $w_0 \in W$ and let $J \subseteq I$. Then the following are equivalent:
PS1 $J \subseteq I^-(w_0)$
PS2 W_J is finite and $l(w_0) = l(w_0 r_J) + l(r_J)$.
PS3 W_J is finite and $P_J(w_0) = w_0 r_J$.

Proof. $PS1 \Rightarrow PS2$: This follows from [5] Theorem 2.16.

$PS2 \Rightarrow PS3$: Let $w_1 \in R_J(w_0)$ then $|l(w_0) - l(w_1)| \leq l(r_J)$. This shows that $l(w_1) \geq l(w_0) - l(r_J) = l(w_0 r_J)$, hence $w_0 r_J = P_J(w_0)$.

$PS3 \Rightarrow PS1$: Since $J \subseteq I^-(r_J)$ and $P_J(w_0) = w_0 r_J$, it follows from lemma 2.6 that $J \subseteq I^-(w_0 r_J^2) = I^-(w_0)$. $\qquad\square$

3. Admissible Partitions

Let I be a set and let M be a Coxeter diagram over I. A partition $\tilde{I}$ of I is called spherical with respect to M if M_α is spherical for all $\alpha \in \tilde{I}$.

We will now fix some notation concerning spherical partitions which will be valid throughout the rest of this section: Let I be a set, let M be a Coxeter diagram over I and let $\tilde{I}$ be a spherical partition of I with respect to M. Let (W, S) be the Coxeter system of type M. We put $R = \{r_\alpha | \alpha \in \tilde{I}\}$ and $\tilde{W} = \langle R \rangle$. Thus $(\tilde{W}, R)$ is a system of involutions over $\tilde{I}$. Let $\tilde{l} : \tilde{W} \longrightarrow N_o$ denote its length function (see section 2) and let $\tilde{M} = (\tilde{m}_{\alpha\beta})_{(\alpha,\beta) \in \tilde{I} \times \tilde{I}}$ be its type. Let $\tilde{F}$ denote the free monoid on $\tilde{I}$. If $\tilde{m}_{\alpha\beta} \neq \infty$ let $\tilde{p}(\alpha, \beta) \in \tilde{F}$ be defined to be the word $\ldots \alpha\beta\alpha\beta$ of length $\tilde{m}_{\alpha\beta}$ and let $\tilde{q}(\alpha, \beta)$ be the word $\ldots \alpha\beta\alpha\beta$ of length $\tilde{m}_{\alpha\beta} - 1$.

Definition 3.1 Let $\tilde{I}$ be a spherical partition with respect to M.
  1. The partition $\tilde{I}$ is said to be admissible if $\tilde{I}$ is admissible at each $\tilde{w} \in \tilde{W}$.
  2. Let $\tilde{w} \in \tilde{W}$ and let $\tilde{f} = \alpha_1\alpha_2 \ldots \alpha_d \in \tilde{F}$ be a representation of $\tilde{w}$ with $\alpha_i \in \tilde{I}$. We say that the representation $\tilde{f}$ is compatible if the following holds:
$$l(\tilde{w}) = \sum_{k=1}^{d} l(r_{\alpha_k})$$

The following three lemmas are immediate.

Lemma 3.2 Let $\tilde{I}$ be an admissible partition of I. Let $\tilde{J} \subseteq \tilde{I}$ and put $J = \cup_{\beta \in \tilde{J}} \beta$. Then the partition $\{\beta | \beta \in \tilde{J}\}$ is an admissible partition of J with respect to M_J.

Lemma 3.3 Assume $|\tilde{I}| = 2$ with $\tilde{I} = \{\alpha, \beta\}$. The following are equivalent:
  a The partition $\{I_\alpha, I_\beta\}$ is admissible.
  b If $\tilde{w}$ is in $\tilde{W}$, then a representation $\tilde{f} \in \tilde{F}$ of $\tilde{w}$ is reduced if and only if it is compatible.

Lemma 3.4 Let M be a spherical diagram over I and let $\tilde{I} = \{\alpha, \beta\}$ be a partition of I. Then $\tilde{m}_{\alpha\beta} \neq \infty$. Moreover: The partition $\tilde{I}$ is admissible if and only if the words $\tilde{p}(\alpha, \beta)$ and $\tilde{p}(\beta, \alpha)$ are compatible representations of r_I.

Proposition 3.5 Let M be a Coxeter diagram over I. Let $\tilde{I}$ be a spherical partition such that for each pair $(\alpha, \beta) \in \tilde{I} \times \tilde{I}$ the partition $\{\alpha, \beta\}$ of $\alpha \cup \beta$ is admissible with respect to $M_{(\alpha \cup \beta)}$. Then we have for all $\tilde{w} \in \tilde{W}$:

A1 A representation $\tilde{f} \in \tilde{F}$ of $\tilde{w}$ is reduced if and only if it is compatible.

A2 Let $\alpha, \beta \in \tilde{I}$ and put $\tilde{v} = P_{(\alpha \cup \beta)}(\tilde{w})$. Then we have: $\tilde{v} \in \tilde{W}$, $\tilde{l}(\tilde{v}) + \tilde{l}(\tilde{v}^{-1}\tilde{w}) = \tilde{l}(\tilde{w})$ and $\tilde{v}^{-1}\tilde{w} \in \langle r_\alpha, r_\beta \rangle$.

A3 The partition $\tilde{I}$ is admissible at $\tilde{w}$.

A4 $\tilde{l}(\tilde{w}r_\alpha) = \tilde{l}(\tilde{w}) - 1$ if and only if $\alpha \subseteq I^-(\tilde{w})$.

A5 $\tilde{l}(\tilde{w}r_\alpha) = \tilde{l}(\tilde{w}) + 1$ if and only if $\alpha \subseteq I^+(\tilde{w})$.

A6 If $\alpha, \beta \in \tilde{I}$ such that $\alpha \cup \beta \subseteq I^-(\tilde{w})$, then there exists a reduced representation of $\tilde{w}$ in $\tilde{F}$ ending with the word $\tilde{p}(\alpha, \beta)$.

Proof. First observe that if $\alpha, \beta \in \tilde{I}$ are such that $M_{(\alpha \cup \beta)}$ is spherical it follows that $\tilde{p}(\alpha, \beta)$ and $\tilde{p}(\beta, \alpha)$ are compatible representations of $r_{(\alpha \cup \beta)}$. In this case we have also that $\tilde{q}(\alpha, \beta)$ (resp. $\tilde{q}(\beta, \alpha)$) is a compatible representation of $r_{(\alpha \cup \beta)}r_\alpha$ (resp. $r_{(\alpha \cup \beta)}r_\beta$) of length $m_{\alpha\beta} - 1$.

The proof goes by induction on $\tilde{l}(\tilde{w})$:

If $\tilde{l}(\tilde{w}) = 0$, then $\tilde{w}$ is the identity. The assertions A1 - A6 are obviously satisfied and the induction starts.

Let $\tilde{w} \in \tilde{W}$ and let $\tilde{f}$ be a reduced representation ending with $\epsilon \in \tilde{I}$. We may write $\tilde{f}$ as $\tilde{f}'\epsilon$. We put $\tilde{w}' = \tilde{w}r_\epsilon$. The word $\tilde{f}'$ is a reduced representation of $\tilde{w}'$. Now $\tilde{l}(\tilde{w}') = \tilde{l}(\tilde{w}) - 1$ and we can apply the induction hypothesis to $\tilde{w}'$. We have $\tilde{l}(\tilde{w}'r_\epsilon) = \tilde{l}(\tilde{w}') + 1$, so the assertion A5 of the induction hypothesis implies that $\epsilon \subseteq I^+(\tilde{w}')$. From lemma 2.4 it is seen that $l(\tilde{w}'r_\epsilon) = l(\tilde{w}') + l(r_\epsilon)$. Now one uses lemma 2.8 to see that $\epsilon \subseteq I^-(\tilde{w})$. Since the representation $\tilde{f}'$ of $\tilde{w}'$ is reduced it must be compatible by the assertion A1 of the induction hypothesis. Combining this with the equality $l(\tilde{w}) = l(\tilde{w}'r_\epsilon) = l(\tilde{w}') + l(r_\epsilon)$ we obtain that $\tilde{f}$ is a compatible representation of $\tilde{w}$.

(I) We have shown that every reduced representation of $\tilde{w}$ is compatible. Moreover, if $\alpha \in \tilde{I}$ we have the implication

$$\tilde{l}(\tilde{w}r_\alpha) = \tilde{l}(\tilde{w}) - 1 \Rightarrow \alpha \subseteq I^-(\tilde{w}).$$

Now let $\delta \in \tilde{I}$. First observe that $P_{(\epsilon \cup \delta)}(\tilde{w}) = P_{(\epsilon \cup \delta)}(\tilde{w}')$; we denote this element in $\tilde{W}$ by $\tilde{v}$. The induction hypothesis shows that $\tilde{v} \in \tilde{W}$ and $\tilde{l}(\tilde{w}') = \tilde{l}(\tilde{v}) + \tilde{l}(\tilde{v}^{-1}\tilde{w}')$. Since $\tilde{l}(\tilde{w}) = \tilde{l}(\tilde{w}') + 1$ and $\tilde{l}(\tilde{v}^{-1}\tilde{w}) = \tilde{l}(\tilde{v}^{-1}\tilde{w}'r_\epsilon) \leq \tilde{l}(\tilde{v}^{-1}\tilde{w}') + 1$ it follows:

$$\tilde{l}(\tilde{w}) \leq \tilde{l}(\tilde{v}) + \tilde{l}(\tilde{v}^{-1}\tilde{w}) \leq \tilde{l}(\tilde{v}) + \tilde{l}(\tilde{v}^{-1}\tilde{w}') + 1 = \tilde{l}(\tilde{w}') + 1 = \tilde{l}(\tilde{w})$$

and equality must hold. Since $\tilde{v}^{-1}\tilde{w}' \in \langle r_\epsilon, r_\delta \rangle$ it follows that $\tilde{v}^{-1}\tilde{w} = \tilde{v}^{-1}\tilde{w}'r_\epsilon \in \langle r_\epsilon, r_\delta \rangle$.

Since $\tilde{v}^{-1}\tilde{w} \in \langle r_\epsilon, r_\delta \rangle$ it follows that ϵ and δ are admissible at $\tilde{v}^{-1}\tilde{w}$. Lemma 2.6 provides now the admissibility of ϵ and δ at $\tilde{w}$.

(II) This proves A3. Moreover, we have shown A2 for all pairs $(\alpha, \beta) \in \tilde{I} \times \tilde{I}$ having the property that there is a reduced representation of $\tilde{w}$ ending with α or β.

Now let $\mu \in \tilde{I}$ be such that $\mu \subseteq I^-(\tilde{w})$. We put $\tilde{z} = P_{(\epsilon \cup \mu)}$. Since we have already shown that $\epsilon \subseteq I^-(\tilde{w})$, it follows $\epsilon \cup \mu \subseteq I^-(\tilde{w})$. From lemma 2.8 it is seen that $M_{(\epsilon \cup \mu)}$ is spherical and that $\tilde{w} = \tilde{z} r_{(\epsilon \cup \mu)}$. Now let $\tilde{f}_z$ be a reduced representation of $\tilde{z}$. Since $\tilde{z} \neq \tilde{w}$ the induction hypothesis shows that $\tilde{f}_z$ is a compatible representation of $\tilde{z}$. On the other hand our first remark above shows that $\tilde{q}(\epsilon, \mu)$ is a compatible representation of $r_{(\epsilon \cup \mu)} r_\epsilon = \tilde{z}^{-1} \tilde{w}'$ of length $\tilde{m}_{\epsilon \mu} - 1$. Combining these considerations we obtain that $\tilde{f}_z \tilde{q}(\epsilon, \mu)$ is a compatible representation of $\tilde{w}'$ of length $\tilde{l}(\tilde{z}) + \tilde{m}_{\epsilon, \mu} - 1$. By the induction hypothesis the representation $\tilde{f}_z \tilde{q}(\epsilon, \mu)$ is reduced. We obtain that $\tilde{f}_z \tilde{p}(\mu, \epsilon)$ is a reduced representation of $\tilde{w}$ and that $\tilde{l}(\tilde{w}) = \tilde{l}(\tilde{z}) + \tilde{m}_{\epsilon \mu}$. We have also that $\tilde{f}_z \tilde{q}(\mu, \epsilon)$ is a representation of $\tilde{w} r_\mu$ of length $\tilde{l}(\tilde{z}) + \tilde{m}_{\epsilon \mu} - 1 = \tilde{l}(\tilde{w}) - 1$. This shows $\tilde{l}(\tilde{w} r_\mu) = \tilde{l}(\tilde{w}) - 1$.

(III) For $\alpha \in \tilde{I}$ we have proved the implication

$$\alpha \subseteq I^-(\tilde{w}) \Rightarrow \tilde{l}(\tilde{w} r_\alpha) = \tilde{l}(\tilde{w}) - 1$$

which accomplishes A4. Moreover we have also shown that A6 is valid.

Let now $\tilde{c}$ be a compatible representation of $\tilde{w}$ ending with γ. So we may write $\tilde{c}$ as $\tilde{c}' \gamma$. Since $\tilde{c}$ is compatible it follows that $\gamma \subseteq I^-(\tilde{w})$. Since we have already shown A4 we can deduce $\tilde{l}(\tilde{w} r_\gamma) = \tilde{l}(\tilde{w}) - 1$. So we may apply the induction hypothesis to $\tilde{w} r_\gamma$. Since $\tilde{c}'$ is a compatible representation of $\tilde{w} r_\gamma$ we can use the induction hypothesis to see that $\tilde{c}'$ is a reduced representation of $\tilde{w} r_\gamma$, hence its length is $\tilde{l}(\tilde{w}) - 1$. This shows that the representation $\tilde{c}$ of $\tilde{w}$ has length $\tilde{l}(\tilde{w})$, so $\tilde{c}$ is a reduced representation of $\tilde{w}$.

(IV) The proof of A1 is now complete.

Let now $\alpha, \beta \in \tilde{I}$. Assume first that $\alpha \cup \beta \subseteq I^+(\tilde{w})$. Then we have $\tilde{v} = P_{(\alpha \cup \beta)}(\tilde{w}) = \tilde{w}$ by lemma 2.4 and the assertions made in A2 are obviously satisfied. Now assume that $\alpha \cup \beta \not\subseteq I^+(\tilde{w})$. Since we have already shown A3 we may assume w.l.o.g. that $\alpha \subseteq I^-(\tilde{w})$. Now, since we have already shown A4 we obtain $\tilde{l}(\tilde{w} r_\alpha) = \tilde{l}(\tilde{w}) - 1$ and A2 follows by (II).

(V) This accomplishes the proof of A2

Now let $\alpha \in \tilde{I}$ be such that $\tilde{l}(\tilde{w} r_\alpha) = \tilde{l}(\tilde{w}) + 1$. From A4 applied to $\tilde{w}$ it follows $\alpha \not\subseteq I^-(\tilde{w})$; now it follows from A3 applied to $\tilde{w}$ that $\alpha \subseteq I^+(\tilde{w})$.

(VI) For $\alpha \in \tilde{I}$ we have shown the implication

$$\tilde{l}(\tilde{w} r_\alpha) = \tilde{l}(\tilde{w}) + 1 \Rightarrow \alpha \subseteq I^+(\tilde{w}).$$

Let $\pi \in \tilde{I}$ be such that $\pi \subseteq I^+(\tilde{w})$. By lemma 2.4 we have $\tilde{w} = P_\pi(\tilde{w})$. Assume that $\tilde{l}(\tilde{w} r_\pi) \neq \tilde{l}(\tilde{w}) + 1$. It follows from A4 applied to $\tilde{w}$ that $\tilde{l}(\tilde{w} r_\pi) = \tilde{l}(\tilde{w}) - 1$

is not possible since $\pi \subseteq I^+(\tilde{w})$. So we obtain $\tilde{l}(\tilde{w}r_\pi) = \tilde{l}(\tilde{w})$. Note that we can now apply A1 - A4 also to $\tilde{w}_1 = \tilde{w}r_\pi$.

We have $\tilde{l}(\tilde{w}_1 r_\pi) = \tilde{l}(\tilde{w}) \neq \tilde{l}(\tilde{w}_1) - 1$. Applying A4 to $\tilde{w}_1$ we obtain $\pi \nsubseteq I^-(\tilde{w}_1)$; applying now A3 to $\tilde{w}_1$ yields $\pi \subseteq I^+(\tilde{w}_1)$. Lemma 2.4 provides $\tilde{w}_1 = P_\pi(\tilde{w}_1)$.

Observe that $R_\pi(\tilde{w}_1) = R_\pi(\tilde{w})$ and therefore $P_\pi(\tilde{w}_1) = P_\pi(\tilde{w})$. Combining all these considerations we obtain

$$\tilde{w}r_\pi = \tilde{w}_1 = P_\pi(\tilde{w}_1) = P_\pi(\tilde{w}) = \tilde{w}$$

which is a contradiction. It follows that $\tilde{l}(\tilde{w}r_\pi) = \tilde{l}(\tilde{w}) + 1$.

(VII) For $\alpha \in \tilde{I}$ we have shown the implication

$$\alpha \subseteq I^+(\tilde{w}) \Rightarrow \tilde{l}(\tilde{w}r_\alpha) = \tilde{l}(\tilde{w}) + 1.$$

This completes the proof of A5 and we are done. $\qquad\square$

Proof of Theorem 1.1: It is obvious that $(\tilde{W}, R)$ is a system of involutions of type $\tilde{M}$. Since $\tilde{I}$ is an admissible partition of I with respect to M it follows from lemma 3.2 that $\{\alpha, \beta\}$ is an admissible partition of $\alpha \cup \beta$ with respect to $M_{(\alpha \cup \beta)}$ for each pair $(\alpha, \beta) \in \tilde{I} \times \tilde{I}$. So we can apply Proposition 3.5. Axiom A of our characterization of Coxeter systems follows now from A3, A4 and A5 of Proposition 3.5. The assertion A6 provides the validity of axiom B.

Proof of Theorem 1.2: The implication (1) $\Rightarrow$ (2) is provided by lemma 3.2. The assertion A3 of Proposition 3.5 shows the other implication.

4. Automorphisms of diagrams

Let I be a set and let M be a Coxeter diagram over I. Let (W, S) be a Coxeter system of type M and let l denote the length function. Let Γ be a group of automorphisms of M and let $\tilde{I}$ be the decomposition of I into the orbits of Γ. Let $\tilde{J} \subseteq \tilde{I}$ be the set of all α such that M_α is spherical and put $J = \bigcup_{\beta \in \tilde{J}} \beta$. We define $R = \{r_\beta | \beta \in \tilde{J}\}$.

Let $g \in \Gamma$, then g induces a bijection $s_i \longrightarrow s_{i^g}$, which extends uniquely to an automorphism of W. Observe that we have $l(w) = l(w^g)$ for all $w \in W$. We put $\tilde{W} = \{w \in W | w^g = w \text{ for all } g \in \Gamma\}$.

Proof of Theorem 1.3: Let $\beta \in \tilde{J}$. Then Γ fixes the subgroup W_β. Since Γ preserves the length, it must leave r_β invariant (cf. lemma 2.7). This shows $\langle R \rangle \subseteq \tilde{W}$.

Let now $\tilde{w} \in \tilde{W}, i \in I^+(\tilde{w})$ and $g \in \Gamma$. It follows that $l(\tilde{w}) + 1 = l(\tilde{w}s_i) = l((\tilde{w}s_i)^g) = l(\tilde{w}^g s_i^g) = l(\tilde{w}s_{i^g})$ showing that $i^g \in I^+(\tilde{w})$. Similarly one deduces that $i \in I^-(\tilde{w})$ implies $i^g \in I^-(\tilde{w})$ for all $g \in \Gamma$. This shows that α is admissible at $\tilde{w}$ for all $\alpha \in \tilde{I}$.

Using the fact that $\tilde{I}$ is admissible at each $\tilde{w} \in \tilde{W}$ and lemma 2.8 one shows by induction on $l(\tilde{w})$ that $\tilde{w} \in \langle R \rangle$. This completes the proof of theorem 1.3.

5. Examples and Applications

Examples: In the following we adopt the notation of [5] for the Coxeter diagrams.

1. The polarity of A_n (resp. D_n, E_6, F_4) provides an admissible partition; the resulting diagram $\tilde{M}$ is $C_{[n+1/2]}$ (resp. $C_{n-1}, F_4, G_2(8)$).

2. The triality of D_4 (resp. $\tilde{E}_6$) provides an admissible partition; the resulting diagram is G_2 (resp. $\tilde{G}_2$).

3. Consider the diagram A_n over the set $I = \{1, 2, \ldots, n\}$ with the natural labelling. Let ω (resp. ϵ) denote the set of odd (resp. even) numbers. Then $\{\omega, \epsilon\}$ is an admissible partition. The resulting diagram $\tilde{M}$ is $G_2(n+1)$.

   We refer to remark 4 of [6], where this example is mentioned in a different context. The proof of the admissibility of the partition described above will be given in [3].

4. The partition $\{\alpha, \beta, \gamma, \delta, \epsilon\}$ with $\alpha = \{a_1, a_2\}$, $\beta = \{b_1, b_2\}$, $\gamma = \{c_1, c_2\}$, $\delta = \{d_1, d_2\}$ and $\epsilon = \{e_1, e_2\}$ of the following diagram is admissible:

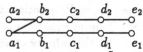

   The resulting diagram $\tilde{M}$ is the following:

$$
\begin{array}{ccccc}
& 5 & & & \\
\circ\!\!-\!\!\!\!&\!\!\!\!-\!\!\circ\!\!-\!\!\!\!&\!\!\!\!-\!\!\circ\!\!-\!\!\!\!&\!\!\!\!-\!\!\circ\!\!-\!\!\!\!&\!\!\!\!-\!\!\circ \\
\alpha & \beta & \gamma & \delta & \epsilon
\end{array}
$$

   This is seen from Theorem 1.2 and our third example.

   The consideration of the corresponding residues in this example provides the well known embeddings (see for instance [6]) of H_3 (resp. H_4) in D_6 (resp. E_8).

Application 1: Example 4 may be seen as a special case of the following result.

Theorem 5.1 *Let $\tilde{I}$ be a set and let $\tilde{M}$ be a Coxeter diagram over $\tilde{I}$ such that $|\{\tilde{m}_{\alpha\beta}|(\alpha, \beta) \in \tilde{I} \times \tilde{I}\}| < \infty$. Then there exists a set I, a Coxeter diagram M over I having only single bonds and an admissible partition $\{I_\alpha | \alpha \in \tilde{I}\}$ of I with respect to M with the property that $\tilde{m}_{\alpha\beta} = |\langle r_{I_\alpha} r_{I_\beta} \rangle|$.*

In particular, every Coxeter group which belongs to a diagram satisfying the above finiteness condition is contained in a Coxeter group which belongs to a diagram having only single bonds.

This theorem will be proved in [3]. As in example 4 its proof uses theorem 1.2 and the admissible partitions of example 3.

Application 2: If M is a diagram over I and $\tilde{I}$ is a admissible partition, providing the diagram $\tilde{M}$, then it is easy to see that M is spherical if and only if $\tilde{M}$ is spherical. Our results can be used to give another proof for the classification of the spherical diagrams. For instance the following is an easy consequence of our considerations: $\tilde{A}_3$ is spherical iff $\tilde{C}_2$ is spherical iff $\tilde{B}_2$ is spherical iff $\tilde{D}_4$ is spherical iff $\tilde{G}_2$ is spherical iff $\tilde{E}_6$ is spherical iff $\tilde{F}_4$ is spherical iff $\tilde{E}_7$ is spherical.

Using the theorem of application 1 above we can reduce the classification of the spherical diagrams to the classification of the spherical diagrams having only single bonds.

As shown in [7] it is easy to see that for instance $\tilde{A}_3$ is not spherical by using only the solution of the word problem, whereas it seems to be very hard to do the same for the diagram $\tilde{E}_7$. Our results can be used to give a proof of the classification, which does not make use of the associated bilinear forms. This will be elaborated in [4].

References

[1] **N. Bourbaki**. *Groupes et Algèbres de Lie, Chapitres 4,5 et 6*. Hermann, Paris, 1968.

[2] **K. Brown**. *Buildings*. Springer Verlag, New York, 1989.

[3] **B. Mühlherr**. Admissible partitions of Coxeter diagrams. In preparation.

[4] **B. Mühlherr**. An alternative proof for the classification of the spherical Coxeter diagrams. In preparation.

[5] **M. A. Ronan**. *Lectures on Buildings*. Academic Press, San Diego, 1989.

[6] **O. P. Shcherbak**. Wavefronts and reflection groups. *Russian Math. Surveys*, 43, pp. 149–194, 1988.

[7] **J. Tits**. Le problème de mots dans les groupes de Coxeter. In *Symposia Mathematica*, volume 1, pages 175–185, London, 1969. INDAM, Rome 1967/68, Academic Press.

[8] **J. Tits**. *Buildings of Spherical Type and Finite BN-pairs*, volume 386 of *Lect. Notes in Math.* Springer Verlag, Berlin, 1974.

B. Mühlherr, Mathematisches Institut, Auf der Morgenstelle 10, 7400 Tübingen, Germany. e-mail: mmgmu01@convex.zdv.uni-tuebingen.de

A local characterization of the graphs of alternating forms

A. Munemasa [*]   S. V. Shpectorov [†]

Abstract

Let Δ be the line graph of $PG(n-1, q)$, $q > 2$, $\mathrm{Alt}(n, q)$ be the graph of the n-dimensional alternating forms over $GF(q)$, $n \geq 4$. It is shown that every connected locally Δ graph, such that the number of common neighbours of any pair of vertices at distance two is the same as in $\mathrm{Alt}(n, q)$, is covered by $\mathrm{Alt}(n, q)$.

1. Introduction

There have been extensive studies in local characterization of graphs. Certain strongly regular graphs are characterized by their local structure. In this paper we shall investigate graphs which are locally a $(q-1)$-clique extension of the Grassmann graph $\begin{bmatrix} V \\ 2 \end{bmatrix}$ over $GF(q)$, $q > 2$. The Grassmann graph $\begin{bmatrix} V \\ 2 \end{bmatrix}$ has as vertices all 2-spaces of an n-dimensional vector space V over $GF(q)$. Two vertices are adjacent whenever they intersect nontrivially. The alternating forms graph $\mathrm{Alt}(n, q)$ is locally a $(q-1)$-clique extension of $\begin{bmatrix} V \\ 2 \end{bmatrix}$. In this paper, we restrict ourselves to the case $\mu = q^2(q^2 + 1)$, i.e., the number of common neighbours of two vertices at distance 2 is always $q^2(q^2 + 1)$. Under the assumption $\mu = q^2(q^2 + 1)$, $\mathrm{Alt}(4, q)$ is the only graph which is locally a $(q-1)$-clique extension of $\begin{bmatrix} V \\ 2 \end{bmatrix}$ with $n = 4$. This result follows from the classification of affine polar spaces due to Cohen and Shult [2]. For large n, however, we cannot expect such a result, since the quotient graphs of $\mathrm{Alt}(n, q)$ over many subgroups of translations have the same local structure and the same value of μ. The purpose of this paper is to show that $\mathrm{Alt}(n, q)$ is universal in the following sense.

Main Theorem. *Let Γ be a graph which is locally a $(q-1)$-clique extension of the Grassmann graph $\begin{bmatrix} V \\ 2 \end{bmatrix}$, where V is an n-dimensional vector space over $GF(q)$, $q > 2$. Suppose that $\mu(\Gamma) = q^2(q^2 + 1)$. Then Γ is covered by $\mathrm{Alt}(n, q)$.*

---

[*]. This research was completed during this author's visit at the Institute for System Analysis, Moscow, as a Heizaemon Honda fellow of the Japan Association for Mathematical Sciences.

[†]. A part of this research was completed during the visit at University of Technology, Eindhoven.

From this theorem we immediately obtain:

Corollary. *If Γ is a distance-regular graph having the same intersection numbers and the same local structure as $\mathrm{Alt}(n,q)$, $q > 2$, then Γ is isomorphic to $\mathrm{Alt}(n,q)$.*

2. Preliminaries

For a vertex v of a graph Γ, denote by $\Gamma_i(v)$ the set of vertices at distance i from v, and denote by $v^\perp$ the set $\{v\} \cup \Gamma_1(v)$. Denote by $\Gamma(u,v,w,\ldots)$ the intersection of the sets $\Gamma_1(u), \Gamma_1(v), \Gamma_1(w), \ldots$, where $u,v,w,\ldots$ are vertices of Γ, hence we denote $\Gamma_1(u) = \Gamma(u)$.

A graph Γ is said to be *locally* Δ, where Δ is a graph, if for each vertex γ of Γ the induced subgraph on $\Gamma(\gamma)$ is isomorphic to Δ. A *μ-graph* for vertices u,v is the set $\Gamma(u,v)$ where u,v are vertices at distance 2. We also often regard a μ-graph as an induced subgraph. If all μ-graphs have the same cardinality, then we denote this cardinality by $\mu(\Gamma)$. The following lemma can be found in [4].

Lemma 2.1 *Let Γ, Δ be graphs, and suppose that Γ is locally Δ. Suppose also that Δ has diameter 2 and any μ-graph of Δ is isomorphic to a graph M. Then any μ-graph of Γ is locally M.*

Two graphs $\Gamma, \tilde{\Gamma}$ are said to *have the same local structure*, if for any vertex v of Γ and any vertex $\tilde{v}$ of $\tilde{\Gamma}$, the subgraphs $\Gamma(v), \tilde{\Gamma}(\tilde{v})$ are isomorphic.

The *Grassmann graph* has as vertices the set of d-dimensional subspaces of an n-dimensional vector space V over a finite field. Two vertices α, β are adjacent whenever $\dim(\alpha \cap \beta) = d - 1$. By abuse of notation we denote by $[\begin{smallmatrix} V \\ d \end{smallmatrix}]$ both the Grassmann graph and the set of vertices of the Grassmann graph. The $n \times m$ *grid* has as vertices $\{(i,j) | i = 1, \ldots, n, \ j = 1, \ldots, m\}$, two distinct vertices $(i_1, j_1), (i_2, j_2)$ are adjacent if and only if $i_1 = i_2$ or $j_1 = j_2$.

The *m-clique extension* of a graph Δ is a graph obtained by replacing vertices of Δ by m-cliques, and joining all pairs of vertices from two m-cliques whenever the corresponding vertices of Δ are adjacent. If $\tilde{\Delta}$ is the m-clique extension of a graph Δ, then an m-clique of $\tilde{\Delta}$ obtained by replacing a vertex of Δ is called a *basic clique*. This definition makes sense if the automorphism group of $\tilde{\Delta}$ leaves the set of basic cliques invariant, and it is the case when Δ is a noncomplete grid or a noncomplete Grassmann graph. A *canonical mapping* is a mapping from the set of vertices of $\tilde{\Delta}$ to the set of vertices of Δ such that a basic clique is mapped to a single vertex, and adjacent basic cliques are mapped to adjacent vertices.

The *alternating forms graph* $\mathrm{Alt}(n,q)$ has as vertices all alternating forms on a vector space V of dimension n over $\mathrm{GF}(q)$. Two vertices γ, δ are adjacent whenever $\text{rank}(\gamma - \delta) = 2$. The graph $\text{Alt}(n, q)$ is locally $\Delta^{(n,q)}$, where $\Delta^{(n,q)}$ is the $(q-1)$-clique extension of the Grassmann graph $\begin{bmatrix} V \\ 2 \end{bmatrix}$.

A *small clique* of the Grassmann graph $\begin{bmatrix} V \\ 2 \end{bmatrix}$ is a clique of the form $\begin{bmatrix} W \\ 2 \end{bmatrix}$, where $W \in \begin{bmatrix} V \\ 3 \end{bmatrix}$, while a *grand clique* of $\begin{bmatrix} V \\ 2 \end{bmatrix}$ is a clique of the form $\{U \in \begin{bmatrix} V \\ 2 \end{bmatrix} | U \ni x\}$, where $0 \neq x \in V$. Both small cliques and grand cliques are maximal cliques of the Grassmann graph, and any maximal clique is either grand or small. Any maximal clique of $\Delta^{(n,q)}$ is the preimage of a maximal clique of $\begin{bmatrix} V \\ 2 \end{bmatrix}$ under a canonical mapping. We call a maximal clique of $\Delta^{(n,q)}$ *small* (resp. *grand*) if it is the preimage of a small (resp. grand) clique of $\begin{bmatrix} V \\ 2 \end{bmatrix}$. A small clique of $\Delta^{(n,q)}$ has size $q^3 - 1$, a grand clique of $\Delta^{(n,q)}$ has size $q^{n-1} - 1$. The intersection of two distinct grand cliques has size $q - 1$, the intersection of two distinct small cliques has size at most $q - 1$, while the intersection of a small clique and a grand clique has size at most $q^2 - 1$.

Lemma 2.2 *Let* Γ *be locally* $\Delta^{(n,q)}$, *let* u, v, w *be vertices of* Γ *with* $u, w \in \Gamma_2(v)$. *If* $\Gamma(u, v, w)$ *contains a* q^2-*clique, then* u *and* w *are adjacent.*

Proof.   Let C be a q^2-clique contained in $\Gamma(u, v, w)$, and fix $x \in C$. Consider the graph $\Gamma(x) \cong \Delta^{(n,q)}$. Suppose that $\{v\} \cup C - \{x\}$ is contained in a grand clique of $\Gamma(x)$. Then $\{u\} \cup C - \{x\}$ must be contained in a small clique, say C_1. Also, $\{w\} \cup C - \{x\}$ must be contained in a small clique, say C_2, but $|C_1 \cap C_2| \geq |C - \{x\}| > q - 1$ forces $C_1 = C_2$. Thus u and w are adjacent. If $\{v\} \cup C - \{x\}$ is contained in a small clique, then C_1 and C_2 must be grand cliques, again we obtain $C_1 = C_2$. $\qquad\square$

Lemma 2.3 *Let* u, v, w *be vertices of* $\Gamma = \text{Alt}(n, q)$ *with* $u, w \in \Gamma_2(v)$, $w \in \Gamma(u)$. *Then* $\Gamma(u, v, w)$ *contains a* q^2-*clique.*

A *singular line* (or simply, a *line*) of a graph Γ is a set of the form $\{u, v\}^{\perp\perp}$, where u, v are adjacent vertices. If Γ is locally Δ for some regular graph Δ, then Γ is edge-regular, so that the incidence system whose points are vertices of Γ, lines are singular lines of Γ, becomes a gamma space ([1] 1.14.1). We call this gamma space the gamma space associated with the graph Γ. A subset of the form $v^{\perp}$, where v is a vertex of Γ, is a subspace of this gamma space, i.e., any line which is incident with at least two points of $v^{\perp}$ is contained in $v^{\perp}$. A hyperplane is a proper subspace which meets every line.

3. μ-graphs

In this section we determine μ-graphs of a graph Γ, where Γ is locally $\Delta^{(n,q)}$, namely, we determine how a μ-graph is embedded in the neighbourhood of a vertex in Γ.

Proposition 3.1 *Suppose that a graph Γ is locally $\Delta^{(n,q)}$, v is a vertex of Γ, and $\sigma : \Gamma(v) \rightarrow \left[\begin{smallmatrix} V \\ 2 \end{smallmatrix}\right]$ is a canonical mapping. If $u \in \Gamma_2(v)$, then either*

    (i) *there exist a $W \in \left[\begin{smallmatrix} V \\ 4 \end{smallmatrix}\right]$ and a nondegenerate alternating form γ on W such that $\sigma(\Gamma(u,v))$ is the set of 2-spaces of W, nonisotropic with respect to γ, or*

    (ii) *there exist a $W \in \left[\begin{smallmatrix} V \\ 4 \end{smallmatrix}\right]$ and $U \in \left[\begin{smallmatrix} V \\ 2 \end{smallmatrix}\right]$ such that $\sigma(\Gamma(u,v))$ is the set of 2-spaces of W which intersect trivially with U.*

    Moreover, if $\mu(\Gamma) = q^2(q^2 + 1)$, then (i) must occur.

Proposition 3.1 is a consequence of [2], Proposition 5.1, after a series of reductions given below.

    It is easy to see that a μ-graph of $\Delta^{(n,q)}$ is isomorphic to $M^{(q)}$, where $M^{(q)}$ is the $(q-1)$-clique extension of the $(q+1) \times (q+1)$ grid. By Lemma 2.1, a μ-graph of Γ is locally $M^{(q)}$. Thus, we wish to describe subgraphs of $\Delta^{(n,q)}$, which are locally $M^{(q)}$. By the following lemma, it suffices to consider a subgraph of the Grassmann graph $\left[\begin{smallmatrix} V \\ 2 \end{smallmatrix}\right]$, which is locally $M^{(q)}$.

Lemma 3.2 *If Σ is a subgraph of $\Delta^{(n,q)}$ and Σ is locally $M^{(q)}$, then Σ contains at most one vertex of each basic clique of $\Delta^{(n,q)}$.*

Proof. Suppose the contrary and let u, v be two vertices of Σ belonging to the same basic clique. Then $u \in \Sigma(v)$ and u is adjacent to any vertex of $\Sigma(v)$ other that u. This contradicts the fact that $\Sigma(v)$ is isomorphic to $M^{(q)}$.   □

    Suppose that Σ is a subgraph of the Grassmann graph $\left[\begin{smallmatrix} V \\ 2 \end{smallmatrix}\right]$ and that Σ is locally $M^{(q)}$. A μ-graph of $M^{(q)}$ is isomorphic to $K^{(q)}$, where $K^{(q)}$ is the disjoint union of two copies of the complete graph on $q - 1$ vertices. Again by Lemma 2.1, a μ-graph of Σ is locally $K^{(q)}$. Recall that $M^{(q)}$ is itself a clique extension.

Lemma 3.3 *Let Θ be a subgraph of $M^{(q)}$, and suppose that Θ is locally $K^{(q)}$. Then Θ contains at most one vertex of each basic clique of $M^{(q)}$. Write $M^{(q)} = \cup_{j=1}^{q+1} C_j$, where C_j are disjoint maximal cliques of $M^{(q)}$. Then there exists an index j_0 such that $|\Theta \cap C_j| = q$ holds for any j with $j = 1, 2, \ldots, q+1$, $j \neq j_0$.*

    Next we reduce the classification of subgraphs of the Grassmann graph $\left[\begin{smallmatrix} V \\ 2 \end{smallmatrix}\right]$ which are locally $M^{(q)}$, to the case $\dim V = 4$.

Lemma 3.4 *Let Σ be a subgraph of the Grassmann graph $\left[\begin{smallmatrix} V \\ 2 \end{smallmatrix}\right]$ and suppose that Σ is locally $M^{(q)}$. Then there exists a subspace $W \in \left[\begin{smallmatrix} V \\ 4 \end{smallmatrix}\right]$ such that $\Sigma \subset \left[\begin{smallmatrix} W \\ 2 \end{smallmatrix}\right]$.*

Proof. Let u, v be two nonadjacent vertices of Σ, and W the unique 4-dimensional subspace of V for which $u, v \in \begin{bmatrix} W \\ 2 \end{bmatrix}$ holds. Clearly, $\Sigma(u, v) \subset \begin{bmatrix} W \\ 2 \end{bmatrix}$. Indeed, more generally, if a vertex z is adjacent to two nonadjacent vertices in $\begin{bmatrix} W \\ 2 \end{bmatrix}$, then z belongs to $\begin{bmatrix} W \\ 2 \end{bmatrix}$. We first want to show that $\Sigma_2(v) \subset \begin{bmatrix} W \\ 2 \end{bmatrix}$. Write $\Sigma(v) = \cup_{i=1}^{q+1} R_i = \cup_{j=1}^{q+1} C_j$, where $\{R_i\}_{i=1}^{q+1}$, $\{C_j\}_{j=1}^{q+1}$ are the two families of disjoint maximal cliques (so that $R_i \cap C_j$ is a basic clique of $\Sigma(v)$). By Lemma 3.3, there exist indices i_0, j_0 such that $|\Sigma(u, v) \cap R_i| = q$ for $i \neq i_0$, $|\Sigma(u, v) \cap C_j| = q$ for $j \neq j_0$. Similarly, if $u' \in \Sigma_2(v)$, then there exist indices i'_0, j'_0 such that $|\Sigma(u', v) \cap R_i| = q$ for $i \neq i'_0$, $|\Sigma(u', v) \cap C_j| = q$ for $j \neq j'_0$. Since $q > 2$, there exist i_1, i_2, j_1, j_2 such that $i_1 \neq i_2$, $j_1 \neq j_2$, and

$$|\Sigma(u, v) \cap R_{i_1}| = |\Sigma(u, v) \cap R_{i_2}| = q,$$

$$|\Sigma(u, v) \cap C_{j_1}| = |\Sigma(u, v) \cap C_{j_2}| = q,$$

$$|\Sigma(u', v) \cap R_{i_1}| = |\Sigma(u', v) \cap R_{i_2}| = q,$$

$$|\Sigma(u', v) \cap C_{j_1}| = |\Sigma(u', v) \cap C_{j_2}| = q.$$

Let B_{ij} denote the basic clique $R_i \cap C_j$. Without loss of generality, we may assume

$$|\Sigma(u, v) \cap B_{i_1 j_1}| = |\Sigma(u, v) \cap B_{i_2 j_2}| = 1,$$

Therefore, $B_{i_1 j_2}, B_{i_2 j_1} \subset \begin{bmatrix} W \\ 2 \end{bmatrix}$, and then $B_{i_1 j_1}, B_{i_2 j_2} \subset \begin{bmatrix} W \\ 2 \end{bmatrix}$. Now

$$|\Sigma(u', v) \cap B_{i_1 j_1}| = |\Sigma(u', v) \cap B_{i_2 j_2}| = 1,$$

or

$$|\Sigma(u', v) \cap B_{i_1 j_2}| = |\Sigma(u', v) \cap B_{i_2 j_1}| = 1.$$

In either case $u' \in \begin{bmatrix} W \\ 2 \end{bmatrix}$. Therefore, we have proved $\Sigma_2(v) \subset \begin{bmatrix} W \\ 2 \end{bmatrix}$, and it follows immediately that $\Sigma(v) \subset \begin{bmatrix} W \\ 2 \end{bmatrix}$. Now it is easy to prove by induction that $\Sigma_i(v) \subset \begin{bmatrix} W \\ 2 \end{bmatrix}$ for all $i = 1, 2, \ldots$. $\qquad \square$

Remark

Lemma 3.4 is not valid when $q = 2$. A locally 3×3 grid subgraph of the Grassmann graph $\begin{bmatrix} V \\ 2 \end{bmatrix}$ over $GF(2)$ may not be contained in a 4-dimensional subspace. The Johnson graph $J(6, 3)$ is one of the two locally 3×3 grid graphs, and the classification of the embedding of $J(6, 3)$ into the Grassmann graph $\begin{bmatrix} V \\ 2 \end{bmatrix}$ over $GF(2)$ is given in [3]. There are two inequivalent embeddings, of dimension 4 and 5. The 5-dimensional embedding occurs as a μ-graph of the graph of the quadratic forms over $GF(2)$.

Now the determination of μ-graphs of Γ has reduced to the classification of the locally $M^{(q)}$ subgraph in the 4-dimensional Grassmann graph. Let W be a 4-dimensional vector space over $GF(q)$, $\mathcal{L}$ the set of pencils (x, π), where

$x \in \begin{bmatrix} W \\ 1 \end{bmatrix}$, $\pi \in \begin{bmatrix} W \\ 3 \end{bmatrix}$, $x \subset \pi$. The incidence system $(\begin{bmatrix} W \\ 2 \end{bmatrix}, \mathcal{L})$, where incidence of $U \in \begin{bmatrix} W \\ 2 \end{bmatrix}$ with the pencil (x, π) is given by $x \subset U \subset \pi$, is called the Grassmannian. Its collinearity graph is the Grassmann graph $\begin{bmatrix} W \\ 2 \end{bmatrix}$.

Lemma 3.5 *If Σ is a subgraph of the Grassmann graph $\begin{bmatrix} W \\ 2 \end{bmatrix}$, and if Σ is locally $M^{(q)}$, then $\begin{bmatrix} W \\ 2 \end{bmatrix} - \Sigma$ is a hyperplane of the Grassmannian $(\begin{bmatrix} W \\ 2 \end{bmatrix}, \mathcal{L})$.*

Proof. Suppose that l is a pencil not contained in $\begin{bmatrix} W \\ 2 \end{bmatrix} - \Sigma$, i.e., there exists a vertex $v \in \Sigma$ incident with l. Let Λ be the subgraph of $\begin{bmatrix} W \\ 2 \end{bmatrix}$ induced by the neighbours of v. Then Λ is a q-clique extension of the $(q + 1) \times (q + 1)$ grid. Since Σ is locally $M^{(q)}$, $\Sigma(v)$ is embedded in Λ in such a way that each basic clique of $\Sigma(v)$ is contained in a basic clique of Λ. Since the set of points incident with l is the union of $\{v\}$ and a basic clique of Λ, there exists a unique point on l not in Σ. In other words, l meets $\begin{bmatrix} W \\ 2 \end{bmatrix} - \Sigma$ at a unique point, proving the assertion. $\qquad \square$

The hyperplanes in the Grassmannian $(\begin{bmatrix} W \\ 2 \end{bmatrix}, \mathcal{L})$ were determined (among others) in [2]. Particularly, Proposition 3.1 follows now from [2], Proposition 5.1.

4. Subgraphs isomorphic to Alt$(4, q)$

In this section, we construct subgraphs isomorphic to Alt$(4, q)$ in our graph Γ, where Γ is assumed to be locally $\Delta^{(n,q)}$ and to satisfy $\mu = q^2(q^2 + 1)$. For any vertex v of Γ, fix a canonical mapping $\sigma_v : \Gamma(v) \rightarrow \begin{bmatrix} V \\ 2 \end{bmatrix}$. For a subspace W of V, define $S(v, W)$ by

$$S(v, W) = \{v\} \cup \tau_v(W) \cup \{u \in \Gamma_2(v) | \Gamma(u, v) \subset \tau_v(W)\},$$

where $\tau_v(W) = \sigma_v^{-1}(\begin{bmatrix} W \\ 2 \end{bmatrix})$. We want to prove the following proposition.

Proposition 4.1 *Let Γ be locally $\Delta^{(n,q)}$ with $\mu = q^2(q^2 + 1)$, let v be a vertex of Γ. Then the subgraph induced by $S(v, W)$ is isomorphic to Alt$(4, q)$ for any $W \in \begin{bmatrix} V \\ 4 \end{bmatrix}$.*

The proof of Proposition 4.1 will be given at the end of this section. The graph Alt$(4, q)$ has intersection array

$$\{(q^2 + 1)(q^3 - 1), q^2(q - 1)(q^3 - 1); 1, q^2(q^2 + 1)\}.$$

First of all, we compute some parameters of $S(v, W)$.

Lemma 4.2 *Let v be a vertex of Γ, and let $W \in \begin{bmatrix} V \\ 4 \end{bmatrix}$. For any $w \in S(v, W) \cap \Gamma(v)$, we have*

$$|\Gamma(w) \cap \Gamma_2(v) \cap S(v, W)| = q^4(q - 1).$$

In particular,

$$|\Gamma_2(v) \cap S(v,W)| = q^2(q-1)(q^3-1),$$

$$|S(v,W)| = q^6.$$

Proof. Note that if $u \in \Gamma(v) \cap \Gamma_2(w) \cap S(v,W)$, then $\Gamma(w,u) \subset S(v,W)$ by Lemma 3.4, since W is the only 4-subspace of V containing both $\sigma_v(w)$ and $\sigma_v(u)$. Counting in two ways the number of edges between $\Gamma(v) \cap \Gamma_2(w) \cap S(v,W)$ and $\Gamma_2(v) \cap \Gamma(w) \cap S(v,W)$, we obtain

$$|\Gamma(v) \cap \Gamma_2(w) \cap S(v,W)|(\mu(\Gamma) - \mu(\Delta^{(n,q)}) - 1)$$
$$= |\Gamma_2(v) \cap \Gamma(w) \cap S(v,W)|(\mu(\Gamma) - (q-1)(q+1)^2 - 1).$$

Here $(q-1)(q+1)^2$ appears as the valency of the μ-graph of Γ. Since $\mu(\Delta^{(n,q)}) = (q-1)(q+1)^2$ as well, and $|\Gamma(v) \cap \Gamma_2(w) \cap S(v,W)| = q^4(q-1)$ the desired equality follows. Counting in two ways the number of edges between $\Gamma(v) \cap S(v,W)$ and $\Gamma_2(v) \cap S(v,W)$, we obtain $|\Gamma_2(v) \cap S(v,W)| = q^2(q-1)(q^3-1)$. $\square$

Lemma 4.3 *Let γ be a nondegenerate alternating form on a 4-dimensional vector space W, and $U \in [{}^W_2]$. Then there exist nonisotropic subspaces $U_1, U_2 \in [{}^W_2]$ such that $U_1 \cap U \neq \emptyset$, $U_2 \cap U \neq \emptyset$, $U_1 \cap U_2 = \emptyset$.*

Lemma 4.4 *Let v be a vertex of Γ, and let $W \in [{}^V_4]$. If $v' \in S(v,W)$, then there exists a subspace $W' \in [{}^V_4]$ such that $S(v,W) = S(v',W')$.*

Proof. Since $S(v,W)$ is connected, we may assume without loss of generality $v' \in \Gamma(v) \cap S(v,W)$. Let $x \in \Gamma(v) \cap \Gamma_2(v') \cap S(v,W)$, and let W' be the unique 4-dimensional subspace of V satisfying $\tau_{v'}(W') \supset \Gamma(v',x)$. The graph $\Gamma(v,v') \cap S(v,W)$ is isomorphic to the join of the $q(q-1)$-clique extension of a $(q+1) \times (q+1)$ grid and a complete graph on $q-2$ vertices. It contains the graph $\Gamma(v,v',x)$ which is isomorphic to $M^{(q)}$. It is easy to see that the graph $\Gamma(v,v',x)$ must be embedded into $\Gamma(v,v') \cap S(v,W)$ in a natural way, i.e., each basic clique of $\Gamma(v,v',x)$ is contained in a basic clique of the $q(q-1)$-clique extension of a $(q+1) \times (q+1)$ grid. For any vertex $u \in \Gamma(v,v') \cap S(v,W)$, one can find two nonadjacent vertices $y, y' \in \Gamma(v,v',x,u)$. Since $\Gamma(v,v',x) \subset \tau_{v'}(W')$, we see that $u \in \tau_{v'}(W')$, i.e., $\Gamma(v,v') \cap S(v,W) \subset \tau_{v'}(W')$. For any vertex $u \in \Gamma(v') \cap \Gamma_2(v) \cap S(v,W)$, there exist by Lemma 4.3 two nonadjacent vertices $y, y' \in \Gamma(v,v',u) \cap S(v,W)$, so that $u \in \tau_{v'}(W')$, i.e., $\Gamma(v') \cap \Gamma_2(v) \cap S(v,W) \subset \tau_{v'}(W')$. Thus we have

$\Gamma(v') \cap S(v, W) \subset \tau_{v'}(W')$, but both sides have the same size by Lemma 4.2, so $\Gamma(v') \cap S(v, W) = \tau_{v'}(W')$.

Next suppose that $u \in S(v, W)$ is not adjacent to v'. There exist two nonadjacent vertices $y, y' \in \Gamma(v', u) \cap S(v, W)$. This is obvious if $u \in \Gamma(v)$, and follows easily from Lemma 4.3 if $u \in \Gamma_2(v)$.

By Lemma 3.4, this forces $\Gamma(u, v') \subset \tau_{v'}(W')$ so that $S(v, W) \subset S(v', W')$. Now Lemma 4.2 forces the equality. □

Now we are ready to prove Proposition 4.1. By Lemma 4.4, the graph $S(v, W)$ is locally $\Delta^{(4,q)}$, and $\mu = q^2(q^2 + 1)$. Given a locally $\Delta^{(4,q)}$ graph, one can construct a rank 4 affine polar space by taking as points the vertices, and as lines the singular lines. All affine polar spaces of rank ≥ 4 were determined in [2]. In particular, $\text{Alt}(4, q)$ is the only locally $\Delta^{(4,q)}$ graph having $\mu = q^2(q^2 + 1)$. Thus we obtain $S(v, W) \cong \text{Alt}(4, q)$.

We conclude the section with the following corollary.

Corollary 4.5 *Whenever* v, u_0 *and* u_1 *are vertices of* Γ *with* $u_0, u_1 \in \Gamma_2(v)$, *one has* $\Gamma(v, u_0) \neq \Gamma(v, u_1)$.

Proof. By Proposition 4.1, it suffices to check the statement for $\Gamma = \text{Alt}(4, q)$, in which case it is obvious. □

5. Isomorphism of the gamma spaces on neighbourhoods

Throughout this section, we let $\Gamma, \tilde{\Gamma}$ be locally $\Delta^{(n,q)}$ with $\mu(\Gamma) = \mu(\tilde{\Gamma}) = q^2(q^2 + 1)$, $q > 2$. In this section, we shall show that the gamma spaces $v^{\perp}$ and $\tilde{v}^{\perp}$ are isomorphic for any vertex v of Γ and any vertex $\tilde{v}$ of $\tilde{\Gamma}$. Moreover, any isomorphism of the gamma spaces $\tilde{v}^{\perp} \to v^{\perp}$ can be extended naturally to a mapping $\tilde{v}^{\perp} \cup \tilde{\Gamma}_2(\tilde{v}) \to v^{\perp} \cup \Gamma_2(v)$, if one takes $\tilde{\Gamma} = \text{Alt}(n, q)$.

Lemma 5.1 *If* v *is a vertex of* Γ *and if* W *is a 4-dimensional subspace of* V, *then* $S(v, W)$ *is a subspace of the gamma space associated with* Γ. *In particular, the set of lines of* Γ *contained in* $S(v, W)$ *is the set of lines of the gamma space associated with the subgraph induced by* $S(v, W)$.

Proof. Let v', w be two adjacent vertices in $S(v, W)$. We aim to show $l = \{v', w\}^{\perp\perp}$ is contained in $S(v, W)$. By Lemma 4.4, we may assume $v' = v$. Then $l = \{v\} \cup \tau_v(U)$ for some $U \in [\begin{smallmatrix} W \\ 2 \end{smallmatrix}]$, hence $l \subset S(v, W)$. □

Lemma 5.2 *If* v *is a vertex of* Γ *and if* U *is a 3-dimensional subspace of* V, *then* $\{v\} \cup \tau_v(U)$ *is a subspace of the gamma space associated with* Γ, *and it* *is isomorphic to the incidence system of points and lines of $AG(3,q)$.*

Proof. Take a 4-dimensional subspace W containing U. We may consider $\{v\} \cup \tau_v(U)$ as a subset of $S(v, W)$. By Proposition 4.1 and Lemma 5.1, the proof reduces to the case $\Gamma = \text{Alt}(4, q)$, for which it is straightforward. □

Lemma 5.3 *Suppose that M is a maximal clique of the graph $\text{Alt}(4, q)$ containing 0. If φ is a bijection of M onto M with $\varphi(0) = 0$, and φ stabilizes all lines contained in M, then there exists a unique extension of φ to an automorphism of $\text{Alt}(4, q)$ stabilizing all lines through 0.*

Proof. There is only one class of maximal cliques in $\text{Alt}(4, q)$ up to automorphisms. That is, M has size q^3 and the lines of $\text{Alt}(4, q)$ contained in M define an incidence system isomorphic to $AG(3, q)$. By the assumption φ must be the scalar multiplication with regard to this affine space structure, hence φ can be extended to an automorphism of $\text{Alt}(n, q)$. On the other hand, the subgroup of $\text{Aut} \, \text{Alt}(n, q)$ fixing 0 and all lines through 0 is precisely the group of scalar multiplications, thus the uniqueness follows. □

With the above lemmas as building blocks, we can construct an isomorphism of the gamma spaces $v^{\perp}$ and $\tilde{v}^{\perp}$, where v is a vertex of Γ and $\tilde{v}$ is a vertex of $\tilde{\Gamma}$. Fix a canonical mapping $\pi : \tilde{\Gamma}(\tilde{v}) \rightarrow \begin{bmatrix} V \\ 2 \end{bmatrix}$ and set $\tilde{\tau}(W) = \pi^{-1}(\begin{bmatrix} W \\ 2 \end{bmatrix})$,

$$\tilde{S}(\tilde{v}, W) = \{\tilde{v}\} \cup \tilde{\tau}(W) \cup \{w \in \tilde{\Gamma}_2(\tilde{v}) | \tilde{\Gamma}(\tilde{v}, w) \subset \tilde{\tau}(W)\},$$

where W is a subspace of V.

Proposition 5.4 *If the graphs Γ and $\tilde{\Gamma}$ are locally $\Delta^{(n,q)}$ with $\mu = q^2(q^2+1)$, and if v is a vertex of Γ and $\tilde{v}$ is a vertex of $\tilde{\Gamma}$, then the gamma spaces $v^{\perp}$ and $\tilde{v}^{\perp}$ are isomorphic.*

Proof. Fix a sequence of subspaces $V = V_n \supset V_{n-1} \supset \cdots \supset V_4$ with $\dim V_j = j$. By Lemma 4.1, we have $\tilde{S}(\tilde{v}, V_4) \cong S(v, V_4) \cong \text{Alt}(4, q)$. Moreover, we can choose an isomorphism between $\tilde{S}(\tilde{v}, V_4)$ and $S(v, V_4)$ in such a way that its restriction φ to $\{\tilde{v}\} \cup \tilde{\tau}(V_4)$ satisfies $\sigma_v \varphi = \pi$ on $\tilde{\tau}(V_4)$. By Lemma 5.1, this mapping φ is an isomorphism of the gamma spaces $\{\tilde{v}\} \cup \tilde{\tau}(V_4)$ and $\{v\} \cup \tau_v(V_4)$. We prove by induction on j, that there exists an isomorphism of gamma spaces $\varphi : \{\tilde{v}\} \cup \tilde{\tau}(V_j) \rightarrow \{v\} \cup \tau_v(V_j)$ satisfying $\sigma_v \varphi = \pi$ on $\tilde{\tau}(V_j)$. Suppose that the assertion holds for some $4 \leq j < n$. We need to define an extension of φ on the basic clique $\tilde{\tau}(U)$ for each $U \in \begin{bmatrix} V_{j+1} \\ 2 \end{bmatrix} - \begin{bmatrix} V_j \\ 2 \end{bmatrix}$. Choose a subspace $W \in \begin{bmatrix} V_{j+1} \\ 4 \end{bmatrix}$ with $U \subset W$. Note that φ is already defined on the subspace $\{\tilde{v}\} \cup \tilde{\tau}(W \cap V_j)$ which is isomorphic to $AG(3, q)$. By Lemma 5.3, φ can be extended to an isomorphism ψ of the graphs (and also of gamma spaces, by Lemma 5.1) $\tilde{S}(\tilde{v}, W)$ to $S(v, W)$, in such a way that $\sigma_v \psi = \pi$ holds on $\tilde{\tau}(W)$. The restriction of ψ on $\tilde{\tau}(U)$ gives the desired extension of φ. In order to check well-definedness, choose another 4-dimensional subspace W' of V_{j+1} containing U. By connectivity we may assume that $\dim(W \cap W') = 3$. Let ψ' be an isomorphism of the graphs $\tilde{S}(\tilde{v}, W')$ to $S(v, W')$ such that $\sigma_v \psi' = \pi$ holds on $\tilde{\tau}(W')$. The restriction of $\psi^{-1} \psi'$ to $\{\tilde{v}\} \cup \tilde{\tau}(W \cap W')$ is an automorphism of $\{\tilde{v}\} \cup \tilde{\tau}(W \cap W') \cong AG(3, q)$ stabilizing all lines through $\{\tilde{v}\}$ and fixing all points on the line $\{\tilde{v}\} \cup \tilde{\tau}(W \cap W' \cap V_j)$. This forces $\psi^{-1} \psi' = 1$ on $\{\tilde{v}\} \cup \tilde{\tau}(W \cap W')$, proving the well-definedness. At the same time we have proved that our extension of φ on $\{\tilde{v}\} \cup \tilde{\tau}(V_{j+1})$ maps lines in $\tilde{\tau}(W)$ to lines in $\tau_v(W)$ for any $W \in \binom{V_{j+1}}{4}$. This implies that the extension of φ to $\{\tilde{v}\} \cup \tilde{\tau}(V_{j+1})$ is an isomorphism of gamma spaces. Therefore, the proof is complete. $\quad\square$

Proposition 5.5 Let v be a vertex Γ and let $\tilde{v}$ be a vertex of $\tilde{\Gamma}$. Let φ be an isomorphism of the gamma spaces $\tilde{v}^\perp \to v^\perp$. Then there exists a bijection $\varphi_2 : \tilde{\Gamma}_2(\tilde{v}) \to \Gamma_2(v)$ such that $\varphi(\tilde{\Gamma}(\tilde{v}, \tilde{u})) = \Gamma(v, \varphi_2(\tilde{u}))$ for any $\tilde{u} \in \tilde{\Gamma}_2(\tilde{v})$. Moreover, if $\tilde{\Gamma} = \mathrm{Alt}(n, q)$, then φ_2 maps edges in $\tilde{\Gamma}_2(\tilde{v})$ to edges in $\Gamma_2(v)$.

Proof. Since $\tilde{\Gamma}_2(\tilde{v})$ is partitioned into sets of the form $\tilde{\Gamma}_2(\tilde{v}) \cap \tilde{S}(\tilde{v}, W)$, where $W \in \binom{V}{4}$, it suffices to prove the existence of φ_2 on each $\tilde{\Gamma}_2(\tilde{v}) \cap \tilde{S}(\tilde{v}, W)$. In other words, we may assume $n = 4$ and $\Gamma = \tilde{\Gamma} = \mathrm{Alt}(4, q)$, and the assertion is easy to verify in this case.

It remains to show that, if $\tilde{\Gamma} = \mathrm{Alt}(n, q)$, then σ_2 maps edges in $\tilde{\Gamma}_2(\tilde{v})$ to edges in $\Gamma_2(v)$. If $\tilde{u}$ and $\tilde{w}$ are adjacent vertices in $\tilde{\Gamma}_2(\tilde{v})$, then by Lemma 2.3, $\tilde{\Gamma}(\tilde{u}, \tilde{v}, \tilde{w})$ contains a q^2-clique. Then $\Gamma(\varphi_2(\tilde{u}), v, \varphi_2(\tilde{w}))$ contains a q^2-clique, so that $\varphi_2(\tilde{u})$ and $\varphi_2(\tilde{w})$ are adjacent by Lemma 2.2. $\quad\square$

Lemma 5.6 For any vertex v of Γ and any $W \in \binom{V}{5}$, $S(v, W) \cong \mathrm{Alt}(5, q)$ holds. Moreover, the set of lines contained in $S(v, W)$ is precisely the set of lines of the subgraph $S(v, W)$.

Proof. It is straightforward to check the assertion for $\Gamma = \mathrm{Alt}(n, q)$. Let $\tilde{\Gamma} = \mathrm{Alt}(n, q)$, $\tilde{v}$ a vertex of $\tilde{\Gamma}$. By Propositions 5.4 and 5.5, there exists a bijection φ from $\tilde{S}(\tilde{v}, W) \cong \mathrm{Alt}(5, q)$ to $S(v, W)$, such that φ maps edges to edges. We want to show that φ does not map a non-edge to an edge. It suffices to check this claim on the subgraph $\Omega = \tilde{S}(\tilde{v}, W) \cap \tilde{\Gamma}_2(\tilde{v})$. Let $u, w \in \Omega$ be non-adjacent, but suppose that $\varphi(u)$ and $\varphi(w)$ are adjacent. Notice that u and w are at distance 2 in $\tilde{\Gamma}$ and that $\tilde{\Gamma}(u, w) \subset \tilde{S}(\tilde{v}, W)$. Choose a vertex $x \in \tilde{\Gamma}(u, \tilde{v}) \cap \tilde{\Gamma}_2(w)$. Then $\Gamma(\varphi(x), \varphi(w)) \supset \varphi(\tilde{\Gamma}(x, w)) \cup \{\varphi(u)\}$, so that $\mu(\Gamma) = |\Gamma(\varphi(x), \varphi(w))| \geq \mu(\tilde{\Gamma}) + 1$. This is a contradiction, since $\mu(\Gamma) = \mu(\tilde{\Gamma}) = q^2(q^2 + 1)$. Thus we have proved that φ is an isomorphism, i.e., $S(v, W) \cong \mathrm{Alt}(5, q)$.

A line of the subgraph $S(v, W)$ is the intersection of two maximal cliques of size q^4, while a line l of Γ is the intersection of two maximal cliques of size q^{n-1}. A clique of size q^4 is contained in a unique maximal clique of size q^{n-1} in Γ, hence a line of $S(v, W)$ coincides with a line of Γ.   □

6.   Triangulability

Recall that a *path* is a sequence of vertices $(v_0, v_1, \ldots, v_s)$ such that v_i is adjacent to v_{i+1} for $0 \leq i < s$. Here s is the length of the path. A subpath of the form (u, v, u) is called a *return*. We do not distinguish paths, which can be obtained from each other by adding or removing returns. Clearly, this gives an equivalence relation on the set of paths. Equivalence classes of this relation are in a natural bijection with paths without returns.

A *closed path* (a *cycle*) is a path with $v_0 = v_s$. For cycles we also do not specially distinguish the starting vertex, i.e., two cycles obtained from one another by a cyclic permutation of vertices are considered as equivalent.

Given two cycles $v = (v_0, v_1, \ldots, v_s = v_0)$ and $w = (w_0, w_1, \ldots, w_t = w_0)$ with $v_0 = w_0$, we may form a cycle $v \cdot w = (v_0, v_1, \ldots, v_s, w_1, \ldots, w_t)$. Iterating this process and adopting to our concept of equivalence, we say that a cycle u can be *decomposed* into a product of cycles $v_1, \ldots, v_k$, whenever there are cycles $u', v'_1, \ldots, v'_k$, equivalent to $u, v_1, \ldots, v_k$, respectively, such that $u' = v'_1 \cdot \ldots \cdot v'_k$.

A graph is called *triangulable* if every cycle in it can be decomposed into a product of triangles (i.e., cycles of length 3). In this section we prove the following proposition, which follows immediately from Lemma 6.2 and Lemma 6.3.

Proposition 6.1 *The graph* $\mathrm{Alt}(n, q)$ *is triangulable.*

Lemma 6.2 *Let* Γ *be a graph. Assume that for any vertex v of Γ and $u_0, u_1 \in \Gamma_j(v)$, $j \geq 2$,*

(i) $\Gamma_{j-1}(u_0) \cap \Gamma(v)$ *is connected,*

(ii) *if u_0 and u_1 are adjacent, then* $\Gamma_{j-1}(u_0) \cap \Gamma_{j-1}(u_1) \cap \Gamma(v) \neq \emptyset$.

Then Γ is triangulable.

Proof.   Suppose $v = (v_0, v_1, \ldots, v_m)$ is a shortest cycle which cannot be decomposed into a product of triangles. Then it cannot be decomposed into a product of cycles of lengths $\leq m$, as well. In particular, no pair of its vertices can be joined in Γ by a path shorter than the distance between these vertices in the cycle. Consider the vertices in v which are at maximal distance from

$v = v_0$. Then, depending on the parity of m, it is either a single vertex, or an edge. Now, if $m \geq 4$, the conditions (i) and (ii), respectively, lead us to a decomposition of v, a contradiction. $\qquad \square$

Lemma 6.3 Let $\Gamma = \mathrm{Alt}(n, q)$, v, u_0, u_1 vertices of Γ with $u_0, u_1 \in \Gamma_j(v)$, $j \geq 2$. Then
> (i) $\Gamma(u_0) \cap \Gamma_{j-1}(v)$ is connected,
> (ii) if u_0 and u_1 are adjacent, then $\Gamma(u_0) \cap \Gamma(u_1) \cap \Gamma_{j-1}(v) \neq \emptyset$.

Proof. This was essentially proved in [1], see the proof of Proposition 9.5.12 for (i), and the proof of Proposition 9.5.13 for (ii). $\qquad \square$

7. Proof of the main theorem

The proof of the main theorem depends on a rather general covering theorem. Let $\Gamma, \tilde{\Gamma}$ be two graphs having the same local structure. Let $v, \tilde{v}$ be vertices of $\Gamma, \tilde{\Gamma}$, respectively. An isomorphism $\varphi : \tilde{v}^{\perp} \to v^{\perp}$ is called *extendable* if there is a bijection $\varphi' : \tilde{v}^{\perp} \cup \tilde{\Gamma}_2(\tilde{v}) \to v^{\perp} \cup \Gamma_2(v)$, mapping edges to edges, such that $\varphi'|_{\tilde{v}^{\perp}} = \varphi$. In this case we call φ' the extension of φ. We will say that Γ has *distinct μ-graphs* if $\Gamma(v, u_0) = \Gamma(v, u_1)$ for $u_0, u_1 \in \Gamma_2(v)$ implies $u_0 = u_1$. Clearly, this property implies that the extension φ' above is unique.

Theorem 7.1 Let $\Gamma, \tilde{\Gamma}$ be graphs having the same local structure. Assume that Γ has distinct μ-graphs and also:
> (i) There exists a vertex v_0 of Γ and a vertex $\tilde{v}_0$ of $\tilde{\Gamma}$, and an extendable isomorphism $\varphi : \tilde{v}_0^{\perp} \to v_0^{\perp}$.
> (ii) if $v, \tilde{v}$ are vertices of $\Gamma, \tilde{\Gamma}$, respectively, $\varphi : \tilde{v}^{\perp} \to v^{\perp}$ is an extendable isomorphism, φ' its extension, and $w \in \tilde{\Gamma}(\tilde{v})$, then $\varphi'|_{w^{\perp}} : w^{\perp} \to \varphi(w)^{\perp}$ is extendable.
> (iii) $\tilde{\Gamma}$ is triangulable.
> Then Γ is covered by $\tilde{\Gamma}$.

Proof. The proof will be given in a series of steps. First, let v be a vertex of Γ and let $\tilde{v}$ be vertex of $\tilde{\Gamma}$, let $\varphi : \tilde{v}^{\perp} \to v^{\perp}$ be an extendable isomorphism, and let φ' be the extension of φ. Let $w \in \tilde{\Gamma}(\tilde{v})$, $\tau = \varphi'|_{w^{\perp}} : w^{\perp} \to \Gamma(\varphi(w))$, τ' the extension of τ. Then $\varphi'(u) = \tau'(u)$ for any $u \in (\tilde{v}^{\perp} \cup \tilde{\Gamma}_2(\tilde{v})) \cap (w^{\perp} \cup \tilde{\Gamma}_2(w))$.

Indeed, by definition $\varphi'(u) = \tau'(u)$ for any $u \in w^{\perp}$. Let $u \in (\tilde{v}^{\perp} \cup \tilde{\Gamma}_2(\tilde{v})) \cap \tilde{\Gamma}_2(w)$. Since φ' maps edges to edges, $\varphi'(\tilde{\Gamma}(w, u)) \subset \Gamma(\varphi'(w), \varphi'(u))$. By definition of τ, we obtain $\tau(\tilde{\Gamma}(w, u)) \subset \Gamma(\varphi'(w), \varphi'(u))$. Since Γ has distinct μ-graphs by assumption, we obtain $\tau'(u) = \varphi'(u)$.

In particular, if $\tau = \varphi'|_{w^{\perp}}$ then $\varphi = \tau'|_{\tilde{v}^{\perp}}$.

Now let T be the covering tree of $\tilde{\Gamma}$, and let the corresponding covering be denoted by χ. If t is an arbitrary vertex of T, then the vertices of T are naturally indexed by the paths (without returns) in $\tilde{\Gamma}$ beginning with $\chi(t)$. Two vertices of T are adjacent if and only if the corresponding paths are of the form $(w_0, \ldots, w_i)$, $(w_0, \ldots, w_i, w_{i+1})$, where $w_0 = \chi(t)$.

We are going to construct a covering $\psi : T \to \Gamma$ such that for any vertex t of T,

(a) $\sigma_t : \chi(t)^\perp \xrightarrow{\chi^{-1}} t^\perp \xrightarrow{\psi} \psi(t)^\perp$ is an extendable isomorphism,

(b) $\psi(r) = \sigma_t'(\chi(r))$, where r is an arbitrary vertex in $T_2(t)$ and σ_t' is the extension of σ_t.

Let us pick a vertex t_0 in T. We will define ψ on $T = \cup_{i=0}^{\infty} T_i(t_0)$ by induction on i. Let v_0 be a vertex of Γ. Let σ be an extendable isomorphism $\chi(t_0)^\perp \to v_0^\perp$, σ' its extension. Define $\psi(t_0) = v_0$, $\psi(r) = \sigma'(\chi(r))$, for every $r \in t_0^\perp \cup T_2(t_0)$. Clearly, the conditions (a) and (b) hold for $t = t_0$.

Now suppose that ψ has been defined on $T^j = \cup_{i=0}^{j} T_i(t_0)$, $j \geq 2$, and suppose the conditions (a) and (b) hold for $t \in T^{j-2}$. Since T is a tree, for every $t \in T_{j+1}(t_0)$ there is exactly one $r \in T_{j-1}(t_0) \cap T_2(t)$. Let us define $\psi(t) = \sigma_r'(\chi(t))$.

Let us now choose an arbitrary $t \in T_{j-1}$. Let $d \in t^\perp \cap T_{j-2}(t_0)$. By induction, $\psi(r) = \sigma_d'(\chi(r))$ for every $r \in d^\perp \cup T_2(d)$. In particular, σ_t is extendable by (ii). Now let $\Omega = t^\perp \cup T_2(t)$. Clearly, $\Omega \cap T^j = \Omega \cap (d^\perp \cup T_2(d))$. By the first step, we have $\psi(r) = \sigma_d'(\chi(r)) = \sigma_t'(\chi(r))$ for $r \in \Omega \cap T^j$. If $r \in \Omega \cap T_{j+1}(t_0)$, then the same conclusion follows by definition. Hence (a) and (b) hold for t, and the covering ψ exists by induction.

Now, at the last step, we are going to prove that $\psi(t_1) = \psi(t_2)$, whenever $\chi(t_1) = \chi(t_2)$. Let $\tilde{v}$ be a cycle in $\tilde{\Gamma}$, t any of its lifting in T, and let v be the image of t in Γ. In such a case we say that v is an image of $\tilde{v}$. We say that a cycle in $\tilde{\Gamma}$ is good if any of its image is a cycle. Since $\chi(t_1) = \chi(t_2)$ if and only if any path between these two vertices maps onto a cycle in $\tilde{\Gamma}$, it suffice to prove that every cycle in $\tilde{\Gamma}$ is good.

Clearly, if a cycle is good, then any equivalent cycle is good as well. Also, a cycle, which can be decomposed into a product of good cycles, is good itself. Hence, by (iii), we have only to prove that every triangle is good. Let (t_0, t_1, t_2, t_3) be a path in T, which maps onto a triangle in $\tilde{\Gamma}$. By the preceding step, $\psi(t_0) = \sigma_{t_1}'(\chi(t_0)) = \sigma_{t_1}'(\chi(t_3)) = \psi(t_3)$.

We have shown that $\psi(t_1) = \psi(t_2)$, whenever $\chi(t_1) = \chi(t_2)$. Thus, $\psi\chi^{-1}$ is a well defined mapping from $\tilde{\Gamma}$ to Γ, which is a covering, since both χ and ψ are such. This completes the proof. $\square$

We are now ready for the final proof.

Proof of main theorem

We want to apply Theorem 7.1 with $\tilde{\Gamma} = \text{Alt}(n, q)$. First of all, Γ has distinct μ-graphs by Corollary 4.5. The hypo-thesis (i) is satisfied by Proposition 5.5, while the hypothesis (iii) by Proposition 6.1. It remains to verify (ii). Let $v, \tilde{v}$ be vertices of $\Gamma, \tilde{\Gamma}$, $\sigma : \tilde{v}^{\perp} \to v^{\perp}$ an extendable isomorphism, σ' its extension. Let $w \in \tilde{\Gamma}(\tilde{v})$. In order to show the extendability of $\sigma'|_{w^{\perp}}$, it suffices to prove that $\sigma'|_{w^{\perp}}$ is an isomorphism of gamma spaces. Let l be a line in $w^{\perp}$. If $w \in l$, then, clearly, $\sigma'(l)$ is a line. Suppose $w \notin l$. Then there exists a subspace $W \in \begin{bmatrix} V \\ 5 \end{bmatrix}$ such that $\{v\} \cup l \subset \tilde{S}(w, W)$. Since $\sigma'|_{w^{\perp}}$ is an isomorphism of graphs, $\sigma'(\tilde{S}(w, W)) = S(\sigma(w), W')$ for some $W' \in \begin{bmatrix} V \\ 5 \end{bmatrix}$. Since σ' maps edges to edges, and $\tilde{S}(w, W) \cong \text{Alt}(5, q) \cong S(\sigma(w), W')$ by Lemma 5.6, it follows that $\sigma'|_{\tilde{S}(w,W)}$ is an isomorphism of graphs. Also from Lemma 5.6, l is a line of the graph $\tilde{S}(w, W)$, so that $\tilde{\sigma}(l)$ is a line of the graph $S(\sigma(w), W')$, and hence $\tilde{\sigma}(l)$ is a line of Γ again by Lemma 5.6. Therefore, we have shown that $\sigma'|_{w^{\perp}}$ maps lines to lines, so $\sigma'|_{w^{\perp}}$ is extendable by Proposition 5.5. Now all hypotheses of Theorem 7.1 are satisfied, so that Γ is covered by $\text{Alt}(n, q)$. $\square$

References

[1] A. E. Brouwer, A. M. Cohen, and A. Neumaier. *Distance-regular Graphs.* Springer Verlag, Berlin, 1989.

[2] A. M. Cohen and E. E. Shult. Affine polar spaces. *Geom. Dedicata,* 35, pp. 43–76, 1990.

[3] A. Munemasa, D. V. Pasechnik, and S. V. Shpectorov. A local characterization of the graphs of alternating forms and the graphs of quadratic forms over $GF(2)$. In A. Beutelspacher, F. Buekenhout, F. De Clerck, J. Doyen, J. W. P. Hirschfeld, and J. A. Thas, editors, *Finite Geometry and Combinatorics,* pages 305–319, Cambridge, 1993. Cambridge University Press.

[4] D. V. Pasechnik. Geometric characterization of the graphs from the Suzuki chain. To appear in *European J. Combin.*

A. Munemasa, Department of Mathematics, Kyushu University, 6-10-1 Hakozaki, Higashi-ku, Fukuoka 812, Japan.
e-mail: munemasa@math.sci.kyushu-u.ac.jp

S. V. Shpectorov, Institute for System Analysis, 9 Pr. 60 Let Oktyabrya, 117312 Moscow, Russia. e-mail: ssh@cs.vniisi.msk.su

A local characterization of the graphs of alternating forms and the graphs of quadratic forms over GF(2)

A. Munemasa [*]   **D.V. Pasechnik** [†]
S.V. Shpectorov [‡]

Abstract

Let Δ be the line graph of PG($n-1$,2), Alt(n,2) be the graph of the n-dimensional alternating forms over GF(2), $n \geq 4$. Let Γ be a connected locally Δ graph such that

1. the number of common neighbours of any pair of vertices at distance two is the same as in Alt(n,2).
2. the valency of the subgraph induced on the second neighbourhood of any vertex is the same as in Alt(n,2).
   It is shown that Γ is covered either by Alt(n,2) or by the graph of $(n-1)$-dimensional GF(2)-quadratic forms Quad($n-1$,2).

1. Introduction

In this paper we investigate graphs which are locally the same as the graph Alt($n,2$) of alternating forms on an n-dimensional vector space V over GF(2). An analogous question for GF(q), $q > 2$, is considered in [4]. The local graph of Alt($n,2$) is isomorphic to the Grassmann graph $\begin{bmatrix} V \\ 2 \end{bmatrix}$, i.e. the line graph of PG($n-1$,2). Thus, we investigate graphs which are locally $\begin{bmatrix} V \\ 2 \end{bmatrix}$. It turns out that, besides Alt($n, 2$), there is another well-known graph which is locally $\begin{bmatrix} V \\ 2 \end{bmatrix}$. It is the graph Quad($n-1, 2$) of quadratic forms on an $(n-1)$-dimensional vector space over GF(2). Both Alt($n, 2$) and Quad($n-1, 2$) are distance regular and have the same parameters, though they are non-isomorphic if $n \geq 5$.

Consider a half dual polar space of type D_n over GF(2). Then the collinearity graph, induced by the complement of a geometric hyperplane,

---
[*]. This research was completed during this author's visit at the Institute for System Analysis, Moscow, as a Heizaemon Honda fellow of the Japan Association for Mathematical Sciences.
[†]. A part of this research was completed when this author held a position at the Institute for System Analysis, Moscow.
[‡]. A part of this research was completed during the visit at University of Technology, Eindhoven.

303

is always locally $\begin{bmatrix} V \\ 2 \end{bmatrix}$. The number of such hyperplanes, even taken up to the action of the automorphism group of the polar space, increases at least exponentially with n, so that in general, classification of locally $\begin{bmatrix} V \\ 2 \end{bmatrix}$ graphs seems to be a hard problem. In this paper, we restrict ourselves to the case $\mu = 20$, i.e., the number of common neighbours of two vertices at distance 2 is always 20. Both graphs $\text{Alt}(n,2)$ and $\text{Quad}(n-1,2)$ possess this property, as well as another property $a_2 = 15 \cdot 2^{n-1} - 105$, which means that the graph induced by the second neighbourhood of every vertex is regular of the shown valency. This latter condition is rather technical and it is used only once in the proof. Hopefully, in further research it may be shown superfluous.

Under the assumption $\mu = 20$, $\text{Alt}(4,2)$ is the only graph which is locally $\begin{bmatrix} V \\ 2 \end{bmatrix}$ with $n = 4$, while $\text{Alt}(5,2)$ and $\text{Quad}(4,2)$ are the only graphs which are locally $\begin{bmatrix} V \\ 2 \end{bmatrix}$ with $n = 5$. For large n, however, we cannot expect the analogous result, since the quotient graphs of $\text{Alt}(n,2)$ or $\text{Quad}(n-1,2)$ by many subgroups of translations have the same local structure and the same values of μ and a_2. The purpose of this paper is to show that $\text{Alt}(n,2)$ and $\text{Quad}(n-1,2)$ are universal in the following sense.

Main Theorem. *Let Γ be a connected graph which is locally the Grassmann graph $\begin{bmatrix} V \\ 2 \end{bmatrix}$, where V is a vector space of dimension n over $\text{GF}(2)$. Suppose that $\mu(\Gamma) = 20$ and $a_2(\Gamma) = 15 \cdot 2^{n-1} - 105$. Then Γ is covered by either $\text{Alt}(n,2)$ or $\text{Quad}(n-1,2)$.*

The following corollary was a motivation for this work.

Corollary. *If Γ is a distance-regular graph having the same intersection numbers and the same local structure as $\text{Alt}(n,2)$, then Γ is isomorphic to either $\text{Alt}(n,2)$ or $\text{Quad}(n-1,2)$.*

As a non-logical consequence of these investigations, new automorphisms of $\text{Quad}(n,q)$, q even, were found, mixing forms of the same rank, but with different Witt indices [3].

The contents of the paper is as follows. In Section 2 we collect definitions and state the notation. In Section 3 we check that the graph $\text{Quad}(n,2)$ has the prescribed local structure and that it is triangulable, that is, every cycle in it can be decomposed into a product of 3-cycles. Similar results for $\text{Alt}(n,2)$ have been known earlier. In Section 4 we determine the possibilities for the μ-graphs, their intersections, vertex-subgraph relationship, etc. This allows us to determine in Section 5 the second neighbourhood of a vertex. Finally, in Section 6 we apply the covering theorem from [4] to show that every graph under consideration is covered either by $\text{Alt}(n,2)$ or by $\text{Quad}(n-1,2)$.

2. Preliminaries

Let Γ, $\bar{\Gamma}$ be connected graphs. We say that Γ is a *cover* of $\bar{\Gamma}$ if there exists a mapping φ from Γ to $\bar{\Gamma}$ which maps edges to edges and, for every $\gamma \in \Gamma$, induces a bijection from $\Gamma(\gamma)$ to $\bar{\Gamma}(\varphi(\gamma))$. If the subgraph $\Gamma_2(\alpha)$ is regular for any vertex α of a graph Γ, and its valency is independent on α, then this valency is denoted by $a_2 = a_2(\Gamma)$.

Let V be an n-dimensional vector space over GF(2), $\begin{bmatrix} V \\ 2 \end{bmatrix}$ the Grassmann graph. A *grand clique* of $\begin{bmatrix} V \\ 2 \end{bmatrix}$ is a set of the form $\{\gamma \in \begin{bmatrix} V \\ 2 \end{bmatrix} | x \in \gamma\}$ where $0 \neq x \in V$. A *small clique* of $\begin{bmatrix} V \\ 2 \end{bmatrix}$ is a set of the form $\begin{bmatrix} W \\ 2 \end{bmatrix}$, where $W \in \begin{bmatrix} V \\ 3 \end{bmatrix}$. Any maximal clique of $\begin{bmatrix} V \\ 2 \end{bmatrix}$ is either grand or small. Two distinct grand cliques have exactly one vertex in common, and two distinct small cliques have at most one vertex in common. A grand clique and a small clique can have at most 3 common vertices. Thus, any 4-clique in $\begin{bmatrix} V \\ 2 \end{bmatrix}$ is contained in a unique maximal clique.

Let V be an n-dimensional vector space over GF(q). The *rank* of an alternating form γ on V is defined by rank $\gamma = \dim(V/\text{Rad}\,\gamma)$, where Rad $\gamma = \{u \in V | \gamma(u,v) = 0 \text{ for any } v \in V\}$. Note that rank γ is always even. The *alternating forms graph* Alt(n, q) has as vertices the alternating forms on V. Two alternating forms γ, δ are adjacent whenever rank $(\gamma - \delta) = 2$.

Let V be as before. A map $\gamma : V \to$ GF(q) is called a *quadratic form* if for any $u, v \in V$ and $a, b \in$ GF(q), $\gamma(au + bv) = a^2\gamma(u) + b^2\gamma(v) + abB_\gamma(u,v)$ for some bilinear form B_γ. We call B_γ the *bilinear form associated with* γ. In case q is even, B_γ is an alternating form. The *rank* of γ is defined by rank $\gamma = \dim(V/\text{Rad}\,\gamma)$, where Rad $\gamma = \{u \in \text{Rad}\,B_\gamma | \gamma(u) = 0\}$. If q is even and rank $\gamma = \dim V$ odd, then the 1-dimensional space Rad B_γ is called the *nucleus* of γ. The *quadratic forms graph* Quad(n, q) has as vertices the quadratic forms on V. Two quadratic forms γ, δ are adjacent whenever rank $(\gamma - \delta) = 1$ or 2. The graph Quad(n, q) is distance-regular and has the same parameters as Alt($n + 1, q$) [1].

3. Some properties of Quad(n-1,2)

It is well-known that Alt($n, 2$) is locally the Grassmann graph $\begin{bmatrix} V \\ 2 \end{bmatrix}$, where V is an n-dimensional vector space over GF(2).

Proposition 3.1 *The graph* $\Gamma = $ Quad($n - 1, 2$) *is locally the Grassmann graph* $\begin{bmatrix} V \\ 2 \end{bmatrix}$, *where* V *is a vector space of dimension* n *over* GF(2).

Proof. Let $V = W \oplus \langle e_0 \rangle$, where Γ is the set of all quadratic forms on W. Define the mapping $\varphi : \Gamma(0) \to \begin{bmatrix} V \\ n-2 \end{bmatrix}$ as follows. For $\gamma \in \Gamma(0)$, define $\varphi(\gamma) = \text{Rad}\,\gamma \in \begin{bmatrix} W \\ n-2 \end{bmatrix} \subset \begin{bmatrix} V \\ n-2 \end{bmatrix}$ if rank $\gamma = 1$. If rank $\gamma = 2$, then write $W = \text{Rad}\,\gamma \oplus \langle x_1, x_2 \rangle$. Define $\varphi(\gamma)$ by $\varphi(\gamma) = \text{Rad}\,\gamma \oplus \langle e_0 + (1 + \gamma(x_2))x_1 +$

$(1 + \gamma(x_1))x_2\rangle \in [^{V}_{n-2}]$. It is easy to see that φ is well-defined and bijective. Since $\Gamma(0)$ and $[^{V}_{n-2}]$ have the same valency, φ is an isomorphism if and only if it preserves adjacency.

Let $\gamma, \delta \in \Gamma(0)$ be adjacent. If $\operatorname{rank} \gamma = \operatorname{rank} \delta = 1$, then clearly $\varphi(\gamma)$ is adjacent to $\varphi(\delta)$. If $\operatorname{rank} \gamma = 1$ and $\operatorname{rank} \delta = 2$, then we have $\operatorname{rank}(\gamma + \delta) = 2$ and $\operatorname{Rad} \delta \subset \operatorname{Rad} \gamma$. Thus $\operatorname{Rad} \delta \subset \varphi(\gamma) \cap \varphi(\delta)$, i.e., $\varphi(\gamma)$ is adjacent to $\varphi(\delta)$. If $\operatorname{rank} \gamma = \operatorname{rank} \delta = 2$ and $\operatorname{rank}(\gamma + \delta) = 1$, then $\operatorname{Rad} \gamma = \operatorname{Rad} \delta \subset \varphi(\gamma) \cap \varphi(\delta)$, i.e., $\varphi(\gamma)$ is adjacent to $\varphi(\delta)$. Finally suppose that $\operatorname{rank} \gamma = \operatorname{rank} \delta = \operatorname{rank}(\gamma + \delta) = 2$. Then $\dim(\operatorname{Rad} \gamma \cap \operatorname{Rad} \delta) = n - 4$. Choose $x, y, z \in W$ in such a way that $W = (\operatorname{Rad} \gamma \cap \operatorname{Rad} \delta) \oplus \langle x, y, z \rangle$, $\operatorname{Rad} \gamma = (\operatorname{Rad} \gamma \cap \operatorname{Rad} \delta) \oplus \langle y \rangle$, and $\operatorname{Rad} \delta = (\operatorname{Rad} \gamma \cap \operatorname{Rad} \delta) \oplus \langle z \rangle$. Then $\operatorname{Rad}(\gamma + \delta) = (\operatorname{Rad} \gamma \cap \operatorname{Rad} \delta) \oplus \langle y + z \rangle$. Since $\gamma(y) = \delta(z) = (\gamma + \delta)(y + z) = 0$, we have $\gamma(z) = \delta(y)$, so that the element $e_0 + (1 + \gamma(z))x + (1 + \delta(x))y + (1 + \gamma(x))z$ belongs to $\varphi(\gamma) \cap \varphi(\delta)$. Thus, $\dim \varphi(\gamma) \cap \varphi(\delta) \le n - 3$, i.e., $\varphi(\gamma)$ is adjacent to $\varphi(\delta)$. $\quad\square$

A graph is called *triangulable* if every cycle in it can be decomposed into a product of 3-cycles (cf. [4]).

Proposition 3.2 *If q is a power of 2, then the graph $\operatorname{Quad}(n, q)$ is triangulable.*

To prove triangulability of a graph Γ it suffices to check the following two conditions (γ a fixed vertex of Γ):

(T1) $\Gamma_{j-1}(\delta) \cap \Gamma(\gamma)$ *is connected for every* $\delta \in \Gamma_j(\gamma)$, $j \ge 2$;

(T2) *if* $\delta_0, \delta_1 \in \Gamma_j(\gamma), j \ge 2$ *are adjacent, then the subsets* $\Gamma_{j-1}(\delta_0) \cap \Gamma(\gamma)$ *and* $\Gamma_{j-1}(\delta_1) \cap \Gamma(\gamma)$ *are at distance at most 1 from each other.*

For a proof of this criterion see the proof of Lemma 6.2 in [4]. From now on we denote by V an n-dimensional vector space over $\operatorname{GF}(q)$, q even. Let $\Gamma = \operatorname{Quad}(n, q), \bar{\Gamma} = \operatorname{Alt}(n, q)$ and let φ be the mapping from Γ to $\bar{\Gamma}$ defined by $\gamma \mapsto B_\gamma$. Clearly, φ takes adjacent vertices to adjacent or equal. Moreover, two quadratic forms are mapped to the same vertex of $\operatorname{Alt}(n, q)$ if and only if their difference has rank 1, i.e., such forms are adjacent.

The following technical facts were taken from [1], see 9.5.5(i) for (i), and 9.6.2 for (ii). A proof for (iii) can be found in [1], page 292.

Lemma 3.3 (i) *Let γ and δ be two alternating forms on V. If $\operatorname{rank}(\gamma + \delta) = \operatorname{rank} \gamma + \operatorname{rank} \delta$, then $\operatorname{Rad} \gamma + \operatorname{Rad} \delta = V$ and $\operatorname{Rad} \gamma \cap \operatorname{Rad} \delta = \operatorname{Rad}(\gamma + \delta)$.*

(ii) *Let γ and δ be quadratic forms on V of rank $2j + 1$ and 2, respectively. If $\operatorname{rank}(\gamma + \delta) = 2j$ then $\operatorname{Rad} \delta \cap \operatorname{Rad} B_\gamma = \operatorname{Rad} \gamma$.*

(iii) *Let γ be a rank 2 quadratic form and δ a rank 2 alternating form on V, such that $\operatorname{rank}(B_\gamma + \delta) = 2$. Then there is a rank 2 quadratic form δ' with $B_{\delta'} = \delta$, such that $\gamma + \delta'$ also has rank 2.*

306

Lemma 3.4 *Let γ be a quadratic form on V of even rank $2j$. Then φ establishes an isomorphism between the subgraphs $\Gamma(0) \cap \Gamma_{j-1}(\gamma)$ and $\bar{\Gamma}(0) \cap \bar{\Gamma}_{j-1}(B_\gamma)$. In particular, $\Gamma(0) \cap \Gamma_{j-1}(\gamma)$ is connected.*

Proof. Clearly, φ maps the former set to the latter. Hence, it suffices to prove that every alternating form from $\bar{\Gamma}(0) \cap \bar{\Gamma}_{j-1}(B_\gamma)$ has exactly one preimage in $\Gamma(0) \cap \Gamma_{j-1}(\gamma)$, and the preimage of an edge is an edge. Let $\beta \in \bar{\Gamma}(0) \cap \bar{\Gamma}_{j-1}(B_\gamma)$. By Lemma 3.3(i), $\operatorname{Rad}\beta \supset \operatorname{Rad}B_\gamma$. Hence, we may assume for simplicity that $\operatorname{Rad}\gamma = 0$. Let us consider V as a symplectic space with respect to the form B_γ. Then $U = \operatorname{Rad}(B_\gamma + \beta)$ is orthogonal to $\operatorname{Rad}\beta$, hence by Lemma 3.3(i), $U = \operatorname{Rad}\beta^\perp$. Let a quadratic form δ be defined by $\operatorname{Rad}\delta = \operatorname{Rad}\beta$ and $\delta|_U = \gamma|_U$. Then $B_\delta = \beta$ and $\operatorname{Rad}(\gamma + \delta) = U$, hence $\gamma + \delta$ has rank $2j - 2$. On the other hand, if for a quadratic form δ one has $B_\delta = \beta$ and $\gamma + \delta$ has rank $2j - 2$ then $\operatorname{Rad}(\gamma + \delta) = \operatorname{Rad}(B_\gamma + \beta) = U$, and hence δ must be defined as above. Hence, φ is indeed a bijection between $\Gamma(0) \cap \Gamma_{j-1}(\gamma)$ and $\bar{\Gamma}(0) \cap \bar{\Gamma}_{j-1}(B_\gamma)$.

Now suppose that δ_1, δ_2 are two rank 2 quadratic forms, such that both $\delta_1' = \gamma + \delta_1$ and $\delta_2' = \gamma + \delta_2$ have rank $2j - 2$. Let $U_i = \operatorname{Rad}(B_{\delta_i'})$, $i = 1, 2$. Suppose $B_{\delta_1} + B_{\delta_2}$ has rank 2. By Lemma 3.3(i), it means that $Z = \operatorname{Rad}B_{\delta_1} \cap \operatorname{Rad}B_{\delta_2}$ has codimension at most 3 in V. In particular, U_1 and U_2, which are both orthogonal to Z, generate at most a 3-dimensional subspace. It follows that $T = U_1 \cap U_2$ is nontrivial. Now the equality $\gamma = \delta_1 + \delta_1' = \delta_2 + \delta_2'$ implies $\delta_1 + \delta_2 = \delta_1' + \delta_2'$. Therefore, both Z and T are in the radical of $\delta_1 + \delta_2$. Since Z has codimension at most 3, and $T \not\subset Z$, we finally have that $\delta_1 + \delta_2$ has rank at most 2. □

Now let us consider the case $\operatorname{rank}\gamma = 2j - 1$. Let, as in [1], $R_s(\gamma)$ denotes the set of quadratic forms δ, such that $\gamma + \delta$ has rank s. Then $\Gamma(0) \cap \Gamma_{j-1}(\gamma)$ consists of three parts, namely, $\Omega_1(\gamma) = R_1(0) \cap R_{2j-2}(\gamma)$, $\Omega_2(\gamma) = R_2(0) \cap R_{2j-3}(\gamma)$ and $\Omega_3(\gamma) = R_2(0) \cap R_{2j-2}(\gamma)$. Clearly, $\Omega_1(\gamma)$ is a clique. Furthermore, $\Omega_2(\gamma)$ consists of all rank 2 forms in the preimage of $\bar{\Gamma}(0) \cap \bar{\Gamma}_{j-2}(B_\gamma)$. In particular, $\Omega_2(\gamma)$ induces a connected subgraph.

Lemma 3.5 *If $\operatorname{rank}\gamma = 2j - 1$ then $\Gamma(0) \cap \Gamma_{j-1}(\gamma)$ is a connected subgraph.*

Proof. Let $\delta \in \Omega_3(\gamma)$. By Lemma 3.3(ii) $\operatorname{Rad}\gamma = \operatorname{Rad}B_\gamma \cap \operatorname{Rad}\delta$. Hence there is a hyperplane U in V, such that $\operatorname{Rad}\delta \subset U$ and $\operatorname{Rad}\gamma = \operatorname{Rad}B_\gamma \cap U$. Define a rank 1 quadratic form α by $\operatorname{Rad}\alpha = U$ and $\alpha|_{\operatorname{Rad}B_\gamma} = \gamma|_{\operatorname{Rad}B_\gamma}$. Clearly, $\alpha \in \Omega_1(\gamma)$ and δ, α are connected by an edge. Since $\Omega_1(\gamma)$ is a clique, it means that $\Omega_1(\gamma) \cup \Omega_3(\gamma)$ induces a connected subgraph.

Since $\Omega_2(\gamma)$ is connected, it remains to find an edge between $\Omega_2(\gamma)$ and $\Omega_3(\gamma)$. Let δ be as above and let $U = \operatorname{Rad}\delta + \operatorname{Rad}B_\gamma$. Then U is a hyperplane in V and, clearly, we can find an alternating form β in $\bar{\Gamma}(0) \cap \bar{\Gamma}_{j-2}(B_\gamma)$, such that $\operatorname{Rad}\beta \subset U$. Then by Lemma 3.3(iii) there is a rank 2 quadratic form α with $B_\alpha = \beta$, such that $\delta + \alpha$ has rank 2. On the other hand, clearly, $\alpha \in \Omega_2(\gamma)$. $\qquad\square$

Lemmas 3.4 and 3.5 give (T1). Next we check (T2).

Lemma 3.6 *If $\gamma, \delta \in \Gamma_j(0)$ are adjacent, then there is an edge between $\Gamma(0) \cap \Gamma_{j-1}(\gamma)$ and $\Gamma(0) \cap \Gamma_{j-1}(\delta)$.*

Proof. Let B_γ and B_δ be of rank $2s$ and $2k$, respectively. We have seen above that $\varphi(\Gamma(0) \cap \Gamma_{j-1}(\gamma))$ contains $\bar{\Gamma}(0) \cap \bar{\Gamma}_{s-1}(B_\gamma)$ and $\varphi(\Gamma(0) \cap \Gamma_{j-1}(\delta))$ contains $\bar{\Gamma}(0) \cap \bar{\Gamma}_{k-1}(B_\delta)$. If $k = s$ then [4], Lemma 6.3(ii) implies that $\bar{\Gamma}(0) \cap \bar{\Gamma}_{s-1}(B_\gamma) \cap \bar{\Gamma}_{k-1}(B_\delta) \neq \emptyset$. If $s < k$ then $\bar{\Gamma}(0) \cap \bar{\Gamma}_{s-1}(B_\gamma) \subset \bar{\Gamma}(0) \cap \bar{\Gamma}_{k-1}(B_\delta)$, and, in particular, we obtain the same conclusion as above. Similarly in the case $s > k$. Hence in any case $\varphi(\Gamma(0) \cap \Gamma_{j-1}(\gamma)) \cap \varphi(\Gamma(0) \cap \Gamma_{j-1}(\delta)) \neq \emptyset$. Since the preimage of any vertex of $\bar{\Gamma}$ is a clique, we obtain the desired edge between $\Gamma(0) \cap \Gamma_{j-1}(\gamma)$ and $\Gamma(0) \cap \Gamma_{j-1}(\delta)$. $\qquad\square$

Since both conditions (T1) and (T2) have been checked, the proof of Proposition 3.2 is complete.

4. μ-graphs

We will say that Γ *has distinct μ-graphs* if $\Gamma(v, u_0) = \Gamma(v, u_1)$ for $u_0, u_1 \in \Gamma_2(v)$ implies $u_0 = u_1$. Assume that the graph Γ is locally $\begin{bmatrix} V \\ 2 \end{bmatrix}$, where V is an n-dimensional vector space over GF(2). Any μ-graph of $\begin{bmatrix} V \\ 2 \end{bmatrix}$ is a 3×3 grid, and $\begin{bmatrix} V \\ 2 \end{bmatrix}$ has distinct μ-graphs. It follows (see [5], Lemma 1) that any μ-graph of Γ is locally a 3×3 grid and that Γ has distinct μ-graphs.

Lemma 4.1 *Suppose $\mu = \mu(\Gamma) = 20$. Then every μ-graph of Γ is isomorphic to $J(6,3)$, and Γ has distinct μ-graphs.*

Proof. It is well-known that there are precisely two connected locally a 3×3 grid graphs, as a reference we can suggest [2], where these graphs appear as a very special case. The Johnson graph $J(6,3)$ is the only locally a 3×3 grid graph with 20 vertices. $\qquad\square$

Now we are going to determine all embeddings of $J(6,3)$ into the Grassmann graph $\begin{bmatrix} V \\ 2 \end{bmatrix}$. We denote by $A(W)$ the set of all nondegenerate alternating forms on W, where W is a 4-dimensional vector space over GF(2). We also denote by $Q(W, x)$ the set of all nondegenerate quadratic forms on W with nucleus $\langle x \rangle$, where W is a 5-dimensional vector space over GF(2), $0 \neq x \in W$. For $\gamma \in A(W)$, we define

$$M_\gamma = \{U \in \begin{bmatrix} W \\ 2 \end{bmatrix} \mid U \text{ is nonisotropic with respect to } \gamma\}.$$

Then $|M_\gamma| = 20$.

For $\gamma \in Q(W, x)$, we define

$$
\begin{aligned}
M_\gamma &= \{U \in [\tbinom{W}{2}] | \gamma(u) = 1 \text{ for all } u \in U, u \neq 0\}. \\
T_\gamma &= \{U \in [\tbinom{W}{2}] | x \notin U, \gamma|_U \neq 0, \gamma \text{ is linear on } U\}. \\
T_\gamma^0 &= \{U \in [\tbinom{W}{2}] | \gamma = 0 \text{ on } U\}.
\end{aligned}
$$

Then $|M_\gamma| = 20$, $|T_\gamma| = 45$, $|T_\gamma^0| = 15$.

The subgraph of $[\tbinom{W}{2}]$ induced by M_γ, where $\gamma \in A(W)$, or $Q(W, x)$, is isomorphic to $J(6, 3)$. We aim to show that these two are the only subgraphs of $[\tbinom{W}{2}]$ isomorphic to $J(6, 3)$ up to automorphisms of $[\tbinom{W}{2}]$.

For a finite set Ω, consider the incidence system $\Omega_{2,3} = ((\tbinom{\Omega}{2}), (\tbinom{\Omega}{3}))$, where $(\tbinom{\Omega}{i})$ denotes the set of i-subsets of Ω, and the incidence is defined by inclusion. A representation φ of $\Omega_{2,3}$ is an injective mapping $\varphi : (\tbinom{\Omega}{2}) \to V - \{0\}$, where V is a vector space over GF(2), satisfying the property

$$
\varphi(\{i, j\}) + \varphi(\{j, k\}) + \varphi(\{j, i\}) = 0
$$

for any $\{i, j, k\} \in (\tbinom{\Omega}{3})$. Since the Johnson graph $J(|\Omega|, 3)$ is the line graph of $\Omega_{2,3}$, we obtain an embedding of the Johnson graph into the Grassmann graph $[\tbinom{V}{2}]$ whenever we have a representation of $\Omega_{2,3}$. One can construct the universal representation of $\Omega_{2,3}$ in the sense that any linear relation between the vectors representing pairs from $(\tbinom{\Omega}{2})$ follows from the above relations. The dimension of the space generated by the image of $(\tbinom{\Omega}{2})$ in the universal representation is the dimension of the left null space of the incidence matrix for $\Omega_{2,3}$ taken over GF(2). If $|\Omega| = 6$, then one checks easily that the dimension of the universal representation is 5.

Lemma 4.2 *Let V be an n-dimensional vector space over* GF(2) *and M a subgraph of $[\tbinom{V}{2}]$ isomorphic to $J(6, 3)$. Then one of the following holds.*

(i) *There exist $W \in [\tbinom{V}{4}]$, $\gamma \in A(W)$ such that $M = M_\gamma$.*

(ii) *There exist $W \in [\tbinom{V}{5}]$, $0 \neq x \in W$, $\gamma \in Q(W, x)$ such that $M = M_\gamma$.*

Proof. The graph $J(6, 3)$ has 30 maximal cliques. It can be shown that there are 15 maximal cliques, each of which is contained in a grand clique of $[\tbinom{V}{2}]$. This establishes a representation of $\Omega_{2,3}$ in V. One checks easily that the embedding (ii) is obtained from the universal representation, while (i) is the only quotient of it. □

In the rest of this section we list some properties of the subgraphs $J(6, 3)$ of $[\tbinom{V}{2}]$, mostly found by a computer program. In every case a check is straightforward. Table 1 is self-explanatory. The Petersen graph is the complement of $J(5, 2)$. By the type of a quadratic form, we mean the rank r together with a

description of γ and δ				$M_\gamma \cap M_\delta$	# of δ
$\gamma \in Q(W,x)$	$\delta \in Q(W,x)$ $\delta \neq \gamma$	type of $\gamma + \delta$	1	K_4	15
			2+	N_8	45
			2−	$\emptyset$	15
			3	K_2	180
			4+	3-claw	120
			4−	$\emptyset$	72
$\gamma \in A(W_1)$	$\delta \in A(W_1)$ $x \in W_1 \in [{}^W_4]$	type of γ on Rad $(B_\gamma|_{W_1} + \delta)$	2−	K_4	4
			2+	K_2	12
		Rad $(B_\gamma|_{W_1} + \delta) = 0$		K_2	12
	$\delta \in Q(W,x)$ $x \in W_1 \in [{}^W_4]$	type of δ on Rad $(B_\delta|_{W_1} + \gamma)$	2−	K_4	64
			2+	K_2	192
		Rad $(B_\delta|_{W_1} + \gamma) = 0$		K_2	192
	$\delta \in A(W_1)$ $\delta \neq \gamma$	rank $(\gamma + \delta) = 2$		N_{12}	15
		rank $(\gamma + \delta) = 4$		Petersen	12
	$\delta \in A(W_2)$ dim $W_1 \cap W_2 = 3$	$\gamma|_{W_1 \cap W_2} = \delta|_{W_1 \cap W_2}$		K_4	4
		$\gamma|_{W_1 \cap W_2} \neq \delta|_{W_1 \cap W_2}$		K_2	24

Table 1. Intersection of two μ-graphs

sign "+" or "−" when r is even. The sign "+" indicates that the Witt index is $r/2$, "−" indicates that it is $r/2 - 1$.

Let N_{12} be the graph with 12 vertices $\{(i,j)|i = 1,2, \ j = 1,\ldots,6\}$ where two distinct vertices $(i_1,j_1),(i_2,j_2)$ are adjacent if and only if $|j_1 - j_2| = 0,1$, or 5. Let N_8 be the induced subgraph of N_{12} on the vertices $\{(i,j)|i = 1,2, \ j = 1,\ldots,4\}$.

Lemma 4.3 Let $\gamma \in A(W)$ with $W \in [{}^V_4]$, or $\gamma \in Q(W,x)$ with $W \in [{}^V_5]$. For $U \in [{}^V_2]$, set $N_{\gamma,U} = \{U_1 \in M_\gamma | U \cap U_1 \neq 0\}$ and regard $N_{\gamma,U}$ as a subgraph of M_γ.

(i) If $\gamma \in A(W)$, $U \in [{}^V_2] - M_\gamma$, and $N_{\gamma,U} \neq \emptyset$, then $N_{\gamma,U}$ is isomorphic to either N_{12} or K_4. Moreover, $N_{\gamma,U} \cong N_{12}$ if and only if $U \in [{}^W_2] - M_\gamma$.

(ii) If $\gamma \in Q(W,x)$, $U \in [{}^V_2]-M_\gamma$, and $N_{\gamma,U} \neq \emptyset$, then $N_{\gamma,U}$ is isomorphic to either N_8 or K_4. Moreover, $N_{\gamma,U} \cong N_8$ if and only if $U \in T_\gamma$.

(iii) The number of $U \in [{}^V_2] - M_\gamma$ with $N_{\gamma,U} \neq \emptyset$ is $15 \cdot 2^{n-1} - 105$ if $\gamma \in A(W)$ and is $15 \cdot 2^{n-1} - 120$ if $\gamma \in Q(W,x)$.

Lemma 4.4 Let $\gamma \in A(W)$ or $Q(W,x)$, $M_\gamma \supset N \cong N_8$. Then N generates W, i.e., there is no proper subspace W_1 of W for which $N \subset [{}^{W_1}_2]$ holds.

Lemma 4.5 Let $\gamma \in Q(W, x)$, $\delta \in Q(W, y)$. If $M_\gamma \cap M_\delta \cong N_8$, then $x = y$.

Lemma 4.6 (i) *The graph with vertex set $A(W)$ whose edges are (γ, δ), where $M_\gamma \cap M_\delta \cong N_{12}$, is a connected graph of valency 15.*

(ii) *The graph with vertex set $Q(W, x)$ whose edges are (γ, δ), where $M_\gamma \cap M_\delta \cong N_8$, is a connected graph of valency 45.*

Lemma 4.7 (i) Let $U_1, U_2 \in \begin{bmatrix} W \\ 2 \end{bmatrix}$, $\dim W = 4$, $U_1 \cap U_2 = 0$. Then

$$|\{\gamma \in A(W)|U_1 \in M_\gamma,\ U_2 \in M_\gamma\}| = 10.$$

(ii) Let $U_1, U_2 \in \begin{bmatrix} W \\ 2 \end{bmatrix}$, $\dim W = 5$, $U_1 \oplus U_2 \oplus \langle x \rangle = W$. Then

$$|\{\gamma \in Q(W, x)|U_1 \in M_\gamma,\ U_2 \in M_\gamma\}| = 10.$$

Lemma 4.8 Suppose that $\gamma, \delta \in Q(W, x)$, the type of $\gamma + \delta$ is $2-$. Then

$$\sum_{W_1 \in \begin{bmatrix} W \\ 4 \end{bmatrix},\ W_1 \ni x} |\{\alpha \in A(W_1)|M_\alpha \cap M_\gamma \supseteq K_4,\ M_\alpha \cap M_\delta \supseteq K_4\}| = 24.$$

5. Determination of the second neighbourhood

In this section, we assume that Γ is locally $\begin{bmatrix} V \\ 2 \end{bmatrix}$ with $\mu = 20$, and $a_2 = 15 \cdot 2^{n-1} - 105$, where $\begin{bmatrix} V \\ 2 \end{bmatrix}$ is the Grassmann graph on a vector space V of dimension n over GF(2). The assumption on a_2 is, however, unnecessary before Lemma 5.5. Let u, v be vertices of Γ at distance 2. Then u is said to be of type 1 (resp. of type 2) with respect to v if the subgraph $\Gamma(u, v)$ in the Grassmann graph $\Gamma(v)$ satisfies Lemma 4.2(i) (resp. (ii)).

Lemma 5.1 Let u, v be vertices of Γ at distance 2. Then u is of type 1 with respect to v if and only if v is of type 1 with respect to u. If u is of type 1 with respect to v, then

$$|\{w \in \Gamma_2(v) \cap \Gamma(u)|\Gamma(u, v, w) \cong N_{12}\}| = 15.$$

If u is of type 2 with respect to v, then

$$|\{w \in \Gamma_2(v) \cap \Gamma(u)|\Gamma(u, v, w) \cong N_8\}| = 45.$$

Proof. If v is of type 2 with respect to u, then by Lemma 4.3 there exists a vertex $w \in \Gamma_2(v) \cap \Gamma(u)$ such that $\Gamma(u,v,w) \cong N_8$. By Lemma 4.4, $\Gamma(u,v)$ and $\Gamma(w,v)$ generate the same subspace in the Grassmann graph $\Gamma(v)$. In particular, u and w are of the same type with respect to v. According to Table 1, N_8 does not occur as the intersection of two μ-graphs of vertices of type 1. Thus, u and w are of type 2 with respect to v, proving the first assertion. The rest of the statements follows from Lemma 4.3. $\square$

For the rest of this section, fix an arbitrary vertex v_0 of Γ and identify $\Gamma(v_0)$ with $\begin{bmatrix} V \\ 2 \end{bmatrix}$. By Lemmas 4.2 and 4.1, one can identify $\Gamma_2(v_0)$ with a subset of $(\bigcup_{W_1} A(W_1)) \cup (\bigcup_{W_2, x} Q(W_2, x))$ in such a way that $U \in \begin{bmatrix} V \\ 2 \end{bmatrix}$ and $\gamma \in \Gamma_2(v_0)$ are adjacent if and only if $U \in M_\gamma$. The goal of this section is to determine $\Gamma_2(v_0)$ as such a subset.

Lemma 5.2 (i) *If $A(W) \cap \Gamma_2(v_0) \neq \emptyset$, then $A(W) \subset \Gamma_2(v_0)$.*
(ii) *If $Q(W, x) \cap \Gamma_2(v_0) \neq \emptyset$, then $Q(W, x) \subset \Gamma_2(v_0)$.*

Proof. (i) Let $\gamma \in A(W) \cap \Gamma_2(v_0)$. Then by Lemma 5.1, we have

$$|\{\delta \in \Gamma_2(v_0) \cap \Gamma(\gamma) | M_\delta \cap M_\gamma \cong N_{12}\}| = 15.$$

Lemma 4.4 implies that if $M_\delta \cap M_\gamma \cong N_{12}$, then $\delta \in A(W)$. By Lemma 4.6(i) we find $A(W) \subset \Gamma_2(v_0)$.
(ii) The proof is analogous, using Lemmas 4.5, 4.6(ii) and Table 1. $\square$

Lemma 5.3 *For any $W \in \begin{bmatrix} V \\ 4 \end{bmatrix}$, one and only one of the following holds.*
(i) $A(W) \subset \Gamma_2(v_0)$,
(ii) $Q(\tilde{W}, x) \subset \Gamma_2(v_0)$ *for exactly one $\tilde{W} \in \begin{bmatrix} V \\ 5 \end{bmatrix}$ and exactly one $x \in \tilde{W}$*
with $\tilde{W} = W \oplus \langle x \rangle$.

Proof. Choose $U_1, U_2 \in \begin{bmatrix} W \\ 2 \end{bmatrix}$ such that $U_1 \cap U_2 = \emptyset$. Since $|\Gamma(U_1, U_2) \cap v_0^\perp| = 10 < \mu(\Gamma)$, we can find a vertex γ in $\Gamma_2(v_0) \cap \Gamma(U_1, U_2)$. If γ is of type 1 with respect to v_0, then $\gamma \in A(W)$, so that (i) holds by virtue of Lemma 5.2(i). If γ is of type 2 with respect to v_0, then $\gamma \in Q(\tilde{W}, x)$ for some $\tilde{W} \in \begin{bmatrix} V \\ 5 \end{bmatrix}$, $x \in \tilde{W}$ with $W \subset \tilde{W}$. It follows easily that $x \notin W$. By Lemma 5.2(ii), $Q(\tilde{W}, x) \subset \Gamma_2(v_0)$. If (i) and (ii) hold simultaneously (resp. if the pair $(\tilde{W}, x)$ is not unique), then $\Gamma(U_1, U_2)$ contains subsets $\Gamma(U_1, U_2) \cap v_0^\perp$, $\Gamma(U_1, U_2) \cap Q(\tilde{W}, x)$, $\Gamma(U_1, U_2) \cap A(W)$, (resp. $\Gamma(U_1, U_2) \cap Q(\tilde{W}', x')$ for another pair $(\tilde{W}', x')$), each having cardinality 10 by Lemma 4.7. Since the intersections are clearly trivial, we obtain a contradiction with $\mu = 20$. $\square$

Lemma 5.4 *Let $\gamma, \delta \in \Gamma_2(v_0)$, $M_\gamma \cap M_\delta \neq \emptyset$. Then γ is adjacent to δ if and only if $M_\gamma \cap M_\delta$ contains a 4-clique.*

Proof. By Lemma 4.3(i–ii), if γ is adjacent to δ, then $M_\gamma \cap M_\delta$ is one of N_{12}, N_8, or K_4, all of which contain K_4. The converse is a special case of [4], Lemma 2.2. $\qquad\square$

Lemma 5.5 *If* $Q(W, x) \subset \Gamma_2(v_0)$, $W_1 \in [^V_4]$, $\dim(W \cap W_1) = 3$ *and* $x \notin W_1$, *then* $A(W_1) \not\subset \Gamma_2(v_0)$.

Proof. Suppose $A(W_1) \subset \Gamma_2(v_0)$. Let $\gamma \in A(W_1)$, and choose $\delta \in Q(W, x)$ in such a way that $M_\gamma \cap [^{W \cap W_1}_2] = M_\delta \cap [^{W \cap W_1}_2]$. Then $M_\gamma \cap M_\delta \cong K_4$, hence γ and δ are adjacent by Lemma 5.4. It is easy to see that there exists an $U_0 \in M_\delta$ such that $U_0 \cap W_1 = \emptyset$. On the other hand, Lemma 4.3(iii) together with the assumption on a_2 implies

$$\begin{aligned}\Gamma_2(\gamma) \cap \Gamma(v_0) &= \{U \in [^V_2] - M_\gamma | N_{\gamma, U} \neq \emptyset\} \\ &= \{U \in [^V_2] - M_\gamma | U \cap W_1 \neq \emptyset\}.\end{aligned}$$

This is a contradiction since $U_0 \in \Gamma_2(\gamma)$. $\qquad\square$

For brevity, let us call a 4-space W of type A (resp. of type Q) if Lemma 5.3 (i) (resp. (ii)) holds.

Lemma 5.6 *Let* $W_0, W_1 \in [^V_4]$, $\dim(W_0 \cap W_1) = 3$. *Suppose that* W_0 *is of type A, and* W_1 *is of type Q, i.e.,* $Q(W_1', x) \subset \Gamma_2(v_0)$ *with* $W_1' = W_1 \oplus \langle x \rangle$. *Then*

    (i) $x \in W_0 \subset W_1'$,
    (ii) *every* $W \in [^{W_1'}_4]$ *with* $x \in W$ *is of type A.*
    (iii) *every* $W \in [^V_4]$ *with* $x \in W$ *and* $\dim(W \cap W_1') = 3$, *is of type A.*

Proof. (i) If $W_0 \not\subset W_1'$, then $W_0 \cap W_1' = W_0 \cap W_1$, which is impossible by Lemma 5.5. Thus, $W_0 \subset W_1'$. If $x \notin W_0$, then $W_1' = W_0 \oplus \langle x \rangle$, which would imply, by Lemma 5.3, that W_0 is of type Q.

(ii) Since W_0 is of type A, we may assume $W \neq W_0$. If W is of type Q, then $Q(W', y) \subset \Gamma_2(v_0)$ for some $y \notin W$, $W' = W \oplus \langle y \rangle$. By (i), we have $y \in W_0 \subset W'$. This implies $W' = W_1'$, and by Lemma 5.3, $y = x \in W$, a contradiction. Thus, W is of type A.

(iii) If W is of type Q, then $Q(W', y) \subset \Gamma_2(v_0)$ for some $y \notin W$, $W' = W \oplus \langle y \rangle$. Let U_1, U_2, U_3 be the 4-spaces of W_1' containing $W \cap W_1'$. Then U_1, U_2, U_3 are of type A by (ii), and hence by (i), we have $U_i \subset W'$ for $i = 1, 2, 3$. This forces $W' = W_1'$, which is a contradiction. $\qquad\square$

Let $\tilde{Q}(W, x) = Q(W, x) \cup (\bigcup_{x \in W_1 \in [^W_4]} A(W_1))$.

Lemma 5.7 *We have either*

(i)
$$\Gamma_2(v_0) = \bigcup_{W \in \left[\begin{smallmatrix} V \\ 4 \end{smallmatrix}\right]} A(W),$$

or

(ii)
$$\Gamma_2(v_0) = \bigcup_{W \in \left[\begin{smallmatrix} V \\ 5 \end{smallmatrix}\right], W \ni x} \tilde{Q}(W, x)$$

for some nonzero element $x \in V$.

Proof. Suppose first that every vertex $u \in \Gamma_2(v_0)$ is of type 1 with respect to v_0. Then $\Gamma_2(v_0)$ is a union of $A(W)$'s, so the first equality holds.

Next suppose that there is a vertex $u \in \Gamma_2(v_0)$ of type 2 with respect to v_0. This means that there exists a $W_1 \in \left[\begin{smallmatrix} V \\ 4 \end{smallmatrix}\right]$ of type Q. Since $|Q(W, x)|$ does not divide $|\Gamma_2(v_0)|$ ($|\Gamma_2(v_0)|$ is known since μ is known), there exists a $W_0 \in \left[\begin{smallmatrix} V \\ 4 \end{smallmatrix}\right]$ of type A. Since the Grassmann graph $\left[\begin{smallmatrix} V \\ 4 \end{smallmatrix}\right]$ is connected, we may assume without loss of generality that $\dim(W_0 \cap W_1) = 3$. By Lemma 5.6(i), we see that $\tilde{Q}(W_1', x) \subset \Gamma_2(v_0)$ for some $x \in W_0 - W_1$, $W_1' = W_1 \oplus \langle x \rangle$. Let $W_2' \in \left[\begin{smallmatrix} V \\ 5 \end{smallmatrix}\right]$, $x \in W_2'$, $\dim(W_1' \cap W_2') = 4$. We want to show that $\tilde{Q}(W_2', x) \subset \Gamma_2(v_0)$. In order to do so, it suffices to prove that $Q(W_2', x) \subset \Gamma_2(v_0)$, by virtue of from Lemma 5.6(ii). Let $W_2 \in \left[\begin{smallmatrix} W_2' \\ 4 \end{smallmatrix}\right]$ with $x \notin W_2 \cap W_1' \in \left[\begin{smallmatrix} W_1' \\ 3 \end{smallmatrix}\right]$. By Lemma 5.5, W_2 must be of type Q. Applying Lemma 5.6(i) for $(W_1' \cap W_2', W_2)$, we see that $\tilde{Q}(W_2', y) \subset \Gamma_2(v_0)$ for some $y \in W_1' \cap W_2' - W_2$. If $y \neq x$, then there exists a $W \in \left[\begin{smallmatrix} W_2' \\ 4 \end{smallmatrix}\right]$ such that $x \in W$, $y \notin W$. Since $W_2' = W \oplus \langle y \rangle$, W is of type Q, while W is of type A by Lemma 5.6(iii). This contradiction proves $x = y$, so that $Q(W_2', x) \subset \Gamma_2(v_0)$. We have shown $\tilde{Q}(W_2', x) \subset \Gamma_2(v_0)$ for any $W_2' \in \left[\begin{smallmatrix} V \\ 5 \end{smallmatrix}\right]$ with $x \in W_2'$, $\dim(W_1' \cap W_2') = 4$. Now it follows from connectivity of the Grassmann graph $\left[\begin{smallmatrix} V/\langle x \rangle \\ 4 \end{smallmatrix}\right]$ that $\tilde{Q}(W_2', x) \subset \Gamma_2(v_0)$ for every $W_2' \in \left[\begin{smallmatrix} V \\ 5 \end{smallmatrix}\right]$ with $x \in W_2'$. Finally, a simple counting shows that the sets $\tilde{Q}(W, x)$ cover the whole of $\Gamma_2(v_0)$. $\square$

It is straightforward to check that Lemma 5.7(i) holds if $\Gamma = \mathrm{Alt}(n, 2)$, and Lemma 5.7(ii) holds if $\Gamma = \mathrm{Quad}(n-1, 2)$. Moreover, two vertices γ, δ of $\mathrm{Quad}(n-1, 2)$ at distance 2 are of type 1 to each other if and only if $\mathrm{rank}(\gamma + \delta) = 3$. If Γ satisfies Lemma 5.7(ii), then the grand clique $C = \{U \in \left[\begin{smallmatrix} V \\ 2 \end{smallmatrix}\right] | U \ni x\}$ of the Grassmann graph $\left[\begin{smallmatrix} V \\ 2 \end{smallmatrix}\right] \cong \Gamma(v_0)$ will be called the *nucleus* with respect to v_0. If $u \in \Gamma_2(v_0)$, then $\Gamma(v_0, u) \cap C = K_4$ or $\emptyset$ depending on whether u is of type 1 or 2 with respect to v_0, and the nucleus is the only such grand clique. The following lemma gives a more convenient characterization of the nucleus.

Lemma 5.8 *Let γ be a type 2 vertex with respect to v_0, and $W \in \left[\begin{smallmatrix} V \\ 5 \end{smallmatrix}\right]$ be the subspace generated by M_γ. Then the nucleus C with respect to v_0 is* *characterized by the properties* $[^W_2] \cap C \neq \emptyset$ *and*

$$\{U \in M_\gamma | U \cap U_0 \neq \emptyset\} \cong K_4 \quad \text{for any } U_0 \in [^W_2] \cap C.$$

Proof. Note that γ can be regarded as a nondegenerate quadratic form on W, and its nucleus x is characterized by the property: for any $U_0 \in [^W_2]$ with $U_0 \ni x$, U_0 contains a unique nonsingular 1-space distinct from $\langle x \rangle$. There are exactly four elements in M_γ containing a given nonsingular 1-space. $\qquad \square$

6. Proof of the main theorem

Let Γ and $\tilde{\Gamma}$ be graphs with the same local structure, let v be a vertex of Γ and let $\tilde{v}$ be a vertex of $\tilde{\Gamma}$. An isomorphism $\sigma : \tilde{v}^\perp \to v^\perp$ with $\sigma(\tilde{v}) = v$ is called *extendable* if there exists a bijection $\sigma' : \tilde{v}^\perp \cup \Gamma_2(\tilde{v}) \to v^\perp \cup \Gamma_2(v)$, mapping edges to edges and satisfying $\sigma'|_{\tilde{v}^\perp} = \sigma$. In this case the mapping σ' is called the *extension* of σ.

Let Γ be locally $[^V_2]$ with $\mu = 20$ and $a_2 = 15 \cdot 2^{n-1} - 105$, where V is a vector space of dimension n over GF(2).

Lemma 6.1 *Suppose v_0 is a vertex of Γ, such that Lemma 5.7(i) holds. Take $\tilde{\Gamma} = \text{Alt}(n, 2)$ and let $\tilde{v} \in \tilde{\Gamma}$. Then every isomorphism $\sigma : \tilde{v}^\perp \to v^\perp$ is extendable.*

Proof. Clearly, every μ-graph $\tilde{\Gamma}(\tilde{v}, \tilde{u})$ in $\tilde{v}^\perp$ is mapped onto a μ-graph $\Gamma(v, u)$ in $v^\perp$. By the second part of Lemma 4.1, such a vertex u is defined uniquely and, hence, we can define the extension σ' by $\sigma'(\tilde{u}) = u$. It remains to check that σ' maps an edge to an edge. Unless both ends of the edge in $\tilde{v}^\perp \cup \tilde{\Gamma}_2(\tilde{v})$ belong to $\tilde{\Gamma}_2(\tilde{v})$, the claim follows by definition. Let $u \in \tilde{\Gamma}_2(\tilde{v})$. Identifying $\tilde{\Gamma}(u)$ with $[^V_2]$, it follows from Lemma 4.3(iii) that the set

$$S = \{w \in \tilde{\Gamma}(u) - \tilde{\Gamma}(u, \tilde{v}) | \tilde{\Gamma}(u, \tilde{v}, w) \supset K_4\}$$

has cardinality $a_2 = 15 \cdot 2^{n-1} - 105$. By Lemma 5.4, S is contained in $\tilde{\Gamma}(u) \cap \tilde{\Gamma}_2(\tilde{v})$, whose cardinality is also a_2. This implies $S = \tilde{\Gamma}(u) \cap \tilde{\Gamma}_2(\tilde{v})$. Now if (u, w) is an edge in $\tilde{\Gamma}_2(\tilde{v})$, then $\tilde{\Gamma}(u, \tilde{v}, w) \supset K_4$, so that $\Gamma(\sigma'(u), v, \sigma'(w)) \supset K_4$. Again by Lemma 5.4, we see that $(\sigma'(u), \sigma'(w))$ is an edge. $\qquad \square$

Lemma 6.2 *Suppose v_0 is a vertex of Γ, such that Lemma 5.7(ii) holds. Take $\tilde{\Gamma} = \text{Quad}(n-1, 2)$ and let $\tilde{v} \in \tilde{\Gamma}$. Then an isomorphism $\sigma : \tilde{v}^\perp \to v^\perp$ is extendable if and only if it maps the nucleus with respect to $\tilde{v}$ to the nucleus with respect to v. In particular, if there exist type 2 vertices $u \in \Gamma_2(v)$ and $\tilde{u} \in \tilde{\Gamma}_2(\tilde{v})$, such that $\sigma(\tilde{\Gamma}(\tilde{v}, \tilde{u})) = \Gamma(v, u)$, then σ is extendable.*

Proof. The "only if" part is obvious, so let us consider an isomorphism $\sigma : \tilde{v}^{\perp} \to v^{\perp}$, which maps the nucleus with respect to $\tilde{v}$ to the nucleus with respect to v. As in Lemma 6.1, the μ-graphs in $\tilde{v}^{\perp}$ are mapped onto μ-graphs in $v^{\perp}$ and this gives us the extension σ'. We only need to check that σ' maps edges from $\tilde{\Gamma}_2(\tilde{v})$ onto edges.

Let $u \in \tilde{\Gamma}_2(\tilde{v})$. If u is of type 1 with respect to $\tilde{v}$ then, as in the previous lemma, we show that for every edge (u, w) in $\tilde{\Gamma}_2(\tilde{v})$ we have $\tilde{\Gamma}(u, \tilde{v}, w) \supset K_4$ and hence $(\sigma'(u), \sigma'(w))$ is an edge by Lemma 5.4. Suppose u is of type 2 with respect to $\tilde{v}$. Consider once again the set $S = \{w \in \tilde{\Gamma}(u) - \tilde{\Gamma}(u, \tilde{v}) | \tilde{\Gamma}(u, \tilde{v}, w) \supset K_4\} \subset \tilde{\Gamma}(u) \cap \tilde{\Gamma}_2(\tilde{v})$. This time, $|S| = a_2 - 15$ by Lemma 4.3 (iii). Identify now $\tilde{\Gamma}(\tilde{v})$ with $\begin{bmatrix} V \\ 2 \end{bmatrix}$ and suppose $u = \gamma \in Q(W, x) \subset \tilde{\Gamma}_2(\tilde{v})$, where $0 \neq x \in W \in \begin{bmatrix} V \\ 5 \end{bmatrix}$. From Table 1, there are 15 elements $\delta \in Q(W, x)$ such that the type of $\gamma + \delta$ is $2-$, and for such a vertex δ, $M_{\gamma} \cap M_{\delta} = \emptyset$ holds. By Lemma 4.8, there are 24 vertices in $\tilde{Q}(W, x)$ which are of type 1 with respect to $\tilde{v}$ and are adjacent to both γ and δ. Since $\mu = 20 < 24$, γ and δ must be adjacent. Therefore, $\tilde{\Gamma}(u) \cap \tilde{\Gamma}_2(\tilde{v})$ consists of the disjoint union of S and the set

$$S_0 = \{\delta \in Q(W, x) | \text{type of } \gamma + \delta \text{ is } 2-\}.$$

Clearly, $\sigma'(w)$ is adjacent to $\sigma'(u)$ if $w \in S$. If $\delta \in S_0$, then by the first part of the proof, the 24 type 1 common neighbours of $\gamma = u$ and δ in $\tilde{Q}(W, x)$ are mapped to common neighbours of $\sigma'(u)$ and $\sigma'(\delta)$, so that $\sigma'(u)$ and $\sigma'(\delta)$ are adjacent since $\mu < 24$.

The last statement of the lemma follows from Lemma 5.8. $\qquad\square$

Proof of Main Theorem. First suppose that for any vertices u, v of Γ at distance 2, u is of type 1 with respect to v. Let $\tilde{\Gamma} = \text{Alt}(n, 2)$. Then for any vertex v of Γ and for any vertex $\tilde{v}$ of $\tilde{\Gamma}$ and for any isomorphism $\sigma : \tilde{v}^{\perp} \to v^{\perp}$, σ is extendable by Lemma 6.1. Let σ' be the extension of σ. The mapping σ' maps edges to edges. Hence, if $w \in \tilde{\Gamma}(\tilde{v})$, then $\sigma'|_{w^{\perp}}$ is also extendable by Lemma 6.1. Now, by [4], Proposition 6.1, all the hypotheses of [4], Theorem 7.1 are satisfied, so Γ is covered by $\text{Alt}(n, 2)$.

Next suppose that there exist vertices u_0, v_0 of Γ which are of type 2 with respect to each other. Again we want to use [4], Theorem 7.1, this time with $\tilde{\Gamma} = \text{Quad}(n-1, 2)$. By Lemma 6.2, there exists an extendable isomorphism $\tilde{v}_0^{\perp} \to v_0^{\perp}$, where $\tilde{v}_0$ is a vertex of $\tilde{\Gamma}$. Suppose that $\tilde{v}$ is a vertex of $\tilde{\Gamma}$ and that v is a vertex of Γ and $\sigma : \tilde{v}^{\perp} \to v^{\perp}$ is an extendable isomorphism with the extension σ'. Then σ' maps edges to edges. Let $w \in \tilde{\Gamma}(\tilde{v})$. We want to show that $\sigma'|_{w^{\perp}}$ is extendable. If $u \in \tilde{v}^{\perp} \cup \tilde{\Gamma}_2(\tilde{v})$, then clearly $\sigma'(\tilde{\Gamma}(w, u)) = \Gamma(\sigma'(w), \sigma'(u))$. If, moreover, u is of type 2 with respect to w, then, as well, $\sigma'(u)$ is of type 2 with respect to $\sigma'(w)$. By Lemma 6.2, it implies that the nucleus with respect to w is mapped to the nucleus with respect to $\sigma'(w)$. Again by Lemma 6.2 we obtain that $\sigma'|_{w^{\perp}}$ is extendable.

316

It remains to find a vertex $u \in \tilde{v}^{\perp} \cup \tilde{\Gamma}_2(\tilde{v})$, which is of type 2 with respect to w. Since $\tilde{\Gamma} = \mathrm{Quad}(n-1, 2)$, it means that we must find, for every quadratic form γ of rank ≤ 2, a quadratic form δ of rank at most 4, such that $\mathrm{rank}\,(\gamma + \delta) = 4$. This is easy to check, so that $\sigma'|_{w^{\perp}}$ is indeed extendable. Triangulability of $\mathrm{Quad}(n-1, 2)$ has been shown in Proposition 3.2. Now all the hypotheses of [4], Theorem 7.1 are satisfied, so Γ is covered by $\mathrm{Quad}(n-1, 2)$.

$\square$

References

[1] **A. E. Brouwer, A. M. Cohen, and A. Neumaier.** *Distance-regular Graphs.* Springer Verlag, Berlin, 1989.

[2] **F. Buekenhout and X. Hubaut.** Locally polar spaces and related rank 3 groups. *J. Algebra,* 45, pp. 391–434, 1977.

[3] **A. Munemasa, D. V. Pasechnik, and S. V. Shpectorov.** Automorphism group of the graph of quadratic forms over a finite field of characteristic 2. Preprint, 1992.

[4] **A. Munemasa and S. V. Shpectorov.** A local characterization of the graphs of alternating forms. In A. Beutelspacher, F. Buekenhout, F. De Clerck, J. Doyen, J. W. P. Hirschfeld, and J. A. Thas, editors, *Finite Geometry and Combinatorics,* pages 289–302, Cambridge, 1993. Cambridge University Press.

[5] **D. V. Pasechnik.** Geometric characterization of the graphs from the Suzuki chain. To appear in *European J. Combin.*

A. Munemasa, Department of Mathematics, Kyushu University, 6-10-1 Hakozaki, Higashi-ku, Fukuoka 812, Japan.
e-mail: munemasa@math.sci.kyushu-u.ac.jp

D.V. Pasechnik, Department of Mathematics, University of Western Australia, Nedlands 6009 WA, Australia.
e-mail: dima@madvax.maths.uwa.edu.au

S. V. Shpectorov, Institute for System Analysis, 9 Pr. 60 Let Oktyabrya, 117312 Moscow, Russia. e-mail: ssh@cs.vniisi.msk.su

On some locally 3-transposition graphs

D. V. Pasechnik *

Abstract

Let Σ_n^ε be the graph defined on the $(+)$-points of an n-dimensional GF(3)-space carrying a nondegenerate symmetric bilinear form with discriminant ε, points are adjacent iff they are perpendicular. We prove that if $\varepsilon = 1$, $n \geq 6$ (resp. $\varepsilon = -1$, $n \geq 7$) then Σ_{n+1}^ε is the unique connected locally Σ_n^ε graph. One may view this result as a characterization of a class of $c^k \cdot C_2$-geometries (or 3-transposition groups). We briefly discuss an application of the result to a characterization of Fischer's sporadic groups.

1. Introduction and results

The study of geometries of classical groups as point-line systems with fixed local structure is very extensive. For instance, see Tits [12]. We refer the reader to a paper of Cohen and Shult [4] for a brief survey on more recent results.

In this paper we use the aforementioned approach to characterize some "nonclassical" geometries, namely 3-transposition graphs that arise from orthogonal GF(3)-groups.

For the concept of a 3-transposition graph and a 3-transposition group, see Fischer [8]. In case of characteristic 2 groups such graphs come from classical geometries. It is worthwhile to mention the work of Hall and Shult [9] characterizing some class of 3-transposition graphs as *locally cotriangular* ones. Note that the graphs Σ_n^ε considered in the present paper are not locally cotriangular (apart from finitely many exceptions for small n).

There is a one-to-one correspondence between 3-transposition graphs and *Fischer spaces* (see, e.g. Buekenhout [2], Cuypers [6], Weiss [13], [14]), namely 3-transposition graphs are complements to the collinearity graphs of Fischer spaces. It turns out that it is possible to exploit this duality and classification of Fischer spaces in the final part of the proof of our theorem 1.1. Note, however, that locally Σ_n^ε graphs not always arise from Fischer spaces.

---

*. A part of this research was completed when this author held a position at the Institute for System Analysis, Moscow.

E.g., there are at least two nonisomorphic locally Σ_6^--graphs, one of them has diameter 4, (i.e. it does not correspond to any Fischer space) see e.g. [1]. The author has shown (in a paper in preparation) that they are the only examples of such graphs.

Throughout the paper we consider undirected graphs without loops and multiple edges. Given a graph Γ, let us denote the set of vertices by $V = V\Gamma$, the set of edges by $E = E\Gamma$. Given two graphs Γ, Δ, the graph $\Gamma \cup \Delta$ (resp. the graph $\Gamma \cap \Delta$) is the graph with the vertex set $V\Gamma \cup V\Delta$ (resp. $V\Gamma \cap V\Delta$) and the edge set $E\Gamma \cup E\Delta$ (resp. $E\Gamma \cap E\Delta$). Given $v \in V\Gamma$, we denote by $\Gamma_i(v)$ the subgraph induced by vertices at distance i from v, and $\Gamma_1(v) = \Gamma(v)$. Furthermore, $\Gamma(X) = \bigcap_{x \in X} \Gamma(x)$. To simplify the notation we use $\Gamma(v_1, \ldots, v_k)$ instead of $\Gamma(\{v_1, \ldots, v_k\})$ and $u \in \Gamma_i(\ldots)$ instead of $u \in V\Gamma_i(\ldots)$. As usual, $v = v(\Gamma) = |V\Gamma|$, $k = k(\Gamma) = v(\Gamma(x))$, where $x \in V\Gamma$. Let $y \in \Gamma_2(x)$. We denote $\mu = \mu(x,y) = \mu(\Gamma) = v(\Gamma(x,y))$. Of course, we use k, μ if it makes sense, i.e. if those numbers are independent on the particular choice of the corresponding vertices. If Δ is a (proper) subgraph of Γ we denote this fact as $\Delta \subseteq \Gamma$ (resp. $\Delta \subset \Gamma$).

We denote the complete multipartite graph with the m parts of equal size n by $K_{n \times m}$. $\mathrm{Aut}(\Gamma)$ denotes the automorphism group of Γ. Our group-theoretic notation is as in [5]. Let Γ, Δ be two graphs. We say that Γ is *locally* Δ if $\Gamma(v) \cong \Delta$ for any $v \in V(\Gamma)$. Let Γ, $\overline{\Gamma}$ be two graphs. We say that Γ is a *cover* of $\overline{\Gamma}$ if there exists a mapping φ from $V\Gamma$ to $V\overline{\Gamma}$ which maps edges to edges. Suppose we have a chain of graphs $\Sigma_1, \ldots, \Sigma_n$, such that Σ_i is locally Σ_{i-1}, $i = 2, \ldots, n$. Then for any complete k-vertex subgraph Υ of Σ_m, $1 < k < m \le n$ we have $\Sigma_m(V\Upsilon) \cong \Sigma_{m-k}$. We say that a graph Γ is a *triple* graph if for each nonedge (u,v) there exist a unique $w \in V\Gamma$ such that $\Gamma(u,v) = \Gamma(u,w) = \Gamma(v,w)$.

We slightly adopt notation and several basic facts from [10]. Let $T = T_n$ be an n-dimensional $\mathrm{GF}(3)$-vector space carrying a nondegenerate symmetric bilinear form $(,)$ with discriminant ε. We say that the point $\langle v \rangle \subset T$ (or a nonzero vector $v \in T$) is of type $(+)$, $(-)$ or *isotropic* according to $(v,v) = 1, -1, 0$ respectively. Since the form is constant on a point, the notation like (p,q) for points p, q will be used freely. The orthogonal complement of $X \subseteq T$ in T is denoted by $X^\perp$.

Define the graph $\Sigma_n^\varepsilon = \Gamma(V,E)$ as follows. Let V be the set of $(+)$-points of T. Define $E = \{(u,v) \subset V \times V | (u,v) = 0\}$, i.e. the edges are pairs of perpendicular $(+)$-points. Given $\Gamma = \Sigma_n^\varepsilon$, we denote by $T(\Gamma)$ the underlying $\mathrm{GF}(3)$-space. Note that Σ_n^ε, $n \ge 5$ if $\varepsilon = 1$, $n \ge 4$ if $\varepsilon = -1$, is a rank 3 graph with automorphism group $GO_n^\mu(3)$, where Witt defect μ is empty if n is odd, otherwise $\mu = (-\varepsilon)^{n/2}$. Since T_n can be represented as the orthogonal direct sum of a $(+)$-point and T_{n-1}, the graph Σ_n^ε is locally Σ_{n-1}^ε. We denote Σ_n^1 by

320

Σ_n^+, and Σ_n^{-1} by Σ_n^-. Note that Σ_n^ε is a triple graph. We refer the interested reader to [10] for more detailed information about Σ_n^ε. We will prove the following theorem.

Theorem 1.1 *Let* $\Theta = \Theta_{n+1}$ *be a connected locally* Σ_n^ε-*graph, where* $\varepsilon = 1$, $n \geq 6$, *or* $\varepsilon = -1$, $n \geq 7$. *Then* Θ *is isomorphic to* Σ_{n+1}^ε.

Remark 1

One may view the graphs Θ_{n+1} as the collinearity graphs of certain $c^k \cdot C_2(s,t)$-geometries $\mathcal{G}(\Theta_{n+1})$ (here $k = n - (\varepsilon + 9)/2$), i.e. rank $k + 2$ geometries with diagram

Conversely, the elements of the geometry may be viewed as i-cliques of the graph with natural incidence, $i = 1, 2, \ldots, k+1, s+k+1$. Meixner has proved the following result [10].

Result 1.2 *Let* $\mathcal{G}$ *be a residually connected flag-transitive* $c^k \cdot C_2$-*geometry. Then if the* $c^1 \cdot C_2$-*residues (resp.* $c^3 \cdot C_2$-*residues) of* $\mathcal{G}$ *are isomorphic to* $\mathcal{G}(\Sigma_6^+)$ *(resp. to* $\mathcal{G}(\Sigma_7^-)$*) then* $\mathcal{G} = \mathcal{G}(\Sigma_{k+5}^+)$ *(resp.* $\mathcal{G} = \mathcal{G}(\Sigma_{k+4}^-)$*).*

Our theorem implies that the flag-transitivity assumption can be replaced to a geometric condition (X) from [10]. (X) states that in the collinearity graph of $\mathcal{G}$ each i-clique is the shadow of some element, $i = 1, 2, \ldots k + 1, s + k + 1$.

Remark 2

The significance of Σ_{i+3}^+ as subgraphs of 3-transposition graphs Δ_{2i} of Fischer's sporadic simple groups Fi_{2i} ($i = 2, 3, 4$) is well-known. Namely, for $\Gamma = \Delta_{2i}$ the subgraph $\Gamma(x, y)$, where x and y are two vertices at distance 2, is isomorphic to Σ_{i+3}^+. The author used theorem 1.1 to prove the following result [11].

Result 1.3 *Any connected locally* Δ_{22} *(resp.* Δ_{23}*) graph is isomorphic to* Δ_{23} *(resp. to* Δ_{24} *or to its 3-fold antipodal cover).*

2. Proof of the theorem

Preliminaries. A proof of the following technical statement is omitted.

Lemma 2.1 *Let* $\Gamma = \Sigma_n^\varepsilon$, $n \geq 5$, *and let* a, b, c *be isotropic points of* $T = T(\Gamma)$.

   If $(a, b) = 0$, $a \neq b$, $(a, c) \neq 0$ *then*
   (i) $a^\perp \cap c^\perp \cap \Gamma \cong \Sigma_{n-2}^{-\varepsilon}$,
   (ii) $a^\perp \cap b^\perp \cap \Gamma$ *is not isomorphic to* $\Sigma_{n-2}^{-\varepsilon}$.
   (iii) Denote $\Upsilon_p = \Gamma \cap p^\perp$. *If* $\Upsilon_a \cap \Upsilon_b$ *contains* $\{v\} \cup \Upsilon_a(v)$ *for some* $v \in V\Upsilon_a$ *then* $a = b$.
   (iv) For any $u \in V\Gamma \setminus V\Upsilon_a$, *the subgraph* $\Gamma(u) \cap \Upsilon_a$ *is isomorphic to* $\Sigma_{n-2}^{-\varepsilon}$.

<u>*Neighbourhood of two vertices at distance two*</u>. We start with a simple general fact. Let Δ be a connected graph satisfying the following property

(*) For any $u \in V\Delta$ and $v \in V\Delta \setminus (V\Delta(u) \cup \{u\})$ the subgraph $\Delta(u, v)$ is isomorphic to some M_Δ, whose isomorphism type is independent on the particular choice of u and v.

Lemma 2.2 *Let* Γ *be locally* Δ *graph, where* Δ *satisfies (*). Then for any* $u \in V\Gamma$, $v \in \Gamma_2(u)$ *the graph* $\Gamma(u, v)$ *is locally* M_Δ.

Note that $\Gamma = \Sigma_n^\varepsilon$ satisfies (*), when $\varepsilon = 1, n \geq 5$ (resp. $\varepsilon = -1, n \geq 4$). Indeed, the stabilizer of $u \in V\Gamma$ in $\mathrm{Aut}(\Gamma)$ acts transitively on $\Gamma_2(u)$. This implies (*). Thus lemma 2.2 holds for locally Γ graphs. The next statement characterizes locally M_Γ-subgraphs of Γ.

Proposition 2.3 *Let* $\Gamma = \Sigma_n^\varepsilon$, *and either* $\varepsilon = 1$, $n \geq 6$, *or* $\varepsilon = -1$, $n \geq 7$. *Let* $\Omega \subset \Gamma$ *be a locally* M_Γ *graph. Then there exists a unique isotropic point* $p \subset T(\Gamma)$ *such that* $\Omega = \Gamma \cap p^\perp$.

Proof. We proceed by induction on n. It is straightforward to check its basis. We leave it to the reader, noticing that for $\varepsilon = 1$, $n = 6$ (resp. $\varepsilon = -1$, $n = 6$) it suffices to classify locally $K_{3\times2} \cup K_{3\times2} \cup K_{3\times2}$ (resp. $K_{3\times4} \cup K_{3\times4} \cup K_{3\times4}$) subgraphs of Γ.
   Now let us check the inductive step. Let Ω be a connected component of a locally M_Γ subgraph of Γ, $v_1 \in V\Omega$, $v_2 \in \Omega(v_1)$. By the inductive hypothesis, $\Omega(v_i) = \Gamma(v_i) \cap p_i^\perp$, where p_i is an isotropic point of $T(\Gamma(v_i))$, hence of $T(\Gamma)$ $(i = 1, 2)$. Since $p_2 \subset v_1^\perp$, it defines a locally $M_{\Gamma(v_1)}$ subgraph Υ of $\Gamma(v_1)$, and $\Upsilon(v_2) = \Omega(v_1, v_2)$. Hence by lemma 2.1 (iii), applied to $\Gamma(v_1)$, $p_1 = p_2$. Therefore $\Omega = \Gamma \cap p_1^\perp$. Finally, it is easy to check that Ω is a unique connected component of the subgraph under consideration. $\square$

Final part of the proof. Let Θ be a connected locally $\Gamma = \Sigma_n^\varepsilon$ graph. Here we assume either $\varepsilon = 1$, $n \geq 6$ or $\varepsilon = -1$, $n \geq 7$. Pick a vertex $u \in V\Theta$.

Lemma 2.4

(i) $\mu(\Theta) = \mu(\Sigma_{n+1}^\varepsilon)$, $v(\Theta) = v(\Sigma_{n+1}^\varepsilon)$.

(ii) Θ is a triple graph.

Proof. (i). The first claim follows from proposition 2.3. Indeed, counting in two ways the edges between $\Theta(u)$ and $\Theta_2(u)$, we obtain the precise value of the number of vertices of Θ. Let $v \in \Theta_2(u)$, $w \in \Theta(v) \setminus \Theta(u)$. By lemma 2.1 (iv) we obtain $\Theta(u,v,w) \cong \Sigma_{n-2}^{-\varepsilon}$, hence nonempty. Thus the diameter of Θ equals two, and we are done.

(ii). Assume that there exist three distinct vertices $v_i \in \Theta_2(u)$, $i = 1, 2, 3$ such that $\Upsilon = \Theta(u,v_1) = \Theta(u,v_2) = \Theta(u,v_3)$. It contradicts the fact that $\Theta(w)$, where $w \in V\Upsilon$, is a triple graph. Observe that $|\Theta_2(u)|$ is exactly twice the number of isotropic points in $T(\Gamma)$. Hence we have no choice determining the edges between $\Theta(u)$ and $\Theta_2(u)$. Now since Σ_{n+1}^ε is a triple graph, the same is true for Θ. □

Let us denote $\Gamma = \Theta(u)$, $\Xi = \Theta_2(u)$. Let Υ be the graph defined on isotropic points of $T(\Gamma)$, two points are adjacent if they are not perpendicular.

Proposition 2.5 Ξ *is a two-fold cover of* Υ.

Proof. Let $(v_1, v_2) \in E\Xi$. By lemma 2.1 (iv) we have $\Theta(u,v_1,v_2) \cong \Sigma_{n-2}^{-\varepsilon}$. Denote by p_i the isotropic point of $T(\Gamma)$ such that $p_i^\perp \cap \Gamma = \Theta(u,v_i)$ $(i = 1, 2)$. By proposition 2.3 such a point p_i exists and is unique $(i = 1, 2)$. By lemma 2.1 (i), (ii), $(p_1, p_2) \neq 0$.

Conversely, assume p_1, p_2 are nonperpendicular isotropic points of $T(\Gamma)$. Denote by Ω_i the locally M_Γ subgraph of Γ, which corresponds to p_i $(i = 1, 2)$. For each Ω_i we have exactly two vertices $v_{ij} \in \Theta_2(u)$ such that $\Omega_i = \Theta(u,v_{ij})$ $(i, j = 1, 2)$. Considering the neighbourhood of v_{11}, we see that both v_{21} and v_{22} cannot be ajacent to v_{11}. On the other hand $k(\Xi) = k(\Upsilon)$. Hence one of v_{21} and v_{22} must be ajacent to v_{11}. We have shown that the mapping $v \mapsto \Theta(u,v)$, where $v \in V\Xi$, is a covering from Υ to Ξ. □

The latter statement implies that Ξ possesses an involutory automorphism g_u which interchanges any $v, w \in V\Xi$ such that $\Theta(u,v) = \Theta(u,w)$, and fixes $\{u\} \cup \Theta(u)$ pointwise.

Consider the subgroup G_u of $\mathrm{Aut}(\Theta)$ generated by g_x, $x \in \Theta(u)$. We have $G_u \cong \mathrm{Aut}(\Gamma)$. Therefore Θ is the collinearity graph of a $c^k \cdot C_2$-geometry satisfying the conditions of result 1.2. Hence our result follows from it. However, we would like to give a complete proof of theorem 1.1 here.

Our claim is that $\{g_v | v \in V\Theta\}$ is a class of 3-transpositions in $\mathrm{Aut}(\Theta)$. Indeed, clearly for any $x \in \Theta(u)$ the involutions g_u and g_x commute. Now let $y \in \Theta_2(u)$. We must prove $\tau = (g_u g_y)^3 = 1$. Note that τ belongs to the kernel of the action of the stabilizer of every $v \in \Theta(u, y)$ on $\Theta(u)$. Therefore it fixes every vertex of Θ. Our claim is proved. The use of the classification of 3-transposition groups given in [8] completes the proof of theorem 1.1. □

Note: Hans Cuypers (personal communication) has suggested another idea how to complete the proof, which is much more geometric. Namely, it may be easily shown that the partial linear space on $V\Theta$, whose lines are triples, is an irreducible Fischer space (see introduction) (or a locally polar geometry with affine planes, see Cuypers and Pasini [7]). Then the use of classification of these objects [6] (resp. [7]) completes the proof.

References

[1] A. E. Brouwer, A. M. Cohen, and A. Neumaier. *Distance-regular Graphs*. Springer Verlag, Berlin, 1989.

[2] F. Buekenhout. La géométrie des groupes de Fischer, 1974. Unpublished notes.

[3] F. Buekenhout. Diagram geometries for sporadic groups. In J. McKay, editor, *Finite Groups Coming of Age*, volume 45 of *A.M.S. Series Contemp. Math.*, pages 1–32, 1985.

[4] A. M. Cohen and E. E. Shult. Affine polar spaces. *Geom. Dedicata*, 35, pp. 43–76, 1990.

[5] J. H. Conway, R. T. Curtis, S. P. Norton, R. A. Parker, and R. A. Wilson. *Atlas of Finite Groups*. Clarendon Press, Oxford, 1985.

[6] H. Cuypers. On a generalization of Fischer spaces. *Geom. Dedicata*, 34, pp. 67–87, 1990.

[7] H. Cuypers and A. Pasini. Locally polar geometries with affine planes. *European J. Combin.*, 13, pp. 39–57, 1992.

[8] B. Fischer. Groups generated by 3-transpositions I. *Invent. Math.*, 13, pp. 232–246, 1971. See also University of Warwick Lecture Notes.

[9] J. I. Hall and E. E. Shult. Locally cotriangular graphs. *Geom. Dedicata*, 18, pp. 113–159, 1985.

[10] T. Meixner. Some polar towers. *European J. Combin.*, 12, pp. 397–416, 1991.

[11] D. V. Pasechnik. Geometric characterization of Fi_{22}, Fi_{23} and Fi_{24} sporadic simple groups, 1993. Submitted to *J. Combin. Theory Ser. A*.

[12] J. Tits. A local approach to buildings. In *The Geometric Vein (Coxeter Festschrift)*, pages 519–547. Springer Verlag, Berlin, 1981.

[13] **R. Weiss.** On Fischer generalizations of $Sp_{2n}(2)$ and $U_n(2)$. *Comm. Algebra*, 11, pp. 2527–2554, 1983.

[14] **R. Weiss.** A uniqueness lemma for groups generated by 3-transpositions. *Math. Proc. Cambridge Philos. Soc.*, 97, pp. 421–431, 1985.

D. V. Pasechnik, Department of Mathematics, University of Western Australia, Nedlands 6009 WA, Australia. e-mail: dima@maths.uwa.edu.au

Coherent configurations derived from quasiregular points in generalized quadrangles

S. E. Payne

Abstract

Let $S = (S^p, T)$ be a translation generalized quadrangle with group T of translations about the point p and having parameters (s, t), with $1 < s \leq t < s^2$. The point p is a quasiregular point (as defined in this paper). Using the points and lines of S away from the point p we construct a coherent configuration with 15 classes. Unfortunately, we leave open the problem of deciding whether or not there are examples with $s < t < s^2$.

1. Introduction and Review

Our standard reference for definitions and results concerning generalized quadrangles (GQ) is the monograph [3] by S. E. Payne and J. A. Thas, and we assume the reader has access to that work. Translation generalized quadrangles (TGQ), which are studied at some length in [3], especially in Chapter 8, provide the motivation for this paper. However, no familiarity with TGQ is required for an understanding of this essay.

Let $S = (P, B, I)$ be a GQ with parameters (s, t), $s > 1$, $t > 1$. If S is a TGQ with base point p (see [3] for the appropriate definitions), and if $s \neq t$, there is a prime power q and an odd integer a for which $s = q^a$ and $t = q^{a+1}$. All known examples have $a = 1$, i.e. $t = s^2$, and when q is a power of 2, this is the only possibility. (If $s = t$, also s must be a prime power, and there are examples with $s = t$ any prime power.) But for q a power of an odd prime, we can state the problem that originally motivated this article.

Problem 1.1 *Is there a TGQ having parameters (s, t) with $s < t < s^2$ and s odd?*

In this paper we want to consider a more general problem. If S is a TGQ with base point p, from results contained in [3] we know that each triad of points (three points, no two collinear) contained in $p^\perp$ has exactly $q + 1$

centers (points collinear with all three points of the triad). And each triad (p, x, y) containing p has either no center or exactly $q + 1$ centers. (Write $x \perp y$ to denote that x and y are collinear points; similarly, $L \perp M$ denotes that L and M are concurrent lines.) It is the existence of a point p such that each triad containing p has 0 or $q + 1$ centers which makes it possible to construct the coherent configuration that we wish to study. So throughout the remainder of this essay that becomes the central hypothesis.

As this work took shape throughout 1990 - 91, we benefitted greatly from conversations with and/or financial support arranged by the following colleagues: S. Tsaranov, J. A. Thas, H. Van Maldeghem, E. E. Shult, R. A. Liebler, D. Nichols and S. Hobart.

2. Quasiregular Points

Let $S = (P, B, I)$ be a GQ with parameters $(s, t), s > 1, t > 1$. For any point x of S let $T_i(x)$ be the set of all triads of points contained in $x^\perp$. The members of $T_i(x)$ are called the *inner triads* of x. Let $T_o(x)$ be the set of all triads containing x. The members of $T_o(x)$ are called the *outer triads* of x. Put

$$\mathrm{spec}_i(x) = \{|T^\perp| : T \in T_i(x)\}, \text{ and } \mathrm{spec}_o(x) = \{|T^\perp| : T \in T_o(x)\}.$$

The union of the *inner spectrum* of x ($\mathrm{spec}_i(x)$) and the *outer spectrum* of x ($\mathrm{spec}_o(x)$) is the *spectrum* of x: $\mathrm{spec}(x) = \{|T^\perp| : T \in T_i(x) \cup T_o(x)\}$.

The point x is said to be *quasiregular* provided there is an integer a such that $\mathrm{spec}(x) \subseteq \{0, 1, 1 + a\}$.

Each point or line of a classical GQ has a fairly small spectrum (see the examples given at the end of this section), and in some cases the existence of sufficiently many points with sufficiently small spectra guarantees that the GQ is classical.

Problem 2.1 *Is there an integer a such that $\mathrm{spec}_i(x) \subseteq \{1, 1 + a\}$ if and only if there is an integer a for which $\mathrm{spec}_o(x) \subseteq \{0, 1, 1 + a\}$ if and only if the point x is quasiregular?*

Let x and y be distinct collinear points. Define a set of triads:

$$T_y(x) = \{T : T \text{ is a triad with } y \in T \subseteq x^\perp\}$$

Let L_0, L_1, L_2 be three distinct lines through x. Fix points y_0 on L_0, y_1 on L_1, y_0 not collinear with y_1. Let $\{y_0, y_1\}^\perp = \{x = x_0, x_1, \ldots, x_t\}$. For $0 \le i \le t$, let a_i be the number of triads (y_0, y_1, y_2) with y_2 on L_2, having exactly i centers from among $x_1, \ldots, x_t$. Then the total number of such triads is $s = \sum a_i$. And the number of pairs (x_j, y_2) with $1 \le j \le t$, y_2 on L_2, $x_j \perp y_2$, is equal to

$t = \sum i a_i$. So $0 = t - t = t/s \sum a_i - \sum i a_i = \sum_{i=0}^{t}(t/s - i)a_i$. Hence

$$\sum_{i<t/s} \left(\frac{t}{s} - i\right) a_i = \sum_{i>t/s} \left(i - \frac{t}{s}\right) a_i. \tag{1}$$

It is immediate from (1) that $a_i = 0$ for all $i < t/s$ iff $a_i = 0$ for all $i > t/s$ iff s divides t and each triad (y_0, y_1, y) with y on L_2 has exactly $1 + t/s$ centers. Also, it is clear that

If each triad in $T_y(x)$ has $1 + b$ centers for some constant b, then

$$b = t/s. \tag{2}$$

In order to make various counts easier, all triads in the next few paragraphs are *ordered triads*.

Theorem 2.2 *If each $T \in T_y(x)$ has $1 + t/s$ centers, then each $T \in T_x(y)$ has $1 + t/s$ centers.*

Proof. $|T_y(x)| = t(t-1)s^2 = |T_x(y)|$. Assume each triad of $T_y(x)$ has $1+t/s$ centers, and let $T_x(y) = \{T_1, \ldots, T_d\}$, $d = t(t-1)s^2$. Let T_i have $1 + r_i$ centers, i.e. T_i has r_i centers different from y. First count the pairs (T_i, w) for which $w \in x^\perp \setminus y^\perp$ is a center of T_i. The number of such pairs is $st \cdot t(t-1) = \sum_1^d r_i$. Second, count the ordered triples (T_i, w, w') for which $w, w' \in x^\perp \setminus y^\perp$ are distinct centers of T_i. Given w, w', $T = (y, w, w') \subseteq x^\perp$ is a triad with $1 + t/s$ centers, t/s of which come from $y^\perp \setminus x^\perp$. So there are $(t/s)(t/s - 1)$ triads T_i having w, w' as centers. So the desired count is $t(t-1)s^2(t/s)(t/s-1) = t^2(t-1)(t-s)$. On the other hand, given one of the d triads T_i with r_i centers, there are $r_i(r_i - 1)$ triples (T_i, w, w'). So $t^2(t-1)(t-s) = \sum r_i(r_i - 1)$. Adding the previous count to this one yields $\sum_1^d r_i^2 = t^2(t-1)(t-s)+t^2(t-1)s = t^3(t-1)$. It is well known that $(\sum_1^d r_i)^2 \leq d \cdot \sum_1^d r_i^2$, with equality iff each r_i equals $\sum r_i/d = t/s$. But $(\sum r_i)^2 = s^2 t^4(t-1)^2$, and $d \cdot \sum r_i^2 = s^2 t(t-1)t^3(t-1) = (\sum r_i)^2$. Hence $r_i = t/s$ for all $i = 1, \ldots, d$. $\square$

By fixing x and letting y vary over the points of $x^\perp \setminus \{x\}$, we obtain the following corollary.

Theorem 2.3 $spec_i(x) = \{1 + t/s\}$ iff $spec_o(x) \subseteq \{0, 1 + t/s\}$.

Recall the following from Section 1.3 of [3]. Let x, y be fixed noncollinear points. For each integer α, $0 \leq \alpha \leq t + 1$, let N_α be the number of triads (x, y, z) having exactly α centers. Now suppose there are three distinct integers α, β, γ with $0 \leq \alpha, \beta, \gamma \leq 1 + t$, for which $\delta \notin \{\alpha, \beta, \gamma\}$ implies that $N_\delta = 0$. Note that we allow $N_\alpha = 0$ also, for example. Then we have the

329

following formula.

$$N_\alpha = \frac{(s^2t - st - s + t)\beta\gamma - (t^2 - 1)s(\beta + \gamma) + (t^2 - 1)(s + t)}{(\alpha - \beta)(\alpha - \gamma)}. \quad (3)$$

The situation in Theorem 2.2 corresponds to $\alpha = 0$, $\beta = 1 + t/s$, $N_\gamma = 0$. In this case we find (putting $q = t/s$):

$$N_0 = \frac{t(s-1)(s^2-t)}{s+t} = \frac{q(s-1)s(s-q)}{q+1}; \text{ and}$$

$$N_{1+t/s} = \frac{(t^2-1)s^2}{s+t} = \frac{s(t^2-s)}{q+1}. \quad (4)$$

Recall (1.2.4 of [3]) that if $t = s^2$, then $\mathrm{spec}_0(x) = \{1 + t/s\}$ for every point x. So in fact we have

$$|\mathrm{spec}_0(x)| = 1 \text{ iff } t = s^2, \text{ in which case } \mathrm{spec}_0(x) = \{1 + t/s\}. \quad (5)$$

To complete this section we survey the classical examples.

Example 2.4 $Q(4, q)$. Here all lines are regular with $s = t = q$. So for any line L, $\mathrm{spec}_i(L) = \{1, 1 + q\} = \mathrm{spec}_0(L)$. For q even, all points are regular. So $\mathrm{spec}_i(x) = \mathrm{spec}_0(x) = \{1, 1+q\}$ for each point x. For q odd, all points are antiregular, i.e. for each point x, $spec_i(x) = \{2\}$, $spec_0(x) = \{0, 2\}$.

Example 2.5 $Q(5, q)$. Here $s = q$, $t = q^2$ and all lines are regular. So for each line L, $\mathrm{spec}_i(L) = \{1, 1+q\}$, $\mathrm{spec}_0(L) = \{0, 1, 1+q\}$. Each point is 3-regular. This means that for each point triad T, $|T^\perp| = 1+q$ and $|T^{\perp\perp}| = 1+q$. Moreover (see 1.4.2 (iii) of [3]), each point $x \in P\setminus(T^\perp \cup T^{\perp\perp})$ is collinear with exactly two points of $T^\perp \cup T^{\perp\perp}$. So $\mathrm{spec}_i(x) = \{1 + q\}$, $\mathrm{spec}_0(x) = \{1 + q\}$.

Example 2.6 $H(4, q^2)$. Here $s = q^2$, $t = q^3$. For each noncollinear pair (x, y) of points, $|\{x, y\}^{\perp\perp}| = 1 + q$. If (x, y, z) is a triad with $z^\perp \cap \{x, y\}^{\perp\perp} \neq \emptyset$, then (x, y, z) has a unique center. If (x, y, z) is a triad with $z^\perp \cap \{x, y\}^{\perp\perp} = \emptyset$, then (x, y, z) has $1 + q$ centers. And any triad T contained in $\{x, y\}^{\perp\perp}$ has $1 + t$ centers. Hence for any point x, $\mathrm{spec}_i(x) = \mathrm{spec}_0(x) = \{1, 1 + q, 1 + t\}$. Now consider a triad $T = (L, M, N)$ of lines with axis K (i.e., $K \in \{L, M, N\}^\perp$). If N lies in the 3-space of $PG(4, q^2)$ spanned by L and M, then T has $1 + q$ axes. If N is not in the 3-space spanned by L and M, then K is the unique axis of T. It follows that $\mathrm{spec}_0(L) = \{0, 1, 1 + q\}$, $\mathrm{spec}_i(L) = \{1, 1 + q\}$, for each line L.

330

3. A Coherent Configuration

For the remainder of this article we let $S = (P, B, I)$ be a GQ with parameters (s, t), $s > 1$, $t > 1$, with a point p having $\text{spec}_i(p) = \{\alpha\}$ for some constant α. Then we know s divides t and $\alpha = 1 + t/s$. And $\text{spec}_o(p) \subseteq \{0, 1 + t/s\}$, with equality if and only if $t < s^2$. Put $q = t/s$ and fix a point $x \in P \setminus p^\perp$. Then we have seen

$$N_0 = \frac{t(s-1)(s-q)}{q+1} \text{ is the number of acentric triads } T = (p, x, y). \quad (6)$$

$$N_{1+q} = \frac{s(t^2 - 1)}{q+1} \text{ is the number of triads } T = (p, x, y) \text{ with } 1 + q \text{ centers.}$$
$$(7)$$

To construct the coherent configuration, we want $N_0 > 0$, so from now on we assume $s \le t < s^2$.

Let $X_1 = P \setminus p^\perp$, so $|X_1| = s^2 t$. Put $X_2 = \{L \in B : p \text{ is not on } L\}$, so $|X_2| = st(1 + t)$. We are going to construct a coherent configuration $\mathcal{C}$ on $X = X_1 \cup X_2$ (using definitions of D. G. Higman [2]) consisting of 15 relations $f_0, \ldots, f_{14}$. Notation: For $L \in B$, $L^* = \{p : p \text{ is on } L\}$.

$$
\begin{aligned}
f_0 &= \{(x, x) : x \in X_1\}. \\
f_1 &= \{(x, y) \in X_1 \times X_1 : x \perp y \ne x\}. \\
f_2 &= \{(x, y) \in X_1 \times X_1 : (p, x, y) \text{ is a triad with } 1 + q \text{ centers}\}. \\
f_3 &= \{(x, y) \in X_1 \times X_1 : (p, x, y) \text{ is an acentric triad}\}. \\
f_4 &= \{(x, L) \in X_1 \times X_2 : x \text{ is on } L\}. \\
f_5 &= \{(x, L) \in X_1 \times X_2 : x \text{ is not on } L \text{ and } x^\perp \cap L^* \in p^\perp\}. \\
f_6 &= \{(x, L) \in X_1 \times X_2 : x \text{ is not on } L \text{ and } x^\perp \cap L^* \notin p^\perp\}. \\
f_7 &= f_4^t \subseteq X_2 \times X_1. \\
f_8 &= f_5^t \subseteq X_2 \times X_1. \\
f_9 &= f_6^t \subseteq X_2 \times X_1. \quad\quad\quad (8) \\
f_{10} &= \{(L, L) : L \in X_2\}. \\
f_{11} &= \{(L, M) \in X_2 \times X_2 : L \ne M \\
&\quad\quad \text{and } L \text{ meets } M \text{ at a point of } p^\perp\}. \\
f_{12} &= \{(L, M) \in X_2 \times X_2 : L \text{ does not meet } M \\
&\quad\quad \text{but } L \text{ and } M \text{ meet the same line through } p\}. \\
f_{13} &= \{(L, M) \in X_2 \times X_2 : L \text{ meets } M \\
&\quad\quad \text{but } L \text{ and } M \text{ meet different lines through } p\}. \\
f_{14} &= \{(L, M) \in X_2 \times X_2 : L \text{ does not meet } M, \\
&\quad\quad \text{and } L \text{ and } M \text{ meet different lines through } p\}.
\end{aligned}
$$

Using the notation of D. G. Higman [2] we have $I = \{0, 1, \ldots, 14\}$; $\Omega = \{0, 10\}$. So $f_0 \cup f_{10} = \Delta$, the diagonal relation on X. Write $f_{i^*} = f_i^t$, so $i = i^*$ for $0 \le i \le 3$ and $10 \le i \le 14$, and $4^* = 7, 5^* = 8, 6^* = 9$, and $(i^*)^* = i$ for $0 \le i \le 14$.

Put $n_0 = |X_1| = s^2 t$; $n_{10} = |X_2| = st(t+1)$. And put $m_i = |f_i|$ for $0 \le i \le 14$. For x in the domain of f_i, put $v_i = |f_i(x)|$. Then it is straightforward to compute the entries in the following table.

i	m_i	v_i
0	$s^2 t$	1
1	$s^2 t(t+1)(s-1)$	$(t+1)(s-1)$
2	$s^3 t(t^2-1)/(q+1)$	$N_{1+q} = s(t^2-1)/(q+1)$
3	$s^2 t^2(s-1)(s-q)/(q+1)$	$N_0 = t(s-1)(s-q)/(q+1)$
4	$s^2 t(t+1)$	$t+1$
5	$s^2 t(t^2-1)$	t^2-1
6	$s^2 t^2(t+1)(s-1)$	$t(t+1)(s-1)$
7	$s^2 t(t+1)$	s
8	$s^2 t(t^2-1)$	$s(t-1)$
9	$s^2 t^2(t+1)(s-1)$	$st(s-1)$
10	$st(t+1)$	1
11	$st(t^2-1)$	$t-1$
12	$st^2(t+1)(s-1)$	$t(s-1)$
13	$s^2 t^2(t+1)$	st
14	$s^2 t^2(t^2-1)$	$st(t-1)$

$$(9)$$

For $0 \le i, j, k \le 14$, if $(x, y) \in f_k$, put

$$p_{ij}^k = |\{z \in X : (x, z) \in f_i, (z, y) \in f_j\}|. \tag{10}$$

The following identities, although lifted straight from [2], are easily verified.

$$p_{i0}^k = \begin{cases} \delta_{ik}, & 0 \le i^* \le 6; \\ 0, & \text{otherwise.} \end{cases}$$

$$p_{i,10}^k = \begin{cases} \delta_{ik}, & 7 \le i^* \le 14; \\ 0, & \text{otherwise.} \end{cases}$$

$$p_{0j}^k = \begin{cases} \delta_{jk}, & 0 \le j \le 6; \\ 0, & \text{otherwise.} \end{cases} \tag{11}$$

$$p_{10,j}^k = \begin{cases} \delta_{jk}, & 7 \le j \le 14; \\ 0, & \text{otherwise.} \end{cases}$$

$$p_{ij}^0 = \begin{cases} \delta_{ij} \cdot v_i, & 0 \le i \le 6; \\ 0, & \text{otherwise.} \end{cases}$$

$$p_{ij}^{10} = \begin{cases} \delta_{ij} \cdot v_i, & 7 \leq i \leq 14; \\ 0, & \text{otherwise.} \end{cases}$$

$$p_{ij}^{k} = p_{j^* i^*}^{k^*}, \quad 0 \leq i, j, k \leq 14. \tag{12}$$

$$p_{ij}^{k} v_k = p_{kj^*}^{i} \cdot v_i, \quad 0 \leq i, j, k \leq 14. \tag{13}$$

$$\begin{aligned} I^{0,0} &= \{0,1,2,3\}; \quad r_{0,0} = |I^{0,0}| = 4. \\ I^{0,10} &= \{4,5,6\}; \quad I^{10,0} = \{7,8,9\}; \quad r_{0,10} = r_{10,0} = 3. \\ I^{10,10} &= \{10,11,12,13,14\}; \quad r_{10,10} = 5; \quad r = |I| = 15. \end{aligned} \tag{14}$$

If $p_{ij}^{k} \neq 0$, there must be $\alpha, \beta, \gamma \in \{0,10\}$ for which

$$i \in I^{\alpha,\beta}, \; j \in I^{\beta,\gamma}, \text{ and } k \in I^{\alpha,\gamma}. \tag{15}$$

$$\begin{aligned} p_{ii^*}^{0} &= \sum_{j} p_{ij}^{k} = v_i, \text{ for } 0 \leq i, k \leq 6. \\ p_{jj^*}^{0} &= \sum_{i} p_{ij}^{k} = v_{j^*} \text{ for } 0 \leq k^*, j^* \leq 6. \\ p_{ii^*}^{10} &= \sum_{j} p_{ij}^{k} = v_i \text{ for } 7 \leq i, k \leq 14. \\ p_{jj^*}^{10} &= \sum_{i} p_{ij}^{k} = v_{j^*} \text{ for } 7 \leq k^*, j^* \leq 14. \end{aligned} \tag{16}$$

Let A_i be an $n \times n$ $(0,1)$-incidence matrix of f_i, $n = n_0 + n_{10}$, corresponding to some ordering of X. Then

$$A_i A_j = \sum_{k} p_{ij}^{k} A_k, \tag{17}$$

and

$$\sum_{v} p_{ki}^{v} p_{vj}^{u} = \sum_{v} p_{ij}^{v} p_{kv}^{u}. \tag{18}$$

Let $M_j = (p_{ij}^{k})$ be the matrix with p_{ij}^{k} in row i and column k. With the aid of (18) it follows that

$$M_i M_j = \sum_{k} p_{ij}^{k} M_k, \tag{19}$$

and $\theta : A_i \mapsto M_i$ induces an algebra isomorphism.

The next major step in the analysis of $\mathcal{C}$ is to compute the intersection matrices M_j, their eigenvalues, and the multiplicities of those as eigenvalues of the incidence matrices A_j.

4. The Intersection Matrices

Before exhibiting any M_j we give some specific counting lemmas that help determine the p_{ij}^k. Let L be a line not incident with p, and define p' to be the point on L collinear with p. Let $y \in X_1, y$ not on L but collinear with p'. Then there are s points x on L for which $(x, y) \in f_2$. This says

$$p_{24}^5 = s = p_{72}^8. \tag{20}$$

Continuing with p, p', L as above, let $y \in X_1$ be chosen so that y is not collinear with p', and define y' to be the point of L collinear with y. So p' and y' are not collinear and we may define the point-line pair (p'', L_1) by $pIL_1Ip''Iyy'$. So $p'' \neq y$. There are $t - 1$ lines $L_2, \ldots, L_t$ through p and different from $L_0 = pp'$ and $L_1 = pp''$. Let w_j be the point of L_j collinear with y, $2 \leq j \leq t$, and x_j the point of L collinear with w_j. So (p, y, x_j) is a triad with center w_j. Since each centric triad containing p has exactly $1 + q$ centers, the points x_j each get labeled $1 + q$ times. Hence

$$L \text{ has } (t - 1)/(q + 1) \text{ points } x \text{ for which } (x, y) \in f_2 \tag{21}$$

and has $(s-1)-(t-1)/(q+1) = (s-q)/(q+1)$ points x' for which $(x', y) \in f_3$.

This proves

$$p_{24}^6 = p_{72}^9 = (t - 1)/(q + 1); \quad p_{34}^6 = p_{73}^9 = (s - q)/(q + 1). \tag{22}$$

Note 4.1 *Both $(t-1)/(q+1)$ and $(s+1)/(q+1)$ must be positive integers.*

It is quite a tedious task to compute all the p_{ij}^k, although relations (11) through (16) of the preceding section help immensely. Possibly the most difficult entry to compute, among those which we computed directly, was p_{22}^3. Hence we record the details in this one case.

Theorem 4.2 $p_{22}^3 = (t + 1)(t - 1)^2/(q + 1)^2.$

Proof. Start with $(x, y) \in f_3$. Let $L_0, \ldots, L_t$ be the lines through p. Let x_i (resp., y_i) be the point of L_i collinear with x (resp., y). By hypothesis $x_i \neq y_i$, $0 \leq i \leq t$. Put $M_i = xx_i$, $N_i = yy_i$. For each i there is a unique i' ($i \neq i'$) for which M_i meets $N_{i'}$ at a point t_i. Each point $z \in \{x_i, y_{i'}\}^{\perp} \setminus \{p, t_i\}$ satisfies $(x, z) \in f_2$ and $(z, y) \in f_2$.

Now for any i, let j be an index different from i, i'.

If w_i is the point of M_i collinear with y_j, and w_i' is the point of N_j collinear with x_i, then all $t - 2$ points of $\{x_i, y_j\}^{\perp} \setminus \{p, w_i, w_i'\}$ are points z for which $(x, z) \in f_2$ and $(z, y) \in f_2$. So in $(t+1)(t-1) + (t+1)(t-1)(t-2) = (t+1)(t-1)^2$ ways a point z is obtained for which $(x, z) \in f_2$ and $(z, y) \in f_2$. But each such z is collinear with $1+q$ x_i's and $1+q$ y_j's, so is counted $(1+q)^2$ times. $\square$

We now exhibit only the most basic M_j's.

$$M_0 = \operatorname{diag}(I_4, O_3, I_3, O_5); \quad M_{10} = \operatorname{diag}(O_4, I_3, O_3, I_5); \quad M_0 + M_{10} = I_{15}. \quad (23)$$

For M_1, if $p_{i1}^k \neq 0$ there is an $\alpha \in \{0, 10\}$ for which $i, k \in I^{\alpha, 0}$. So the nonzero blocks of M_1 are

$$M_1^0 = (p_{i1}^k), \ 0 \le i, k \le 3, \quad \text{and} \quad M_1^{10} = (p_{ik}^k), \ 7 \le i, k \le 9.$$

So

$$M_1 = \begin{pmatrix} M_1^0 & O_{4\times3} & O_{4\times3} & O_{4\times5} \\ \hline O_{3\times4} & O_3 & O_3 & O_{3\times5} \\ \hline O_{3\times4} & O_3 & M_1^{10} & O_{3\times5} \\ \hline O_{5\times4} & O_{5\times3} & O_{5\times3} & O_{5\times5} \end{pmatrix}. \quad (24)$$

Columns 0 and 1 of M_1^0 are easy to determine using (22). And $p_{01}^2 = 0$, $p_{11}^2 = t - q$, $p_{21}^2 = (q+1)(s-1) + (t-q)[(t-1)/(q+1)] - 1) = (s-1)(qt+1)/(q+1)$.

Since column sums equal v_1, p_{31}^2 is easily determined. Finally, $p_{01}^3 = 0$ and $p_{11}^3 = t+1$. With the help of (22), $p_{21}^3 = (t^2-1)/(q+1)$, and M_1^0 is easily completed.

$$M_1^0 = \begin{pmatrix} 0 & 1 & 0 & 0 \\ (t+1)(s-1) & s-2 & t-q & t+1 \\ 0 & \dfrac{t(t-1)}{q+1} & \dfrac{(s-1)(qt+1)}{q+1} & \dfrac{t^2-1}{q+1} \\ 0 & \dfrac{t(s-q)}{q+1} & \dfrac{(t-q)(s-q)}{q+1} & \dfrac{(t+1)(s-2q-1)}{q+1} \end{pmatrix}. \quad (25)$$

The eigenvalues of M_1^0 are

$$\theta_0 = (t+1)(s-1); \quad \theta_1 = s - t - 1; \quad \theta_2 = s - 1; \quad \theta_3 = -(t+1). \quad (26)$$

In a similar fashion we may proceed to determine M_i^{10}, and all the other M_j along with their eigenvalues and their corresponding multiplicities. For the sake of brevity, and since the other M_j are really determined by M_1 anyway, these computations are suppressed in this published version.

5. Eigenvalue Multiplicities of A_1

Specializing eq. (17), if (a_0, a_1, a_2, a_3) is row 1 of M_1^0, we have

$$A_1^2 = \sum_0^3 p_{11}^k A_k = \sum_0^3 a_k A_k. \quad (27)$$

And if (b_0, b_1, b_2, b_3) is row 1 (just below row 0) of $(M_1^0)^2$,

$$A_1^3 = \sum_0^3 p_{11}^k A_1 A_k = \sum_{j=0}^3 \left(\sum_{k=0}^3 p_{11}^k p_{1k}^j \right) A_j = \sum_{j=0}^3 b_j A_j. \qquad (28)$$

We want to find α, β, γ, δ such that if

$$p_0(x) = \alpha x^3 + \beta x^2 + \gamma x + \delta, \text{ then } p_0(A_1) = J. \qquad (29)$$

Put $k = (t+1)(s-1) = p_{11}^0$, and

$$p(x) = (x - k)p_0(x). \qquad (30)$$

It will turn out, of course, that the four roots of $p(x) = 0$ are the eigenvalues of M_1^0, since clearly $p(A_1) = 0$. Since $I = A_0$, A_1, A_2, A_3 are linearly independent,

$$\alpha \sum b_j A_j + \beta \sum a_j A_j + \gamma A_1 + \delta I = J = I + A_1 + A_2 + A_3 \qquad (31)$$

yields a system of four equations in the variables $\alpha, \beta, \gamma, \delta$, with the following solution (put $\Delta = a_2 b_3 - a_3 b_2$):

$$\begin{aligned}
\alpha &= (a_2 - a_3)/\Delta \\
\beta &= (b_3 - b_2)/\Delta \\
\gamma &= [\Delta - a_1(b_3 - b_2) - b_1(a_2 - a_3)]/\Delta \\
\delta &= [\Delta - a_0(b_3 - b_2) - b_0(a_2 - a_3)]/\Delta
\end{aligned} \qquad (32)$$

Of course we already know the a_j and b_j.

$$\begin{aligned}
a_0 &= k = (t+1)(s-1) & b_0 &= k(s-2) \\
a_1 &= s-2 & b_1 &= k + (s-2)^2 + st(t - 2q + 1) \\
a_2 &= t-q & b_2 &= (t-q)(st - 2t + 3s - 3) \\
a_3 &= t+1 & b_3 &= (t+1)(st + 2s - 3t - 3)
\end{aligned} \qquad (33)$$

From eq. (32) and (33) we have

$$\begin{aligned}
\Delta &= -qk(s+t) \\
\alpha\Delta &= -(q+1) \\
\beta\Delta &= -(q+1)(2t - 2s + 3) \\
\gamma\Delta &= -(q+1)(t^2 - 3st + s^2 - 4s + 4t - 3) \\
\delta\Delta &= (q+1)(t+1)(s-1)(t-s+1)
\end{aligned} \qquad (34)$$

Now we can write out $p(x)$ and factor it:

$$p(x) = \alpha(x - k)(x + t + 1)(x - s + 1)(x - s + t + 1). \qquad (35)$$

Put $f(x) = p(x)/\alpha = \prod_{i=0}^{3}(x - \theta_i)$, with θ_i labeled as in eq. (26).

Let d_i be the multiplicity of θ_i as an eigenvalue of the real symmetric (and hence diagonalizable) matrix A_1. So we have

$$d_i = \text{tr}(f_i(A_1))/f_i(\theta_i), \tag{36}$$

where $\text{tr}(C)$ is the trace of C, and $f_i(x) = f(x)/(x-\theta_i)$. We need the following:

$$\text{tr}(I) = s^2 t; \;\; \text{tr}(A_1) = 0; \;\; \text{tr}(A_1^2) = ks^2 t; \;\; \text{tr}(A_1^3) = ks^2 t(s - 2). \tag{37}$$

Since $f_3(x) = (x - \theta_0)(x - \theta_1)(x - \theta_2)$, we have $\text{tr}(f_3(A_1)) = \text{tr}(A_1^3) - (\theta_0 + \theta_1 + \theta_2)\text{tr}(A_1^2) + (\theta_0\theta_1 + \theta_0\theta_2 + \theta_1\theta_2)\text{tr}(A_1) - \theta_0\theta_1\theta_2\text{tr}(I) = ks^2 t(t - s^2)$. And $f_3(\theta_3) = -s^2(t+1)(s+t)$. So eventually we find

$$d_3 = t(s^2 - t)(s - 1)/(s + t). \tag{38}$$

Proceeding in the same manner, we find the remaining entries in the following array.

i	θ_i	d_i
0	$(t+1)(s-1)$	1
1	$s-t-1$	$(t+1)(s-1)$
2	$s-1$	$s(t^2-1)/(q+1) = N_{1+q}$
3	$-(t+1)$	$q(s^2-t)(s-1)/(q+1) = N_0$

(39)

Unfortunately all the d_i are evidently positive integers!

Let $\mathcal{A} = \{I, A_1, A_2, A_3\}$ and $\mathcal{B} = \{I, A_1, A_1^2, A_1^3\}$. So $\mathcal{A}$ and $\mathcal{B}$ are two (ordered) bases for the Bose-Mesner algebra they span. Let C be the 4×4 matrix whose jth column is the coordinate matrix $[A_1^j]_{\mathcal{A}}$ of A_1^j with respect to the basis $\mathcal{A}$, $0 \leq j \leq 3$. So

$$C = \begin{pmatrix} 1 & 0 & a_0 & b_0 \\ 0 & 1 & a_1 & b_1 \\ 0 & 0 & a_2 & b_2 \\ 0 & 0 & a_3 & b_3 \end{pmatrix} = \begin{pmatrix} I_2 & A \\ 0_2 & B \end{pmatrix}, \tag{40}$$

and

$$C^{-1} = \begin{pmatrix} I_2 & -AB^{-1} \\ 0_2 & B^{-1} \end{pmatrix}, \tag{41}$$

with

$$B^{-1} = \Delta^{-1} \begin{pmatrix} b_3 & -b_2 \\ -a_3 & a_2 \end{pmatrix} = \begin{pmatrix} \dfrac{st+2s-3t-3}{(s+t)(t-q)} & \dfrac{st-2t+3s-3}{(s+t)(t+1)} \\[2mm] \dfrac{1}{(s+t)(t-q)} & \dfrac{-1}{(s+t)(t+1)} \end{pmatrix}.$$

337

And the jth column of C^{-1} is $[A_j]_B$. So A_2 and A_3 can be calculated explicitly as polynomials in A_1. Hence A_1, A_2, A_3 have common eigenspaces. Let W_i be the eigenspace (of column vectors) of A_1 belonging to θ_i. So $d_i = v_i = \dim(W_i)$. Hence W_i is an eigenspace of A_j, $0 \le i, j \le 3$.

Let $\theta_i(A_j) = \theta_{ij}$ be the eigenvalue of A_j on W_i.

$$\Theta = (\theta_{ij}) = \begin{pmatrix} 1 & (t+1)(s-1) & \dfrac{s(t^2-1)}{q+1} & \dfrac{t(s-1)(s-q)}{q+1} \\ 1 & s-t-1 & \dfrac{s(t-1)}{q+1} & \dfrac{-t(s-q)}{q+1} \\ 1 & s-1 & -s & 0 \\ 1 & -(t+1) & 0 & t \end{pmatrix}. \quad (42)$$

Note 5.1 $\theta_{0j} = \dim(W_j) = d_j = v_j$.

Put $\vec{x}_i = [\theta_{0i}, \theta_{1i}, \theta_{2i}, \theta_{3i}]^t$. Then

$$(M_j^0)^t \vec{x}_i = \theta_{ij} \vec{x}_i \implies \Theta M_j^0 \Theta^{-1} = \operatorname{diag}(\theta_{0j}, \theta_{1j}, \theta_{2j}, \theta_{3j}) \quad (43)$$

and $\Theta^2 = s^2 t I$.

The incidence matrices A_k were all obtained from some ordering of $X_1 = P \setminus p^{\perp} = \{y_1, \ldots, y_{s^2 t}\}$. Choose a point y in X_1 and put $\Gamma_k = \{y_j \in X_1 : (y, y_j) \in f_k\}$. Then $\Gamma = (\Gamma_0, \Gamma_1, \Gamma_2, \Gamma_3)$ partitions X_1, and $|\Gamma_i| = v_i$. Put $\gamma_{ij}^k = |\{(u, v) \in \Gamma_i \times \Gamma_j : (u, v) \in f_k\}|$. Then $\gamma_{ij}^k / v_i = p_{jk}^i \implies A_k^{\Gamma} = (\gamma_{ij}^k / v_i) = (M_k^0)^t$ is the matrix obtained by blocking A_k according to the partition Γ and replacing each block by its average row sum. We say that the vector $\vec{x} = (x_0, x_1, x_2, x_3)^t$ blows up to $\Gamma(\vec{x}) = (z_1, \ldots, z_{s^2 t})^t$ where $z_j = x_i$ iff $y_j \in \Gamma_i$, $0 \le i \le 3$, $1 \le j \le s^2 t$.

Note 5.2 $\vec{x}_j$ blows up to a vector in W_j, $0 \le j \le 3$.

In setting up the partition Γ, etc., we could have started with any point z of X_1 in place of y.

Conjecture 5.3 *The vectors in W_i blown up from the vectors obtained from the v_i points $z \in f_i(y)$ form a basis for W_i, $0 \le i \le 3$.*

A little thought shows that for a point $y_k \in \Gamma_j = f_j(y)$, the eigenvector $\vec{x}_j = (\theta_{0j}, \theta_{1j}, \theta_{2j}, \theta_{3j})^t$ of $(M_j^0)^t$, when blown up with respect to the point y_k, gives the kth column of $\sum_0^3 \theta_{ij} A_j$. And the eigenvalue of this latter matrix on W_k is $\sum_i \theta_{ij} \theta_{ki} = s^2 t \cdot \delta_{jk}$, since $\Theta^2 = s^2 t I$. So the rank of $\sum_i \theta_{ij} A_i$ equals $\dim(W_j)$. This means the above Conjecture could possibly be true. In fact, if

S is the GQ with $s = t = 3$ and with all points antiregular, a hand calculation shows that the conjecture is indeed true.

6. Conclusion

The goal of this investigation was to find restrictions on the parameters of those GQ having a quasiregular point, and if possible to find restrictions on the parameters of a translation generalized quadrangle. In fact, the only restriction we have found is that if $s < t < s^2$ and $q = t/s$, then $q + 1$ divides $t - 1$, or equivalently, $q + 1$ divides $s - q$. And this restriction was found early in the computations without any real need for the coherent configuration. One other thing to compute would be the Krein parameters (cf. [1]). We did. They are all nonnegative. One then might consider which ones are equal to zero. We did that also. Unfortunately, those that are equal to zero are precisely the ones that are predicted to be zero by the general theory (cf.[4]).

On the one hand it may be possible that some other approach to the study of the coherent configuration will eventually turn up additional restrictions. On the other hand, if the corresponding GQ are ever discovered, we have worked out here many details concerning the associated coherent configuration.

References

[1] **E. Bannai and T. Ito.** *Algebraic Combinatorics I.* Benjamin Cummings, Menlo Park, California, 1984.

[2] **D. G. Higman.** Coherent algebras. *Linear Algebra Appl.*, 93, pp. 209–239, 1987.

[3] **S. E. Payne and J. A. Thas.** *Finite Generalized Quadrangles*, volume 104 of *Research Notes in Mathematics*. Pitman, Boston, 1984.

[4] **P. Terwilliger.** The subconstituent algebra of an association scheme. Unpublished, 1991.

S. E. Payne, CU-Denver Dept. of Math., Campus Box 170, P.O. Box 173364, Denver, CO 80217-3364, USA. e-mail: spayne@cudnvr.denver.colorado.edu

On Veldkamp planes

E. E. Shult

Abstract

If Veldkamp lines exist for a point-line geometry Γ, then the Veldkamp space $\mathcal{V}$ is a linear space. In addition to the ordinary poset structure of the subspaces of $\mathcal{V}$, one automatically obtains in addition an Aut (Γ)-invariant subposet of *flats*. This is because $\mathcal{V}$ has the structure of a partial matroid. It is shown that these two poset structures diverge at the level of Veldkamp planes, for all embeddable proper Grassmann spaces, but coincide at this level for all half-spin geometries.

1. Introduction

Let $\Gamma = (\mathcal{P}, \mathcal{L})$ be a rank 2 incidence geometry of points $(\mathcal{P})$ and lines $(\mathcal{L})$. A **geometric hyperplane** is a proper subspace of Γ which intersects the point-shadow of each line in at least one point. In a previous study ([5]) conditions implying a partial matroid structure on the collection $\mathcal{V}$ of all geometric hyperplanes of a point-line geometry were given. For each positive integer r, the condition is

(1.1) (VELDKAMP $(r-1)$-SPACES EXIST) For any collection $\{A_i\}$ of k geometric hyperplanes of a point-line geometry $\Gamma = (\mathcal{P}, \mathcal{L}), 1 \leq k \leq r$, and any further hyperplane A,

$$A_1 \cap A_2 \cap \cdots \cap A_k \subseteq A \quad \text{implies}$$
$$A_1 \cap A_2 \cap \cdots \cap A_{k-1} \subseteq A \quad \text{or} \quad A_1 \cap A_2 \cap \cdots \cap A_{k-1} \cap A \subseteq A_k.$$

When $k = 1$, the intersection $A_1 \cap A_2 \cap \cdots \cap A_{k-1}$ is understood to represent the set $\mathcal{P}$ of all points. In this case (1.1) becomes

(1.2) (VELDKAMP POINTS EXIST) If for hyperplanes A_1, A of Γ, we have $A_1 \subseteq A$, then $A \subseteq A_1$, – that is, no hyperplane is properly contained in another.

If $k = 2$, the condition is

(1.3) (VELDKAMP LINES EXIST) Veldkamp points exist for Γ and if A_1, A_2, and A are hyperplanes of Γ, then $A_1 \cap A_2 \subseteq A$ implies $A_1 = A$ or $A_1 \cap A \subseteq A_2$.

The import of the second condition is that if we let $\mathcal{V}_2$ be the set of intersections of pairs of distinct geometric hyperplanes (the Veldkamp lines), then the incidence system $(\mathcal{V}, \mathcal{V}_2, \supseteq)$ is a linear space, for the point shadow of any Veldkamp line is determined by any two of its distinct Veldkamp points. This linear space is called the **Veldkamp space** and is only defined when Veldkamp lines exist.

Under certain conditions the Veldkamp space can be shown to be a projective space. We mention here two criteria which insure this:

CRITERION 1. $\Gamma = (\mathcal{P}, \mathcal{L})$ possesses Veldkamp planes and Teirlinck's condition holds: namely, if A and B are hyperplanes and p is a point not in $A \cup B$, then there is a hyperplane C containing p and $A \cap B$.

CRITERION 2. (Peter Johnson) Γ possesses Veldkamp lines and there exists an embedding $e : \Gamma \to \mathbf{P}$, where $\mathbf{P}$ is a projective space, such that every geometric hyperplane H of Γ arises from this embedding – i.e. there exists a projective hyperplane $\mathbf{H}$ of $\mathbf{P}$ such that $H = e^{-1}\left(\mathbf{H} \cap e(\mathcal{P})\right)$.

(For the definitions of (projective) embeddings, the reader may consult any one of [1], [6], or [4].) In the previous study on Veldkamp lines ([5]), the author was able to exploit criterion 2 along with previous results on hyperplanes arising from an embedding to show that the Veldkamp space of the Grassmann spaces $A_{n,d}(D)$, the half-spin geometries $D_{n.n}(F)$, and the exceptional geometries $E_{6,1}(F)$ (where D is a division ring, F is a field) are all projective spaces.

Aside from the *subspace structure* of the Veldkamp space, there is also a natural dependence relation on $\mathcal{V}$ which defines a *partial-matroid structure*. Let $\{A_i \| i = 1, \dots, k\}$ be a collection of hyperplanes of Γ – that is, a finite subset of $\mathcal{V}$. A hyperplane A is said to **depend on** $\{A_i \| i = 1, \dots, k\}$, if and only if A contains the intersection $A_1 \cap A_2 \cap \cdots \cap A_k$. Under this definition, $\mathcal{V}$ satisfies the first two axioms of a *dependence theory* (or matroid), namely

(Reflexivity) Any element of $\mathcal{V}$ depends on any finite subset of $\mathcal{V}$ which contains it as a member.

(Transitivity) If A depends on the finite subset X of $\mathcal{V}$, and every element of X depends on the same finite subset Y of $\mathcal{V}$, then A depends on Y.

What is missing, of course, is the famous "exchange axiom". The condition (1.1), that Veldkamp $(r-1)$-spaces exist, can now be interpreted as a limited version of the exchange axiom. It says

(r-exchange axiom) If k is a positive integer not exceeding r, and A depends on $\{A_1, A_2, \dots, A_k\}$ but not on $\{A_1, A_2, \dots, A_{k-1}\}$, then A_k depends on $\{A_1, A_2, \dots, A_{k-1}, A\}$.

We call a dependence relation on $\mathcal{V} \times \{\text{finite subsets of } \mathcal{V}\}$ satisfying the three axioms just listed, an r-**partial matroid**.

Of course, even without any particular version of the exchange axiom, one can still define independent sets and flats. A collection of hyperplanes $\{A_1, A_2, \dots, A_s\}$ is **independent** if and only if the intersection $A_1 \cap A_2 \cap$

342

$\cdots \cap A$, cannot be the intersection over a proper subset of $\{A_i\}$. Given any subset X of $\mathcal{V}$, the **flat generated by** X is the subset $F(X)$ of all elements of $\mathcal{V}$ which depend on a finite subset of X.

At this stage, we see that if Veldkamp lines at least exist, there are two sorts of partially ordered structures defined on certain subsets of $\mathcal{V}$: (1) **the partially ordered system of subspaces of the Veldkamp space**, and (2) **the partially ordered system of flats of the r-partial matroid**.

To what extent do these two partially ordered structures coincide? There are two immediate elementary observations that can be made:

(1.4) (i) Every flat is a subspace of the Veldkamp space.

(ii) For each integer $k \leq r$, every subspace of the Veldkamp space generated by k points is a flat.

But for values of k larger than r, many Veldkamp subspaces generated by k points are not flats. So in fact $\mathcal{V}$ supports two distinct dependence relations (the one given above, and *subspace dependence* in which a point of $(\mathcal{V}, \mathcal{V}_1)$ depends on a finite set of points if and only if it is in the *subspace* generated by the latter set). Yet both relations (or partially ordered systems, if one takes that point of view) admit the action of the group Aut(Γ).

We give here a classical example at low rank. Consider the (embedded) classical rank 3 polar space defined by a non-degenerate quadric Q of maximal Witt index on $PG(5, F)$. One may represent this projective space as $\mathbf{P}(V)$ where $V = F^{(6)}$, and q is defined by $Q(\alpha_1, \ldots, \alpha_6) = \alpha_1 \alpha_2 + \alpha_3 \alpha_4 + \alpha_5 \alpha_6$. Then any functional $f \in V^*$ defines a hyperplane $H = H_f$ as the set of all Q-singular 1-spaces in ker(f). Thus we can find hyperplanes A_1 and A_2 which are generalized quadrangles of type $Q(4, F)$ for which $A_1 \cap A_2$ is an elliptic quadric $Q^-(3, F)$. We can choose hyperplanes A_3 and A_4 such that A_4 meets $A_1 \cap A_2$ at a conic C, and $A_3 = p^\perp$ for a point p in C. Then $A_1 \cap A_2 \cap A_3 = \{p\}$. Thus

(1.5) (i) $A_1 \cap A_2 \cap A_3$ is properly contained in $A_1 \cap A_2 \cap A_4$, and

(ii) $A_1 \cap A_2 \not\subseteq A_4$.

Comparing with (1.1) this means the following property fails:

(1.6) (Veldkamp planes exist).

(1) Veldkamp lines exist.

(2) If $A_1, \ldots, A_4$ are hyperplanes such that $A_1 \cap A_2 \cap A_3 \subseteq A_4$ then either $A_1 \cap A_2 \cap A_3 = A_1 \cap A_2 \cap A_4$ or else $A_1 \cap A_2 \subseteq A_4$.

In fact, it is property (2) which fails, for Veldkamp lines exist (as they do for all polar spaces of rank ≥ 3 having thick lines (see, for example [5])). In fact it is well known that all hyperplanes of this polar space arise from its natural embedding in $PG(5, F)$, so, by Criterion 2, the Veldkamp space is a projective space. Yet, since Veldkamp planes don't exist, it carries with it a proper $\Omega^+(6, F)$-invariant subcollection of flats which is somehow a natural

343

invariant since it is defined entirely by the polar space itself. We will show that this phenomenon persists for all proper embeddable Grassmann spaces in

Theorem 1.1 *If* $\Gamma = \mathcal{G}(k, V)$ *is the Grassmann space whose points and lines are respectively the k-subspaces and $(k-1, k+1)$-subspace flags of an n-dimensional vector space V over a field F ($2 \le k \le \dim V - 2$), then Veldkamp planes do not exist for Γ.*

There is also an adjunct result for the non-embeddable case.

Theorem 1.2 *Suppose Γ is the corresponding Grassmann space of the k-subspaces of a vector space V over a non-commutative division ring D (k is an integer greater than 1 and less than $\dim V - 1$; V may have infinite dimension here). Then the Veldkamp space is not projective.*

By way of contrast we have

Theorem 1.3 *Veldkamp planes exist for the half-spin geometries (the Lie incidence geometries of type $D_{n,n}$, $n \ge 4$).*

(For the definition of this geometry see [1].) As we have already remarked, the Veldkamp space of a half-spin geometry is a projective space. The theorem says that if there is any disparity between the Veldkamp subspace structure and the matroidal structure, it must occur at higher rank. This result has at least one interesting corollary.

Corollary 1.4 *Any subspace of codimension at most three in the 2^{n-1}-dimensional half-spin module for $\Omega^+(2n, F)$ is spanned by the pure spinors within it.*

2.   Proofs of theorems 1 and 2

The **Grassmann space** $\mathcal{G}_k(V)$ is the point-line geometry $(\mathcal{P}, \mathcal{L})$, whose set of points $\mathcal{P}$ is the collection of all k-dimensional subspaces of the vector space V, and whose lines are the pairs (A, B) where A is a $(k-1)$-dimensional subspace of the $(k+1)$-subspace B – that is, the set of $(k-1, k+1)$-subspace flags. A "point" C is incident with a "line" (A, B) if and only if $A \subset C \subset B$ as subspaces of V. Evidently $k < \dim(V)$. If $k = 1$ or if $k = \dim(V) - 1$ when the latter is finite, then $\mathcal{G}_k(V)$ is just the projective space $\mathrm{P}(V)$ or $\mathrm{P}(V^*)$. Otherwise we say that $\mathcal{G}_k(V)$ is a **proper Grassmann space**. We say that

$\mathcal{G}_k(V)$ has **finite rank** if and only if V is finite-dimensional.

(1) The embeddable case. Here D is a field and there is an injective mapping $e : \mathcal{P} \to \mathbf{P}$, where $\mathbf{P} := \mathbf{P}\,(\Lambda^{(k)}(V))$, the k-fold wedge product of V with itself, taking $\mathcal{P}$ bijectively onto the set of pure 1-spaces of $U = \Lambda^{(k)}(V)$, so that the point shadow of each line is mapped onto the set of 1-spaces of a suitable totally pure 2-subspace depending on that line. This produces an embedding $e : \Gamma \to \mathbf{P}$ of the point-line geometry $(\mathcal{P}, \mathcal{L})$ into the projective space $\mathbf{P}(U)$.

Let $\{x_1, \ldots, x_{n+1}\} = \mathcal{B}$ be a basis for the vector space V. Then the set $\mathcal{B} \wedge \mathcal{B} := \{x_{i_1} \wedge \cdots \wedge x_{i_k} \| 1 \le i_1 \le \cdots \le i_k \le n + 1\}$ is a basis for the vector space $U = \Lambda^{(k)}(V)$. Let W be the 4-space $\langle x_1, x_2, x_3, x_4 \rangle$ and let $u = x_5 \wedge \cdots \wedge x_{k+2}$. Then $S := e^{-1}(e(\mathcal{P}) \cap \mathbf{P}(\Lambda^{(2)}(W) \wedge u))$ is a symplecton of Γ of type $A_{3,2}$.

Let A be any geometric hyperplane of S. Then by [4] there is a functional f of $\Lambda^{(2)}(W)$ such that all pure 1-spaces of $\Lambda^{(2)}(W)$ which are in the kernel of f are precisely the images of the hyperplane A. We form the direct sum

(2.1) $\Lambda^{(k)}(V) = \Lambda^{(2)}(W) \wedge u \oplus BF$

where B is the set of b in $\mathcal{B} \wedge \mathcal{B}$ which are not of the form of exactly two factors from the set $\{x_1, x_2, x_3, x_4\}$ being wedged with u (there are exactly $\binom{n}{k} -$ 6 such basis elements b). Let π be the projection $\Lambda^{(k)}(V) \to \Lambda^{(2)}(W) \wedge u$ with respect to the direct sum (2.1). We define a functional $\hat{f}$ of $\Lambda^{(k)}(V)$ by declaring its value on each basis element of $\mathcal{B} \wedge \mathcal{B}$ as follows:

$$\hat{f}(b) = 0 \text{ if } b \text{ is in } B,$$

$$\hat{f}(x_i \wedge x_j \wedge u) = f(x_i \wedge x_j) \text{ for } 1 \le i \le j \le 4.$$

Then $\hat{A} := e^{-1}(e(\mathcal{P}) \cap \mathbf{P}(\ker(\hat{f})))$ is a geometric hyperplane of Γ satisfying
(2.2) $\hat{A} \cap S = A$.
Now by the Klein correspondence, the polar space S is of type $Q^+(5, F)$. Thus, by the remarks of the introduction, there exist four hyperplanes $A_1, A_2,$ A_3, A_4 of S such that
(2.3) (i) $A_1 \cap A_2 \cap A_4$ properly contains $A_1 \cap A_2 \cap A_3$, and
   (ii) $A_1 \cap A_2 \not\subset A_4$.

We now form the hyperplanes $\hat{A}_i$ defined by functionals $\hat{f}_i$ of $\Lambda^{(k)}(V)$ as above, for $i = 1, 2, 3, 4$. We claim that
(2.4) $\hat{A}_i \cap \hat{A}_2 \cap \hat{A}_3 \subseteq \hat{A}_4$.
Suppose $v = \sum \alpha_j b_j$ (b_j ranging over $\mathcal{B} \wedge \mathcal{B}$) was a vector of $\Lambda^{(k)}(V)$ satisfying $\hat{f}_i = 0$, for $i = 1, 2, 3$. We must show $\hat{f}_4 = 0$. Now each element of $\Lambda^{(2)}(W) \wedge u$ is uniquely expressible in the form $w \wedge u$ where $w \in \Lambda^{(2)}(W)$. Thus we can write $\pi(v) = v_\pi \wedge u$ where v_π is in $\Lambda^{(2)}(W)$. Now from the definition of the $\hat{f}_k$

we see that $\hat{f}_k(b_j) = 0$ for all but the six values $b_{ij} = x_i \wedge x_j \wedge u$, $1 \leq i \leq j \leq 4$
of $\mathcal{B} \wedge \mathcal{B} - \mathcal{B}$. Thus

(2.5) $\hat{f}_k(v) = \hat{f}_k(\pi(v)) = \hat{f}_k(v_\pi \wedge u) = f(v_\pi)$, $k = 1, \ldots, 4$.

But $A_1 \cap A_2 \cap A_3 \subseteq A_4$ means

(2.6) $f_k(v_\pi) = 0$ for $k = 1, 2, 3$ implies $f_4(v_\pi) = 0$.

Thus from (2.5) and (2.6), our claim follows.

Now we see from the claim (2.4) and (2.2) that

(2.7) $\hat{A}_1 \cap \hat{A}_2 \cap \hat{A}_4$ properly contains $\hat{A}_1 \cap \hat{A}_2 \cap \hat{A}_3$, and $\hat{A}_1 \cap \hat{A}_2 \not\subseteq A_4$.

Thus Veldkamp planes do not exist, proving Theorem 1.

(2) The non-embeddable case. Here D is a non-commutative division
ring. In this case, the hyperplanes of $\mathcal{G}_k(V)$ are completely described by the
Corollary to the Main Theorem of [2]. Let R be a subspace of V of codimension
k. Set $H(R)$ to be the set of k-dimensional subspaces V which intersect R
non-trivially. Then it is a Lemma of [2] that $H(R)$ is a hyperplane of $\mathcal{G}_k(V)$.
The Corollary says that in this non-embeddable case, every hyperplane of
$\mathcal{G}_k(V)$ has the form $H(R)$.

Now we can calculate the Veldkamp space $\mathcal{V}(\Gamma)$ directly. We know its
points; we only need to know its lines, and, since Veldkamp lines exist, these
can be determined by examining intersections of the $H(R)$.

First suppose R_1 and R_2 are two subspaces of codimension k in V.
What are the $H(R)$ which contain $H(R_1) \cap H(R_2)$?

Assume then $H(R) \supseteq H(R_1) \cap H(R_2)$. If R did not contain $R_1 \cap R_2$,
one could find a vector r in $R_1 \cap R_2 - R$, form a $(k-1)$-space complement
U to $R \oplus \langle r \rangle$ in V, and thus obtain a k-space $\langle r \rangle \oplus U$ which meets R_1 and
R_2 non-trivially, but meets R trivially. Since this violates our assumption, we
have

(2.8) R contains $R_1 \cap R_2$.

Also, suppose we could find non-zero vectors $r_i \in R_i - R$, $i = 1, 2$, such that
$E = \langle R, r_1, r_2 \rangle = R \oplus \langle r_1 \rangle \oplus \langle r_2 \rangle$ is a direct sum. Then if U is a $(k-2)$-space
complement to E in V, then $\langle r_1 \rangle \oplus \langle r_2 \rangle \oplus U$ is a k-subspace of V meeting
the R_i non-trivially, but meeting R trivially. Thus there is no such choice of
$r_i \in R_i - R$. This implies

(2.9) Either

    (i) R is one of the R_i, or

    (ii) $R_1 + R = R_2 + R = R_1 + R_2$ and this subspace contains R as a
         hyperplane.

In case (ii) of (2.9), $R_1 \cap R_2$ has codimension 1 in each R_i, and R
is a space of codimension k incident with both $R_1 \cap R_2$ and $R_1 + R_2$ of
codimension $k-1$. These values of R (along with R_1 and R_2) form the "points"
of the Grassmann line of $\mathcal{G}_k(V^*)$ determined by R_1 and R_2. The corresponding

Veldkamp line is thick.

Otherwise we have $R_1 \cap R_2$ of codimension greater than $k+1$ and case (i) of (2.9) lists the only possibilities for R. The corresponding Veldkamp line is thin; its point shadow is $\{R_1, R_2\}$.

Thus we have a bijection $\mathcal{V} \to \mathcal{G}_k(V^*)$ given by $H(R) \to R$, where R is regarded as a k-subspace of the dual space. The thick lines of $\mathcal{L}(V)$ are mapped onto the lines of the Grassmann space $\mathcal{G}_k(V^*)$, all other lines of $\mathcal{L}(\mathcal{V})$ are thin. Since the Grassmann space contains two intersecting lines not in a singular subspace, the Veldkamp space contains a plane with exactly two thick lines, and all remaining lines thin. Such a plane cannot be a generalized projective plane, and so the Veldkamp space is not a generalized projective space in this case.

3.   Some preliminary results needed for theorem 1.3 and corollary 1.4

We should first standardize some notation regarding point-line geometries. First $d_\Gamma(p, q)$ denotes the distance in the point-collinearity graph from point p to point q. For each non-negative integer k we let $\Delta_k^*(p)$ be the set of points at distance *at most* k from point p: thus $\Delta_0^*(p) = \{p\}$ and $\Delta_1^*(p) = p^\perp$ in the usual notation. A subspace S of $\Gamma = (\mathcal{P}, \mathcal{L})$ is **convex** if it contains the intermediate vertices of all geodesic paths of the point-collinearity graph connecting any two of its points. For any subset X of $\mathcal{P}$, $\langle X \rangle_\Gamma$, the **convex closure** of X, is the intersection of all convex subspaces containing X.

The following three results were proved in [5].

Lemma 3.1 *To show that Veldkamp $(r-1)$-spaces exist for some point-line geometry $\Gamma = (\mathcal{P}, \mathcal{L})$, it is sufficient to show for any convex subspace S of Γ, and any collection $\{A_1, A_2, \ldots, A_s\}$ of hyperplanes of S, $s \leq r$, that the point set $(A_1 \cap A_2 \cap \cdots \cap A_{s-1}) - (A_1 \cap A_2 \cap \cdots \cap A_s)$ has a connected collinearity graph. (Graphs with empty vertex set are considered connected.)*

Corollary 3.2 *The sufficient condition of Lemma 3.1 holds for any non-degenerate polar space of rank at least r. Hence Veldkamp $(r-1)$-spaces exist for these polar spaces.*

Lemma 3.3 *Suppose $\Gamma = (\mathcal{P}, \mathcal{L})$ is a point-line geometry for which Veldkamp $(r-1)$-spaces exist. Suppose further that every geometric hyperplane of Γ arises from an embedding $e : \Gamma \to \mathbf{P}$ for some projective space $\mathbf{P}$. Then every subspace of $\mathbf{P}$ of codimension at most r is spanned by the points of* *e(P) which are contained in it.*

In the previous study of Veldkamp lines ([5]) certain classes of point-line geometries $\mathcal{E}_n$ were defined for each integer $n \geq 2$. If $\Gamma = (\mathcal{P}, \mathcal{L})$ is in $\mathcal{E}_n$, then Γ must satisfy the following three axioms:

(E1) Γ has thick lines and is connected (i.e. has either a connected point collinearity graph, or equivalently a connected bipartite incidence graph).

(E2) (i) For any positive integer $k \leq n$, every geodesic path of length k extends to one of length n.

(ii) The point-collinearity graph has diameter n, and for each point p, the set $\Delta_{n-1}^*(p) := \{q \in \mathcal{P} \| d_\Gamma(p, q) \leq n - 1\}$ is a geometric hyperplane of Γ.

(E3) If p and q are distinct points of Γ with $d_\Gamma(p, q) \leq k$, $k \geq 2$, then the convex closure $\langle p, q \rangle_\Gamma$ is a subspace geometry belonging to $\mathcal{E}_k$.

If Γ is in $\mathcal{E}_n$, and $2 \leq k \leq n$, the symbol $\mathcal{E}_k(\Gamma)$ will denote the collection of subspaces of the form $\langle p, q \rangle_\Gamma$ where $d_\Gamma(p, q) = k$. It is easy to see that the geometries in $\mathcal{E}_2$ are non-degenerate polar spaces. The members of $\mathcal{E}_2(\Gamma)$ are called **symplecta**.

Examples of geometries of $\mathcal{E}_n$

Geometry	diameter	symplecton type
dual polar spaces $C_{n,n}$	n	classical GQ
Grassmann spaces $A_{2n-1,n}$	n	$\Omega^+(6, F)$
half spin geometries $D_{n,n}$, n even	$n/2$	$\Omega^+(8, F)$
exceptional geometry $E_{7,1}$	3	$\Omega^+(12, F)$

The geometries of $\mathcal{E}_n$ were introduced in [5] to give a general setting for a string of lemmata which could then be applied to any of its members. Of fundamental interest to us here is

Lemma 3.4 *Suppose Γ is a geometry belonging to $\mathcal{E}_n$. Suppose (i) $X = \{x_1, x_2, \ldots, x_r\}$ are points of a symplecton S which span a (singular) $(r-1)$-dimensional projective space, or (ii) $X = \{x_1, x_2, x_3, x_4\}$ is a set of four points of a symplecton, any two of which are collinear except for the pair $\{x_1, x_4\}$. Then the sets $\{H_x = \Delta_{n-1}^*(x) \| x \in X\}$ are independent hyperplanes of Γ.*

Proof: This is Corollary 6.4 of [5].

4. Half-spin geometries have Veldkamp planes

The half-spin geometries of type $D_{n,n}$ form a diagram geometry with diagram

$$\underset{\mathcal{D}_{n-1}}{\circ}\!-\!\underset{\mathcal{D}_{n-2}}{\circ}\!-\!\underset{\mathcal{D}_{n-3}}{\circ}\,-\,-\,-\,\underset{\mathcal{S}}{\circ}\!-\!\underset{\mathcal{M}_3}{\circ}\!-\!\underset{\mathcal{L}}{\circ}\!\!<\!\!\begin{array}{c}\circ^{\mathcal{P}}\\[4pt]\circ_{\mathcal{M'}}\end{array}$$

having n nodes. Here $\mathcal{P}$ and $\mathcal{L}$ are the points and lines of the geometry, $\mathcal{M'}$ is a class of maximal singular subspaces of projective dimension $n-1$, $\mathcal{M}_3$ is a class of maximal singular $PG(3)$'s, and $\mathcal{S}$ is the class of symplecta, which are type D_4. The diameter of the point-collinearity graph is $[n/2]$. The $\mathcal{D}_k$ are half-spin subspaces of type $D_{k,k}$ and the convex closure $\langle p, q\rangle_\Gamma$ for two points at distance $d \geq 2$, is a member of the class $\mathcal{D}_{2d}$ (that is, members of $\mathcal{E}_d(\Gamma)$ in the sense of the last section).

Suppose now, Γ is a half-spin geometry of type $D_{n,n}$, where n is even. Then, as already remarked, Γ is a member of the class $\mathcal{E}_m$, where $m = n/2$. Suppose D is a geometry in $\mathcal{D}_{n-1}$. Then for each point x in $\mathcal{P} - D$, $M_x = x^\perp \cap D$ is a maximal singular subspace of $\mathcal{M'}(D)$, the collection of maximal singular subspaces playing the role of $\mathcal{M'}$ in the diagram $\mathcal{D}_{n-1}$ for D. The singular subspace generated by x and M_x is then a member of $\mathcal{M'}$. Now the point-collinearity graph of the geometry D of type $D_{n-1,n-1}$ has diameter $m - 1 = [(n-1)/2]$. For each point y in D, and singular subspace M of $\mathcal{M'}(D)$, $\Delta_{m-2}^*(y) \cap M$ is either (i) all of M, (ii) of codimension two in M or (iii) empty. The set

$$N_D(M) := \{y \in D \| \Delta_{m-2}^*(y) \cap M \neq \emptyset\}$$

is called the set of **points near** M, and is a geometric hyperplane of D in $\mathcal{D}_{2m-1}$. For any point x, as noted, $\Delta_{m-1}^*(x)$ is a hyperplane of Γ, and if x is not in D,

(4.1) $\Delta_{m-1}^*(x) \cap D = N_D(M_x)$

where $M_x = x^\perp \cap D$.

Now if D_1 and D_2 are two members of $\mathcal{D}_{n-1}$, then $D_1 \cap D_2$ is either a subspace in $\mathcal{D}_{n-2}$ (where by convention $\mathcal{D}_4 := \mathcal{S}$ and $\mathcal{D}_3 := \mathcal{M}_3$ for the cases $n = 6$ and 5) or $D_1 \cap D_2$ is empty. In the latter case D_1 and D_2 are said to be **opposite**. In that case there are isomorphisms

(4.2) $\begin{array}{l}\psi_{12} : (D_1, \mathcal{L}(D_1)) \rightarrow (\mathcal{M'}(D_2), \mathcal{L}(D_2))\\[4pt]\psi_{21} : (D_2, \mathcal{L}(D_2)) \rightarrow (\mathcal{M'}(D_1), \mathcal{L}(D_1))\end{array}$

where $\mathcal{L}(D_i) := \mathcal{L} \cap \operatorname{Res}(D_i)$, the lines of Γ incident with D_i, $i = 1, 2$. (Noting that each maximal singular subspace belonging to $\mathcal{M'}(D)$ is a hyperplane of a unique member of $\mathcal{M'}$ incident with D, and, conversely, that each member of $\mathcal{M'} \cap \operatorname{Res}(D)$ meets D at a member of $\mathcal{M'}(D)$, the sets $\mathcal{M'}(D)$ and $\mathcal{M'} \cap \operatorname{Res}(D)$ may be identified in (4.2) as far as incidence is concerned.)

349

By considering three pairwise opposite members of $\mathcal{D}_{n-1}$ (they must exist since classical polar spaces of reduced rank at least 2 possess 3-cocliques) one infers from (4.2)

(4.3) (i) Any two members of $\mathcal{D}_{n-1}$ are isomorphic.

(ii) For any D in $\mathcal{D}_{n-1}$, there is a twisting automorphism

$$\sigma : (D, \mathcal{L}(D)) \to (\mathcal{M}'(D), \mathcal{L}(D)).$$

We exploit the dualities between opposite members of $\mathcal{D}_{n-1}$ as well as the self-duality of (4.3)(ii) in the proof of the following lemma.

Lemma 4.1 *Let D be a half-spin geometry of type $D_{n,n}$ where $n = 2m+1$ is odd. Suppose $\{M_1, \ldots, M_4\}$ is a set of four members of $\mathcal{M}'(D)$ (the singular subspaces of D corresponding to the objects of type $\mathcal{M}'$ in the diagram $\mathcal{D}_n$).*

We suppose that $M_i \cap M_j = L_{ij}$ is a line for any $\{i, j\}$ a 2-set of $\{1, 2, 3, 4\}$ distinct from $\{1, 4\}$, where $L_{12} \neq L_{23} \neq L_{34}$. In the case that $M_1 \cap M_4$ is a line L_{14} we assume it is not in the plane $\langle L_{12}, L_{23} \rangle$; otherwise, of course, $M_1 \cap M_4$ is empty.

Then the four sets $\{H_i = N_D(M_i) \| i = 1, \ldots, 4\}$ are independent in the sense of section 2.

Proof: The condition $L_{12} \neq L_{23} \neq L_{34}$ means the triplets $\{H_1, H_2, H_3\}$ and $\{H_2, H_3, H_4\}$ are independent. It suffices to show that for every 3-set J of the index set $I = \{1, 2, 3, 4\}$, there is a point of D in the intersection of the H_i for $i \in J$ which is not in H_k where $\{k\} = I - J$. So, in turn, it suffices to prove the "twisted dual" version of this last assertion, namely:

(4.4) Let $X = \{x_1, \ldots, x_4\}$ be a set of four points of D which either (i) form a 4-clique generating a $PG(3)$, or (ii) yield the induced collinearity subgraph

with $\langle x_1, x_2, x_3 \rangle$ and $\langle x_2, x_3, x_4 \rangle$ planes. Then for any 3-subset $X_0 = \{x_i, x_j, x_k\}$ of X, there exists a maximal singular subspace M_0 of $\mathcal{M}'(D)$, such that $N_D(M_0) \cap X = X_0$.

To prove (4.4), we first embed D in the canonical way in a half-spin geometry E of type $D_{n+1,n+1}$ where $n + 1 = 2m + 2$ is even. The word "canonical" here means that the embedding $D \to E$ (which we regard as an inclusion) is such

that D belongs to $\mathcal{D}_n(E)$. Now as previously noted, E is a member of $\mathcal{E}_{m+1}$ satisfying axioms (E1)-(E3). By Lemma 3.3, the sets $\Delta_m^*(x_i)$ are independent, so, for any 3-subset $J = \{i, j, k\}$ of I, there is a point u_J in E such that
$$(4.5) \quad \Delta_m^*(u_J) \cap X = X_J := \{x_S \| S \in J\}.$$
Since u_J has distance $m + 1$ from one member of X, u_J cannot be in the subgeometry D for the point-collinearity graph of D has diameter m. Now, setting $u_J^\perp \cap D = M_J \in \mathcal{M}'(D)$, (4.1) and (4.5) yield

$$N_D(M_J) \cap X = X_J.$$

Thus M_J meets the requirements of M_0 in the conclusion of (4.4). This completes the proof.

We now begin the proof of Theorem 3, that Veldkamp planes exist for the half-spin geometries. By Lemma 3.2, it suffices to show that for any three hyperplanes A_1, A_2 and A_3 of a half-spin geometry E of type $D_{n,n}$, the collinearity graph induced on the set

$$Z = A_1 \cap A_2 - A_1 \cap A_2 \cap A_3$$

is connected. By Corollary 3.2, this is true if $n = 4$, since geometries of type $D_{4,4}$ are polar spaces of rank 4. By way of contradiction we assume $n > 4$ chosen minimally so that Z is not connected, and choose x and y in distinct connected components of Z so that their mutual distance $d_E(x, y)$ in E is minimal. Then

(4.6) (i) $d_E(x, y) = m = \operatorname{diam}(E)$, $E = \langle x, y \rangle_E \in \mathcal{E}_m$, forcing $n = 2m$ to be even.

(ii) There is no point z in Z whose distance from both x and y is less than m.

First choose D in $\mathcal{D}_{n-1}(E)$ on point y, and let D' be a subgeometry of $\mathcal{D}_{n-1}(E)$ which is opposite D and lying on point x (one exists). Now, since D and D' are of type $D_{n-1,n-1}$ where $n - 1 \geq 5$, they must have singular rank at least 5. In fact, if M is chosen in $\mathcal{M}'(D')$ so that it lies on x, then $A_1 \cap A_2 \cap M$ has codimension at most 2 in M, a $PG(n - 1)$, and so it is at least a plane and contains $A_1 \cap A_2 \cap A_3 \cap M$ as a proper subspace. Thus there is a choice of three points x_1, x_2 and x_3 in

$$A_1 \cap A_2 \cap M - A_1 \cap A_2 \cap A_3 \cap M = Z \cap M$$

spanning a plane π of M. Let L be the line on x_2 and x_3. We set $M = M_1$, $\pi = \pi_1$ and let M_2 be a second maximal singular subspace of $\mathcal{M}'(D')$ on line L. We can then find a fourth point x_4 in $Z \cap M_2$ so that $\langle x_2, x_3, x_4 \rangle$ is a plane π_2 in M_2. We have $M_1 \cap M_2 = L$ and at this point there are two possibilities:

(4.7) (i) $\langle x_1, x_2, x_3, x_4 \rangle$ is isomorphic to a $PG(3)$ or

    (ii) the collinearity graph on $X = \{x_1, x_2, x_3, x_4\}$ is

Then setting $M_i = x_i^{\perp} \cap D$, we see from Lemma 4.1 that the sets $H_i := N_D(M_i)$, $i = 1, 2, 3, 4$, are independent hyperplanes of D.

On the other hand, if z is any point of $Z \cap D'$, then $d_E(z, x) \le m - 1$, since D' is convex in E and has point-diameter $m - 1$. By the minimality of n, z lies in the same connected component of Z as does x. Then z can be placed in the role of x in (4.6)(ii) to force the fact that no point of Z can lie in $\Delta_{m-1}^*(z) \cap \Delta_{m-1}^*(y)$. This means $\Delta_{m-1}^*(z) \cap D \cap Z = \emptyset$ so by (4.1)

(4.8) $N_D(M_x) \cap A_1 \cap A_2 \subseteq A_3$.

Now, because $y \in D \cap Z$, $A_3 \cap D$ does not contain $A_1 \cap A_2 \cap D$ – that is, $A_3 \cap D$ does not depend on $A_1 \cap D$ and $A_2 \cap D$. Since, by induction, Veldkamp planes exist for D (minimality of n), an exchange axiom for 3-sets holds for hyperplanes of D. Thus (4.8) and the fact that $A_3 \cap D$ does not depend on $\{A_1 \cap D, A_2 \cap D\}$, yields

(4.9) $A_1 \cap A_2 \cap A_3 \cap D \subseteq N_D(M_x)$.

We now apply (4.9) to $x = x_i$, $i = 1, 2, 3, 4$. This produces the result that each of the hyperplanes $H_i := N_D(M_i)$ of D depends on the set of three hyperplanes, $\{A_i \cap D \| i = 1, 2, 3\}$. Since Veldkamp planes exist for D, we see that this implies that the four hyperplanes $H_1, \ldots, H_4$ are dependent. This contradicts the conclusion that the H_i are independent obtained above. The proof is complete.

We prove now Corollary 1.4. We have three facts at hand. (i) All geometric hyperplanes of the half-spin geometries arise from their embedding in the half-spin module ([3]). (ii) The pure spinor 1-spaces are the embedded points of the half-spin geometry in the half-spin module. (iii) Veldkamp planes exist for the half-spin geometries. The Corollary now follows from these three facts and Lemma 3.3.

References

[1] **B. N. Cooperstein and E. E. Shult.** Geometric hyperplanes of embeddable Lie incidence geometries. In J. W. P. Hirschfeld, D. R. Hughes, and J. A. Thas, editors, *Advances in Finite Geometries and Designs*, pages 81–91, Oxford, 1991. Oxford University Press.

[2] **J. I. Hall and E. E. Shult.** Geometric hyperplanes of non-embeddable Grassmannians, to appear in *European J. Combin.*

[3] **E. E. Shult**. Geometric hyperplanes of the half-spin geometries. Unpublished, 1990.

[4] **E. E. Shult**. Geometric hyperplanes of embeddable Grassmannians. *J. Algebra*, 145, pp. 55–82, 1992.

[5] **E. E. Shult**. On Veldkamp lines. Preprint, 1992.

[6] **E. E. Shult and J. A. Thas**. Hyperplanes of dual polar spaces and the spin module. *Arch. Math.*, 59, pp. 610–623, 1992.

E. E. Shult Department of Mathematics, Kansas State University, Manhattan KS 66502, USA. e-mail: shult@ksuvm.bitnet

The Lyons group has no distance-transitive representation

L. H. Soicher

Abstract

We show that the Lyons group Ly has no distance-transitive representation, and that the only faithful multiplicity-free permutation representations of Ly are those on the cosets of $G_2(5)$ and of $3 \cdot McL: 2$.

One area which is still open in the classification of primitive distance-
transitive graphs is the determination of the primitive distance-transitive representations of certain sporadic simple groups and their automorphism groups
(see [1]). In this note we show that the Lyons group $Ly \cong \mathrm{Aut}(Ly)$ has no
distance-transitive representation. In the process, we find that the only faithful multiplicity-free permutation representations of Ly are those on the cosets
of $G_2(5)$ and of $3 \cdot McL: 2$.

Let G be a permutation group on a finite set Ω, and Γ a connected graph
with vertex set Ω. (Throughout this note all graphs are undirected, with no
loops and no multiple edges.) We say that G acts *distance-transitively* on Γ
if for each $i = 0, \ldots, \mathrm{diam}(\Gamma)$, G is transitive on the set of ordered pairs of
vertices at distance i in Γ. The graph Γ is called *distance-transitive* if $\mathrm{Aut}(\Gamma)$
acts distance-transitively on Γ. A *distance-transitive representation* (DTR)

$$\rho : X \to \mathrm{Sym}(\Omega)$$

of a group X is a faithful permutation representation such that $\rho(X)$ acts
distance-transitively on some connected graph with vertex set Ω.

Let X be a finite group, and $\rho : X \to \mathrm{Sym}(\Omega)$ a DTR. It is well-known
that ρ must be *multiplicity-free*, that is, the sum of distinct (complex) irreducible representations of X (in fact these distinct representations must be
real (see [1]), but it seems worthwhile to classify all multiplicity-free permu-
tation representations of Ly). Thus, $|\Omega| \le D_X$, where D_X is defined to be the
sum of the degrees of the irreducible representations of X. From the character
table of Ly (see [2] or [3, 4]), we have that

$$D_{Ly} = 1297168312.$$

From R.A. Wilson's classification [7, 8] of the maximal subgroups of Ly, we see that the only proper subgroups of index $\leq D_{Ly}$ are $G_2(5)$, $3 \cdot McL : 2$, $5^3 \cdot L_3(5)$, and $2 \cdot A_{11}$. (We use ATLAS notation [2] throughout for group structures and conjugacy classes.)

Proposition 1 *Let $H \cong 5^3 \cdot L_3(5)$ or $2 \cdot A_{11}$, and ρ be the permutation representation of Ly acting on the (right) cosets of H. Then ρ is not multiplicity-free.*

Proof. We make use of the rational character table of Ly, supplied in computer form by the GAP 3.1 group theory system [6].

Let π be the character of ρ, and σ the sum of the irreducible characters of Ly. We suppose that π is the sum of distinct irreducible characters and find a contradiction by examining the possibilities for $\sigma - \pi$, which must be a rational character not having the trivial character as a constituent.

Suppose $H \cong 2 \cdot A_{11}$, and let $k = \sigma(1) - \pi(1) = 341437$. A very easy computer search, using a PASCAL program, shows that there is no sum τ of nontrivial distinct irreducible characters of Ly, such that $\tau(1) = k$ and τ is a rational character.

Now suppose $H \cong 5^3 \cdot L_3(5)$, and let $k = \sigma(1) - \pi(1) = 183938656$. Using the same PASCAL program, and about three minutes of CPU-time on a SUN Sparcstation 2, we have that there are exactly 2325 sums τ of nontrivial distinct irreducible characters of Ly, such that $\tau(1) = k$ and τ is a rational character, but for each such sum τ we find that $\sigma - \tau$ has a negative value on some element of Ly, and so cannot be the permutation character π.

$\square$

We are now left to consider the permutation character π of Ly on the cosets of H, where $H \cong G_2(5)$ or $H \cong 3 \cdot McL : 2$. In each case, π is the sum of just five distinct characters of irreducible real representations (see [2] or [3, 4]). We shall show in each case that the permutation representation corresponding to π is not a DTR.

Lemma 2 *Let G be a permutation group on a finite set Ω, $\omega \in \Omega$, and $\Omega_0, \ldots, \Omega_d$ the distinct orbits of the point stabilizer G_ω on Ω. Suppose that $G^* \leq G$ acts primitively on $\Omega^* = \omega^{G^*}$, and that G_ω^* has exactly $d+1$ distinct orbits $\Omega_0^*, \ldots, \Omega_d^*$ on Ω^*, such that $\Omega_i^* \subseteq \Omega_i$ for $i = 0, \ldots, d$.*

If G acts distance-transitively on a connected graph Γ with vertex set Ω, then G^* acts distance-transitively on the graph Γ^* induced by Γ on Ω^*.

Proof. If G acts distance-transitively on Γ, then without loss of generality we may assume that the Ω_i are ordered so that $\Omega_i = \Gamma_i(\omega)$, the set of vertices

356

at distance i from ω. Since G^* acts primitively on Γ^*, we see that Γ^* is connected and that $\Omega_i^* = \Gamma_i^*(\omega)$. The result follows. $\qquad\square$

Proposition 3 *The permutation representation of $G \cong Ly$ on the cosets of $H \cong G_2(5)$ is not a DTR.*

Proof. Let t be a $3A$-element of H, $G^* = N_G(t) \cong 3 \cdot McL{:}2$, $\omega = H$, $\Omega = \{Hg \,|\, g \in G\}$, and $\Omega^* = \{Hg \,|\, g \in G^*\}$. Now $G_\omega^* \cong 3 \cdot U_3(5){:}2$ has nontrivial orbits of lengths

$$252, \quad 750, \quad 2625, \quad 3500$$

on Ω^* (see [5]), and so the corresponding two point stabilizers A, B, C, D have respective orders

$$3000, \quad 1008, \quad 288, \quad 216.$$

The two point stabilizers corresponding to the nontrivial orbits of H on Ω are of shapes

$$5_+^{1+4}{:}4.S_4, \quad U_3(3), \quad 2.(A_5 \times A_4).2, \quad (3 \times L_2(7)){:}2$$

(see [2] or [3, 4]). Since 5^3 divides $|A|$, we see that the only possible inclusion for A in such a two point stabilizer is $A \le 5_+^{1+4}{:}4.S_4$. We also see that $B \cong (3 \times L_2(7)){:}2$, the only possible inclusion for C is $C \le 2.(A_5 \times A_4).2$ ($U_3(3)$ has no subgroup of order 288), and the only possible inclusion for D is $D \le U_3(3)$ (since 3^3 divides $|D|$).

Now suppose that the representation of G acting on Ω is a DTR. The action of G^* on Ω^* is the primitive action of $McL{:}2$ on the cosets of $U_3(5){:}2$, and so by the preceding discussion and Lemma 2, it follows that the representation of $McL{:}2$ on the cosets of $U_3(5){:}2$ is a DTR. However, this is not the case (see [5]), and this contradiction establishes the result. $\qquad\square$

Proposition 4 *The permutation representation of $G \cong Ly$ on the cosets of $H \cong 3 \cdot McL{:}2$ is not a DTR.*

Proof. We identify the cosets of H with the $3A$-generated subgroups of order 3 of G. Let $\omega = O_3(H)$, and $\Omega = \omega^G$. Now H has just four nontrivial orbits on Ω, which can each be characterised by the group

$$3^2, \quad 2 \cdot A_4, \quad 2 \cdot A_5, \quad \text{or} \quad 5_+^{1+2}{:}3, \qquad\qquad (*)$$

generated by ω and an element of that orbit (see [3, 4]).

Now let $\alpha \in \Omega$, such that $\langle \alpha, \omega \rangle \cong 3^2$, and $G^* = N_G(\alpha) \cong 3 \cdot McL{:}2$. Then $\Omega^* = \omega^{G^*}$ is a conjugacy class (of size 15400) of subgroups of G^*. Now

G^* acts on Ω^* as the primitive action of $McL\!:\!2$ on its $3A$-generated subgroups of order 3, and this action is rank 5 (see [5]). In McL, each of the groups in the list (*) can be generated by a pair of $3A$-elements (this can be seen in $2\cdot A_8$ and $5^{1+2}_+\!:\!3\!:\!8$), and so this is also true for G^*.

We may now apply the preceding discussion and Lemma 2 to assert that if the representation of G on Ω is a DTR, then so must be the representation of $McL\!:\!2$ on its $3A$-generated subgroups of order 3. However, from [5] we know that the latter representation is not a DTR.   □

References

[1] A. E. Brouwer, A. M. Cohen, and A. Neumaier. *Distance-regular Graphs*. Springer Verlag, Berlin, 1989.

[2] J. H. Conway, R. T. Curtis, S. P. Norton, R. A. Parker, and R. A. Wilson. *Atlas of Finite Groups*. Clarendon Press, Oxford, 1985.

[3] R. Lyons. Evidence for a new finite simple group. *J. Algebra*, 20, pp. 540–569, 1972.

[4] R. Lyons. Evidence for a new finite simple group. errata. *J. Algebra*, 34, pp. 188–189, 1975.

[5] C. E. Praeger and L. H. Soicher. Permutation representations and orbital graphs for the sporadic simple groups and their automorphism groups: rank at most five. In preparation.

[6] M. Schönert et al. *GAP – Groups, Algorithms, and Programming*. Lehrstuhl D für Mathematik, Rheinisch Westfälische Technische Hochschule, Aachen, Germany, first edition, 1992. 700 pages.

[7] R. A. Wilson. The subgroup structure of the Lyons group,. *Math. Proc. Cambridge Philos. Soc.*, 95, pp. 403–409, 1984.

[8] R. A. Wilson. The maximal subgroups of the Lyons group. *Math. Proc. Cambridge Philos. Soc.*, 97, pp. 433–436, 1985.

L. H. Soicher, School of Mathematical Sciences, Queen Mary and Westfield College, Mile End Road, London E1 4NS, U.K. email: L.H.Soicher@qmw.ac.uk

Intersection of arcs and normal rational curves in spaces of odd characteristic

L. Storme *   T. Szőnyi [†]

Abstract

We study arcs K in $\mathrm{PG}(n,q)$, $n \geq 3$, q odd, having many points common with a given normal rational curve L. In particular, we show that, if $0.09q + 2.09 \geq n \geq 3$, q large, then $(q+1)/2$ is the largest possible number of points of K on L, improving on the bound given in [11], [12], [14]. When $|K \cap L| = (q+1)/2$, we show that the points of $K \cap L$ are invariant under a cyclic linear collineation of order $(q \pm 1)/2$. The corresponding questions for q even are discussed in [13].

1. Introduction

Let $\Sigma = \mathrm{PG}(n,q)$ denote the n-dimensional projective space over the field $\mathrm{GF}(q)$. A k-arc in Σ, with $k \geq n+1$, is a set K of k points such that no $n+1$ points of K belong to a hyperplane of Σ. A point r of $\mathrm{PG}(n,q)$ extends a k-arc K, in $\mathrm{PG}(n,q)$, to a $(k+1)$-arc if and only if $K \cup \{r\}$ is a $(k+1)$-arc. A k-arc K of $\mathrm{PG}(n,q)$ is complete if and only if K is not contained in a $(k+1)$-arc of $\mathrm{PG}(n,q)$. Otherwise, K is called incomplete.

A normal rational curve K of $\mathrm{PG}(n,q)$, $2 \leq n \leq q-2$, is a $(q+1)$-arc which is projectively equivalent to the set $L = \{(1,t,\ldots,t^n)\|t \in \mathrm{GF}(q)^+\}$ ($\mathrm{GF}(q)^+ = \mathrm{GF}(q) \cup \{\infty\}$; ∞ defines the point $(0,\ldots,0,1)$). More information about arcs and normal rational curves can be found in [2, Chap. 21] and [3, Chap. 27]. In $\mathrm{PG}(2,q)$, $L = \{(1,t,t^2)\|t \in \mathrm{GF}(q)^+\}$ is the conic $C : X_1^2 = X_0 X_2$. This $(q+1)$-arc is complete when q is odd. A normal rational curve of $\mathrm{PG}(3,q)$ is also called a twisted cubic. For results concerning the completeness of normal rational curves we refer to [11], [12], [14].

Given a conic C and a point $r \notin C$, it is easy to construct k-arcs containing approximately half the points of C by choosing one of the two points of C on each secant line passing through r. This construction, sometimes

---

* . Senior Research Assistant of the National Fund for Scientific Research Belgium
[†]. Research of this author was supported by the National Fund for Scientific Research Belgium and by the M.H.B. Fund for the Hungarian Science

called the *Segre construction*, has been studied by several authors. The difficulty is to prove the completeness of the arcs, and various methods such as group theory, algebraic geometry and elementary field theory have been used [5], [9].

For more details about the history of the problem we consider, the reader is referred to [13]. Here we only repeat the main problems:

(1) Consider a $(k+1)$-arc of $PG(n,q)$, $2 \leq n \leq q-2$, which has k points in common with a given normal rational curve. What is the maximum value of k for which this occurs?

(2) Consider a normal rational curve L in $PG(n,q)$. Fix a point r of $PG(n,q) \backslash L$. Characterize the largest k-arcs K, contained in L, such that $K \cup \{r\}$ is a $(k+1)$-arc and investigate the completeness of $K \cup \{r\}$.

For problem (1), there is an easy bound for the cardinality of $K \cap L$ [11], obtained by induction on n, namely

$$|K \cap L| \leq \frac{1}{2}(q+3) + (n-2). \qquad (1)$$

Our results improve this bound and, as for q even, we would like to stress that our bound is independent of the dimension n; namely we show that

$$|K \cap L| \leq \frac{1}{2}(q+1) \quad \text{if} \quad 3 \leq n \leq 0.09q + 2.09.$$

Moreover our bound is attained and we can characterize the arcs K for which $|K \cap L|$ is maximal. We show that the points of $K \cap L$ are invariant under a cyclic linear collineation of order $(q \pm 1)/2$. This key observation allowed us to use induction on the dimension n without having to increase the bound simultaneously with the dimension when $3 \leq n \leq 0.09q + 2.09$. This is done in Section 4.

The method applied is based on a careful inspection of arcs and normal rational curves in three dimensions. Let $r \in K \backslash L$. Projecting the arc K from r we get a subset of a rational plane cubic curve, which allows us to translate everything to the language of abelian groups using 3-independent subsets. Section 2 is devoted to the study of 3-independent subsets in abelian groups, while in Section 3 we characterize the arcs K in $PG(3,q)$ having $(q+1)/2$ points on a normal rational curve.

Historically, there was another source to our problem: coding theory. Arcs are related to MDS codes, subsets of normal rational curves to (Generalized Doubly Extended) Reed–Solomon (GD)RS codes. MDS extensions of GDRS codes were studied by Seroussi and Roth [11]. The bound (1), together with an improvement for q even, is due to them. Later on, Roth and Lempel

[10] used a normal rational curve and a point lying on a tangent of it to construct long non-Reed–Solomon type MDS codes. Their construction is based on subgroups of index two in the group $(\mathrm{GF}(q), +)$, q even; so it is one case in the characterization in [13]. For more on the coding theoretic background and the connection of normal rational curves with MDS codes and with GDRS codes, we refer to [7], [10], [11].

2. The group problem

We study 3-independent subsets of cyclic and elementary abelian groups. The following results and remarks show how these abelian groups are used in the study of arcs. This section is based on Theorems 2.1 to 2.6 of [13] where all the necessary statements and theorems are proved in detail. We restrict ourselves to a survey of the results that will be used in the subsequent sections.

Consider a twisted cubic L in $\mathrm{PG}(3, q)$. A *chord* to L is either a tangent line to L, a (real) bisecant to L through 2 different points of L in $\mathrm{PG}(3, q)$ or a (imaginary) bisecant to L through 2 conjugate points of L in $\mathrm{PG}(3, q^2)\backslash\mathrm{PG}(3, q)$. Every point of $\mathrm{PG}(3, q)\backslash L$ belongs to exactly one chord of L [2, p. 240]. Projecting L from a point r, $r \notin L$, one gets a rational cubic curve Δ. By the tangent-chord law, one can define an abelian group G on its non-singular points. This group is elementary abelian if r lies on a tangent of L, cyclic of order $q - 1$ if r is on a real bisecant, and cyclic of order $q + 1$ if r belongs to an imaginary bisecant of L. There is an element δ in G such that three non-singular points of Δ, with parameters x, y, z, are collinear if and only if $x + y + z = \delta$.

The following theorem, due to Kneser, plays a crucial role in the characterization of our arcs in terms of abelian groups.

Theorem 2.1 (Kneser [8, p. 6]) *Let A and B be finite subsets of an abelian group G. There exists a subgroup H of G such that $A + B = A + B + H$ and $|A + B| \geq |A + H| + |B + H| - |H|$.*

As in [13], q_0 always denotes the smallest integer q for which $(1 + \sqrt{3})r_3(n) < 0.01n, n \geq q - 1$, where $r_3(n)$ is the maximum cardinality of a subset $A \subset N = \{1, \ldots, n\}$ that contains no three-term arithmetic progression.

Definition 2.2 Let G be an abelian group and δ a fixed element of it. A subset A of G is called a 3-independent subset of G with respect to δ if and only if $x + y + z \neq \delta$ for all sets $\{x, y, z\}$ of pairwise distinct elements of A. If $\delta = o$, the identity of G, then we simply say that A is 3-independent.

The concept of 3-independency is used for studying arcs K' contained in a plane rational cubic curve Δ. Let G be the abelian group defined on Δ and S be the set of parameters of the non-singular points of K'. Let x, y, z be three distinct parameters of S. Since the corresponding points of K' are not collinear, $x + y + z \neq \delta$, where δ is defined in the beginning of Section 2. So, S is a 3-independent subset with respect to δ.

Lemma 2.3 [13, Lemma 2.6] *Let G be the cyclic group C_{q-1} or C_{q+1} with q odd, $q \geq q_0$, and let A be a 3-independent subset with $|A| \geq 0.41|G|$. Then A is contained in $G \setminus H$ or H, according as $\delta \in H$ or not, where H is the unique subgroup of index 2 of G. Furthermore, the cyclic group C_q, q prime, $q \geq q_0$, contains no 3-independent subset A of size $|A| \geq 0.41q$.*

Lemma 2.4 *An elementary abelian group $G = C_p^n$, $p > 2$, which contains a non-trivial subgroup H for which $|G/H| \geq 121$ and $|H| \geq 121$, contains no 3-independent subsets T of cardinality at least $0.41|G|$.*

Proof. Let $G = C_p^n$, $p > 2$, and $H \leq G$. Define $k = \max_{a \in G} |T \cap (a + H)|$ and fix an a with $|T \cap (a + H)| = k \geq 0.41|H|$. Pair elements $\bar{b}, \bar{c} \in G/H$ and fix their coset representatives in such a way that $a + b + c = \delta$. We call the pairs $\{\bar{b}, \bar{c}\}$ for which $\bar{b} = \bar{c}$ or $\bar{a} \in \{\bar{b}, \bar{c}\}$ the exceptional pairs since in these cases $\bar{a}, \bar{b}, \bar{c}$ are not pairwise distinct.

Let $T_x = \{t - x \,||\, t \in T \cap (x + H)\}$. If the sets T_b, T_c are non-empty, Theorem 2.1 gives us a subgroup K of H for which $|T_a + T_b| \geq |T_a| + |T_b| - |K|$. As $(T_a + T_b) \cap (-T_c) = \emptyset$, we have

$$|T_a| + |T_b| + |T_c| \leq |H| + |K| \leq |H| \frac{1+p}{p}. \tag{2}$$

From now on, let $p \geq 5$.

Part 1. $k \leq |H|(1+p)/(2p)$.

In this case, we have, from (2), that for all b, c, with $a + b + c = \delta$ and $\bar{a}, \bar{b}, \bar{c}$ pairwise distinct, including those for which $T_b = \emptyset$ or $T_c = \emptyset$, that $|T_b| + |T_c| \leq |H|(1+p)/p - k$. Hence $|T| \leq 3k + (|G/H| - 3)(|H|(1+p)/p - k)/2$. The right hand side is linear in k and reaches its maximum at $k = 0.41|H|$, hence $|T| \leq 0.045|H| + 0.395|G|$, which is false if $|G/H| \geq 11$.

Part 2. $|H|(p+1)/(2p) < k \leq 2|H|(p+1)/(3p)$.

In this case, for all b, c for which $T_b \neq \emptyset \neq T_c$ and $\bar{a}, \bar{b}, \bar{c}$ distinct, from (2), $|T_b| + |T_c| \leq |H|(p+1)/p - k \leq k$. This upper bound is also valid if $T_b = \emptyset$ or $T_c = \emptyset$. It is easy to check that $((\delta - 2a) + H) \cap T = \emptyset$ and $|((\delta - a)/2 + H) \cap T| \leq (|H| + 1)/2$ if $\bar{\delta} \neq 3\bar{a}$. Using this we get $|T| \leq$

$(|G/H| - 3)k/2 + k + (|H| + 1)/2 \le 0.4|G| + 0.1|H| + 1/2$, which is smaller than $0.41|G|$ if $|G : H| \ge 121$.

Part 3. $2|H|(p+1)/(3p) < k \le |H|$.

Let α and β be the number of cosets of H with $|(a' + H) \cap T| \ge (|H| + 5)/2$ and $(|H| + 3)/2 \ge |(a' + H) \cap T| \ge |H|(1 + p)/p - k$. Let A and B be the set of representatives of the cosets of H corresponding to α and β. Then $\bar{\delta} \notin (A \cup B) + A + A$; hence, as with (2), by Theorem 2.1, we get $3\alpha + \beta \le |G/H|(1 + p)/p$.

Let $b + H$ be a coset in $A \cup B$ and $c + H$ its paired coset, i.e., $a + b + c = \delta$. If $\bar{b}$ does not belong to an exceptional pair, $(c + H) \cap T = \emptyset$, whence we get $\alpha + \beta \le (|G/H| + 1)/2$ and there are at least $\alpha + \beta - 1$ cosets having empty intersection with T.

Therefore $|T| \le \alpha k + \beta(|H| + 3)/2 + (|G/H| - 2\alpha - 2\beta + 1)(|H|(1+p)/p - k)$ where $\alpha \ge 1, \beta \ge 0, \alpha + \beta \le (|G/H| + 1)/2, 3\alpha + \beta \le |G/H|(1 + p)/p$. Therefore the right hand side reaches its extremum at one of the following points: $(\alpha, \beta) = (1, 0), (|G/H|(1 + p)/(3p), 0), (1, (|G/H| - 1)/2)$ and $(|G/H|(p + 2)/(4p) - 1/4, |G/H|(p - 2)/(4p) + 3/4)$. This extremum is always smaller than $0.41|G|$. For instance, if $(\alpha, \beta) = (|G/H|(p + 1)/(3p), 0)$, then we have $|T| \le \alpha|H|(p+1)/p < 0.4|G|$ if $p \ge 11$ since $|G/H| - 2\alpha + 1 \le \alpha$. Similarly, if $(\alpha, \beta) = (1, 0)$, then $|T| \le 4|G|/11 + |H|(p - 1)/p < 0.41|G|$ if $p \ge 11$ and $|G/H| \ge 121$.

Part 4. The values $p = 3, 5$ and 7 are treated separately. Since the case $p = 5$ is proved in the same way as $p = 7$, this case is omitted.

(1)$p = 3$: Let $H \le G$ with $|G : H| = 3$. Select H such that $\delta \in H$. Let $T_i = T \cap (i + H), i = 0, 1, 2$. For all i, $|T_i| \le (|H| + 5)/2$, otherwise δ can be written as the sum of 3 distinct elements of T_i. This can be proved by using the method described in [11] and [14, Lemma 20]. Hence $T_i \ne \emptyset$, $i = 0, 1, 2$ or else $|T| \le |H| + 5 < 0.41|G|$.

Now $\delta \notin T_0 + T_1 + T_2$ or $\delta - (t_0 + t_1 + t_2) \notin T_0' + T_1' + T_2'$ where $t_i \in T_i$ and $T_i = T_i' + t_i$. From (2), $|T_0'| + |T_1'| + |T_2'| \le |K| + |H|$ for a subgroup $K < H$. If $|K| \le |H|/9$, then $|T| \le 10|H|/9 < 0.41|G|$ which is false.

So $|K| = |H|/3$ and $|T_0' + K| + |T_1' + K| + |T_2' + K| \le 4|H|/3$. Moreover equality must hold, or else $|T| \le |H| < 0.41|G|$. So for some i, $|T_i' + K| = 2|H|/3$ and $|T_i'| \le (|H| + 5)/2$. Equivalently, $|T| \le (|H| + 5)/2 + 2|H|/3 < 0.41|G|$.

(2)$p = 7$: We can assume that $\delta = 0$. Select $H \le G$ with $|G : H| = 7$, let $T_i = T \cap (H + i), i = 0, \ldots, 6$, let $k = \max |T_i|$. Since Parts 1 and 2 are true for $p = 7$, assume $k > 16|H|/21$. But $|T \cap H| \le (|H| + 5)/2$, so suppose that $k = |T_1|$. Then $T_{-2} = \emptyset$. Consider the pairs $\{T_0, T_{-1}\}, \{T_2, T_{-3}\}, \{T_3\}$.

If $T_3 \ne \emptyset$, then $(T_3 + T_3) \cap (-T_1) = \emptyset$ with $T_3 + T_3 = \{x + y || x, y \in$

$T_3, x \neq y\}$. So $|T_3| \leq |H| - k$. This is also true if $T_3 = \emptyset$.

<u>Case 1.</u> $T_0, T_{-1}, T_2, T_{-3} \neq \emptyset$. Then, from (2), $|T_0| + |T_1| + |T_{-1}| \leq 8|H|/7$ and $|T_2| + |T_1| + |T_{-3}| \leq 8|H|/7$. So $|T| \leq 16|H|/7 - k + |H| - k < 0.41|G|$.

<u>Case 2.</u> Exactly one coset $T_i = \emptyset$, $i = 0, -1, 2, -3$. For instance $T_0 = \emptyset$. Then $|T| \leq 8|H|/7 + |T_{-1}| + |T_3| \leq 15|H|/7 < 0.41|G|$.

<u>Case 3.</u> For at least one $i \in \{0, -1\}$ and $j \in \{2, -3\}$, $T_i = T_j = \emptyset$; for instance $T_0 = T_2 = \emptyset$. Then $T = T_1 \cup T_3 \cup T_4 \cup T_6$ and $|T_1| + |T_3| \leq |H|$. No other coset $T_r = \emptyset$, $r = -1, -3$. Since $0 \notin (T_4 \dot{+} T_4) + T_6$, from [15], either $|T_4| \leq |H|/7$ or $|T_4| \leq (|H| - |T_6|)2/3$. Both cases imply $|T| < 0.41|G|$. $\square$

To sum up the geometric meaning of the results of this section, we can formulate the following theorem.

Theorem 2.5 *Let L be a twisted cubic, K be an arc of $PG(3, q)$ such that $K \backslash L = \{r\}$. Suppose that $|K \cap L| \geq 0.41(q + 1) + 1$ and $q \geq q_0$. Then r lies on a bisecant M of L and there are at most two arcs $K' \supset K$ with $|K' \backslash L| = 1$ intersecting L in exactly $(q + 1)/2$ points. The arc K' is unique if r lies on an imaginary bisecant, and $K' \backslash (M \cap L)$ is unique if r lies on a real bisecant.*

Proof. Projecting L from r we get a rational plane cubic curve Δ. The projection of K minus possibly the singular point of Δ, considered as a subset of the abelian group G defined on the non-singular points of Δ, is a 3-independent subset with respect to some $\delta \in G$ (see 2.2). At most one point of $K \cap L$ is projected onto the singular point of Δ, so this 3-independent subset has at least size $0.41|G|$.

If r belongs to a tangent to L, then G is isomorphic to $(GF(q), +)$. But this group has no 3-independent subsets of size greater than or equal to $0.41|G|$ (Lemmas 2.3 and 2.4). This means that r belongs to a real or imaginary bisecant M to L and G is isomorphic to the cyclic group C_{q-1} or C_{q+1}. Therefore, by Lemma 2.3, we find a subgroup H of size $|G|/2$ with the property that either H or $G \backslash H$ contains the non-singular points of the projection of K. Suppose H contains the non-singular points of the projection of K; the case $G \backslash H$ is treated in a similar way. From Lemma 2.3, $\delta \notin H$. So no three points of H add up to δ. Hence these points of H define an arc in the plane which contains Δ. Let K_1 be the set of points of L which are projected onto the points of H. Then $K_1 \cup \{r\}$ is a $(\frac{1}{2}|G| + 1)$-arc of $PG(3, q)$.

If r belongs to an imaginary bisecant, then $K' = K_1 \cup \{r\}$. If r belongs to a real bisecant M, we can add one of the 2 points of $L \cap M$ to $K_1 \cup \{r\}$. This then gives a $\frac{1}{2}(q + 3)$-arc K' and $K' \backslash (M \cap L)$ is uniquely defined. $\square$

The previous theorem shows that $K \cap L$ is essentially unique but there might be points outside L that extend the arc K. This possibility will be investigated in the second part of the next section.

3.   Three dimensions

In this section we study in detail $(k+1)$-arcs K of PG$(3, q)$, $q \geq q_0$, containing k points of the twisted cubic L. As we remarked in Section 2, we can translate the extendability of arcs, contained in L, by r $(r \notin L)$ into the language of abelian groups. The structure of the abelian group depends on the type of chord of L the point r belongs to.

Theorem 2.5 shows that it is enough to describe $\frac{1}{2}(q+3)$-arcs having $(q+1)/2$ points on L, since all the arcs K with $k \geq 0.41(q+1)+1$ are embedded in such a big arc. These arcs correspond to subgroups of index 2 in the abelian group of the cubic curve Δ obtained by projecting L from r. Another important ingredient of the proof is to show that these groups can be interpreted as subgroups of the projective group PGL$(2, q)$ of L. In the next two sections we frequently use some properties of the linear collineation group PGL$(2, q)$ fixing a normal rational curve L in PG(n, q). The results used here can be found in [2, Chap. 21.1] and [3, Chap. 27.5].

Let us first consider a point r on a real bisecant of $L = \{(1, t, t^2, t^3) \| t \in \mathrm{GF}(q)^+\}$. As PGL$(2, q)$ is 3-transitive on the points of L, we can assume that r belongs to the bisecant through $(1, 0, 0, 0)$ and $(0, 0, 0, 1)$. Then $r = (1, 0, 0, a)$, $a \neq 0$. Three different points of L with parameters t_i, $i = 1, 2, 3$, are coplanar with r if and only if

$$t_1 t_2 t_3 - a = 0. \tag{3}$$

Of course, this also corresponds to the fact that in this case the projection of L is a cubic curve having an ordinary double point and its group is isomorphic to the multiplicative group of the field GF(q). So, the unique subgroup of index 2 of this group consists of the non-zero squares. Combining this with Theorem 2.5 we immediately get the following theorem.

Theorem 3.1 *Let* $r = (1, 0, 0, a)$, $a \neq 0$, *be a point on a real bisecant of* L, ℓ *a fixed non-square and* $K = \{(1, t, t^2, t^3) \| t = \ell a u^2, \ u \in \mathrm{GF}(q)\} \cup \{r\}$. *Then,*

(i) there are precisely two $\frac{1}{2}(q+3)$-arcs containing r and $(q+1)/2$ points of L, namely the arcs K and $(K \backslash \{(1, 0, 0, 0)\}) \cup \{(0, 0, 0, 1)\}$.

(ii) Moreover, precisely $(q-1)/2$ points $(1, 0, 0, b)$, b/a is a non-zero square, extend the arc $K \backslash \{r\} \subset L$.

(iii) Finally, $K \backslash \{(1, 0, 0, 0), r\}$, which is contained in both arcs mentioned in (i), is an orbit of a cyclic subgroup H, of order $(q-1)/2$, of PGL$(2, q)$.

Proof. By the remarks before the theorem we only have to prove (ii) and (iii). Part (ii) follows immediately from (3), namely the points of the form $(1, 0, 0, b)$ that extend the arc $K \backslash \{r\} \subset L$ are precisely the points with b/a a square and Part (iii) is also clear as the subgroup H consists of the mappings

$t \mapsto v^2 t$ where v runs over the non-zero elements of $\mathrm{GF}(q)$. $\qquad\square$

Later on we will show that there are no other points outside L extending $K \backslash \{r\}$ than the ones mentioned in (ii). In particular, this will imply that K is complete.

Let us turn to points belonging to *imaginary bisecants*. So, again let $L = \{(1, t, t^2, t^3) \parallel t \in \mathrm{GF}(q)^+\}$. As we also use elements of $\mathrm{GF}(q^2)$, let us generate this field by an element i that satisfies $i^2 = \ell$, where ℓ is a fixed non-square in $\mathrm{GF}(q)$. An imaginary bisecant is defined by two conjugate points of L with parameters $\omega, \omega^q = \overline{\omega}$ in $\mathrm{GF}(q^2) \backslash \mathrm{GF}(q)$. If $\omega = a + bi$, then the transformation $t \mapsto (t - a)/b$ maps ω onto i and $\overline{\omega}$ onto $-i$. Therefore we can suppose that our imaginary bisecant M is the line joining the points $(1, i, \ell, \ell i)$ and $(1, -i, \ell, -\ell i)$. It is worthwhile to mention that we shall frequently use the fact that an imaginary bisecant can be mapped onto this M using an element of $\mathrm{PGL}(2, q)$ fixing the point with parameter ∞.

Theorem 3.2 *Let r be a point on the imaginary bisecant M joining $(1, i, \ell, \ell i)$ $\in L$ and $(1, -i, \ell, -\ell i) \in L$.*

(i) Then there is a unique $\frac{1}{2}(q+1)$-arc on L that can be extended by r to a $\frac{1}{2}(q+3)$-arc, and it is an orbit $\mathcal{O}$ under a cyclic subgroup H of $\mathrm{PGL}(2, q)$ of order $(q+1)/2$. The group H fixes $(1, i, \ell, \ell i)$ and $(1, -i, \ell, -\ell i)$.
(ii) The orbit mentioned in (i) or its complement consists of the points with parameters in $\{(1 + \ell t^2)/(2t) \parallel t \in \mathrm{GF}(q)^+\}$.
(iii) There are exactly $(q+1)/2$ points on M that extend the arc $\mathcal{O} \subset L$ and the same applies for $L \backslash \mathcal{O}$.

Proof. We know from Theorem 2.5 that the $\frac{1}{2}(q+1)$-arc on L mentioned in this theorem is unique. So, we will consider an orbit of a cyclic group of order $(q+1)/2$ and show that a certain point of M can be added to it to obtain a $\frac{1}{2}(q+3)$-arc.

First, let us consider the transformation $\beta : t \mapsto (t - i)/(t + i)$. This maps the points of L with parameter i onto 0, $-i$ to ∞, and ∞ onto 1. A point on the bisecant joining $(1, i, \ell, \ell i)$ and $(1, -i, \ell, -\ell i)$ is mapped onto a point $(1, 0, 0, a)$ with $a \in \mathrm{GF}(q^2)$. Now β can be extended to a collineation of $\mathrm{PG}(3, q^2)$ [2, p. 233], namely to

$$B : \begin{pmatrix} v_0 \\ v_1 \\ v_2 \\ v_3 \end{pmatrix} \mapsto \begin{pmatrix} i\ell & 3\ell & 3i & 1 \\ -\ell i & -\ell & i & 1 \\ i\ell & -\ell & -i & 1 \\ -i\ell & 3\ell & -3i & 1 \end{pmatrix} \begin{pmatrix} v_0 \\ v_1 \\ v_2 \\ v_3 \end{pmatrix}.$$

Then B maps $i_1 = (1, i, \ell, \ell i)$ to $(8i\ell, 0, 0, 0)$ and $i_2 = (1, -i, \ell, -\ell i)$ to $(0, 0, 0, -8i\ell)$. A point of $\mathrm{PG}(3, q)$ on the line joining i_1 and i_2 can be written

366

as $\alpha i_1 + \overline{\alpha} i_2$, so B maps this point onto the point $a = (\alpha, 0, 0, -\overline{\alpha})$. Now this point is linearly dependent on 3 different points of L with parameters t_1, t_2, t_3 over $\mathrm{GF}(q^2)$ if and only if

$$\alpha t_1 t_2 t_3 + \overline{\alpha} = 0, \text{ i.e., } t_1 t_2 t_3 = -\alpha^{q-1}. \tag{4}$$

We are going to show that a suitable orbit of a cyclic group of order $(q+1)/2$, together with a, forms an arc; then applying B^{-1} we show that the inverse images of the points of that orbit are points of $\mathrm{PG}(3,q)$ because originally we only know that the coordinates belong to $\mathrm{GF}(q^2)$.

Let us start by proving that from a point with parameter $z \in \mathrm{GF}(q^2)$ satisfying $z^{q+1} = 1$ we get a point of $\mathrm{PG}(3,q)$ applying B^{-1}. If we apply B^{-1} on the points of L we just have to apply β^{-1}, which is the mapping $t \mapsto (-it - i)/(t - 1)$. The elements satisfying $z^{q+1} = 1$ can be written as $z = w^{k(q-1)}$ for some k, where w is a primitive element of $\mathrm{GF}(q^2)$. We have to show that $\beta^{-1}(z) \in \mathrm{GF}(q)^+$, that is $\beta^{-1}(z) = \overline{\beta^{-1}(z)}$, when $z \neq 1$. This is equivalent to $(w\overline{w})^{k(q-1)} = 1$; but that is obvious since $w\overline{w} = w^{q+1}$ and $w^{q^2-1} = 1$. Now we see that the parameters of the points of the cyclic orbits should be subsets of the elements satisfying $z^{q+1} = 1$. Indeed, the two orbits are $\mathcal{O}_1 = \{(1, z, z^2, z^3) \| z^{(q+1)/2} = 1\}$ and $\mathcal{O}_2 = \{(1, z, z^2, z^3) \| z^{(q+1)/2} = -1\}$.

We now show that for any α, $\mathcal{O}_i \cup \{a\}$ is an arc for $i = 1$ or 2. First of all recall the condition of linear independence for the point a with respect to points of $\mathcal{O}_i$. Condition (4) tells us that this is $t_1 t_2 t_3 \neq -\alpha^{q-1}$. Notice that $(-\alpha^{q-1})^{(q+1)/2}$ is ± 1. If this is -1 then we can add a to $\mathcal{O}_1$, since raising (4) to the power $(q+1)/2$ we get $+1$ on the left hand side and -1 on the right hand side. Similarly, if $(-\alpha^{q-1})^{(q+1)/2} = +1$ then we can add a to $\mathcal{O}_2$ and we get an arc.

Finally, let us remark that the cyclic transformation of order $(q+1)/2$ that maps $B^{-1}(\mathcal{O}_1)$ into itself is indeed an element of $\mathrm{PGL}(2,q)$ as it maps four points of $\mathrm{PG}(3,q)$ to four points of $\mathrm{PG}(3,q)$. This completes the proof of (i).

To prove (ii), let us parametrize $\mathcal{O}_1$ in another way. The elements for which $z^{(q+1)/2} = 1$ can also be written as $z = (1 + ti)^{2(q-1)}$ or $z = i^{2(q-1)} = \ell^{q-1} = 1$. We also obtain $z = 1$ when we substitute $t = 0$ in $z = (1 + ti)^{2(q-1)}$; so we only consider $z = (1 + ti)^{2(q-1)}$. Easy computation shows that

$$(1 + ti)^{2(q-1)} = \frac{1 + 6t^2\ell + t^4\ell^2 - 4ti - 4t^3 i\ell}{1 - 2\ell t^2 + t^4\ell^2}.$$

The arc which is extendable by a point of M can be obtained from $\mathcal{O}_1$ by applying B^{-1}. So the parameters of the arc are obtained by β^{-1}. Putting the right hand side of the preceding equation into $\beta^{-1}: t \mapsto (-ti - i)/(t - 1)$, gives $(1 + \ell t^2)/(2t)$. This proves (ii).

Concerning (iii) recall that the points of M have the form $\alpha(1, i, \ell, \ell i) + \bar{\alpha}(1, -i, \ell, -\ell i)$ and we could add such a point to $\mathcal{O}_1$ or $\mathcal{O}_2$ according as

$$(-\alpha^{q-1})^{(q+1)/2} = -1 \text{ or } +1.$$

Obviously, any one of these two equations has exactly $(q+1)/2$ solutions in α, which proves (iii). □

Remark 3.3 In the parametric form $(1 + \ell t^2)/(2t)$, we get the same value for t and $(\ell t)^{-1}$, if $t \neq 0$, and for $t = 0$ and $t = \infty$.

The following three theorems, with a corollary, will be the cornerstones of the induction argument of Section 4. They are also very useful for investigating the completeness of the arcs constructed in Theorems 3.1 and 3.2.

Theorem 3.4 *Consider the set of non-zero squares K. Suppose that there is a transformation $\gamma : t \mapsto (at+b)/(ct+d)$, $ad - bc \neq 0$, for which $|K \cap \gamma(K)| > (q+3\sqrt{q})/4$. Then γ is one of the mappings $t \mapsto v^2/t$ or $t \mapsto v^2 t$, $v \neq 0$, which fix K.*

Proof. We have to estimate the number of solutions to the equation

$$v^2 = \frac{au^2 + b}{cu^2 + d}, \qquad ad - bc \neq 0.$$

Instead of the quotient of the two expressions on the right hand side we can consider their product since the product of two elements is a square if and only if their quotient is square. So we consider

$$v^2 = (au^2 + b)(cu^2 + d), \qquad ad - bc \neq 0.$$

For this equation we can apply [6, Theorem 5.41], which shows that

$$\left| \sum_{u \in \mathrm{GF}(q)} \omega((au^2 + b)(cu^2 + d)) \right| \leq 3\sqrt{q},$$

where ω denotes the quadratic character; that is $\omega(x) = +1$ if x is a square, -1 if x is a non-square and $\omega(0) = 0$. In other words, the number of elements u of $\mathrm{GF}(q)$ for which $(au^2 + b)(cu^2 + d)$ is a non-zero square, is at most $(q + 3\sqrt{q})/2$ if the polynomial $(au^2 + b)(cu^2 + d)$ is not a constant times the square of another polynomial. In this case, as u and $-u$, $u \neq 0$, give the same value v^2 one sees that the number of non-zero squares that can be obtained as $(au^2 + b)(cu^2 + d)$ is at most $(q + 3\sqrt{q})/4$.

The excluded case, when $(au^2 + b)(cu^2 + d)$ is a constant times the square of another polynomial, implies that $au^2 + b$ and $cu^2 + d$ are both a constant times squares of other polynomials or $au^2 + b$ is a constant multiple of $cu^2 + d$. The second possibility contradicts $ad - bc \neq 0$, while the first possibility implies that $ab = 0$ and $cd = 0$, and that clearly implies that γ is either a mapping $t \mapsto v^2/t$ or $t \mapsto v^2 t$. $\qquad\Box$

Theorem 3.5 *Consider the set $\mathcal{O}$ of points of L that have parameters $(1 + \ell t^2)/(2t)$ where ℓ is a fixed non-square and t runs over $GF(q)^+$. Apply a transformation $\gamma : t \mapsto at + b$, $a \neq 0$, $a, b \in GF(q)$, on $\mathcal{O}$. Then, if $a \neq \pm 1$ or $b \neq 0$,*

$$\frac{q + 4 - 2\sqrt{q}}{4} \leq |\mathcal{O} \cap \gamma(\mathcal{O})| \leq \frac{q + 4 + 2\sqrt{q}}{4}.$$

Proof. If $(a, b) = (1, 0)$ or $(a, b) = (-1, 0)$, then γ fixes $\{i, -i\}$ with $i^2 = \ell$. Since the cyclic group of order $(q+1)/2$ which fixes $\mathcal{O}$ is uniquely determined by $\{i, -i\}$ (see the proof of 3.2), $\gamma(\mathcal{O}) = \mathcal{O}$ or $\gamma(\mathcal{O}) = GF(q)^+ \setminus \mathcal{O}$.

The parameters $(1 + \ell t^2)/(2t)$ are mapped by γ onto $(a + a\ell t^2 + 2bt)/(2t)$. If a parameter belongs to a point of $\mathcal{O} \cap \gamma(\mathcal{O})$, then

$$\frac{1 + \ell v^2}{2v} = \frac{a + a\ell t^2 + 2bt}{2t}, \text{ i.e., } t + \ell t v^2 = a\ell v t^2 + 2btv + av.$$

This is the equation of a cubic curve $\Gamma : XZ^2 + \ell XY^2 - a\ell X^2 Y - 2bXYZ - aYZ^2 = 0$ which is absolutely irreducible and so we can apply the Hasse–Weil bound [1, p. 228]. This gives $q + 1 - 2\sqrt{q} \leq |\Gamma| \leq q + 1 + 2\sqrt{q}$.

Now we show that any point of the intersection $\mathcal{O} \cap \gamma(\mathcal{O})$ yields 4 points of Γ. First of all, as we saw in Remark 3.3, for any v there is a unique element $v_1 = (\ell v)^{-1}$ that satisfies $(1 + \ell v^2)/(2v) = (1 + \ell v_1^2)/(2v_1)$. Similarly, for any t there is a unique $t_1 = (\ell t)^{-1}$ for which $(a + a\ell t^2 + 2bt)/(2t) = (a + a\ell t_1^2 + 2bt_1)/(2t_1)$, since this is obtained by applying γ to $(1 + \ell t_1^2)/(2t_1) = (1 + \ell t_1^2)/(2t_1)$. One of the solutions is $(x, y) = (0, 0)$ which is counted once. Therefore $|\mathcal{O} \cap \gamma(\mathcal{O})| \leq (|\Gamma| - 1)/4 + 1$, which completes the proof. $\qquad\Box$

Corollary 3.6 *Let $\mathcal{O}_1$ and $\mathcal{O}_2$ be orbits of different cyclic subgroups H_1 and H_2 of order $(q+1)/2$ or $(q-1)/2$ in $PGL(2, q)$. Then $|\mathcal{O}_1 \cap \mathcal{O}_2| \leq (q + 3\sqrt{q})/4$.*

Proof. (i) Suppose that $|\mathcal{O}_1| = |\mathcal{O}_2| = (q + 1)/2$. As $PGL(2, q)$ is 3-transitive on L [2, p. 234], we can suppose that $\mathcal{O}_1$ is just the orbit $\mathcal{O}$ of points with parameters $(1 + \ell t^2)/(2t)$, $t \in GF(q)^+$. Then the other orbit $\mathcal{O}_2$, or its complement, can be obtained from $\mathcal{O}$ using a transformation of the form $\gamma : t \mapsto at + b$. Indeed, $\mathcal{O}_2$ is determined by a pair of conjugate points

$\{w = b + ai, \overline{w} = b - ai\}$ while $\mathcal{O}$ is determined by the pair of conjugate points $\{i, -i\}$. If we use this a and b to define γ, then we indeed get an orbit belonging to the subgroup fixing $\{w, \overline{w}\}$. Now the corollary follows immediately from Theorem 3.5.

(ii) Suppose $\mathcal{O}_1$ and $\mathcal{O}_2$ are orbits of cyclic subgroups H_1 and H_2 which fix 2 points of L in $PG(3, q)$. Using a transformation $\gamma : t \mapsto (at + b)/(ct + d), ad - bc \neq 0$, we can map the fixed points of H_1 onto the fixed points of H_2 and $\gamma(\mathcal{O}_1)$ is $\mathcal{O}_2$ or the second orbit of size $(q-1)/2$ of H_2 on L. From Theorem 3.4, $|\mathcal{O}_1 \cap \mathcal{O}_2| \leq (q + 3\sqrt{q})/4$. $\qquad\square$

Similarly, we can bound the intersection of an orbit of a cyclic group of size $(q-1)/2$ and $(q+1)/2$. Again, these two orbits, of different type, intersect in roughly halve their points.

Theorem 3.7 *Let $\mathcal{O}_1$ be an orbit of a cyclic subgroup of order $(q-1)/2$ and $\mathcal{O}_2$ be an orbit of a cyclic subgroup of order $(q+1)/2$ of $PGL(2, q)$. Then $|\mathcal{O}_1 \cap \mathcal{O}_2| \leq (q + 3\sqrt{q})/4$.*

Proof. As $PGL(2, q)$ is 3-transitive on L, we can suppose that $\mathcal{O}_1$ consists of the points of L having square non-zero parameters. As in the proof of Corollary 3.6, $\mathcal{O}_2$ or its complement is obtained from the orbit $\mathcal{O}$ consisting of the points with parameters $(1 + \ell t^2)/(2t)$ using a transformation $\gamma : t \mapsto at + b$. The points of $\gamma(\mathcal{O})$ have parameters $(a + a\ell t^2 + 2bt)/(2t)$ (cf. the proof of Theorem 3.5). Therefore we have to estimate the number of solutions to

$$u^2 = \frac{a + a\ell t^2 + 2bt}{2t}, \tag{5}$$

with $u \neq 0$. As in the proof of Theorem 3.4, this can be replaced by $u^2 = (a + a\ell t^2 + 2bt) \cdot 2t$, and then we can use [6, Theorem 5.41]. The expression on the right hand side is a constant multiple of a square of another polynomial if and only if $a = 0$, but that is impossible.

So $|\sum_{t \in GF(q)} \omega((a + a\ell t^2 + 2bt)2t)| \leq 3\sqrt{q}$ where ω is the quadratic character defined in the proof of 3.4. For at most $(q + 3\sqrt{q})/2$ values t in $GF(q)$ is $(a + a\ell t^2 + 2bt)2t$ a non-zero square. Again four pairs (t, u) give the same equality (5); so continuing as in the proof of 3.5 implies our theorem. $\qquad\square$

Using the last three theorems we can easily prove the completeness of the arcs constructed in Theorems 3.1 and 3.2.

Theorem 3.8 *Let M be a real or imaginary bisecant of the twisted cubic L. Let $\mathcal{O}$ be an orbit of a cyclic subgroup of $PGL(2, q)$ of order $(q-1)/2$ or $(q+1)/2$ fixing the points of $L \cap M$. Then the arc $\mathcal{O}$ can only be extended by some points of $L \cup M$.*

Proof. Suppose that a point s, $s \notin L$, extends $\mathcal{O}$. Then, by Theorem 2.5, $\mathcal{O}$ is contained in a $\frac{1}{2}(q+1)$-arc on L extendable by s, and s lies on a real or imaginary bisecant M' of L. Theorems 3.1 and 3.2 show that $\mathcal{O}$, with at most one exception, is an orbit corresponding to the bisecant M'. As $\mathcal{O}$ is an orbit with respect to M, we have two different orbits intersecting in at least $(q-3)/2$ points. On the other hand, Corollary 3.6 and Theorem 3.7 imply that two different orbits intersect in at most $(q+3\sqrt{q})/4$ points. This shows that $M = M'$. Now Theorems 3.1(ii) and 3.2(iii) tell us exactly which points of M can be added to $\mathcal{O}$. $\qquad\square$

Let us conclude this section with a theorem about arcs intersecting the twisted cubic L in at least $0.41(q+1)+1$ points. We formulate the results in such a way that the proof also works in n dimensions.

Theorem 3.9 *Let L be a twisted cubic in $PG(3,q)$, q odd, $q \geq q_0$. Let $K \neq L$ be a complete arc with $|K \cap L| \geq 0.41(q+1)+1$. Then*

(1) $|K \backslash L| \leq 2$ and the point(s) of $K \backslash L$ lie on a (real or imaginary) bisecant M of L;
(2) $|K \cap L| = (q \pm 1)/2$.

Proof. Let $r \in K \setminus L$. By Theorem 2.5, r lies on a bisecant M to L and $(K \setminus M) \cap L$ is contained in an arc $\mathcal{O}$ of size $(q \pm 1)/2$. This follows from Theorems 2.5, 3.1 and 3.2.

Suppose a point s, $s \neq r$ and $s \notin L$, extends $K \cap L$. Then s belongs to a bisecant M' of L (Theorem 2.5) and s determines an orbit $\mathcal{O}'$ of a cyclic subgroup of $PGL(2,q)$. If $\mathcal{O}' \neq \mathcal{O}$, then $\mathcal{O}'$ intersects $\mathcal{O}$ in at most $(q+3\sqrt{q})/4$ points (Corollary 3.6 and Theorem 3.7), which is less than $0.41(q+1)-1$. Therefore, r and s determine the same orbit $\mathcal{O}$, which implies $M = M'$. So $K \cap L$ is only extendable by points of M, which shows that $|K \setminus L| \leq 2$ and this proves (1).

We can always assume that $|K \cap M| = 2$ since all hyperplanes of $PG(3,q)$ that contain M can contain at most $n - 2$ points of $L \setminus M$. If M is a real bisecant and K contains one of the 2 points r_1 of $M \cap L$, then $K \subseteq \mathcal{O} \cup \{r_1, r\}$ where $\mathcal{O} \cup \{r_1, r\}$ is a complete $\frac{1}{2}(q+3)$-arc of $PG(3,q)$ containing $(q+1)/2$ points of L. If however K does not contain a point of $L \cap M$, then we can assume that $K \cap M$ contains 2 of the $(q-1)/2$ points of $M \setminus L$ that extend $\mathcal{O}$ to a $\frac{1}{2}(q+1)$-arc (3.1(ii)).

If M is an imaginary bisecant, then we can assume that $K \cap M$ contains 2 of the $(q+1)/2$ points of M that extend $\mathcal{O}$ to a $\frac{1}{2}(q+3)$-arc (3.2(iii)). So $|K \cap L| = (q+1)/2$ and K is a complete $\frac{1}{2}(q+5)$-arc. $\qquad\square$

Remark 3.10 Actually, the previous theorem yields a complete classification of complete arcs K having at least $0.41(q+1)+1$ points on L. Namely, these arcs are contained in an arc consisting of an orbit $\mathcal{O}$ of size $(q\pm 1)/2$ of a cyclic subgroup of $PGL(2,q)$, and two points on the bisecant M. If M is an imaginary bisecant, then $K \cap M$ contains 2 of the $(q+1)/2$ points of M that extend $\mathcal{O}$ to a $\frac{1}{2}(q+3)$-arc (3.2(iii)), while in case of a real bisecant, K either contains 2 points of $M \setminus L$ or one point of $L \cap M$ and a point of $M \setminus L$.

4. Arbitrary dimensions

This section shows how the results in $PG(3,q)$ imply results in $PG(n,q)$ using induction. The method we use is based on the ideas of [13].

Theorem 4.1 *Consider the normal rational curve* $L = \{(1,t,\dots,t^n)\|t \in GF(q)^+\}, q = p^h, q$ *odd,* $q \geq q_0, n \geq 3$. *Let* K *be a* $(k+1)$-*arc of* $PG(n,q)$ *which has k points in common with L. Assume that $\frac{1}{2}(q+1) \geq k \geq 0.41(q+1)+n-2$. Let r be the unique point of $K \setminus L$. Then r belongs to a real or imaginary bisecant M of L.*

(a) If r belongs to a real bisecant of L, then $K \cap L$ is contained in a $\frac{1}{2}(q+1)$-arc $K' \subseteq L$ which contains one point of the bisecant and the set of the remaining $(q-1)/2$ points is fixed by a cyclic subgroup of order $(q-1)/2$ of the automorphism group of L fixing the 2 points of the bisecant.
(b) If r belongs to an imaginary bisecant of L, then $K \cap L$ is contained in a $\frac{1}{2}(q+1)$-arc $K' \subseteq L$. Also K' is fixed by a cyclic subgroup of order $(q+1)/2$ of the automorphism group of L fixing the 2 conjugate points of L on the imaginary bisecant.

Proof. The proof uses induction on n.
$\underline{n=3.}$ This is Remark 3.10.
$\underline{n>3.}$ Assume that this theorem is valid in $n-1$ dimensions. Project from each point p_j of $K \cap L$ onto a hyperplane α_j skew to it. Let r be projected onto the point r_j of α_j and $L \setminus \{p_j\}$ be projected onto the normal rational curve L_j of α_j. The points of $K \setminus \{p_j\}$ are projected onto the points of a k-arc K_j of α_j which has at least $k-1$ points in common with L_j.

At most one point r_j belongs to the normal rational curve L_j [4]. Select 2 points p_1, p_2 of $K \cap L$ for which $r_1 \notin L_1$ and $r_2 \notin L_2$. It follows from the induction hypothesis that r_i belongs to a (real or imaginary) bisecant M_i of L_i, $i = 1, 2$. Select p_1 and p_2 so that r_1 and r_2 belong at the same time to real or imaginary bisecants of L_1 and L_2.

$\underline{\text{Case 1:}}$ (a) r_1 and r_2 belong to imaginary bisecants M_1 and M_2 of L_1 and L_2 respectively. Using the 3-transitivity of the automorphism group of L we

372

can suppose that $p_1 = (1, 0, \ldots, 0)$ and $p_2 = (0, \ldots, 0, 1)$. Choose $\alpha_1 : X_0 = 0$ and $\alpha_2 : X_n = 0$. Then L is projected onto $L_1 = \{(0, 1, t, \ldots, t^{n-1}) || t \in \mathrm{GF}(q)^+\}$. Notice that the parameter t of a point of $L \backslash \{p_1\}$ does not change when we project it from p_1 onto $X_0 = 0$. Then $K \backslash \{p_1\}$ is projected onto K_1. It follows from the induction hypothesis that there is a $\frac{1}{2}(q+1)$-arc $K_1' \subset L_1$ in $X_0 = 0$ containing $K_1 \cap L_1$ and K_1' is fixed by a cyclic group H_1 of order $(q+1)/2$. This group H_1 fixes the 2 conjugate points of L_1 on the bisecant M_1.

(b) Lift everything to n dimensions and consider the set $K' \subset L$ in $\mathrm{PG}(n, q)$ whose points have the same set of parameters as the set $K_1' \subset L_1$. Using the natural action of H_1 on the parameters of the points of L and L_1, H_1 can be regarded as a subgroup of the automorphism group $\mathrm{PGL}(2, q)$ of L and it fixes K' setwise and also two conjugate points of L. The set $K \backslash \{p_1, r\}$ is contained in K' and

$$|K' \backslash (K \backslash \{p_1, r\})| \leq 0.09(q+1) + 3 - n \leq 0.09q.$$

(c) Proceed now in the same way for p_2. Project from p_2 onto $X_n = 0$. Then L is projected onto the normal rational curve $L_2 = \{(1, t, \ldots, t^{n-1}, 0) || t \in \mathrm{GF}(q)^+\}$ of $X_n = 0$, the point r is projected onto a point r_2 of $X_n = 0$ where $r_2 \notin L_2$. The arc $K \backslash \{p_2\}$ is projected onto a k-arc K_2 containing $k-1$ points of L_2. From the induction hypothesis it follows that there is a $\frac{1}{2}(q+1)$-arc $K_2' \subset L_2$ containing $K_2 \cap L_2$ and K_2' is fixed by a cyclic group H_2. This group fixes the two conjugate points of L_2 on the bisecant M_2. Again, observe that the parameter t of a point of $L \backslash \{p_2\}$ does not change when we project it from p_2 onto $X_n = 0$.

(d) We proceed as in (b) and lift everything to n dimensions. This yields a $\frac{1}{2}(q+1)$-arc $K'' \subset L$ having the same parameters as K_2'. This arc K'' is fixed by the group H_2, regarded as a linear group in n dimensions, which fixes 2 conjugate points of L. Now $K \backslash \{p_2, r\}$ is contained in K'' and

$$|K'' \backslash (K \backslash \{p_2, r\})| \leq 0.09q.$$

(e) As $K \backslash \{p_1, p_2, r\} \subset K' \cap K''$, by Corollary 3.6, we have that $K' = K''$. Therefore the two conjugate points corresponding to K' and K'' are also equal and so the groups H_1 and H_2 are equal. Let s, $\bar{s} = s^q \in \mathrm{GF}(q^2)$ be the parameters of these conjugate points of L.

(f) It follows from (a) that r_1 belongs to the bisecant through the points $(0, 1, s, \ldots, s^{n-1})$ and $(0, 1, \bar{s}, \ldots, \bar{s}^{n-1})$ of L_1 in $X_0 = 0$. So $r \in \langle (1, 0, \ldots, 0), (1, s, \ldots, s^n), (1, \bar{s}, \ldots, \bar{s}^n) \rangle = \gamma_1$. From the corresponding facts for p_2 we get that $r \in \langle (0, \ldots, 0, 1), (1, s, \ldots, s^n), (1, \bar{s}, \ldots, \bar{s}^n) \rangle = \gamma_2$. So $r \in \gamma_1 \cap \gamma_2$, which is the imaginary bisecant through $s_1 = (1, s, \ldots, s^n)$ and $\overline{s_1} = (1, \bar{s}, \ldots, \bar{s}^n)$.

(g) Let G be the cyclic linear collineation group of order $q+1$ on L which fixes the two conjugate points s_1, $\overline{s_1}$. Then there exists a unique subgroup H_1 of G of order $(q+1)/2$ and H_1 defines two orbits on L, one of which is $K' = K''$. As $K\backslash\{p_1, r\} \subset K'$ and $K\backslash\{p_2, r\} \subset K'' = K'$, we get that $K\backslash\{r\} \subseteq K'$.

Case 2: (a) The points r_1 and r_2 belong to real bisecants M_1 and M_2 of L_1 and L_2 respectively. We can more or less repeat what we did in Case 1. So we get that the arc $K\backslash\{p_1, r\}$ is contained in a $\frac{1}{2}(q+1)$-arc $K' \subset L$ in $PG(n, q)$ and the points of K' have the same parameters as the points of $K_1' \subset L_1$. Here K_1' is fixed by a cyclic group H_1 of order $(q-1)/2$ fixing also the 2 points of $M_1 \cap L_1$, K_1' contains one of the points of $L_1 \cap M_1$ and the remaining points constitute an orbit of H_1. Analogously, $K\backslash\{p_2, r\}$ is contained in a $\frac{1}{2}(q+1)$-arc $K'' \subset L$ in $PG(n, q)$ and the points of K'' have the same parameters as the points of $K_2' \subset L_2$; K_2' is fixed by a cyclic group H_2 of order $(q-1)/2$ fixing the 2 points of $M_2 \cap L_2$, K_2' contains one of the points of $L_2 \cap M_2$ and the remaining points constitute an orbit of H_2.

(b) The only difference with respect to Case 1 is that the points of L_1 and L_2 on the real bisecants M_1 and M_2 play a symmetric role; they can replace one another. As this point does not change the arc at all, we can select any of the 2 points of $M_1 \cap L_1$ and $M_2 \cap L_2$.

(c) As in Case 1, $|K' \cap K''| \geq 0.41(q+1) + n - 4$, so by Corollary 3.6, the orbits of H_1 and H_2, lifted to $PG(n, q)$, are the same and so $H_1 = H_2$. Therefore the groups fix the same points whose parameters are denoted by s, s'.

(d) As in Case 1, $r \in \langle (1, 0, \ldots, 0), (0, 1, s, \ldots, s^{n-1}), (0, 1, s', \ldots, s'^{n-1}) \rangle$ $= \gamma_1$ and $r \in \langle (0, \ldots, 0, 1), (1, s, \ldots, s^{n-1}, 0), (1, s', \ldots, s'^{n-1}, 0) \rangle = \gamma_2$.

If $s, s' \notin \{0, \infty\}$ then $\gamma_1 \neq \gamma_2$, hence $r \in \gamma_1 \cap \gamma_2$, which is the line joining $(1, s, \ldots, s^n)$ and $(1, s', \ldots, s'^n)$ and it is the desired real bisecant M of L.

If $s = 0$, $s' \neq \infty$, then again $r \in \gamma_1 \cap \gamma_2$, which is the line joining $(1, 0, \ldots, 0)$ and $(1, s', \ldots, s'^n)$. But if we project from $p_1 = (1, 0, \ldots, 0)$ onto $X_0 = 0$, then r is projected onto the point $r_1 = (0, 1, s', \ldots, s'^{n-1})$ and this point belongs to L_1. This contradicts the assumption $r_i \notin L_i$, $i = 1, 2$. The other cases in which at least one of the parameters is 0 or ∞ can be treated similarly, so these cases cannot occur.

(e) As in (g) of Case 1 it is easy to see, using the results of (b), (c) and (d), that the arc K' contains one of the points $(1, s, \ldots, s^n), (1, s', \ldots, s'^n)$ and the remaining points form an orbit of a cyclic linear collineation group of L of order $(q-1)/2$ fixing the 2 points of $M \cap L$. $\qquad\square$

We can obtain an explicit characterization of the $\frac{1}{2}(q+3)$-arcs K containing $(q+1)/2$ points of L, since we know that these arcs contain an orbit of a cyclic subgroup of $PGL(2, q)$ of order $(q \pm 1)/2$. As the action of $PGL(2, q)$ on L, or on $GF(q)^+$, does not depend on the dimension, we know that $K \cap L$

or its complement is projectively equivalent to the set of points with parameter $t = u^2$, $u \in GF(q)$, or $t = (1 + \ell u^2)/(2u)$, $u \in GF(q)^+$, see Theorem 3.1 and Theorem 3.2 (ii). Therefore the only thing we have to check is the extendability of $K \cap L$ by the points of the bisecant M.

Theorem 4.2 *Let* $L = \{(1, t, \ldots, t^n)\|t \in GF(q)^+\}$ *and let* $r = (1, 0, \ldots, 0, a)$, $a \neq 0$, *be a point of* $PG(n, q)$, $n \geq 3$, $q \geq q_0$, q *odd, on the real bisecant of* L *joining* $(1, 0, \ldots, 0)$ *and* $(0, \ldots, 0, 1)$. *Let* $K \subseteq L$ *be a* k-*arc,* $\frac{1}{2}(q+1) \geq k \geq 0.41(q+1) + n - 2$, *which is extendable by* r.

(a) If n *is even,* $-a$ *is a non-square and* $K \backslash \{(1, 0, \ldots, 0), (0, \ldots, 0, 1)\}$ *is contained in* $\mathcal{O}_1 = \{(1, u^2, \ldots, (u^2)^n)\|u \in GF(q)^* = GF(q) \setminus \{0\}\}$ *or in* $\mathcal{O}_2 = \{(1, \ell u^2, \ldots, (\ell u^2)^n)\|u \in GF(q)^*\}$ *with* ℓ *a fixed non-square. Both cases can occur.*
(b) If n *is odd,* $K \subseteq \mathcal{O}_1$ *if* a *is non-square and* $K \subseteq \mathcal{O}_2$ *if* a *is square.*

Proof. (a) n even: It follows from Theorem 4.1 that $K \setminus \{(1, 0, \ldots, 0), (0, \ldots, 0, 1)\}$ is contained in $\mathcal{O}_1$ or $\mathcal{O}_2$. The point $r = (1, 0, \ldots, 0, a)$ is linearly dependent on n distinct points $(1, t_i, \ldots, t_i^n)$, $t_i \in GF(q)$, if and only if $\prod_{i=1}^{n} t_i = (-1)^{n-1} a = -a$ (n even).
Let S and O_i, $i = 1, 2$, be the set of parameters of the points of K and $\mathcal{O}_i$, $i = 1, 2$. Then $S \backslash \{0, \infty\} \subseteq O_i$, for some $i = 1, 2$. If $-a$ is square, then by using the method of the proof of [14, Lemma 20], it is possible to write $-a$ as the product of n distinct elements of S. This is impossible so $-a$ is non-square.
The product of n distinct elements of O_j, $j = 1, 2$, is always square since n is even, so is never equal to $-a$. Therefore $\mathcal{O}_1 \cup \{r\}$, $\mathcal{O}_2 \cup \{r\}$ are $\frac{1}{2}(q+1)$-arcs. We can add one of the points $(1, 0, \ldots, 0), (0, \ldots, 0, 1)$ to extend $\mathcal{O}_i \cup \{r\}$, $i = 1, 2$, to a $\frac{1}{2}(q+3)$-arc.
(b) n odd: The point $(1, 0, \ldots, 0, a)$ is linearly dependent on n distinct points $(1, t_i, \ldots, t_i^n)$, $t_i \in GF(q)$, if and only if $\prod_{i=1}^{n} t_i = (-1)^{n-1} a = a$ since n is odd. The difference between (a) and (b) follows from the fact that the product of n elements of O_i, $i = 1, 2$, now always belongs to O_i, $i = 1, 2$. $\square$

Theorem 4.3 *Let* $L = \{(1, t, \ldots, t^n)\|t \in GF(q)^+\}$ *and let* r *be a point of* $PG(n, q)$, $n \geq 3, q \geq q_0$, *which is on the imaginary bisecant* M *of* L *joining* $i_1 = (1, i, \ldots, i^n)$ *to* $i_2 = (1, -i, \ldots, (-i)^n)$ *where* $i^2 = \ell$ *is non-square in* $GF(q)$. *Let* $\mathcal{O}_1 = \{(1, u, \ldots, u^n)\|u = (1 + \ell t^2)/(2t), t \in GF(q)^+\}$ *and* $\mathcal{O}_2 = L \setminus \mathcal{O}_1$.
Suppose that $K \subset L$ *is a* k-*arc,* $\frac{1}{2}(q+1) \geq k \geq 0.41(q+1) + n - 2$, *extendable by* r *to a* $(k+1)$-*arc.*

(a) *If n is even, then $K \subseteq \mathcal{O}_1$ or $K \subseteq \mathcal{O}_2$ and both cases can occur. Exactly $(q+1)/2$ points of M extend $\mathcal{O}_1$ and $\mathcal{O}_2$ to $\frac{1}{2}(q+3)$-arcs.*

(b) *If n is odd, then $K \subseteq \mathcal{O}_1$ or $K \subseteq \mathcal{O}_2$ and for a fixed point r on M, exactly one case can occur. Each point of M extends exactly one of $\mathcal{O}_i, i = 1, 2$, to a $\frac{1}{2}(q+3)$-arc.*

Proof. From Theorem 4.1, $K \subseteq \mathcal{O}_1$ or $K \subseteq \mathcal{O}_2$. Embed $PG(n,q)$ in $PG(n, q^2)$ and proceed as in Theorem 3.2. The mapping $\beta : t \mapsto (t - i)/(t + i)$ maps the line M to $\langle (1, 0, \ldots, 0), (0, \ldots, 0, 1) \rangle$. This mapping can be extended to a collineation of $PG(n, q^2)$, $\beta(1, i, \ldots, i^n) = (2^n i^n, 0, \ldots, 0)$, $\beta(1, -i, \ldots, (-i)^n) = (0, \ldots, 0, 2^n (-i)^n)$.

So $\beta(\alpha i_1 + \bar{\alpha} i_2) = (2^n i^n \alpha, 0, \ldots, 0, 2^n (-i)^n \bar{\alpha}) = (1, 0, \ldots, 0, (-1)^n \alpha^{q-1})$. A point $(1, 0, \ldots, 0, v)$ is linearly dependent on n different points $(1, t_i, \ldots, t_i^n)$, $t_i \in GF(q^2)$, $1 \le i \le n$, if and only if $v = (-1)^{n-1} \prod_{i=1}^{n} t_i$. Therefore $(1, 0, \ldots, 0, (-1)^n \alpha^{q-1})$ is linearly dependent on n such points if and only if $-\alpha^{q-1} = t_1 \cdots t_n$.

The points of L in $PG(n, q)$ are mapped by β onto the points $(1, z, \ldots, z^n)$ with z in $\mathcal{O}_1 \cup \mathcal{O}_2$ where $\mathcal{O}_1 = \{z || z^{(q+1)/2} = 1, z \in GF(q^2)\}$ and where $\mathcal{O}_2 = \{z || z^{(q+1)/2} = -1, z \in GF(q^2)\}$. So $-\alpha^{q-1} \in \mathcal{O}_1 \cup \mathcal{O}_2$.

(a) *n even*: The product of n elements of $\mathcal{O}_1$ or $\mathcal{O}_2$ always belongs to $\mathcal{O}_1$ since n is even. Since r extends K, $\beta^{-1}(r)$ extends $\beta^{-1}(K)$. As in the proof of 4.2, it is first shown that $-\alpha^{q-1}$ belongs to $\mathcal{O}_2$. If $-\alpha^{q-1} \in \mathcal{O}_2$, no product $\prod_{i=1}^{n} t_i$ of n distinct elements of $\mathcal{O}_1$ or $\mathcal{O}_2$ equals $-\alpha^{q-1}$, hence $(1, 0, \ldots, 0, (-1)^n \alpha^{q-1})$ extends $\mathcal{O}_1' = \{(1, z, \ldots, z^n) || z \in \mathcal{O}_1\}$ and $\mathcal{O}_2' = \{(1, z, \ldots, z^n) || z \in \mathcal{O}_2\}$ to $\frac{1}{2}(q+3)$-arcs. Applying β^{-1} proves this theorem.

(b) *n odd*: The product of n elements of $\mathcal{O}_2$ now belongs to $\mathcal{O}_2$ since n is odd. This explains the difference between (a) and (b). This difference is analogous to the difference between (a) and (b) in 4.2. $\square$

Finally, let us mention that the proof of the completeness of the arcs in three dimensions, Theorem 3.9, was formulated so that it can be used in n dimensions. The completeness of the arcs is stated in the following theorem.

Theorem 4.4 *Let L be a normal rational curve in $PG(n, q)$, q odd, $q \ge q_0$, $n \ge 3$, and let $K \ne L$ be a complete k-arc with $\frac{1}{2}(q+1) \ge |K \cap L| \ge 0.41(q+1) + n - 2$. Then*

(1) *$|K \setminus L| \le 2$ and the point(s) of $K \setminus L$ lie on a real or imaginary bisecant M of L;*

(2) *$|K \cap L| = (q \pm 1)/2$.* $\square$

Theorem 4.5 Let n_0 denote the integral part of $0.09q + 2.09$, $q \geq q_0$. Let L be a normal rational curve in $PG(n, q)$, q odd, where $n \geq n_0$, and let $K \not\subset L$ be an arc in $PG(n, q)$. Then

$$|K \cap L| \leq \frac{q+1}{2} + n - n_0.$$

Proof. The proof is similar to [13, Theorem 4.7] and to [14]. $\qquad\square$

References

[1] **J. W. P. Hirschfeld**. *Projective Geometries over Finite Fields.* Clarendon Press, Oxford, 1979.

[2] **J. W. P. Hirschfeld**. *Finite Projective Spaces of Three Dimensions.* Oxford University Press, Oxford, 1985.

[3] **J. W. P. Hirschfeld and J. A. Thas**. *General Galois Geometries.* Oxford University Press, Oxford. Oxford Science Publications, 1991.

[4] **H. Kaneta and T. Maruta**. An elementary proof and an extension of Thas' theorem on k-arcs. *Math. Proc. Cambridge Philos. Soc.*, 105, pp. 459–462, 1989.

[5] **G. Korchmáros**. Osservazioni sui risultati di B. Segre relativi ai k-archi contenenti $k - 1$ punti di un ovale. *Atti Accad. Naz. Lincei Rend. Cl. Sci. Fis. Mat. Natur. (8)*, 56, pp. 541–549, 1974.

[6] **R. Lidl and H. Niederreiter**. *Finite Fields*, volume 20 of *Encyclopedia Math. Appl.* Addison-Wesley, Menlo Park, California, 1983.

[7] **F. J. MacWilliams and N. J. A. Sloane**. *The Theory of Error-Correcting Codes.* North Holland, Amsterdam, 1977.

[8] **H. B. Mann**. *Addition Theorems.* Wiley-Interscience, London, New York, Sydney, 1965.

[9] **G. Pellegrino**. Sugli archi completi dei piani di Galois di ordine dispari, contenenti $(q + 3)/2$ punti di un ovale. *Rend. Sem. Mat. Brescia*, 7, pp. 495–523, 1984.

[10] **R. M. Roth and A. Lempel**. A construction of non-Reed-Solomon type MDS codes. *IEEE Trans. Inform. Theory*, IT-35, pp. 655–657, 1989.

[11] **G. Seroussi and R. M. Roth**. On M.D.S. extensions of generalized Reed-Solomon codes. *IEEE Trans. Inform. Theory*, IT-32, pp. 349–354, 1986.

[12] **L. Storme**. Completeness of normal rational curves. *J. Algebraic Combin.*, 1, pp. 197–202, 1992.

[13] **L. Storme and T. Szőnyi**. Intersection of arcs and normal rational curves in spaces of even characteristic. To appear in *J. Geom.*

[14] **L. Storme and J. A. Thas**. Generalized Reed-Solomon codes and normal rational curves: an improvement of results by Seroussi and Roth. In J. W. P. Hirschfeld, D. R. Hughes, and J. A. Thas, editors, *Advances in Finite Geometries and Designs*, pages 369–389, Oxford, 1991. Oxford University Press.

[15] **T. Szőnyi and F. Wettl**. On complexes in a finite abelian group, I, II. *Proc. Japan Acad. Ser. A. Math. Sci.*, 64, pp. 245–248; 286–287, 1988.

L. Storme, University of Gent, Galglaan 2, B-9000 Gent (Belgium). e-mail: ls@cage.rug.ac.be

T. Szőnyi, Dept. of Computer Science, Eötvös Loránd University, Múzeum krt. 6-8, H-1088 Budapest (Hungary). e-mail: sztomi@ludens.elte.hu

Flocks and partial flocks of the quadratic cone in PG$(3, q)$

J. A. Thas   C. Herssens   F. De Clerck

Abstract

Let K be the quadratic cone with vertex v in PG$(3, q)$. A partition of $K - \{v\}$ into q disjoint irreducible conics is called a flock of K. A set of disjoint irreducible conics of K is called a partial flock of K. With each flock of K there corresponds a translation plane of order q^2 and also a generalized quadrangle of order (q^2, q). In this paper a partial flock F of size 11 is constructed for any $q \equiv -1 \bmod 12$. For $q = 11$ F is a flock not isomorphic to any previously known flock. The subgroup G of PGL$(4, q)$ fixing K and F is isomorphic to $C_2 \times S_3$, hence has order 12. Finally, with the aid of a computer, it was shown by De Clerck and Herssens [3] that for $q = 11$ the cone K has exactly 4 mutually non-isomorphic flocks.

1. Introduction

1.1. Flocks of quadratic cones

Let K be the quadratic cone with vertex v in PG$(3, q)$. A partition of $K - \{v\}$ into q disjoint irreducible conics is called a flock of K. The flock F is called linear if the q planes of the conics of F all contain a common line L.

A set of disjoint irreducible conics of the cone K is called a partial flock of K.

If F is a flock of the quadratic cone K of PG$(3, q)$, q odd, then Bader, Lunardon and Thas [1] have shown that from F there arise q other flocks $F_1, F_2, \ldots, F_q$, some or all of them possibly projectively equivalent to F. We say that those flocks $F_1, F_2, \ldots, F_q$ are derived from F. From the construction it follows that $F, F_1, F_2, \ldots, F_{i-1}, F_{i+1}, \ldots, F_q$ are derived from F_i, $i = 1, 2, \ldots, q$.

We now list all the known non-linear flocks of the quadratic cone K. Let K be represented by the equation $X_0 X_1 = X_2^2$, and let the planes of the elements of the flock be represented by $a_i X_0 + b_i X_1 + c_i X_2 + X_3 = 0$, $i = 1, 2, \ldots, q$.

(1) The flocks FTW of Fisher, Thas and Walker (see Thas [17]):

here $q \equiv -1 \bmod 3$ and

$$\{(a_i, b_i, c_i)|i = 1, 2, \ldots, q\} = \{(t, 3t^3, 3t^2)|t \in \mathrm{GF}(q)\}.$$

This flock is linear if and only if $q = 2$.

(2) The flocks $K1$ of Kantor (see Thas [17]):
here q is odd, m is a given nonsquare, σ is an automorphism of $\mathrm{GF}(q)$ and

$$\{(a_i, b_i, c_i)|i = 1, 2, \ldots, q\} = \{(t, -mt^\sigma, 0)|t \in \mathrm{GF}(q)\}.$$

This flock is linear if and only if $\sigma = 1$.

(3) The flocks $K2$ of Kantor (see Thas [17]):
here q is odd with $q \equiv \pm 2 \bmod 5$, and

$$\{(a_i, b_i, c_i)|i = 1, 2, \ldots, q\} = \{(t, 5t^5, 5t^3)|t \in \mathrm{GF}(q)\}.$$

This flock is linear if and only if $q = 3$.

(4) The flocks $P1$ of Payne (see Thas [17]):
here $q = 2^e$ with e odd, and

$$\{(a_i, b_i, c_i)|i = 1, 2, \ldots, q\} = \{(t, t^5, t^3)|t \in \mathrm{GF}(q)\}.$$

This flock is linear if and only if $q = 2$.

(5) The flocks $K3$ of Kantor (see Gevaert and Johnson [5]):
here $q = 5^h$, k is a given nonsquare and

$$\{(a_i, b_i, c_i)|i = 1, 2, \ldots, q\} = \{(t, k^{-1}t + 2t^3 + kt^5, t^2)|t \in \mathrm{GF}(q)\}.$$

This flock is always non-linear.

(6) The flocks G of Ganley (see Gevaert and Johnson [5] and Payne [13]):
here $q = 3^h$, n is a given nonsquare, and

$$\{(a_i, b_i, c_i)|i = 1, 2, \ldots, q\} = \{(t, -(nt + n^{-1}t^9), t^3)|t \in \mathrm{GF}(q)\}.$$

This flock is linear if and only if $q = 3$.

(7) The flocks Fi of Fisher (see Gevaert and Johnson [5], Payne [12] and Thas [17]):
let q be odd, let ζ be a primitive element of $\mathrm{GF}(q^2)$, so $w = \zeta^{q+1}$ is a primitive element of $\mathrm{GF}(q)$ and hence a nonsquare in $\mathrm{GF}(q)$; put $i = \zeta^{(q+1)/2}$, so $i^2 = w, i^q = -i$; put $z = \zeta^{q-1} = a + bi$, so z has order $q + 1$ in the multiplicative group of $\mathrm{GF}(q^2)$; then the triples (a_i, b_i, c_i) are given by

$$(t, -wt, 0)$$

with $t \in GF(q)$ and $t^2 - 2(1+a)^{-1}$ a square in $GF(q)$, and

$$(-a_{2j}, -wa_{2j}, 2b_{2j})$$

with $j = 0, 1, \ldots, (q-1)/2$, $a_k = (z^{k+1} + z^{-k})/(z+1)$ and $b_k = i(z^{k+1} - z^{-k})/(z+1)$. This flock is linear if and only if $q = 3$.

(8) The non-linear flocks BLT, discovered by Bader, Lunardon and Thas [1]; see also Johnson [6]. These flocks are derived from the flocks $K3$, where $q = 5^h$ with $h > 1$.

(9) The non-linear flocks JP, discovered by Johnson [6] and Payne [14], applying derivation to the flocks $K2$ with q odd, $q \equiv \pm 2 \bmod 5$, $q \geq 13$.

(10) The non-linear flocks $PTJLW$, discovered by Payne and Thas [16] and Johnson, Lunardon and Wilke [7] (see also Johnson [6]) applying derivation to the flocks G of Ganley with $q = 3^h$, $h > 2$.

(11) The non-linear flocks $P2$ discovered by Payne [14] "re-coordinatizing" the generalized quadrangles arising from the flocks $P1$ with $q = 2^{2h+1}$, $h > 1$ (cf. 1.3.).

1.2. Flocks for small q

In Thas [17] it is shown that for $q = 2, 3, 4$ any flock of the quadratic cone is linear, in De Clerck, Gevaert and Thas [2] it is proved that for any $q \in \{5, 7, 8\}$ there exists exactly one non-linear flock, and by De Clerck and Mylle (see Mylle [9]) it is shown with the aid of a computer that for $q = 9$ there exist exactly two non-linear flocks.

1.3. Flocks, translation planes and generalized quadrangles

Independently, Thas and Walker [18] discovered that with each flock of an irreducible quadric of $PG(3, q)$, in particular a quadratic cone, there corresponds a translation plane of order q^2 and dimension at most two over its kernel; see also Bader, Lunardon and Thas [1], Fisher and Thas [4], Johnson, Lunardon and Wilke [7] and Thas [17].

Payne [10] (see also Kantor [8] and Payne [11]) showed that with a set of q upper triangular 2×2 - matrices over $GF(q)$ of a certain type, there corresponds a generalized quadrangle of order (q^2, q). In Thas [17] it is proved that with such a set of q matrices there corresponds a flock of the quadratic cone of $PG(3, q)$, and conversely that with each flock of the quadratic cone there corresponds such a set of matrices. Hence with each flock of the quadratic cone of $PG(3, q)$ there corresponds a generalized quadrangle of order (q^2, q).

By Payne and Rogers [15] the process of derivation never produces new generalized quadrangles.

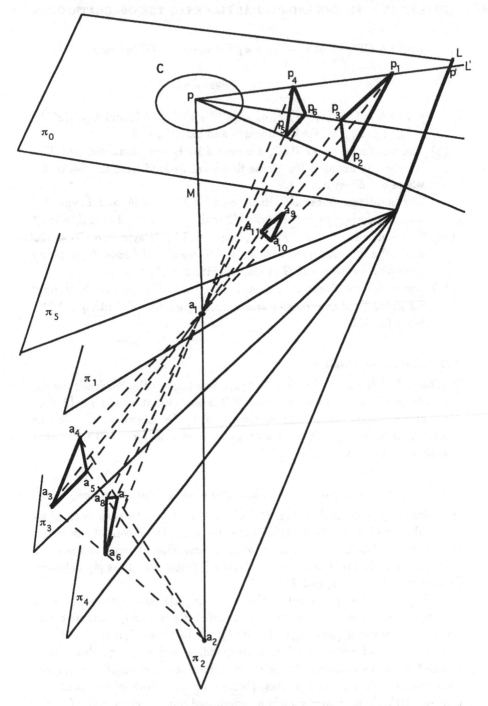

Figure 1: Partial flock for $q \equiv -1 \mod 12$

2. A partial flock for $q \equiv -1 \bmod 12$ and a new flock for $q = 11$

The notion of flock of the quadratic cone K of $PG(3, q)$, q odd, is dualized as follows. With K there corresponds an irreducible conic C in a plane $PG(2, q)$ of $PG(3, q)$. With the q planes of the elements of a flock F of K, there correspond q points $a_1, a_2, \ldots, a_q$ of $PG(3, q) - PG(2, q)$ with the property that any line $a_i a_j$, $i \neq j$ and $i, j \in \{1, 2, \ldots, q\}$, intersects $PG(2, q)$ in an interior point a_{ij} of C. The set $\{a_1, a_2, \ldots, a_q\}$ will also be called a flock of C and will also be denoted by F. A line containing exactly t points of F will be called a t-secant of F.

Lemma 2.1 *Let C be an irreducible conic of $PG(2, q)$ and let L be a line of $PG(2, q)$ having no point in common with C. If $q \equiv -1 \bmod 3$, then there exists an unique subgroup G of order 3 of $PGL(3, q)$ fixing C and L. Also, for any $\alpha \in G - \{1\}$ the line L is the only fixed line, implying that the polar point p of L with respect to C is the only fixed point.*

Proof. Let C be the conic with equation $X_0 X_1 = X_2^2$ and let L be any line having no point in common with C. Further, let $q \equiv -1 \bmod 3$. As the group $PGO(3, q)$ of C acts transitively on the interior points of C, we may assume that L has equation $X_0 + X_1 - X_2 = 0$. The element

$$\alpha : \begin{pmatrix} x_0 \\ x_1 \\ x_2 \end{pmatrix} \longmapsto \begin{pmatrix} 0 & 1 & 0 \\ 1 & 1 & -2 \\ 0 & 1 & -1 \end{pmatrix} \begin{pmatrix} x_0 \\ x_1 \\ x_2 \end{pmatrix}$$

fixes C, L and generates a group G of order 3. Also, the only fixed point of α (resp. α^2) is $p(2, 2, 1)$; hence the only fixed line is L. Since $(1, 0, 0)^\alpha = (0, 1, 0)$, $(0, 1, 0)^\alpha = (1, 1, 1)$, $(1, 1, 1)^\alpha = (1, 0, 0)$ and since $PGO(3, q)$ acts sharply 3 - transitive on C, it is clear that for any three distinct points p_1, p_2, p_3 of C the unique $\gamma \in PGO(3, q)$ with $p_1^\gamma = p_2, p_2^\gamma = p_3, p_3^\gamma = p_1$ has order 3 and fixes a unique interior point of C. Let x_M be the number of such γ's fixing a given line M, with $C \cap M = \emptyset$; by the foregoing $x_M \geq 2$. Then counting the number of pairs (γ, M), with M fixed by γ, we obtain

$$2\begin{pmatrix} q+1 \\ 3 \end{pmatrix} \Big/ \begin{pmatrix} q+1 \\ 3 \end{pmatrix} = \sum_M x_M, \tag{1}$$

where the sum runs over all lines M with $M \cap C = \emptyset$. As there are exactly $q(q-1)/2$ such lines M, and as the first member of (1) equals $q(q-1)$, it follows that each M is fixed by exactly two elements $\gamma, \gamma^{-1} = \gamma^2 \in PGO(3, q)$ of order 3. $\square$

Let $\Sigma = \mathrm{PG}(3,q)$, with $q \equiv -1 \bmod 12$, and let C be an irreducible conic of the plane $\pi_0 = \mathrm{PG}(2,q)$. Further, let p be an interior point of C and let L be the polar line of p with respect to C. By Lemma 2.1 there exist two elements $\alpha, \alpha^{-1} = \alpha^2 \in \mathrm{PGL}(3,q)$ of order 3 fixing C and L.

Let M be a line containing p but not contained in π_0, and let a_1, a_2 be distinct points of M different from p. The plane La_i will be denoted by π_i, $i = 1, 2$. Further, let π_3 be the plane through L defined by $\{\pi_0, \pi_3; \pi_1, \pi_2\} = -1$.

Let L' be a line of π_0 through p having no point in common with C and let $L' \cap L = \{p'\}$. Choosing coordinates in such a way that $\pi_0 : X_3 = 0$, $C : X_3 = X_0 X_1 - X_2^2 = 0$ and $p(2,2,1)$, it is readily seen that p' is an interior point of C if and only if 3 is a square. As $q \equiv -1 \bmod 4$, this condition is satisfied, hence p' is an interior point of C. Let $\gamma, \gamma^{-1} = \gamma^2$ be the elements of order 3 of $\mathrm{PGL}(3,q)$ fixing C and L'. Since γ and γ^2 induce no involution on L', they fix the common points z_1, z_2 of C and L' over $\mathrm{GF}(q^2)$. As $\{z_1, z_2; p, p'\} = -1$ and neither γ nor γ^2 induces an involution on L', we have $p^\gamma \neq p' \neq p^{\gamma^2}$ and so $\{p, p^\gamma, p^{\gamma^2}\} \cap \{p', p'^\gamma, p'^{\gamma^2}\} = \emptyset$. Now put $p_1 = p'^\gamma, p_2 = p_1^\alpha, p_3 = p_2^\alpha$ and $p_4 = p^\gamma, p_5 = p_4^\alpha, p_6 = p_5^\alpha$; clearly $p_1 p_4, p_2 p_5$ and $p_3 p_6$ all contain p. Then $\{z_1, z_2; p_1, p_4\} = \{z_1^\alpha, z_2^\alpha; p_2, p_5\} = \{z_1^{\alpha^2}, z_2^{\alpha^2}; p_3, p_6\} = -1$.

Now put (see Figure 1)

$$\pi_3 \cap a_1 p_i = \{a_{i+2}\}, \ i = 1, 2, 3,$$
$$a_2 a_{i+2} \cap a_1 p_{i+3} = \{a_{i+5}\}, \ i = 1, 2, 3,$$
$$p a_{i+5} \cap a_1 p_i = \{a_{i+8}\}, \ i = 1, 2, 3.$$

Question. Is every point $a_j a_k \cap \pi_0$, $j \neq k$ and $j, k = 1, 2, \ldots, 11$, an interior point of C ?

Lemma 2.2 The plane $a_6 a_7 a_8 = \pi_4$ contains L, the plane $a_9 a_{10} a_{11} = \pi_5$ contains L, and $\{\pi_0, \pi_3; \pi_4, \pi_5\} = -1$.

Proof. Let $p_1 p_3 \cap p_4 p_6 = \{l\}$. Then $p_1^\alpha p_3^\alpha \cap p_4^\alpha p_6^\alpha = p_2 p_1 \cap p_5 p_4 = \{l^\alpha\}$, and $p_2^\alpha p_1^\alpha \cap p_5^\alpha p_4^\alpha = p_3 p_2 \cap p_6 p_5 = \{l^{\alpha^2}\}$. Since $p_1 p_4, p_2 p_5, p_3 p_6$ are concurrent, the points $l, l^\alpha, l^{\alpha^2}$ are collinear. Consequently α fixes the line $l l^\alpha$. As L is the only line fixed by α, we have $l l^\alpha = L$. Hence L contains the points $l, l^\alpha, l^{\alpha^2}$.

The common point $\bar{l}$ of the lines $a_3 a_5$ and $a_6 a_8$ is projected from a_1 onto the common point l of the lines $p_1 p_3$ and $p_4 p_6$. Hence $\bar{l}$ is the common point of $a_1 l$ and $a_3 a_5$. As $a_3 a_5 \cap p_1 p_3$ is a point of $\pi_0 \cap \pi_3 = L$, we necessarily have $l \in a_3 a_5$. Hence $\{\bar{l}\} = a_1 l \cap a_3 a_5 = \{l\}$. Consequently $a_6 a_8$ contains l. Analogously $l^\alpha \in a_7 a_6$ and $l^{\alpha^2} \in a_8 a_7$. We conclude that $a_6 a_7 a_8 = \pi_4$ contains the line L.

The common point $\bar{\bar{l}}$ of the lines $a_6 a_8$ and $a_9 a_{11}$ is projected from a_1 onto the common point l of the lines $p_4 p_6$ and $p_1 p_3$. Hence $\bar{\bar{l}}$ is the common point of

384

$a_1 l$ and $a_6 a_8$. By the preceding section $l \in a_6 a_8$. Hence $\{\bar{\bar{l}}\} = a_1 l \cap a_6 a_8 = \{l\}$. Consequently $a_9 a_{11}$ contains l. Analogously $l^\alpha \in a_{10} a_9$ and $l^{\alpha^2} \in a_{11} a_{10}$. We conclude that $a_9 a_{10} a_{11} = \pi_5$ contains L.

Consider the plane $a_1 L' = \zeta$. This plane ζ contains the points $a_1, a_2, p, p_1, p_4, a_3, a_6, a_9, p'$. Let $\{m\} = \pi_3 \cap a_1 a_2$ and $\{n\} = a_6 a_9 \cap mp'$. We have $-1 = \{\pi_0, \pi_3; \pi_2, \pi_1\} = \{p, m; a_2, a_1\} = \{a_3 p, a_3 m; a_3 a_2, a_3 a_1\} = \{p, n; a_6, a_9\} = \{L', p'n; p'a_6, p'a_9\} = \{\pi_0, \pi_3; \pi_4, \pi_5\}$. Hence $\{\pi_0, \pi_3; \pi_4, \pi_5\} = -1$. □

Lemma 2.3 *The harmonic homology σ with axis π_3 and centre p fixes C, fixes a_3, a_4, a_5, interchanges a_1, a_2, and interchanges also a_i and a_{i+3}, $i = 6, 7, 8$.*

Proof. The homology σ induces in π_0 the harmonic homology σ' with axis L and centre p. Since L is the polar line of p with respect to C, σ' fixes C. Also, since $\{\pi_0, \pi_3; \pi_1, \pi_2\} = \{\pi_0, \pi_3; \pi_4, \pi_5\} = -1$, σ interchanges a_1, a_2 and interchanges a_i, a_{i+3}, $i = 6, 7, 8$. Clearly a_3, a_4, a_5 are fixed by σ. □

Lemma 2.4 *If R is a line of π_0 through p and if $r \in R$, $r \neq p$ and $r \notin L$, then the line $r^\alpha r^{\alpha^2}$ intersects L in the point R^β, with β the polarity defined by C.*

Proof. We use the notations of Lemma 2.1. Let R be the line with equation $(X_0 - 2X_2) + \lambda(X_1 - 2X_2) = 0$, $\lambda \in GF(q) \cup \{\infty\}$. Further, let $r(2 + \lambda + 2a, -2\lambda - 1 + 2a, -\lambda + 1 + a) \in R$, with $a \neq 0, \infty$. By the proof of Lemma 2.1 we have $r^\alpha(-2\lambda - 1 + 2a, \lambda - 1 + 2a, -\lambda - 2 + a)$ and $r^{\alpha^2}(\lambda - 1 + 2a, \lambda + 2 + 2a, 2\lambda + 1 + a)$. Hence $r^\alpha r^{\alpha^2} \cap L$ is the point $(\lambda, 1, 1 + \lambda)$, that is, the point R^β. □

Lemma 2.5 *All the points $a_i a_j \cap \pi_0 = \{a_{i,j}\}$ with $\{i, j\} \neq \{3, 7\}, \{3, 8\}, \{3, 10\}, \{3, 11\}, \{4, 6\}, \{4, 8\}, \{4, 9\}, \{4, 11\}, \{5, 6\}, \{5, 7\}, \{5, 9\}, \{5, 10\}$, are interior points of C.*

Proof. The points $a_{1,i}$, with $i = 2, \ldots, 11$, are the points $p, p_1, \ldots, p_6$, so are interior points of C by Lemma 2.1.

By Lemma 2.3 the harmonic homology σ interchanges a_1 and a_2, hence also the points $a_{2,i}$, $i = 3, \ldots, 11$, are interior points of C.

The point $a_3 a_5 \cap \pi_0$ is the point $p_1 p_3 \cap L = \{l\}$ (see Lemma 2.2). By Lemma 2.4 the point l is the pole of pp_2 with respect to C. Since $pp_2 = L'^\alpha$, the line pp_2 has no point in common with C. Hence l is an interior point of C. Consequently $a_{3,5}$ is an interior point of C. Analogously, $a_{3,4}, a_{4,5}, a_{6,7}, a_{6,8}, a_{7,8}, a_{9,10}, a_{9,11}, a_{10,11}$ are interior points of C.

Since $a_{3,6} = a_{2,3}$ and $a_{3,9} = a_{1,3}$, also $a_{3,6}$ and $a_{3,9}$ are interior points of C. Analogously, $a_{4,7}, a_{4,10}, a_{5,8}, a_{5,11}$ are interior points of C.

As $a_{6,9} = a_{7,10} = a_{8,11} = p$, also these points are interior points of C.

Now we consider the points $a_{6,10}, a_{6,11}, a_{7,9}, a_{7,11}, a_{8,9}, a_{8,10}$. Let $\tilde{\alpha}$ be the element of $\mathrm{PGL}(4,q)$ defined by $a_1^{\tilde{\alpha}} = a_1, a_2^{\tilde{\alpha}} = a_2, p_1^{\tilde{\alpha}} = p_2, p_2^{\tilde{\alpha}} = p_3, p_3^{\tilde{\alpha}} = p_1$. Then $\tilde{\alpha}$ induces α in the plane π_0 and fixes the set $\{a_1, a_2, \ldots, a_{11}\}$. Clearly $a_{6,10}^{\tilde{\alpha}} = a_{7,11}, a_{7,11}^{\tilde{\alpha}} = a_{8,9}, a_{7,9}^{\tilde{\alpha}} = a_{8,10}, a_{8,10}^{\tilde{\alpha}} = a_{6,11}$. Also, the homology σ interchanges $a_{6,10}$ and $a_{7,9}$. Hence it is sufficient to prove that $a_{6,10}$ is an interior point of C.

Projecting a_6 and a_{10} from a_7 onto the plane π_0, we see that $a_{6,10}$ is on the line pl^α; projecting a_6 and a_{10} from a_1 onto the plane π_0, we see that $a_{6,10}$ is on the line $p_2 p_4$.

By Lemma 2.4, with $R = L'$, we have $l^{\alpha^2}(\lambda, 1, 1+\lambda)$; so by Lemma 2.1 we have $l^\alpha(1+\lambda, -\lambda, 1)$. Hence the line pl^α has equation $(\lambda+2)X_0 + (\lambda-1)X_1 - (4\lambda+2)X_2 = 0$.

Let z_1, z_2 be the common points of C and L' over $\mathrm{GF}(q^2)$. Further, let $\{z_1, z_2; p, p_4\} = \xi$. As γ fixes z_1 and z_2, and $p^\gamma = p_4$, we have $\{z_1, z_2; p, p^\gamma\} = \{z_1, z_2; p^\gamma, p^{\gamma^2}\} = \{z_1, z_2; p^{\gamma^2}, p\} = \xi$. Consequently,

$$1 = \{z_1, z_2; p, p^\gamma\} \cdot \{z_1, z_2; p^\gamma, p^{\gamma^2}\} \cdot \{z_1, z_2; p^{\gamma^2}, p\} = \xi^3.$$

It follows that $\xi^2 + \xi + 1 = 0$. The coordinates of the points z_1 and z_2 are given by

$$z_1(\lambda^2, 2\lambda^2 + 3\lambda + 2 + 2(\lambda+1)\theta, \lambda(\lambda+1+\theta)),$$
$$z_2(\lambda^2, 2\lambda^2 + 3\lambda + 2 - 2(\lambda+1)\theta, \lambda(\lambda+1-\theta)),$$

with $\theta^2 = \lambda^2 + \lambda + 1$ and $\theta \in \mathrm{GF}(q^2) - \mathrm{GF}(q)$. If p_4 has coordinates y_0, y_1, y_2, then from $\{z_1, z_2; p, p_4\} = \xi$ follows that

$$y_0 = \lambda + 2 + 2\theta - \xi(\lambda + 2 - 2\theta),$$
$$y_1 = -2\lambda - 1 + 2\theta + \xi(2\lambda + 1 + 2\theta),$$
$$y_2 = -\lambda + 1 + \theta - \xi(-\lambda + 1 - \theta).$$

If p_1 has coordinates t_0, t_1, t_2, then from $\{z_1, z_2; p_1, p_4\} = -1$ follows that

$$t_0 = 2\theta + \lambda + 2 + \xi(\lambda + 2 - 2\theta),$$
$$t_1 = 2\theta - 2\lambda - 1 + \xi(-2\theta - 2\lambda - 1),$$
$$t_2 = \theta - \lambda + 1 + \xi(-\theta - \lambda + 1).$$

Now by Lemma 2.1 the point $p_2 = p_1^\alpha$ has coordinates

$$u_0 = 2\theta - 2\lambda - 1 + \xi(-2\theta - 2\lambda - 1),$$
$$u_1 = 2\theta + \lambda - 1 + \xi(-2\theta + \lambda - 1),$$
$$u_2 = \theta - \lambda - 2 + \xi(-\theta - \lambda - 2).$$

Hence the line p_2p_4 has equation

$$(2\theta + \xi(\theta - \lambda + 2))X_0 + (2\theta + \xi(\theta + 3\lambda + 1))X_1 - (2\theta + \xi(\theta + 4\lambda + 6))X_2 = 0.$$

Since $a_{6,10}$ is the intersection of pl^α and p_2p_4, this point has coordinates

$$\begin{aligned}
r_0 &= 3(\lambda + 1)(\xi + 2) + 8\theta\xi, \\
r_1 &= -3\lambda(\xi + 2) + 8\theta\xi, \\
r_2 &= 3(\xi + 2) + 4\theta\xi.
\end{aligned}$$

Let $GF(q^2) = \{x + \theta y \mid x, y \in GF(q)\}$, $\theta^2 = \lambda^2 + \lambda + 1$. If $\xi = x' + \theta y'$, then $(x' + \theta y')^2 + (x' + \theta y') + 1 = 0$, so

$$\begin{aligned}
x'^2 + \theta^2 y'^2 + x' + 1 &= 0, \text{ and} \\
2x'y' + y' &= 0.
\end{aligned}$$

Since $\xi \notin GF(q)$, we have $y' \neq 0$, and consequently $x' = -1/2$. Hence

$$\theta^2 y'^2 = -3/4. \tag{2}$$

Now

$$\frac{\theta\xi}{\xi + 2} = \frac{\theta(-\frac{1}{2} + \theta y')}{\frac{3}{2} + \theta y'} \overset{(2)}{=} \frac{-\frac{\theta}{2} - \frac{3}{4y'}}{\frac{3}{2} + \theta y'} = -\frac{1}{2y'}. \tag{3}$$

Hence

$$a_{6,10}(3(\lambda + 1) - 4/y', -3\lambda - 4/y', 3 - 2/y').$$

The polar line $A_{6,10}$ of $a_{6,10}$ with respect to C has coordinates

$$\begin{aligned}
\beta_0 &= -3\lambda - 4/y', \\
\beta_1 &= 3(\lambda + 1) - 4/y', \\
\beta_2 &= -2(3 - 2/y').
\end{aligned}$$

Further,

$$\beta_2^2 - 4\beta_0\beta_1 = 36(\lambda^2 + \lambda + 1) - 48/y'^2 = 12(3\theta^2 y'^2 - 4)/y'^2.$$

By (2)

$$\beta_2^2 - 4\beta_0\beta_1 = (-3)(5/y')^2 = \not\square.$$

Consequently, the line $A_{6,10}$ has no point in common with C, hence $a_{6,10}$ is an interior point of C. $\qquad\square$

Lemma 2.6 *All the points $a_i a_j \cap \pi_0 = \{a_{i,j}\}$, with $\{i, j\} = \{3, 7\}, \{3, 8\}, \{3, 10\},$ $\{3, 11\}, \{4, 6\}, \{4, 8\}, \{4, 9\}, \{4, 11\}, \{5, 6\}, \{5, 7\}, \{5, 9\}, \{5, 10\},$ are interior points of C.*

Proof. If σ is the harmonic homology with axis π_3 and centre p, then $a_{3,7}^\sigma = a_{3,10}, a_{3,8}^\sigma = a_{3,11}, a_{4,6}^\sigma = a_{4,9}, a_{4,8}^\sigma = a_{4,11}, a_{5,6}^\sigma = a_{5,9}, a_{5,7}^\sigma = a_{5,10}$. With $\tilde{\alpha}$ the element of $\mathrm{PGL}(4,q)$ introduced in the proof of Lemma 2.5, we have $a_{3,7}^{\tilde{\alpha}} = a_{4,8}, a_{4,8}^{\tilde{\alpha}} = a_{5,6}, a_{5,6}^{\tilde{\alpha}} = a_{3,7}, a_{3,10}^{\tilde{\alpha}} = a_{4,11}, a_{4,11}^{\tilde{\alpha}} = a_{5,9}, a_{5,9}^{\tilde{\alpha}} = a_{3,10}, a_{3,8}^{\tilde{\alpha}} = a_{4,6}, a_{4,6}^{\tilde{\alpha}} = a_{5,7}, a_{5,7}^{\tilde{\alpha}} = a_{3,8}, a_{3,11}^{\tilde{\alpha}} = a_{4,9}, a_{4,9}^{\tilde{\alpha}} = a_{5,10}, a_{5,10}^{\tilde{\alpha}} = a_{3,11}$.

Let η_1 be the harmonic homology with as axis the plane $a_1 a_2 p_1$ and as centre the point l^{α^2}, where $\{l^{\alpha^2}\} = p_2 p_3 \cap p_5 p_6$. By Lemma 2.4 the point l^{α^2} is the pole of the line L' with respect to C. Clearly η_1 fixes C, l^{α^2}, the points of L', the line L, the planes π_i with $i = 0, 1, \ldots, 5$, and the points a_1, a_2, a_9, a_3, a_6. As the lines L'^α, L'^{α^2} are uniquely defined by C, L and L', and as η_1 fixes C, L and L', it also fixes the set $\{L', L'^\alpha, L'^{\alpha^2}\}$. Hence η_1 interchanges L'^α and L'^{α^2}. Since l^{α^2} is on the lines $p_2 p_3$ and $p_5 p_6$, these lines are fixed by η_1, and so $p_2^{\eta_1} = p_3$ and $p_5^{\eta_1} = p_6$. It follows that $a_{10}^{\eta_1} = a_{11}, a_4^{\eta_1} = a_5$ and $a_7^{\eta_1} = a_8$. Hence η_1 fixes the set $F = \{a_1, a_2, \ldots, a_{11}\}$. Also, $a_{3,7}^{\eta_1} = a_{3,8}$.

From the preceding two sections it is now clear that the group of all elements of $\mathrm{PGL}(4,q)$ fixing π_0, C, p and F acts transitively on the set of the 12 points in the statement of the theorem. Hence it is sufficient to prove that $a_{3,7}$ is an interior point of C.

Projecting a_3 and a_7 from a_1 onto the plane π_0, we see that $a_{3,7}$ is on the line $p_1 p_5$; projecting a_3 and a_7 from a_2 onto the plane π_0, and using Lemma 2.3, we see that $a_{3,7}$ is on the line $p_1^\sigma p_2^\sigma$.

By the proof of Lemma 2.5 the point p_1 has coordinates

$$t_0 = (\lambda + 2)(\xi + 1) + 2\theta(1 - \xi),$$
$$t_1 = (-2\lambda - 1)(\xi + 1) + 2\theta(1 - \xi),$$
$$t_2 = (-\lambda + 1)(\xi + 1) + \theta(1 - \xi).$$

Now

$$\frac{1 - \xi}{1 + \xi}\theta = -\frac{3\xi\theta}{\xi + 2} = \frac{3}{2y'}, \text{ by (3)}.$$

Hence

$$p_1(\lambda + 2 + 3/y', -2\lambda - 1 + 3/y', -\lambda + 1 + 3/(2y')).$$

The coordinates of p_4 were determined in the proof of Lemma 2.5. Since $p_5 = p_4^\alpha$, the coordinates of p_5 are given by (cf. proof of Lemma 2.1)

$$s_0 = (2\lambda + 1)(\xi - 1) + 2\theta(\xi + 1),$$
$$s_1 = (-\lambda + 1)(\xi - 1) + 2\theta(\xi + 1),$$
$$s_2 = (\lambda + 2)(\xi - 1) + \theta(\xi + 1).$$

Now

$$\frac{\xi + 1}{\xi - 1}\theta = \frac{\xi\theta}{-\xi - 2} = \frac{1}{2y'}, \text{ by (3)}.$$

Hence

$$p_5(2\lambda + 1 + 1/y', -\lambda + 1 + 1/y', \lambda + 2 + 1/(2y')).$$

If $\gamma_0, \gamma_1, \gamma_2$ denote the coordinates of the line $p_1 p_5$, then

$$\gamma_0 = -3(\lambda^2 + \lambda + 1) + 9\lambda/(2y') + 3/y',$$
$$\gamma_1 = -3(\lambda^2 + \lambda + 1) - 3\lambda/(2y') - 9/(2y'),$$
$$\gamma_2 = 3(\lambda^2 + \lambda + 1) - 6\lambda/y' + 3/y'.$$

Since $L' \cap L = \{p'(2 + \lambda, -2\lambda - 1, -\lambda + 1)\}$ and $\{p, p'; p_1, p_1^\sigma\} = -1$, we have

$$p_1^\sigma(-6 + 4y' + 2\lambda y', -6 - 2y' - 4\lambda y', -3 + 2y' - 2\lambda y').$$

The line $p_1^\sigma p_2^\sigma$ contains the point $p_1 p_2 \cap L$, that is, the point $l^\alpha(-1 - \lambda, \lambda, -1)$. Hence the coordinates $\delta_0, \delta_1, \delta_2$ of the line $p_1^\sigma p_2^\sigma$ are given by

$$\delta_0 = 12y'^2(1 + \lambda + \lambda^2) + 18y'(\lambda + 2),$$
$$\delta_1 = 12y'^2(1 + \lambda + \lambda^2) + 18y'(\lambda - 1),$$
$$\delta_2 = -12y'^2(1 + \lambda + \lambda^2) - 36y'(2\lambda + 1).$$

Since $a_{3,7}$ is the intersection of $p_1 p_5$ and $p_1^\sigma p_2^\sigma$, this point has coordinates

$$d_0 = y'(7\lambda + 5) + 6,$$
$$d_1 = y'(-5\lambda + 2) + 6,$$
$$d_2 = y'(2\lambda + 7) + 3.$$

The polar line $A_{3,7}$ of $a_{3,7}$ with respect to C has coordinates

$$\zeta_0 = y'(-5\lambda + 2) + 6,$$
$$\zeta_1 = y'(7\lambda + 5) + 6,$$
$$\zeta_2 = -2(y'(2\lambda + 7) + 3).$$

Further,

$$\zeta_2^2 - 4\zeta_0\zeta_1 = 12(13y'^2(\lambda^2 + \lambda + 1) - 9) = 12(13y'^2\theta^2 - 9).$$

By (2)

$$\zeta_2^2 - 4\zeta_0\zeta_1 = -9.25 = \not p.$$

Consequently, the line $A_{3,7}$ has no point in common with C, hence $a_{3,7}$ is an interior point of C. $\quad\square$

Theorem 2.7 *The set* $F = \{a_1, a_2, \ldots, a_{11}\}$ *is a partial flock of the conic* C. *In particular, for* $q = 11$ *the set* F *is a flock of* C.

Proof. This follows immediately from Lemma's 2.5 and 2.6. $\quad\square$

Theorem 2.8 For $q = 11$ the flock F is new.

Proof. For $q = 11$, the previously known non–linear flocks of C are the FTW flock and the Fi flock. The Fi flock has a 5-secant (corresponding to the triples $(t, -wt, 0)$ in (7) of 1.1); for the FTW flock there does not exist a plane containing 4 points of the flock.

If F admits a 5-secant N, then N necessarily contains a_1, a_2 and a point of each of $F \cap \pi_i$, $i = 3, 4, 5$. This yields a contradiction as $a_1 a_2 \cap F = \{a_1, a_2\}$. Also, the points a_6, a_7, a_9, a_{10} are contained in a common plane through p. We conclude that F is new. $\qquad\square$

Problem. Is the partial flock F extendable to a flock for any $q \equiv -1 \bmod 12$? For $q = 23$ the partial flock seems to be uniquely extendable to a full flock. This is one of the computer results obtained by De Clerck and Herssens [3], see section 4 for a summary of these results.

3. The group of the partial flock

Theorem 3.1 The group G of all elements of $PGL(4, q)$, with $q \equiv -1 \bmod 12$, leaving C and F fixed is isomorphic to $C_2 \times S_3$; hence G has order 12.

Proof. Suppose that $\eta \in PGL(4, q)$ fixes C and F. So η fixes the plane π_0. Assume by way of contradiction that $\{a_1, a_2\}$ is not fixed by η. Each of the points a_1, a_2 is on exactly three 3-secants of F. If e.g. $a_1^\eta = a_i$, $i > 2$, then a_i is on exactly three 3-secants of F, and consequently a_i is on a 3-secant N containing neither a_1 nor a_2. Let $N \cap F = \{a_i, a_j, a_k\}$. Then necessarily any of the planes π_3, π_4, π_5 contains exactly one of the points a_i, a_j, a_k. Projecting a_i, a_j, a_k from a_1 onto π_0, we see that some side $p_{i'} p_{j'}$ of the triangle $p_1 p_2 p_3$ must contain a vertex $p_{k'}$ of the triangle $p_4 p_5 p_6$, with $p p_{i'} \neq p p_{k'} \neq p p_{j'}$. Applying α, we may assume that p_1, p_3, p_5 are collinear. Now using the coordinates of $p_3 = p_1^{\alpha^2}$ and the equation of $p_1 p_5$ (see proof of Lemma 2.6), we immediately check that p_3 is not on the line $p_1 p_5$. This contradiction proves that η fixes $\{a_1, a_2\}$.

In Lemma 2.3 we proved that the harmonic homology σ with axis π_3 and centre p fixes C, fixes a_3, a_4, a_5, interchanges a_1, a_2, and interchanges also a_i and a_{i+3}, $i = 6, 7, 8$.

Now assume that $\eta \in PGL(4, q)$ fixes C, F, a_1 and a_2. As η fixes the line $a_1 a_2$ and the plane π_0, it also fixes p. Hence it fixes every point of the line $a_1 a_2$. Also the polar line L of p with respect to C is fixed. For any plane π containing L, the line L and the point $\pi \cap a_1 a_2$ are fixed, and so π is fixed. As η fixes $\pi_3 \cap F = \{a_3, a_4, a_5\}$ and L, also the set $\{l, l^\alpha, l^{\alpha^2}\}$ is fixed.

If η fixes each of $l, l^\alpha, l^{\alpha^2}$, then it necessarily fixes each of a_3, a_4, a_5, each of a_6, a_7, a_8, and each of a_9, a_{10}, a_{11}, that is, η is the identity.

Now assume that η fixes l^{α^2} and interchanges l and l^α. Then a_3, a_6, a_9 are fixed, and $a_4^\eta = a_5, a_5^\eta = a_4, a_7^\eta = a_8, a_8^\eta = a_7, a_{10}^\eta = a_{11}, a_{11}^\eta = a_{10}$. Hence η is the harmonic homology η_1 with axis $a_1 a_2 p_1$ and centre l^{α^2}, introduced in the proof of Lemma 2.6. Analogously, if η fixes l and interchanges l^α and l^{α^2}, then it is the harmonic homology η_2 with axis $a_1 a_2 p_2$ and centre l. Finally, if η fixes l^α and interchanges l^{α^2} and l, it is the harmonic homology η_3 with axis $a_1 a_2 p_3$ and centre l^α.

Next, assume that $l^\eta = l^\alpha, l^{\alpha \eta} = l^{\alpha^2}, l^{\alpha^2 \eta} = l$. Then $p_1^\eta = p_2, p_2^\eta = p_3, p_3^\eta = p_1$. Hence η is the unique element $\tilde{\alpha}$ of $\mathrm{PGL}(4, q)$ which fixes a_1, a_2 and induces α in the plane π_0. As $a_3^{\tilde{\alpha}} = a_4, a_4^{\tilde{\alpha}} = a_5, a_5^{\tilde{\alpha}} = a_3, a_6^{\tilde{\alpha}} = a_7, a_7^{\tilde{\alpha}} = a_8, a_8^{\tilde{\alpha}} = a_6, a_9^{\tilde{\alpha}} = a_{10}, a_{10}^{\tilde{\alpha}} = a_{11}, a_{11}^{\tilde{\alpha}} = a_9$, the projectivity $\tilde{\alpha}$ fixes indeed F. Analogously, if $l^\eta = l^{\alpha^2}, l^{\alpha^2 \eta} = l^\alpha, l^{\alpha \eta} = l$, then η is the unique element $\tilde{\tilde{\alpha}}$ of $\mathrm{PGL}(4, q)$ which fixes a_1, a_2 and induces $\alpha^2 = \alpha^{-1}$ in the plane π_0.

Consequently $G' = \{1, \eta_1, \eta_2, \eta_3, \tilde{\alpha}, \tilde{\tilde{\alpha}}\}$ is the group of all elements of $\mathrm{PGL}(4, q)$ leaving C, F, a_1 and a_2 fixed. Clearly G' induces S_3 on the set $\{a_3, a_4, a_5\}$.

It readily follows that $G \cong S_3 \times C_2$. $\qquad\square$

Remark. The orbits of G on F are $\{a_1, a_2\}, \{a_3, a_4, a_5\}, \{a_6, a_7, \ldots, a_{11}\}$.

4. Computer results

With the aid of a computer, the following results were obtained by De Clerck and Herssens [3].

Result 1. For $q = 11$, the only flocks of the quadratic cone are the linear flock, the FTW flock, the Fi flock, and the new flock described in this paper.

Result 2. For $q = 13$, there are at least 3 non-linear flocks, namely the $K2$ flock, the Fi flock and the JP flock. If other flocks exist they should have the same geometric structure as $K2$ or as JP.

Result 3. For $q = 16$, there is exactly one non–linear flock of the quadratic cone. This is the first non–linear flock for q of the form 2^{2h}.

Result 4. For $q = 17$, there is a non–linear flock neither isomorphic to one of the flocks (1) to (7), nor obtainable by derivation from one of these. Hence for $q = 17$ there are at least 6 mutually non–isomorphic flocks of the quadratic cone.

Result 5. For $q = 23$, the partial flock F with 11 elements is uniquely extendable to a full flock. This flock is a non–linear flock neither isomorphic to one of the flocks (1) to (7), nor obtainable by derivation from one of these. Moreover this flock yields a new flock by using the technique of derivation.

Hence for $q = 23$ there are at least 7 mutually non–isomorphic flocks of the quadratic cone.

References

[1] **L. Bader, G. Lunardon, and J. A. Thas**. Derivation of flocks of quadratic cones. *Forum Math.*, 2, pp. 163–174, 1990.

[2] **F. De Clerck, H. Gevaert, and J. A. Thas**. Flocks of a quadratic cone in $PG(3, q)$, $q \leq 8$. *Geom. Dedicata*, 26, pp. 215–230, 1988.

[3] **F. De Clerck and C. Herssens**. Flocks of the quadratic cone in $PG(3, q)$, for q small. *The CAGe Reports*, 8, pp. 1–74, 1992.

[4] **J. C. Fisher and J. A. Thas**. Flocks in PG(3, q). *Math. Z.*, 169, pp. 1–11, 1979.

[5] **H. Gevaert and N. L. Johnson**. Flocks of quadratic cones, generalized quadrangles and translation planes. *Geom. Dedicata*, 27, pp. 301–317, 1988.

[6] **N. L. Johnson**. Derivation of partial flocks of quadratic cones. To appear.

[7] **N.L. Johnson, G. Lunardon, and F.W. Wilke**. Semifield skeletons of conical flocks. *J. Geom.*, 40, pp. 105–112, 1991.

[8] **W. M. Kantor**. Some generalized quadrangles with parameters (q^2, q). *Math Z.*, 192, pp. 45–50, 1986.

[9] **F. Mylle**. Flocks van kwadrieken in PG(3, q). Master's thesis, Universiteit Gent, 1991.

[10] **S. E. Payne**. Generalized quadrangles as group coset geometries. *Congr. Numer.*, 29, pp. 717–734, 1980.

[11] **S. E. Payne**. A new family of generalized quadrangles. *Congr. Numer*, 49, pp. 115–128, 1985.

[12] **S. E. Payne**. The Thas-Fisher generalized quadrangles. *Ann. Disrete Math.*, 37, pp. 357–366, 1988.

[13] **S. E. Payne**. An essay on skew translation generalized quadrangles. *Geom. Dedicata*, 32, pp. 93–118, 1989.

[14] **S. E. Payne**. Collineations of the generalized quadrangles associated with q-clans. *Ann. Discrete Math.*, 52, pp. 449–461, 1992.

[15] **S. E. Payne and L. A. Rogers**. Local group actions on generalized quadrangles. *Simon Stevin*, 64, pp. 249–284, 1990.

[16] **S. E. Payne and J. A. Thas**. Conical flocks, partial flocks, derivation, and generalized quadrangles. *Geom. Dedicata*, 38, pp. 229–243, 1991.

[17] **J. A. Thas**. Generalized quadrangles and flocks of cones. *Europ. J. Combinatorics*, 8, pp. 441–452, 1987.

[18] **M. Walker**. A class of translation planes. *Geom. Dedicata*, 5, pp. 135–146, 1976.

J. A. Thas, F. De Clerck, University of Ghent, Galglaan 2, B-9000 Gent (Belgium). e-mail: jat@cage.rug.ac.be, fdc@cage.rug.ac.be

C. Herssens Limburgs Universitair Centrum, Vakgroep Zuivere Wiskunde, Campus Building D, B-3590 Diepenbeek (Belgium).
e-mail: herc@bdiluc01.bitnet

Some extended generalized hexagons

J. van Bon

Abstract

We determine the flag-transitive extended generalized hexagons, such
that the group 7 : 6 is induced on the residue of each point and a
certain geometric condition (∗) holds.

1. Introduction.

Recently R. Weiss [1], [2], classified flag-transitive extended generalized hexagons, under an additional geometric condition (see (∗) below). These geometries belong to a Buekenhout diagram

Here the G_2-residue is either one of the thick generalized hexagons associated with the groups $G_2(q)$ or $^3D_4(q)$, or the point-thin generalized hexagon
(i.e., each point is incident with exactly two lines) associated with $L_3(q) : 2$.
Let Γ be such a $c.G_2$-geometry. In the special case where G_2-residue is isomorphic to the generalized hexagon associated to $L_3(2) : 2$, Weiss assumed
that the stabilizer of a flag $\mathcal{F}$ corresponding to a G_2-residue induces $L_3(2) : 2$
on $res(\mathcal{F})$, the residue of $\mathcal{F}$ in Γ. The purpose of this paper is to remove this
last assumption.

Let Π be the projective plane over $GF(2)$ and denote by $\mathcal{P}$, $\mathcal{L}$ and $\mathcal{F}$ the
set of points, lines and maximal flags of Π, respectively. Define Π_o to be the
geometry $(\mathcal{F}, \mathcal{P} \cup \mathcal{L}, *)$, where $*$ is the natural incidence relation. Then Π_o is
a generalized hexagon having two lines through every point and three points
on each line. This geometry admits besides $L_3(2) : 2$ one other flag-transitive
group of automorphisms; namely a group isomorphic to 7 : 6.

Let $\Gamma = (\mathcal{B}_1, \mathcal{B}_2, \mathcal{B}_3)$ be a connected $c.\Pi_o$-geometry satisfying condition

(∗) There exist triples of pairwise collinear points not lying on any circle,

where the points, lines and circles are the elements of $\mathcal{B}_1, \mathcal{B}_2$ and $\mathcal{B}_3$,
respectively. We shall prove the following theorem.

Main Theorem *Let $\Gamma = (\mathcal{B}_1, \mathcal{B}_2, \mathcal{B}_3)$ be a connected $c.\Pi_o$-geometry with a flag-transitive group $G \leq aut(\Gamma)$ satisfying condition $(*)$. Suppose that for each $P \in \mathcal{B}_1$, the group G_P^o is isomorphic to $7 : 6$, where G_P^o denotes the group induced by G_P on $res(P)$. Then $|\mathcal{B}_1| = 24$ and $G \cong GL(2,7)/\langle -I \rangle$ or $|\mathcal{B}_1| = 64$ and $G \cong 2^6 : 7 : 6$.*

In [1] and [2] one of the first steps in the proof shows that two points are incident with at most one line. However in our situation we do have some examples occurring for which this fails. Besides the geometry obtained from the first example by taking a quotient with respect to the centre, there is one more flag-transitive $c.G_2$-geometry such that there exist two points that are incident with at least two lines. This one has $|\mathcal{B}_1| = 4$ and admits a flag-transitive group $G \cong 7 : (2 \times A_4)$. In both cases any 3 points are incident with a circle so condition $(*)$ fails, but we mention them as they will appear during the coset enumerations.

One can try to extend the geometry over and over again by a circle-geometry and thus obtain geometries of type $c^k.G_2$, with $k \geq 2$. In fact, in the two previously mentioned papers, Weiss also classified these geometries where the $c.G_2$-residue is one of the examples there found. In this paper we do not attempt to classify $c^k.G_2$-geometries, where $k \geq 2$ and the $c.G_2$-residue is as in the Main Theorem. This is left as an open problem.

The organization of this paper is as follows. In section 2 we shall list some properties of Π_o and the action of the group $7 : 6$ on it. Further we shall state some information about the examples. In section 3 the theorem will be proved by reducing the possible presentations for the group G to a finite number and then performing coset enumerations using CAYLEY. Throughout this paper we will use Atlas-notation for our groups.

2. The geometry Π_o and the examples.

In this section we shall study the geometry Π_o and the examples in some more detail. Recall the definitions from section 1. Let Π be the projective plane over $GF(2)$ and denote by $\mathcal{P}$, $\mathcal{L}$ and $\mathcal{F}$ the set of points, lines and maximal flags of Π, respectively. Define Π_o to be the geometry $(\mathcal{F}, \mathcal{P} \cup \mathcal{L}, *)$, where $*$ is the natural incidence relation. Then Π_o is a generalized hexagon having two lines through every point and three points on each line. This geometry admits a flag-transitive group of automorphisms H isomorphic to $7 : 6$, which is a subgroup of the full automorphism group $L_3(2) : 2$. Let Ψ denote the collinearity graph of Π_o. The points and lines of Π_o can be identified with the vertices and maximal cliques of Ψ. For the remainder of this section we shall freely use this identification without further reference. For an involution $\sigma \in H$ let $fix(\sigma)$ denote the set of points fixed by σ. The following omnibus lemma collects some facts of H and Π_o.

Lemma 2.1 *Let Π_o and H be as above.*

1. *There are 21 points and 14 lines;*
2. *For any point P and line l we have $|H_P| = 2$ and $|H_l| = 3$;*
3. *An involution of H fixes 3 points and no lines;*
4. *For any involution $\sigma \in H$ and any two points $P, Q \in fix(\sigma)$ the distance in Ψ between P and Q is 3.*

Proof. Straightforward. $\square$

The group H can be presented by generators x, y and t subject to the relations:

$$x^7 = y^3 = t^2 = [y,t] = 1, \ x^y = x^4 \text{ and } x^t = x^{-1}.$$

Let $T = \langle t \rangle$, $e = y^{x^4}$ and Ψ' be the coset graph of H on T obtained by calling two cosets Ta and Tb adjacent if and only if $ab^{-1} \in TeT \cup Te^{-1}T$. Hence $\Psi'_1(T) = \{Te, Te^{-1}, Tet, Te^{-1}t\}$. Observe that for each coset Th there are unique $i \in \{0, 1, ..., 6\}$ and $j \in \{0, 1, 2\}$ with $Th = Tx^i y^j$. Now two cosets Ta and Tb, with $a, b \in \langle x, y \rangle$, are adjacent if and only if $ab^{-1} \in \{x^4 y, x^5 y^2, x^3 y, x^2 y^2\}$. From this fact it follows that Ψ' has maximal cliques of size 3 and any two maximal cliques intersect in at most one point. The group H contains 7 involutions presented by $t^{x^i} = tx^{2i}$, where $i \in \{0, 1, ..., 6\}$. Using the action of H on Ψ' one finds that $\Psi'_2(T)$ contains the 4 cosets Tx, Tx^3, Tx^4 and Tx^6 and $\Psi'_3(T)$ contains the 2 cosets Tx^2 and Tx^5 which are both adjacent to Ty. Moreover, the latter set also contains the two cosets Ty and Ty^2 which are fixed by T. In fact it is not too hard to prove the following lemma, whose proof is left to the reader.

Lemma 2.2 *The graphs Ψ and Ψ' are isomorphic.* $\square$

We now embark on the construction of the examples.

The group $GL(2,7)$ in its natural representation acts transitively on the 48 nonzero vectors of a 2-dimensional vector space over $GF(7)$. Let $\tilde{H}$ be the stabilizer of such a vector. Let $G = GL(2,7)/\langle -I \rangle$ and $H = \tilde{H}\langle -I \rangle/\langle -I \rangle$. Then $H \cong 7 : 6$. Let A be a subgroup of G with $A \cong A_4$ and $A \cap H \neq \{1\}$. Calculations in G reveal that such a group A exists and that $|A \cap H| = 3$. Let t be an involution in H. Then there exists an elementary Abelian 2-group L of order four containing t. Call the elements of the set $\mathcal{B}_1 = \{Hh | h \in G\}$ points, those of the set $\mathcal{B}_2 = \{Lg | g \in G\}$ lines and those of the set $\mathcal{B}_3 = \{Ah | h \in G\}$ circles. Let $P, Q \in \{H, L, A\}$ with $P \neq Q$, then a coset Pa is called incident with Qb if and only if $Pa \cap Qb \neq \emptyset$. Let $\Gamma_1 = (\mathcal{B}_1, \mathcal{B}_2, \mathcal{B}_3)$ be the geometry constructed above. There are 24 points, 252 lines and 84 circles; moreover, each point is incident with 21 lines and 14 circles and each circle is incident with 6 lines. Hence each line is incident with 2 points and 2 circles. Straightforward calculations show that Γ_1 is a connected $c.\Pi_o$-geometry satisfying condition $(*)$. Let Z be the centre of G. It can be seen that in the collinearity graph each point Ha is adjacent to all others except Haz and Haz^2, where $\langle z \rangle = Z$. A second example can be obtained as a quotient from this example by factoring out the centre of $GL(2,7)$.

A third example can be found in the following way. The group $L_3(2) : 2$ has an 8-dimensional representation over $GF(2)$ (this is better known as the adjoint representation). Let V be this vector space. There exists a $v \in V$ whose orbit $\mathcal{O}$ under $L_3(2) : 2$ has length 14. Call two vectors $u, w \in V$ adjacent if and only if $u - w \in \mathcal{O}$. There exist maximal cliques of size 4, corresponding to the translates of 2-dimensional subspaces contained in $\mathcal{O} \cup \{0\}$. The group $L_3(2) : 2$ contains a subgroup H isomorphic to $7 : 6$. This group H leaves a 2-dimensional subspace invariant whose intersection with $\mathcal{O}$ is empty. Denote by $\bar{V}$ the quotient of V by this subspace and restrict the adjacency relation to $\bar{V}$. We obtain a graph on which the semidirect product of $\bar{V}$ with H acts. If one takes $\mathcal{B}_1 = \{\bar{v} | v \in V\}$, $\mathcal{B}_2 = \{\{\bar{u}, \bar{w}\} | u - w \in \mathcal{O}$ for some $u \in \bar{u}, w \in \bar{w}\}$ and $\mathcal{B}_3$ to be the set of 4-cliques. Then $\Gamma_2 = (\mathcal{B}_1, \mathcal{B}_2, \mathcal{B}_3)$ is a connected $c.\Pi_o$-geometry and has the desired structure. In particular, straigthforward calculations show that the collinearity graph of $res(P)$ has the following properties. Let $\alpha \in G_P$ and Q be an element of $res(P)$ not fixed by α. Then Q is adjacent to Q^α if and only if Q and Q^α are at distance three in $res(P)$. Moreover, these two vertices, S_1 and S_2, are the only vertices at distance three from Q in $res(P)$, to which Q is adjacent. Furthermore, S_1 and S_2 have a common neighbour in $res(P)$ that is fixed by the unique involution fixing Q.

Finally consider the group $7 : (2 \times A_4)$, that is the group generated by four symbols a, b, c, d subject to the relations

$$a^7 = b^2 = c^2 = d^3 = 1, a^b = a^{-1}, a^d = a^2, [a, c] = [b, c] = [b, d] = 1, c * c^d = c^{d^2}.$$

Call the cosets of $< a, b, d >$ points, those of $< b, c >$ lines and the $< c, d >$-cosets circles. If we choose incidence as before, then straight forward computations show us that we get an example of an extended hexagon, but this time condition $(*)$ fails.

3. The proof of the Main Theorem.

Let G and $\Gamma = (\mathcal{B}_1, \mathcal{B}_2, \mathcal{B}_3)$, be as before. The elements of $\mathcal{B}_1$, $\mathcal{B}_2$ and $\mathcal{B}_3$, will be called points, lines and circles, respectively. Suppose that $P \in \mathcal{B}_1$, $l \in \mathcal{B}_2$ and $\gamma \in \mathcal{B}_3$ form a maximal flag of Γ. Then $\mathcal{B}_1{}^l = \mathcal{B}_1{}^{\{l,\gamma\}}$ since Γ_l is a generalized 2-gon, so $|\mathcal{B}_1{}^l| = 2$ since Γ_γ is a geometry of type c. Also $|\mathcal{B}_2{}^{\{P,\gamma\}}| = 3$ since $\Gamma_P \cong \Pi_o$ and hence $|\mathcal{B}_1{}^\gamma| = 4$. Let Δ be the collinearity graph on $\mathcal{B}_1$. Since Γ is connected, so is Δ. Let $Q \in \Delta(P)$ (where $\Delta(P)$

denotes the set of all points adjacent to P in Δ). Notice that each circle is incident with 4 points so induces a 4-clique in the collinearity graph.

If there is more than one line incident with P and Q, then the collection of lines incident with P and Q form a block of imprimitivity for the action of the group $7:6$ on $res(P)$. Hence the blocks have either size three or seven. In case the block size is three, then the blocks are the 3 lines on P fixed by an involution of $7:6$. Hence $|\Delta(P)| = 7$. The group $G_{P,Q}$ induces a group of order six on $res(P)$ which acts transitively on the six blocks not containing a line incident with Q. It follows that Δ is an 8-clique. In case the block size is seven it follows that $|\Delta(P)| = 3$ and thus Δ is a 4-clique.

Lemma 3.1 *If P, l, γ is a maximal flag, then $G_{P,l,\gamma} = 1$; in particular $G_P \cong 7:6$.*

Proof. Notice that in $res(P)$ the stabilizer of a flag is trivial. Let $g \in G_{P,l,\gamma}$. Then g fixes all lines and circles incident with P and also all neighbours of P in the collinearity graph. Let l', γ' be a flag incident with P. Let Q be a point, different from P, incident with this flag. Then $g \in G_{Q,l',\gamma'}$, hence G fixes all lines and circles incident with Q. By connectedness it now follows that G fixes all points, lines and circles.   □

If Δ is an 8-clique, then G is a subgroup of S_8 acting 2-transitively on eight points, hence $G \cong GL(2,7)$. If Δ is an 4-clique, then the elements of order seven and two fix Δ point wise. It follows that $G \cong 7:(2 \times A_4)$

From now on we will identify an element $x \in \mathcal{B}_2 \cup \mathcal{B}_3$ with the set $\mathcal{B}_1{}^x$. For any circle γ, let N_γ denote the kernel of the action of G_γ on γ. Then $G_P \cong 7:6$, $|G_{P,Q}| = 2$ and $|G_{P,\gamma}/N_\gamma| = 3$. From this it immediately follows that $G_P \cap G_{Q,R,S} = \{1\}$ for each circle $\{P, Q, R, S\}$ of Γ.

Lemma 3.2 *For all circles γ of Γ we have $N_\gamma = \{1\}$.*

Proof. Let P, Q, R, S be the four points incident with γ. Each element of N_γ fixes the points P, Q, R and S, so $N_\gamma \subseteq G_P \cap G_Q \cap G_R \cap G_S = \{1\}$.   □

Corollary 3.3 *For all circles γ of Γ we have $G_\gamma \cong A_4$.*

Proof. Let $\gamma = \{P, Q, R, S\}$ be a circle. Then $|G_P \cap G_{\{Q,R,S\}}| = 3$, so G_γ acts 2-transitively on a set of four vertices with a point stabilizer of order 3. Hence $G_\gamma \cong A_4$.   □

Lemma 3.4 *Let γ be a circle incident with the line $\{P,Q\}$, $\langle t \rangle = G_{P,Q}$ and $\langle f \rangle = G_{\{P,Q\},\gamma}$ then $[t, f] = 1$.*

Proof. Both f and t are involutions so $[f, t] = (ft)^2$. The line $\{P, Q\}$ is incident with exactly two circles, the element f fixes both of them whereas t interchanges them. Now $(ft)^2$ fixes P, Q and the each of the two circles incident with $\{P, Q\}$, hence $(ft)^2 = 1$. $\square$

Let $l = \{P, Q\}$. For $1 \leq i \leq 3$, let X_i denote the set of elements of $\mathcal{B}_2{}^P$ at distance $2i$ from l in the graph $(\mathcal{B}_2{}^P \cup \mathcal{B}_3{}^P, *)$ and $Y_i = \{T \in \mathcal{B}_1 | \{P, T\} \in X_i\}$. The elements of Y_1 are precisely the points co-circular with P and Q. By condition $(*)$, $\Delta(Q) \cap Y_i \neq \emptyset$ for at least one of $i = 2$ or 3.

Lemma 3.5 *Either* $|\Delta(Q) \cap Y_2| = 0$, 4 *or* 8.

Proof. Let $R \in \Delta(Q) \cap Y_2$. Let $\{P, Q, S_1, S_2\}$ and $\{P, S_2, S_3, R\}$ be the two unique circles incident with P and connecting the line $\{P, Q\}$ with $\{P, R\}$ in $res(P)$. These two circles are the only ones incident with both P and S_2. The involution in $G_{\{P, S_2, S_3, R\}}$ which interchanges P with S_2 and S_3 with R also fixes $\{P, Q, S_1, S_2\}$ and interchanges Q with S_1. So we have $R \in \Delta(Q)$ if and only if $S_3 \in \Delta(S_1)$. Now let $e \in G_P$ be the element of order 3 fixing $\{Q, S_1, S_2\}$ with $S_1^e = Q$. Thus we have $S_3 \in \Delta(S_1)$ if and only if $S_3^e \in \Delta(Q)$. In $res(P)$ the circle $\{P, S_2, S_3, R\}$ is not fixed by e. Hence $\{P, S_1, S_3^e, R^e\}$ is a different circle, and so $S_3^e \in Y_2$. Using the action of the group $G_{P,Q} = \langle t \rangle$ we now find that $\{R, S_3^e, R^t, S_3^{et}\}$ is a subset of Y_2 with the property that $|\{R, S_3^e, R^t, S_3^{et}\} \cap \Delta(Q) \cap Y_2|$ is equal to either 0 or 4. Similarly we find that $|\{S_3, R^e, S_3^t, R^{et}\} \cap \Delta(Q) \cap Y_2|$ is equal to either 0 or 4. As these subsets form a partition of Y_2 the lemma follows. $\square$

From now on fix $P, Q, R \in \mathcal{B}_1$ with $\{P, Q\}$, $\{P, R\}$ and $\{Q, R\}$ lines but $\{P, Q, R\}$ not incident with a circle. In other words we have $Q, R \in \Delta(P)$ and $R \in Y_2 \cup Y_3$. Such a set exists by our assumption $(*)$. The group $G_P \cong 7 : 6$ has generators x, y and t such that

$$x^7 = y^3 = t^2 = [y, t] = 1, \ x^y = x^4 \text{ and } x^t = x^{-1}.$$

Without loss of generality we may assume that Q can be identified with the coset $\langle t \rangle$ in G_P. Set $e = y^{x^4}$ then e normalizes a circle in $res(P)$ incident with Q. Let γ the circle incident with P and Q and normalized by e. As $G_\gamma \cong A_4$ there exist an involution $f \in G_\gamma$ acting fix point freely on γ with $\langle f, e \rangle \cong A_4$ and $P^f = Q$. Hence, with help of Lemma 3.4, we find

$$f^2 = (ef)^3 = [t, f] = 1.$$

As P, Q and R are pairwise collinear and $P^f = Q$ it follows that R^f is collinear with P and Q, but P, Q and R^f are not co-circular. We shall investigate, in $res(P)$, the possible positions of the lines $\{P, R\}$ and $\{P, R^f\}$ with respect to the line $\{P, Q\}$.

Lemma 3.6 *If* $R = Q^\alpha$ *for some involution* $\alpha \in G_P$, *then* Γ *is isomorphic to either* Γ_1 *or* Γ_2.

Proof. Suppose that $R = Q^\alpha$ for some involution $\alpha \in G_P$. Then $P^{f\alpha f} = R^f = Q^\beta$ for some $\beta \in \langle x, y \rangle$. Hence $P^{f\alpha f \beta^{-1}} = Q$ and, as $\beta \in G_P$, we find that the element $\beta f \alpha f \beta^{-1}$ interchanges P with Q. Consequently, we have either $\beta f \alpha f \beta^{-1} = f$ or ft. Hence

$$\alpha^f = (ft^k)^{(x^j y^i)}$$

for some $i \in \{1, 2, 3\}$, $j \in \{1, 2, ..., 7\}$ and $k \in \{0, 1\}$. If $R \in Y_2$, then it follows by the results of Section 2 and Lemma 3.5 that all four elements of $\{R, S_3^e, R^t, S_3^{et}\}$ are the image of Q under some involution of G_P. Hence we may assume that $\alpha = t^{x^2}$; whereas if $R \in Y_3$, then we may assume that $\alpha = t^x$. Moreover, as $\beta \in G_P$ and $Q^\beta = R^f$ it follows that $R^f \in \Delta(P)$, so $R \in \Delta(Q)$. Hence $\{P, Q, R\}$ is a triangle not incident with a circle, thus Γ satisfies (∗). Performing a coset enumeration with CAYLEY proves the lemma. □

Remark From section 2 it follows that the automorphism group of Γ_1 contains elements satisfying both possibilities for α; whereas that of Γ_2 only contains elements satisfying $\alpha = t^x$.

From now on we shall assume that, if P_1, P_2 and P_3 are points of Γ pairwise collinear, but the three not co-circular, then P_i and P_j are not interchanged by an involution of G_{P_k}, where $\{i, j, k\} = \{1, 2, 3\}$.

Lemma 3.7 We have $R^f \notin \{R, R^t\}$.

Proof. Both involutions ft and f interchange P with Q. As P, Q and R are not co-circular we see that in $res(R)$ the lines $\{R, P\}$ and $\{R, Q\}$ are not on a circle but in Δ are joint by an edge. Thus if R is fixed by either f or ft, then we violate the assumption just made. □

Lemma 3.8 If S is a point collinear with both P and Q, but not co-circular, then S is not fixed by t; in particular $R^f \neq R^{ft}$.

Proof. The involution t fixes in $\Delta(P)$ besides Q only Q^y and Q^{y^2}. Thus if $S^t = S$, then S is either Q^y or Q^{y^2}. Hence in Δ we find that Q is joint by an edge to one of Q^y or Q^{y^2}. Using the action of $\langle y \rangle$, we see that $C = \{P, Q, Q^y, Q^{y^2}\}$ is a clique of Δ. Any involution in G_P fixes exactly 3 neighbours of P in Δ. Hence the connected component containing P of the vertices fixed by t is equal to the clique C. As f commutes with t and interchanges P with Q it follows that f leaves C invariant. Again we may assume that f does not fix Q^y, so f interchanges Q^y with Q^{y^2} and P with Q. Straightforward calculations show that $(fy)^3$ fixes C point wise, hence $(fy)^3 = 1$ or t. Adding this relation to the previous ones and using CAYLEY gives the desired result. (The coset enumeration yields 4 as an answer, which corresponds to a quotient of Π_o.) □

Lemma 3.9 If P, R and R^ϕ are co-circular, where $\phi \in \{f, ft\}$, then $\Delta(Q) \cap Y_2 \neq \emptyset$.

Proof. Let $\{P, R, R^\phi, U\}$ be a circle. Then $U \neq Q$ and $\{Q, R, R^\phi, U^\phi\}$ is a

401

circle too. Hence in $res(R)$ we have that $\{R, P\}$ and $\{R, Q\}$ are not collinear but $\{R, R^\phi\}$ is collinear with both $\{R, P\}$ and $\{R, Q\}$, so $\{R, P\}$ and $\{R, Q\}$ are at distance two. Recall that P and Q are joint by an edge and that G acts flag-transitively on Γ, thus we can find a similar structure in $res(P)$ and the lemma follows. $\square$

Corollary 3.10 We have $\Delta(Q) \cap Y_2 \neq \emptyset$; in particular $|\Delta(Q) \cap Y_2| = 4$.
Proof. Suppose that $\Delta(Q) \cap Y_2 = \emptyset$. Then the vertices R, R^t, R^f and R^{ft} are all in $\Delta(Q) \cap Y_3$. By Lemma 3.7 and 3.8 these vertices are all different, and are neither fixed by t nor the image of Q under some involution of G_P. From the structure of Y_3 follows that $\{P, R\}$ is co-circular with either R^f or R^{ft}, a contradiction with the previous lemma. As the set Y_2 contains four vertices which are the image of Q under some involution it follows from Lemma 3.5 that $|\Delta(Q) \cap Y_2| = 4$. $\square$

Lemma 3.11 Either $|\Delta(Q) \cap Y_3| = 2$ or $|\Delta(Q) \cap Y_3| = 4$.
Proof. First observe that t acts fixed point freely on the set $\Delta(Q) \cap Y_3$. By the various assumptions we are making, it follows that neither the two points fixed by t nor the two points which are the image of Q under some involution of G_P are in $\Delta(Q) \cap Y_3$. Consequently it is sufficient to show that $\Delta(Q) \cap Y_3 \neq \emptyset$. To this end let $R \in \Delta(Q) \cap Y_2$ and $\{P, Q, S_1, S_2\}$ and $\{P, S_2, S_3, R\}$ be the two unique circles incident with P and connecting $\{P, Q\}$ with $\{P, R\}$ in $res(P)$. Clearly Q is not collinear with S_3 and $\{P, Q, R\}$ is not incident with a circle. Similarly we find in $res(S_2)$ that $\{R, Q, S_2\}$ is not incident with a circle. Now in $res(R)$ we have the circle $\{R, P, S_2, S_3\}$ and neither line of it is co-circular with $\{R, Q\}$. Hence at least one of the lines $\{R, P\}$ and $\{R, S_2\}$ is at distance 3 from $\{R, Q\}$. As both P and S_2 are collinear with Q and G acts flag-transitively the lemma follows. $\square$

By Corollary 3.10 we may take $R = Q^{ete}$, so $R = P^{fete}$. Write $R^f = Q^\beta$. Then $pR^f = P^{fetef} = P^{f\beta}$, hence $fetef\beta^{-1}f \in G_P$ If $R^f \in Y_2$ then, as $R^f \neq R^t$, either $R^f = Q^{e^2te^2} = Q^{xy}$ or $R^f = Q^{e^2te^2t} = Q^{x^{-1}y}$. If $R^f \in Y_3$ then either $R^f = Q^{xy^{-1}}$, $R^f = Q^{x^{-1}y^{-1}}$, $R^f = Q^{yx}$ or $R^f = Q^{yx^{-1}}$. Consequently $\beta \in \{xy, x^{-1}y, xy^{-1}, x^{-1}y^{-1}, yx, yx^{-1}\}$. Thus we obtain the additional relation $fetef\beta^{-1}f = x^i y^j t^k$ for some $i \in \{0, 1, ..., 6\}$, $j \in \{0, 1, 2\}$ and $k \in \{0, 1\}$. Remains to show that a possible outcome will satisfy condition $(*)$.

Lemma 3.12 There are triangles not incident with circles.
Proof. Let $\alpha = ete$, β as above and $\gamma \in G_P$ such that $f\alpha f\beta^{-1}f\gamma = 1$. Then P is adjacent to $P^{f\alpha}$, $P^{f\beta^{-1}}$ and $P^{f\gamma}$. Clearly $P^{f\gamma} \neq P^{f\beta^{-1}f\gamma}$ and $P^{f\beta^{-1}f\gamma} \neq P^{f\alpha f\beta^{-1}f\gamma} = P$. Hence P, $P^{f\gamma} = Q^\gamma$ and $P^{f\beta^{-1}f\gamma} = Q^{\beta^{-1}f\gamma}$ are three different mutually adjacent vertices. These are incident with a circle if

and only if P, Q and $Q^{\beta^{-1}f}$ are. Applying f and β to this triple we see that this is true if and only if P, Q and $Q^\beta = R^f$ are incident with a circle, which is not so. □

Performing coset enumerations with help of CAYLEY, Theorem 1.1 follows.

Remark The coset enumerations yield also 4 and 8 as an answer, these correspond to the geometries determined at the beginning of this section.

References

[1] **R. Weiss**. Extended generalized hexagons. *Math. Proc. Cambridge Philos. Soc.*, 108, pp. 7–19, 1990.

[2] **R. Weiss**. A geometric characterization of the groups McL and Co_3. *J. London Math. Soc.*, 44, pp. 261–269, 1991.

J. van Bon, Department of Mathematics, Tufts University, Medford, MA 02155, U.S.A. e-mail: jvanbon@Jade.Tufts.edu

Nuclei in finite non-Desarguesian projective planes

F. Wettl [*]

Abstract

Let K be a point set in a finite projective plane. A point P is called a nucleus of K if no line through P intersects $K \setminus \{P\}$ in more than one point. There are several results on the number of nuclei in a finite Desarguesian plane. This paper deals with the non-Desarguesian case. An application on strong representative systems is given.

1. Introduction

Let K be a set of k points in a finite projective plane π. A point P is called a *nucleus* of K if no line through P intersects $K \setminus \{P\}$ in more than one point. P is called an *external nucleus* if $P \notin K$, and P is called an *internal nucleus* if $P \in K$. If K has an external nucleus, then $|K| \leq q+1$, if K has an internal nucleus, then $|K| \leq q+2$. The set of external nuclei of K is denoted by $E(K)$, the set of internal nuclei is denoted by $I(K)$. Results on external nuclei can be found in [3], [5], on internal nuclei in [12], [10], [13], [14]. Blokhuis and Wilbrink [5] proved the next theorem:

Result 1 [5] *In $\pi = PG(2, q)$ the lines are the only $(q + 1)$-sets admitting at least q external nuclei.*

This means, that if the points of a $(q + 1)$-set are not on a line, then the number of external nuclei is at most $q - 1$, that is $|E(K)| \leq q - 1$. The aim of this paper is to find an upper bound for the number of external nuclei of a $(q + 1)$-set, when π is not a Desarguesian plane. The next theorem will be proved.

Theorem 1.1 *If K is a $(q+1)$-set in the finite projective plane π of order q and if the points are not on a line, then*

$$|E(K)| \leq \frac{q\sqrt{4q+1} - q + \sqrt{4q+1} + 1}{2} = n^3 + n^2 + n + 1,$$

---

[*]. This work was supported by the Hungarian Research Foundation for Scientific Research (OTKA) # 326-0313.

where $n = \frac{1}{2}(\sqrt{4q+1} - 1)$, that is $q = n(n+1)$.

The sharpness of this theorem is not known. If q is a prime power, and $q > 2$, then the right side of the inequality is not an integer. Equality may occur, only if n is an integer, and a plane of order $n(n + 1)$ exists, with a subplane of order n.

As an application of Theorem 1.1, we prove a result on maximal strong representative systems. A *flag* of a finite projective plane π is an incident point-line pair (P, L). A set of flags

$$S = \{(P_1, L_1), \ldots, (P_k, L_k)\}$$

is called a *strong representative system*, if $P_i \in L_j$ holds if and only if $i = j$. S is said to be *maximal* if it is maximal subject to inclusion as a set of flags. It was proved in [8], that

$$q + 1 \leq k \leq q\sqrt{q} + 1$$

holds for a maximal strong representative system S of k flags. A paper of Blokhuis and Metsch [4] in this book deals with the upper bound. The next configuration theorem was proved in [8] about the lower bound of k.

Result 2 [8] Let $S = \{(P_1, L_1), \ldots, (P_{q+1}, L_{q+1})\}$ be a maximal strong representative system of $q + 1$ flags in $\pi = PG(2, q)$. Then either $P_1, \ldots, P_{q+1}$ are on a line, or $L_1, \ldots, L_{q+1}$ are concurrent lines.

In this paper we generalize this theorem for the non-Desarguesian planes.

Theorem 1.2 Let $S = \{(P_1, L_1), \ldots, (P_{q+1}, L_{q+1})\}$ be a maximal strong representative system of $q + 1$ flags in a finite projective plane π. Then either $P_1, \ldots, P_{q+1}$ are on a line, or $L_1, \ldots, L_{q+1}$ are concurrent lines.

Let us study next the case of internal nuclei. We know the next results:

Result 3 [2] Let $|K| = q + 2$ in $\pi = PG(2, q)$. If $|I(K)| \geq 3$, then q is even, and in this case $I(K) = K$, or $|I(K)| \leq \frac{q}{2}$.

Result 4 [13] Let $|K| = q + 1$ in $\pi = PG(2, q)$. If $I(K) \neq K$, then $|I(K)| \leq \lfloor \frac{q+1}{2} \rfloor$.

The internal nuclei of a point set are necessarily on an arc. In the extremal constructions of the results above, the points of $K \setminus I(K)$ are on a line. Point sets with this condition have an application in cryptography (see [9]). The next theorem is about this type of point sets.

Theorem 1.3 *Suppose* $|K|$ *is a* $(q+2)$-set *in a projective plane of order* q, q *even. If the set* $K \setminus I(K)$ *is a nonempty subset of the points on a line then*

$$|I(K)| \leq q - \frac{\sqrt{4q+1} - 1}{2} = n^2.$$

The form of this inequality is

$$|I(K)| \leq q - \sqrt{q},$$

if q *is a square, and this bound is sharp. If* K *satisfies the same conditions, but* q *is odd, then* $|I(K)| \leq q - 2$.

2.   Proofs

For the proof of Theorem 1.1, we first study the case where K is a $(q+1)$-set, which contains at least $q - \sqrt{q}$ collinear points.

Lemma 2.1 *If* K *is a* $(q+1)$-set *in a finite projective plane* π *of order* q *and if* K *has* $q+1-s$ *collinear points* $(s \geq 2)$, *then*

$$|E(K)| \leq s(s-1).$$

Proof of Lemma: Let L be that line which contains $q + 1 - s$ collinear points of K. Let us denote the points of $L \setminus K$ by $Q_1, Q_2, \ldots, Q_s$ and the points of $K \setminus L$ by $R_1, R_2, \ldots, R_s$ $(s \geq 2)$. If P is a nucleus of K, then a permutation ϕ of the subscripts $1, 2, \ldots, s$ can be given by the collinearities, such that $\phi(i) = j$ if and only if P, Q_i and R_j are collinear points. Every permutation ϕ belonging to a nucleus is determined uniquely by the effect of ϕ on any two different subscripts. This means that there are no two different permutations which both map i_1 to j_1 and i_2 to j_2 $(i_1 \neq i_2)$. Let us denote the maximum number of such a permutation set on s points by $p(s)$. It is clear that $p(s) \leq s(s-1)$, because the subscripts $1, 2$ have $s(s-1)$ possible images. This proves our lemma.   □

Applying this lemma we get the next corollary.

Corollary 2.2 *If* K *is a* $(q+1)$-set *in a projective plane* π *of order* q *and if* K *has at least* $q - \sqrt{q} - 1$ *and at most* q *collinear points, then*

$$|E(K)| \leq q + 3\sqrt{q} + 2.$$

Proof of Theorem 1.1: First we follow [7]. Let $l_1, \ldots, l_k$ be the secants of K. Let e_i denote the number of points of K on l_i. Clearly

$$\sum_{i=1}^{k} e_i(e_i - 1) = (q+1)q. \qquad (1)$$

Let $m = \max\{ e_i \,\|\, i = 1, 2, \ldots, k\}$, let $\mathcal{B}$ be the set of lines disjoint from K and let b be the number of these lines. A standard computation shows that

$$b = \sum_{i=1}^{k} (e_i - 1) - q$$

and so from (1)

$$(q+1)q = \sum_{i=1}^{k} e_i(e_i - 1) \leq m \sum_{i=1}^{k} (e_i - 1) = m(b+q),$$

that is

$$b \geq \frac{q^2 + q}{m} - q. \qquad (2)$$

It is also clear that

$$b \geq (q + 1 - m)(m - 1). \qquad (3)$$

We show that

$$b \geq \frac{q\sqrt{4q+1} - 3q + \sqrt{4q+1} - 1}{2} = n^3, \quad \text{if} \quad m \leq q + 1 - n.$$

This follows from (2) if $m \leq n + 1$ and from (3) if $n + 1 \leq m \leq q + 1 - n$.

A line disjoint from K is not incident with a nucleus of K, that is $E(K)$ is a subset of those points which are not covered by the lines of B. Let $\mathcal{B}'$ be an arbitrary subset of $\mathcal{B}$ for which $|\mathcal{B}'| = n^3$. Let t_i denote the number of points which are incident with exactly i lines of $\mathcal{B}'$. This means that $|E(K)| \leq t_0$. Counting in two different ways the points of the plane, the flags (P, L), with P a point incident with a line L of $\mathcal{B}'$, the triples (L, L', P) with L and L' different lines of $\mathcal{B}'$ incident with P, we get the following equations:

$$\sum_{i=1}^{q+1} t_i = q^2 + q + 1 - t_0,$$

$$\sum_{i=1}^{q+1} i t_i = |\mathcal{B}'|(q+1),$$

$$\sum_{i=1}^{q+1} i(i-1) t_i = |\mathcal{B}'|(|\mathcal{B}'| - 1).$$

line. Point sets with this condition have an application in cryptography (see [9]). The next theorem is about this type of point sets.

Theorem 1.3 *Suppose* $|K|$ *is a* $(q+2)$-*set in a projective plane of order* q, q *even. If the set* $K \setminus I(K)$ *is a nonempty subset of the points on a line then*

$$|I(K)| \leq q - \frac{\sqrt{4q+1} - 1}{2} = n^2.$$

The form of this inequality is

$$|I(K)| \leq q - \sqrt{q},$$

if q *is a square, and this bound is sharp. If* K *satisfies the same conditions, but* q *is odd, then* $|I(K)| \leq q - 2$.

2. Proofs

For the proof of Theorem 1.1, we first study the case where K is a $(q+1)$-set, which contains at least $q - \sqrt{q}$ collinear points.

Lemma 2.1 *If* K *is a* $(q+1)$-*set in a finite projective plane* π *of order* q *and if* K *has* $q + 1 - s$ *collinear points* $(s \geq 2)$, *then*

$$|E(K)| \leq s(s-1).$$

Proof of Lemma: Let L be that line which contains $q + 1 - s$ collinear points of K. Let us denote the points of $L \setminus K$ by Q_1, $Q_2, \ldots$, Q_s and the points of $K \setminus L$ by R_1, $R_2, \ldots$, R_s $(s \geq 2)$. If P is a nucleus of K, then a permutation ϕ of the subscripts $1, 2, \ldots, s$ can be given by the collinearities, such that $\phi(i) = j$ if and only if P, Q_i and R_j are collinear points. Every permutation ϕ belonging to a nucleus is determined uniquely by the effect of ϕ on any two different subscripts. This means that there are no two different permutations which both map i_1 to j_1 and i_2 to j_2 $(i_1 \neq i_2)$. Let us denote the maximum number of such a permutation set on s points by $p(s)$. It is clear that $p(s) \leq s(s-1)$, because the subscripts 1, 2 have $s(s-1)$ possible images. This proves our lemma. $\square$

Applying this lemma we get the next corollary.

Corollary 2.2 *If* K *is a* $(q+1)$-*set in a projective plane* π *of order* q *and if* K *has at least* $q - \sqrt{q} - 1$ *and at most* q *collinear points, then*

$$|E(K)| \leq q + 3\sqrt{q} + 2.$$

Proof of Theorem 1.1: First we follow [7]. Let $l_1, \ldots, l_k$ be the secants of K. Let e_i denote the number of points of K on l_i. Clearly

$$\sum_{i=1}^{k} e_i(e_i - 1) = (q+1)q. \tag{1}$$

Let $m = \max\{ e_i \| i = 1, 2, \ldots, k \}$, let $\mathcal{B}$ be the set of lines disjoint from K and let b be the number of these lines. A standard computation shows that

$$b = \sum_{i=1}^{k} (e_i - 1) - q$$

and so from (1)

$$(q+1)q = \sum_{i=1}^{k} e_i(e_i - 1) \leq m \sum_{i=1}^{k} (e_i - 1) = m(b+q),$$

that is

$$b \geq \frac{q^2 + q}{m} - q. \tag{2}$$

It is also clear that

$$b \geq (q+1-m)(m-1). \tag{3}$$

We show that

$$b \geq \frac{q\sqrt{4q+1} - 3q + \sqrt{4q+1} - 1}{2} = n^3, \quad \text{if} \quad m \leq q+1-n.$$

This follows from (2) if $m \leq n+1$ and from (3) if $n+1 \leq m \leq q+1-n$.

   A line disjoint from K is not incident with a nucleus of K, that is $E(K)$ is a subset of those points which are not covered by the lines of $\mathcal{B}$. Let $\mathcal{B}'$ be an arbitrary subset of $\mathcal{B}$ for which $|\mathcal{B}'| = n^3$. Let t_i denote the number of points which are incident with exactly i lines of $\mathcal{B}'$. This means that $|E(K)| \leq t_0$. Counting in two different ways the points of the plane, the flags (P, L), with P a point incident with a line L of $\mathcal{B}'$, the triples (L, L', P) with L and L' different lines of $\mathcal{B}'$ incident with P, we get the following equations:

$$\sum_{i=1}^{q+1} t_i = q^2 + q + 1 - t_0,$$

$$\sum_{i=1}^{q+1} it_i = |\mathcal{B}'|(q+1),$$

$$\sum_{i=1}^{q+1} i(i-1)t_i = |\mathcal{B}'|(|\mathcal{B}'| - 1).$$

Adding up the last two equations we get

$$\sum_{i=1}^{q+1} i^2 t_i = |\mathcal{B}'|(q + |\mathcal{B}'|).$$

To estimate of t_0 we consider the next inequality:

$$\sum_{i=1}^{q+1} (i - n)^2 t_i \geq 0.$$

Substituting the previous equations into this inequality, we get:

$$|\mathcal{B}'|(q + |\mathcal{B}'|) - 2n|\mathcal{B}'|(q + 1) + n^2(q^2 + q + 1 - t_0) \geq 0,$$

from which

$$t_0 \leq n^3 + n^2 + n + 1 = \frac{q\sqrt{4q + 1} - q + \sqrt{4q + 1} + 1}{2},$$

and this completes the proof. An other end of this proof is possible by the inequality between the quadratic and arithmetic means, from which we obtain

$$\left(\sum_{i=1}^{q+1} it_i / \sum_{i=1}^{q+1} t_i \right)^2 \leq \sum_{i=1}^{q+1} i^2 t_i / \sum_{i=1}^{q+1} t_i,$$

that is, after the substitutions,

$$\left(|\mathcal{B}'|(q + 1)/(q^2 + q + 1 - t_0) \right)^2 \leq |\mathcal{B}'|(q + |\mathcal{B}'|)/(q^2 + q + 1 - t_0),$$

and this leads to the same result. $\qquad\square$

Proof of Theorem 1.2: Here we follow the proof of Result 2, using the result of Theorem 1.1. As before, let $n = \frac{1}{2}(\sqrt{4q + 1} - 1)$, so that $q = n(n + 1)$. Let $K = \{P_1, \ldots, P_{q+1}\}$ and suppose that the points of K are not all on a line. Similarly suppose that $\{L_1, \ldots, L_{q+1}\}$ is not a pencil. Hence there exists a point P which is not covered by $L_1 \cup \ldots \cup L_{q+1}$. Such a point P belongs to $E(K)$, because otherwise there would be a line L through P which is disjoint from K, and thus S would be extendible by the flag (P, L). The number of these non-covered points is at least

$$q^2 + q + 1 - \left((q + 1 - m)^2 + mq + 1 \right) = m(q + 2 - m) - q - 1,$$

where m denotes the maximum number of concurrent lines of S. This is true because m concurrent lines cover $mq + 1$ points, and each other line covers at most $q + 1 - m$ new points, i.e. the number of the covered points is at most $(q + 1 - m)^2 + mq + 1$. Let L_i and L_j are two different lines of S, and denote by Q the intersection of them. Let us suppose that $Q \notin E(K)$, then there is a line L through Q, which has no point of K. At least one point R of this line is not covered by the lines L_k ($k = 1, \ldots, q+1$), because Q is covered by at least two lines. It means that S is extendible by the flag (R, L), and this contradiction proves that $Q \in E(K)$. The number of this type of points is at least $q + 1$, by Fischer's inequality, because the points of K are not all on one line. Comparing these lower bounds with the upper bound of Theorem 1, we get the following inequality:

$$m(q + 2 - m) \leq \frac{q\sqrt{4q+1} - q + \sqrt{4q+1} + 1}{2} = qn + n + 1.$$

It follows that $m \leq n+1$ or $m \geq q+1-n$. Let us now derive another bound for the size of $E(K)$. We saw that all points that are on no line of S are in $E(K)$ as well as all points that are on more than one line of S. Since there are at most m lines in S concurrent we see that every line in S contains at most $q + 1 - q/(m-1)$ non-nuclei. It follows that the total number of nuclei is at least

$$q^2 + q + 1 - (q+1)(q + 1 - \frac{q}{m-1}) = \frac{q(q+1)}{m-1} - q.$$

Using again the upper bound of Theorem 1 we obtain

$$\frac{q(q+1)}{m-1} - q \leq qn + n + 1,$$

and this implies $m \geq n+1$. Combining this with the above we see that either $m = n + 1$ and we have equality in all estimates, or $m \geq q + 1 - n$. Let us first get rid of the second possibility, since this is the easiest. Consider the point P that is the intersection of at least $q + 1 - n$ lines of S. Let M be another line through P. At most n points of M lie on exactly one line of S, so M contains at least $q + 1 - n > q - \sqrt{q}$ nuclei. By a result of Cameron and Fisher [6], this means that all nuclei are on the line M, so M is unique, and there are q lines of S passing through P. One can easily verify that such a set can not be maximal.

Next we consider the case $m = n + 1$. Using the fact that we have equality in all estimates, we get that the lines in S form a (dual) subplane Π_1 (of order n) and the points in K are the points of a disjoint subplane Π_2. Note that every point in Π_1 is a nucleus of K, the points of Π_2, since it lies on more than one line of S, on the other hand no point in Π_2 is nucleus of Π_1. Since both subplanes have the same number of nuclei, there is a point P that is a nucleus of Π_1 and not of Π_2. This means that the point is not on a line in S and on a line missing K. It follows that the strong representative system is not maximal. □

Proof of Theorem 3: Let q be even, $|K| = q + 2$, assume that the elements of $K \setminus I(K)$ are on the line L (so $|L| \geq 3$), $|L \cap K| = k$, so $|K \setminus L| = q + 2 - k$. There are k tangents to every point of the arc $K \setminus L$, these points are nuclei, so every tangent meets L in a point of K. This means that $q + 2 - k$ and so k must be even, otherwise there would be at least one tangent through every point of L. Let P be a point of $L \cap K$, and let $\mathcal{L}$ be the set of lines, which contains the next lines:

the line L,

those lines, which intersect K in P and in a point of $K \setminus L$,

those lines, which intersect K in a point of $(K \cap L) \setminus \{P\}$.

The number of these lines is clearly $1 + (q + 2 - k) + (k - 1)(k - 2)$. We prove that $\mathcal{L}$ is a dual blocking set, that is every point of the plane is covered by a line of $\mathcal{L}$, and $\mathcal{L}$ has no $q + 1$ concurrent lines. This last statement is clear, so indirectly, let Q be a point, which is not covered by $\mathcal{L}$. This means, that the line PQ has no more points of K, and the lines PR, where $R \in (K \cap L) \setminus \{P\}$, intersect $K \setminus L$ in one point. In this way we have $k - 1$ tangents of $K \setminus L$ through Q, but $k - 1$ is odd, so we have at least one more tangent, but it intersects $L \setminus K$, which is a contradiction.

Let L' be a line, for which $L \cap L' = \{P\}$. The points of $L' \setminus \{P\}$ are covered by the lines of type (3), so

$$(k - 1)(k - 2) \geq q,$$

which implies the given inequalities. The sharpness of the estimation in the case of square q follows from the construction in Hall planes given by Szőnyi [11].

A good estimation in the case of odd q seems to be more difficult. To exclude the cases $|I(K)| = q$, $q - 1$ is simple. If $|I(K)| = q - 2$, that is 4 non-nuclei are collinear, then $K \setminus L$ together with two of the collinear points is a complete q-arc. The complete q-arcs are studied in a paper of Beutelspacher [1]. At the moment the existence of a complete q-arc is undecided, for which there is a point Q, such that there are $q - 1$ tangents and a secant of the arc through Q. Adding one or two of this type of points to a complete q-arc, we may get a set of $q + 1$ or $q + 2$ points K, such that $|I(K)| = q - 2$. □

Acknowledgement. The author thanks T. Szőnyi for fruitful discussions and the referee for correcting the proof of Theorem 1.2 adding new steps to it.

References

[1] **A. Beutelspacher**. On complete q-arcs in projective planes of order q. *Boll. Un. Mat. Ital. A (6)*, 5, pp. 449–454, 1986.

[2] A. Bichara and G. Korchmáros. Note on $(q + 2)$-sets in a Galois plane of order q. Ann. Discrete Math., 14, pp. 117–122, 1982.

[3] A. Blokhuis and F. Mazzocca. On maximal sets of nuclei in $PG(2, q)$ and quasi-odd sets in $AG(2, q)$. In J. W. P. Hirschfeld, D . R. Hughes, and J. A. Thas, editors, Advances in Finite Geometries and Designs, pages 35–46, Oxford, New York, Tokyo, 1991. Oxford University Press.

[4] A. Blokhuis and K. Metsch. Large minimal blocking sets, strong representative systems and partial unitals. In A. Beutelspacher, F. Buekenhout, F. De Clerck, J. Doyen, J. W. P. Hirschfeld, and J. A. Thas, editors, Finite Geometry and Combinatorics, pages 37–52, Cambridge, 1993. Cambridge University Press.

[5] A. Blokhuis and H. A. Wilbrink. A characterization of exterior lines of certain sets of points in $PG(2, q)$. Geom. Dedicata, 23, pp. 253–254, 1987.

[6] P. J. Cameron. Nuclei. Handwritten notes, 1985.

[7] P. Erdős and L. Lovász. Problems and results on 3-chromatic hypergraphs and some related questions. In Infinite and Finite Sets, volume 10 of Colloq. Math. Soc. János Bolyai, pages 607–627, Amsterdam, 1975. North-Holland.

[8] T. Illés, T. Szőnyi, and F. Wettl. Blocking sets and maximal strong representative systems in finite projective planes. Mitt. Math. Sem. Giessen, 201, pp. 97–107, 1991.

[9] G. Simmons. Sharply focused sets of lines on a conic in $PG(2, q)$. Congr. Numer., 73, pp. 181–204, 1990.

[10] T. Szőnyi. Arcs and k-sets with large nucleus-set in Hall planes. J. Geom., 34, pp. 187–194, 1989.

[11] T. Szőnyi. Complete arcs in Galois planes: a survey. Quad. Sem. Geom. Comb. Univ. Roma "La Sapienza", 94, 1989.

[12] T. Szőnyi. k-sets in $PG(2, q)$ having a large set of internal nuclei. In A. Barlotti, G. Lunardon, F. Mazzocca, N. Melone, D. Olanda, A. Pasini, and G. Tallini, editors, Combinatorics '88, pages 449–458, Rende (I), 1991. Mediterranean Press.

[13] F. Wettl. On the nuclei of a finite projective plane. J. Geom., 30, pp. 157–163, 1987.

[14] F. Wettl. Internal nuclei of k-sets in finite projective spaces of three dimensions. In J. W. P. Hirschfeld, D. R. Hughes, and J. A. Thas, editors, Advances in Finite Geometries and Designs, pages 407–419, Oxford, 1991. Oxford University Press.

Ferenc Wettl, Technical University Budapest, Transport Engineering Faculty, Department of Mathematics, Budapest, Műegyetem rkp. 9. H ép. V.4. 1111 Hungary. e-mail: h2817wet@ella.hu

Printed in the United States
By Bookmasters